CW00971868

Object Model Notation — Basic Concepts

(This page is in the public domain.)

Object:

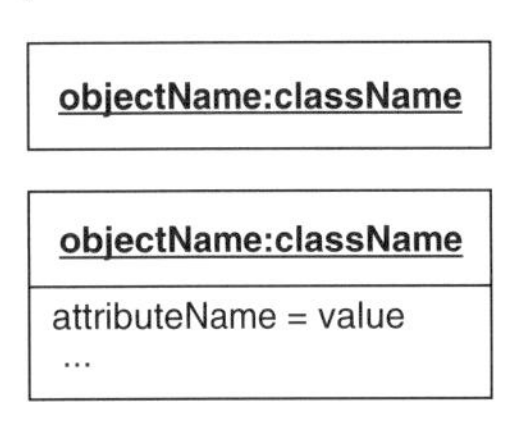

Class:

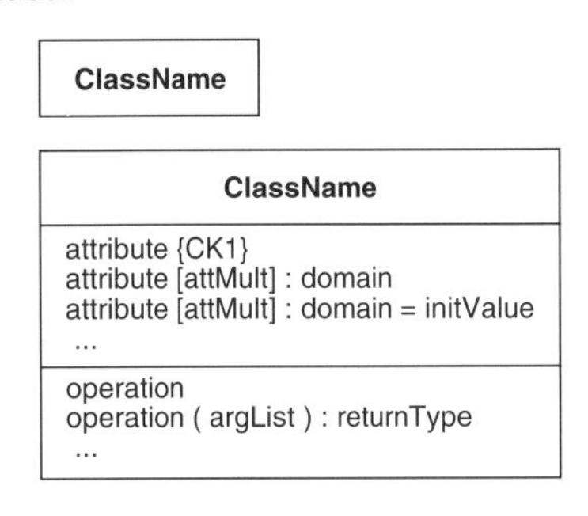

Link Attribute:

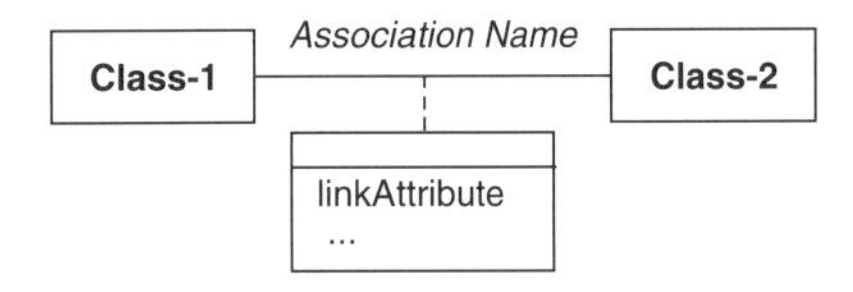

Association Class:

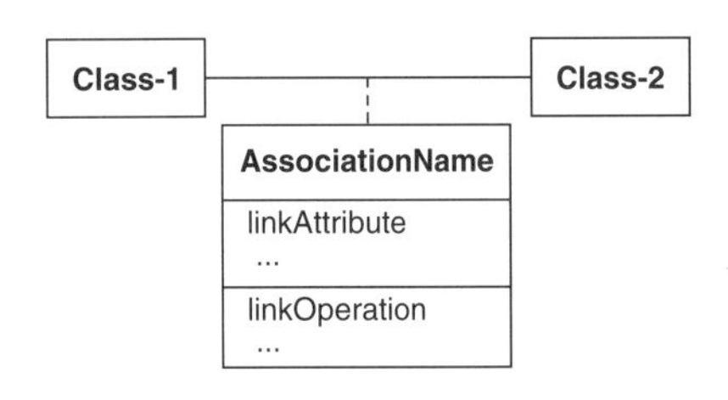

Link:

Association:

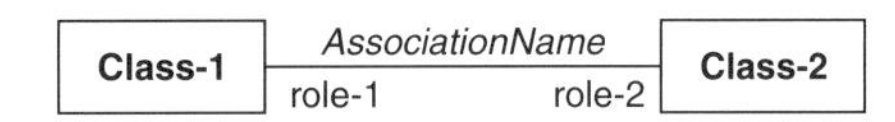

Multiplicity of Associations:

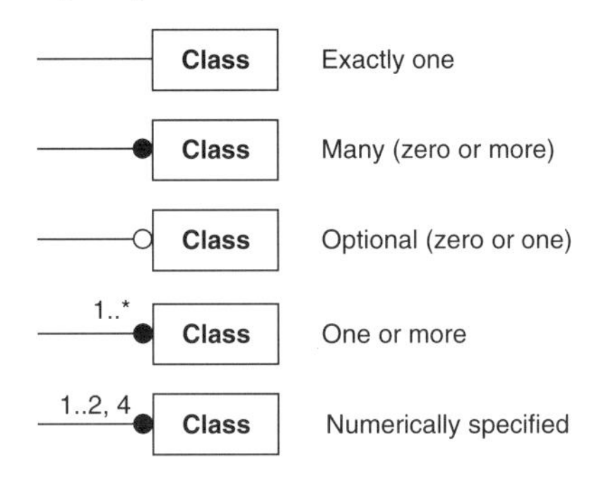

Ordering:

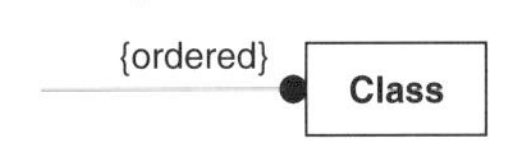

Qualified Association:

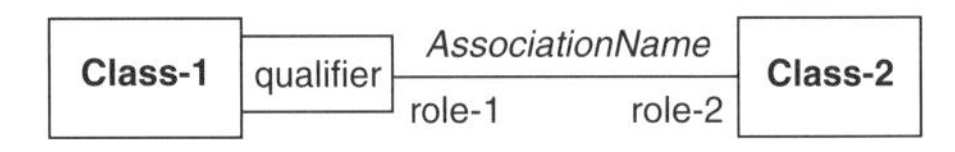

Generalization (Inheritance):

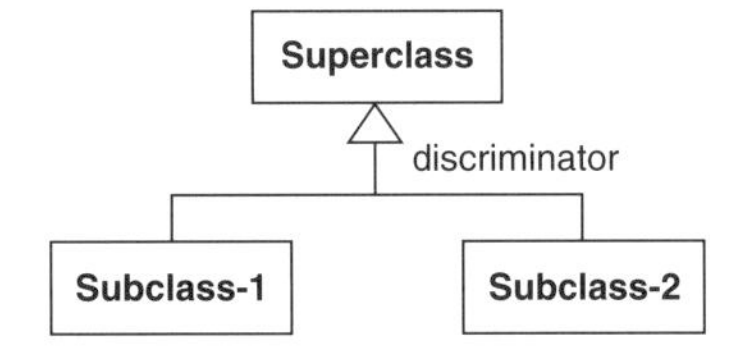

Object Model Notation — Advanced Concepts

Instantiation:

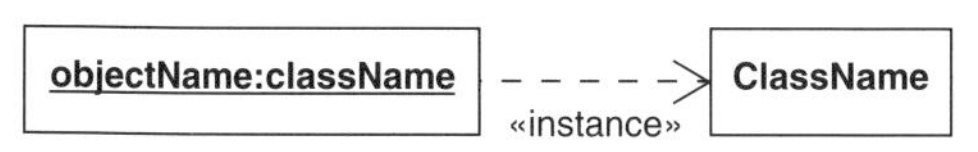

Ternary Association:

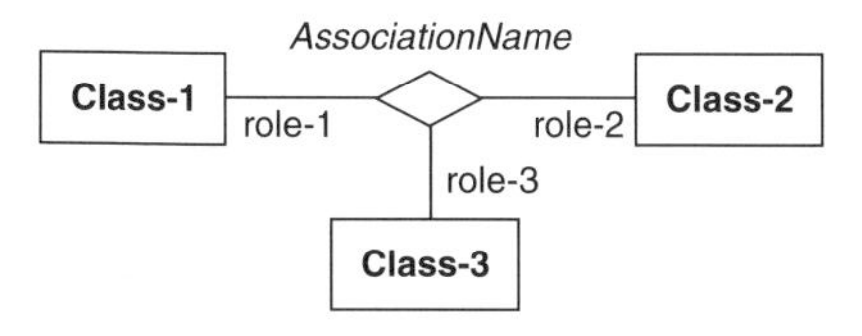

Concrete class:

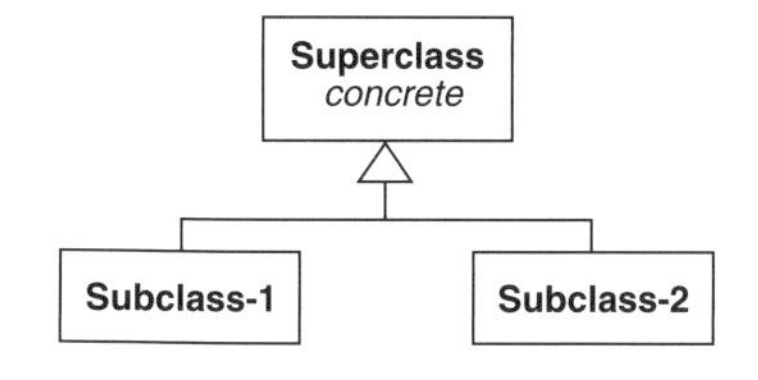

Abstract class:

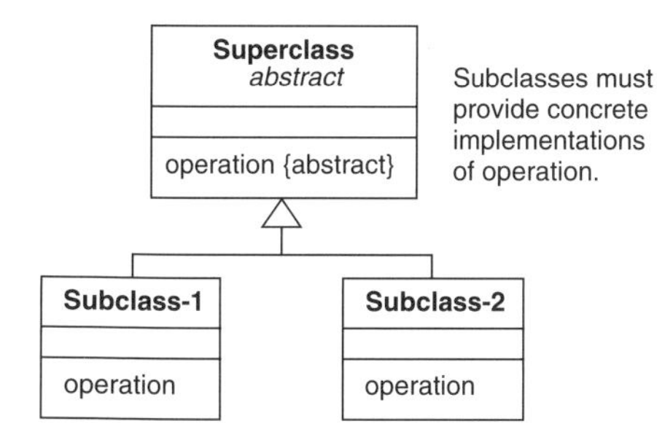

Multiple Inheritance:

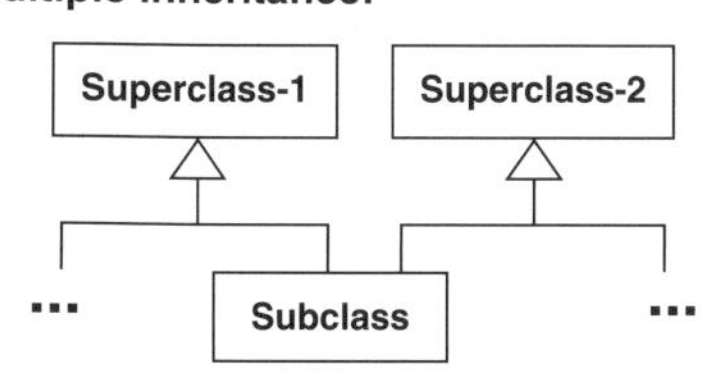

Class Attributes and Class Operations:

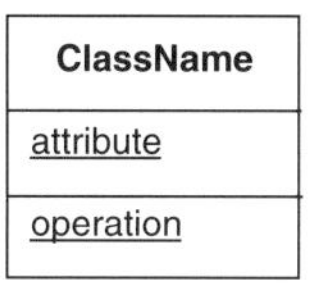

Exclusive-Or Association

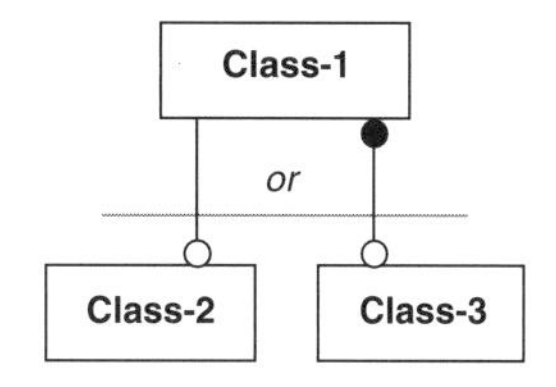

Aggregation:

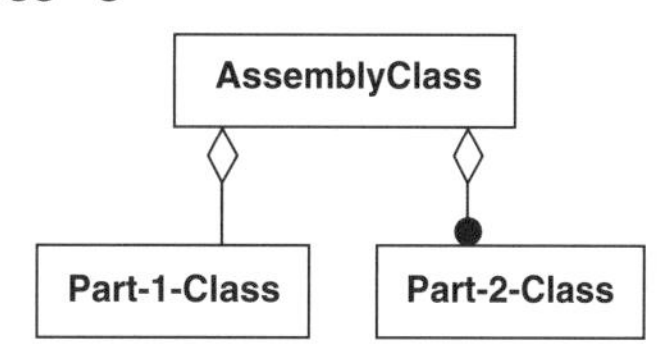

Package:

Constraints on Objects:

Class
attrib-1
attrib-2
{ attrib-1 ≥ 0 }

Constraint between Associations:

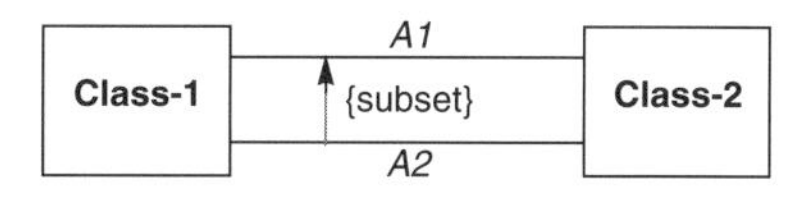

Derived Class:

Derived Association:

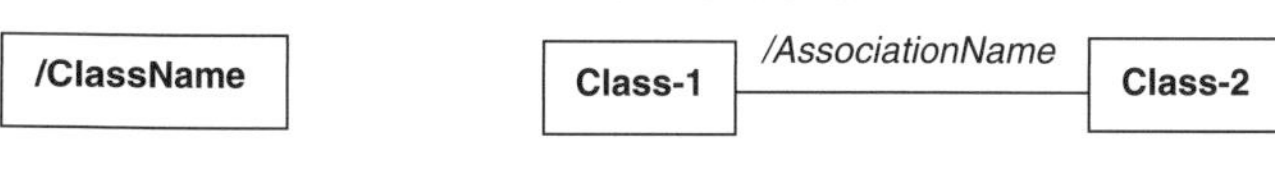

Derived Attribute:

ClassName
/attribute

Object-Oriented Modeling and Design for Database Applications

MICHAEL BLAHA
OMT Associates Inc.
Chesterfield, Missouri

WILLIAM PREMERLANI
General Electric Research and Development Center
Schenectady, New York

Prentice-Hall International, Inc.

Editor-in-chief: **MARCIA HORTON**
Publisher: **ALAN APT**
Editor: **LAURA STEELE**
Director of production and manufacturing: **DAVID W. RICCARDI**
Managing editor: **BAYANI MENDOZA DE LEON**
Creative director: **PAULA MAYLAHN**
Production editor: **KATHARITA LAMOZA**
Art director: **HEATHER SCOTT**
Manufacturing buyer: **DONNA SULLIVAN**
Editorial assistant: **TONI HOLM**

Simon & Schuster / A Viacom Company
Upper Saddle River, New Jersey 07458

Printed in the United States of America

10 9 8 7 6 5 4 3 2 1

ISBN 0-13-899089-1

Prentice-Hall International (UK) Limited, London
Prentice-Hall of Australia Pty. Limited, Sydney
Prentice-Hall Canada Inc., Toronto
Prentice-Hall Hispanoamericana, S.A., Mexico
Prentice-Hall of India Private Limited, New Delhi
Prentice-Hall of Japan, Inc., Tokyo
Simon & Schuster Asia Pte. Ltd., Singapore
Editora Prentice-Hall do Brasil, Ltda., Rio de Janeiro
Prentice-Hall, Inc., Upper Saddle River, New Jersey

To Jean, Brian, Karen, and my parents, Robert and Ann.
M. R. B.

To my family.
W. J. P.

Contents

FOREWORD xi

PREFACE xiii

What You Will Find, xiv
Who Should Read This Book?, xiv
Acknowledgments, xv

CHAPTER 1 INTRODUCTION 1

1.1 Why Object Orientation?, 1
1.2 Object-Oriented Software Engineering, 2
1.3 The OMT Methodology, 5
1.4 Organization of This Book, 8
1.5 Font Conventions, 9
1.6 Chapter Summary, 9
Bibliographic Notes, 10
References, 11

Part 1: Modeling Concepts

CHAPTER 2 BASIC OBJECT MODELING 12

2.1 Object and Class Concepts, 13
2.2 Link and Association Concepts, 17
2.3 Generalization Concepts, 26
2.4 Sample Diagrams, 27
2.5 Practical Tips, 31
2.6 Chapter Summary, 33
Changes to OMT Notation, 34
Bibliographic Notes, 35
References, 35
Exercises, 36

CHAPTER 3 ADVANCED OBJECT MODELING **42**
3.1 Object and Class Concepts, 42
3.2 Link and Association Concepts, 48
3.3 Aggregation, 51
3.4 Generalization, 56
3.5 Multiple Inheritance, 58
3.6 Packages, 61
3.7 Derived Data and Constraints, 67
3.8 Advanced Practical Tips, 68
3.9 Chapter Summary, 69
Changes to OMT Notation, 70
Bibliographic Notes, 71
References, 71
Exercises, 72

CHAPTER 4 OBJECT METAMODELING **75**
4.1 Metadata and Metamodels, 75
4.2 Frameworks, 80
4.3 Patterns, 81
4.4 Chapter Summary, 88
Bibliographic Notes, 88
References, 89
Exercises, 89

CHAPTER 5 FUNCTIONAL MODELING **96**
5.1 Pseudocode, 96
5.2 Pseudocode with the Object Navigation Notation, 98
5.3 ONN Constructs, 98
5.4 Combining ONN Constructs, 105
5.5 Additional ONN Properties, 108
5.6 Other Paradigms, 110
5.7 Practical Tips, 113
5.8 Chapter Summary, 114
Bibliographic Notes, 115
References, 116
Exercises, 116

Part 2: Analysis and Design Process

CHAPTER 6 PROCESS PREVIEW **118**
Reference, 120

CHAPTER 7 CONCEPTUALIZATION **121**
7.1 Overview, 121
7.2 The Portfolio Manager Case Study, 122

7.3 Chapter Summary, 124
Bibliographic Note, 124
CHAPTER 8 ANALYSIS **125**
8.1 Overview, 125
8.2 Problem Statement, 126
8.3 Object Model, 126
8.4 The Data Dictionary, 145
8.5 Dynamic Model, 148
8.6 Functional Model, 148
8.7 Lessons Learned, 156
8.8 Chapter Summary, 157
Bibliographic Notes, 158
References, 159
Exercises, 160
CHAPTER 9 SYSTEM DESIGN **169**
9.1 Overview, 169
9.2 Devising an Architecture, 170
9.3 Choosing an Implementation for External Control, 175
9.4 Choosing a Data Management Approach, 177
9.5 Choosing a Database Management Paradigm, 182
9.6 Determining Opportunities for Reuse, 187
9.7 Choosing a Strategy for Data Interaction, 190
9.8 Choosing an Approach to Object Identity, 193
9.9 Dealing with Temporal Data, 195
9.10 Dealing with Secondary Aspects of Data, 196
9.11 Specifying Default Policies for Detailed Design, 198
9.12 Chapter Summary, 200
Bibliographic Notes, 202
References, 202
Exercises, 202
CHAPTER 10 DETAILED DESIGN **210**
10.1 Overview, 210
10.2 Object Model Transformations, 210
10.3 Elaborating the Object Model, 222
10.4 Elaborating the Functional Model, 224
10.5 Evaluating the Quality of a Design Model, 230
10.6 Chapter Summary, 230
Bibliographic Notes, 231
References, 232
Exercises, 233

CHAPTER 11 PROCESS REVIEW **236**

11.1 Conceptualization, 236
11.2 Analysis, 236
11.3 System Design, 237
11.4 Detailed Design, 238
11.5 Overview of Implementation, 238

Part 3: Implementation

CHAPTER 12 FILES **239**

12.1 Introduction to Files, 239
12.2 Implementing the Object Model, 240
12.3 Organizing Data into Files, 241
12.4 Selecting a File Approach, 243
12.5 Implementing Identity, 246
12.6 Implementing Domains, 246
12.7 Implementing Classes, 248
12.8 Implementing Attributes, 249
12.9 Implementing Simple Associations, 250
12.10 Implementing Advanced Associations, 253
12.11 Implementing Generalizations, 255
12.12 Summary of Object Model Mapping Rules, 256
12.13 Implementing the Dynamic Model, 256
12.14 Implementing the Functional Model, 256
12.15 Other Implementation Issues, 263
12.16 Chapter Summary, 264
Bibliographic Notes, 264
References, 265
Exercises, 265

CHAPTER 13 RELATIONAL DATABASES: BASICS **269**

13.1 Introduction to Relational Databases, 270
13.2 Implementing the Object Model, 276
13.3 Implementing Identity, 276
13.4 Implementing Domains, 278
13.5 Implementing Classes, 282
13.6 Implementing Simple Associations, 282
13.7 Implementing Advanced Associations, 285
13.8 Implementing Single Inheritance, 288
13.9 Implementing Multiple Inheritance, 291
13.10 Summary of Object Model Mapping Rules, 291
13.11 Implementing the Dynamic Model, 292

13.12 Implementing the Functional Model, 295
13.13 Chapter Summary, 301
Bibliographic Notes, 303
References, 304
Exercises, 304

CHAPTER 14 RELATIONAL DATABASES: ADVANCED **307**

14.1 Implementing the Object Model, 307
14.2 Defining Constraints on Tables, 307
14.3 Defining Indexes, 309
14.4 Allocating Storage, 311
14.5 Creating a Schema, 312
14.6 Implementing the Dynamic Model, 314
14.7 Implementing the Functional Model, 318
14.8 Other Functional Modeling Issues, 326
14.9 Physical Implementation Issues, 330
14.10 Lessons Learned with the Portfolio Manager, 330
14.11 Chapter Summary, 331
Bibliographic Notes, 332
References, 332
Exercises, 332

CHAPTER 15 OBJECT-ORIENTED DATABASES: BASICS **334**

15.1 Introduction to ObjectStore, 334
15.2 Implementing the Object Model, 339
15.3 Implementing Domains, 340
15.4 Implementing Classes, 342
15.5 Implementing Generalizations, 344
15.6 Implementing Simple Associations, 344
15.7 Implementing Advanced Associations, 348
15.8 Summary of Object Model Mapping Rules, 349
15.9 Implementing the Dynamic Model, 349
15.10 Implementing the Functional Model, 350
15.11 Mapping ONN Constructs, 351
15.12 Creation and Deletion Methods, 356
15.13 Chapter Summary, 360
Bibliographic Notes, 361
References, 362
Exercises, 363

CHAPTER 16 OBJECT-ORIENTED DATABASES: ADVANCED **368**

16.1 Keys, 368
16.2 Extents, 369

16.3 Folding Attributes into a Related Class, 372
16.4 Promoting Associations, 376
16.5 Using ObjectStore Queries, 385
16.6 Software Engineering Issues, 387
16.7 Chapter Summary, 390
References, 391
Exercises, 392

CHAPTER 17 IMPLEMENTATION REVIEW 394
17.1 Implementing the Object Model, 394
17.2 Implementing the Functional Model, 397
17.3 Overview of Large System Issues, 397

Part 4: Large System Issues

CHAPTER 18 DISTRIBUTED DATABASES 399
18.1 Introduction to Distributed Databases, 399
18.2 Client-Server Computing, 401
18.3 Distributing Data, 406
18.4 Chapter Summary, 413
Bibliographic Notes, 413
References, 414

CHAPTER 19 INTEGRATION OF APPLICATIONS 415
19.1 Overview, 415
19.2 Enterprise Modeling, 417
19.3 Integration Techniques, 425
19.4 Integration Architecture, 429
19.5 Data Warehouse, 432
19.6 Chapter Summary, 433
Bibliographic Notes, 434
References, 435

CHAPTER 20 REVERSE ENGINEERING 436
20.1 Overview, 436
20.2 Hierarchical Databases, 439
20.3 Network Databases, 443
20.4 Relational Databases, 451
20.5 Chapter Summary, 457
Bibliographic Notes, 458
References, 458

APPENDIX A GLOSSARY 459
APPENDIX B BNF GRAMMAR FOR THE ONN 470
INDEX 473

Foreword

The OMT method began in 1987 when I worked with Mary Loomis (then at Calma Company) to combine the programming-language notation that I had invented for my DSM programming language with the database notation that Mary had used in her *Database Book*. We came to the realization that a good modeling notation could span both programming languages and databases by incorporating the best aspects of both media. Later that year Mary, Ashwin Shah, and I published the first paper on OMT at ECOOP '87 and I published my paper at OOPSLA '87 on the advantages of relations in programming languages. Mary and Ashwin implemented a code generator for the DSM language and Calma Corporation became the world's leading user of DSM (indeed, the only users outside of GE).

At the same time I began working with my colleagues Mike Blaha and Bill Premerlani at the GE Research & Development Center in Schenectady, New York. Mike and Bill learned DSM and OMT and began applying them to database applications. They expanded OMT from its programming language focus by showing how to use it to generate schemas for relational databases. We published a paper in the *Communications of the ACM* in 1988 on "Relational Database Design Using an Object-Oriented Methodology." I don't know that this paper revolutionized the database world, but it did attract the attention of a publisher who asked us to write a book on object-oriented database design. We replied that we didn't think we could write a whole book on database design, but what about a book on object-oriented design?

Two and one-half years later that proposal resulted in the book *Object-Oriented Modeling and Design*, whose success exceeded our wildest expectations. That book contained one chapter on database design, primarily written by Mike and Bill; not much, but all that we had room and material for at the time. We always felt that database design deserved a whole book. You now have that book in your hands. This book is no mere academic exercise. It is the result of years of work by Mike Blaha and Bill Premerlani in building real databases for real problems.

Some critics have complained that OMT is data focused and influenced by databases. I accept these characterizations, but I do not find in them the limitations that the critics intend.

We need to remove the barriers of the past and adopt good ideas without regard to artificial labels. It is a great shame that the programming language community and the database community still hold themselves so far apart and have resisted adopting more ideas from each other. There are many things that are far easier in databases, such as operations on sets and complicated queries; programming languages would do well to incorporate them. There are many things that are far easier in programming languages, such as implementing algorithms and special cases; database programmers have often been forced to perform awkward translations between database and programming language implementations. As Mike Blaha and Bill Premerlani show in this book, a good object modeling technique transcends both programming languages and databases by abstracting the essence of a system. Such an approach accepts that real systems are neither completely behavioral (as some object-oriented purists would have it) nor completely data (as some database purists would have it) but a balanced mixture of both aspects.

I worked with Mike and Bill for almost ten years at the GE R&D Center. Our partnership was always highly productive. Occasionally someone would ask us to close the office door during one of our discussions. They would ask "what are you fighting about?" "Fighting? We're not fighting. This is just a normal technical discussion." Those were great times. We accomplished a lot because no one was afraid of hard-hitting honest feedback. Mike and Bill both combine a no-nonsense pragmatism with a high ability to abstract the essential aspects of a situation.

This book fills an important need. There are many books on programming language design, from language tricks to idioms to software patterns. There are many books on database mechanics and theory, but there are few books on database design for the actual practitioner, although the number of database applications is substantial and they are crucial to many businesses. This book combines both a high-level view of abstract design with the pragmatic details that are necessary to use real technology. I am pleased that this book is finally available and I hope that it can begin to heal the gap between databases and programming languages.

James Rumbaugh
Santa Clara, California

Preface

Object-oriented modeling techniques are no longer new. Every day brings a new success story: lower costs, shorter time to market, better quality systems, new levels of customer satisfaction. The benefits are many. The success of the book we coauthored with Jim Rumbaugh, Fred Eddy, and Bill Lorensen (*Object-Oriented Modeling and Design*, Prentice Hall, 1991) is evidence of the enthusiasm for this technology.

However, despite the general agreement that object-oriented approaches are a good thing, we still see people hesitating to apply them in some areas. Database design and implementation is one of these. We believe this hesitation is the result of a hole in the literature: Although many excellent books cover object-oriented concepts and database concepts individually, we have not yet found a book that systematically shows how to use object-oriented principles to design and implement database applications. We wrote this book in part to address this gap.

We have also experienced a need for the book first-hand. Although database management state of the art is quite good and experts understand applications well, we have often been disappointed when we have reverse engineered databases from existing applications. Many databases have deficient schema that are confusing to understand and difficult to program against. It seems to us that many programmers are expected to design a database, even if they don't know how. For example, we have encountered some applications that are several orders of magnitude slower than they could be with a proper database design and implementation. We have found other databases that are just plain wrong; they are incapable of storing the desired data. Less extreme, perhaps, we have seen many developers underutilize database management system (DBMS) capabilities and cause themselves additional work. We hope our book can help avoid situations like this. We have found the Object Modeling Technique (OMT) methodology extremely helpful in developing database applications.

WHAT YOU WILL FIND

This book is about the OMT methodology. It is essentially a sequel to *Object-Oriented Modeling and Design* (Prentice Hall, 1991), but one targeted to data management applications. As such, you can expect the same emphasis on straightforward organization and clear explanations, as well as the provision of additional reading and practical exercises at the end of the chapters.

The structure of the book parallels the major software development activities:

- Introduction. Gives a refresher of OMT models (object, dynamic, and functional), analysis and design.
- Part 1: Modeling Concepts. Includes both basic and advanced issues.
- Part 2: Analysis and Design. Gives step by step recommendations from conceptual analysis through detailed design in the context of a case study.
- Part 3: Implementation. Follows the case study through three possible implementations and gives recommendations and practical tips. We have made all the code from the case study implementations available on the World Wide Web.
- Part 4: Large-System Issues. Gives a flavor of more advanced technologies.

We use the Unified Modeling Language (UML) developed by Grady Booch, Jim Rumbaugh, and Ivar Jacobson as our notation for the object model. We believe the UML will become an influential standard and we want to coordinate advances in database technology with those of programming technology.

WHO SHOULD READ THIS BOOK?

We see many ways to use this book. Practitioners should find it useful to learn a systematic approach to developing data management applications. This book addresses files, relational databases, and object-oriented databases. We authors are both practitioners, and this book describes the techniques that we actually use. We have taught these techniques to other software engineers and they have also found them helpful.

Software developers and teachers will find the exercises and complete case study useful in elucidating abstract concepts. (The complete solutions to the exercises are available separately from Prentice Hall.) Several courses could be devised around this book:

- **Basic object-oriented modeling course**. A one-semester course could introduce concepts and modeling with Chapters 1–3 and 5–8.
- **Graduate Management Information Systems (MIS) course**. The computing budgets of large corporations are dominated by support for information systems—the primary focus of this book. An instructor could teach the basic object-oriented modeling course in the first semester and cover Chapters 9–14 and 17–20 in a second-semester MIS course.

- **Advanced modeling course**. This could be another second-semester course following the basic object-oriented modeling course. The instructor could emphasize Chapters 4, 9, 10, and 19.
- **Undergraduate software engineering course**. Students could be organized into teams for a one-semester course, during which they develop software together. Chapters 1, 2, and 5–9 would be appropriate for such a course. The instructor could also cover the appropriate implementation chapters—Chapter 12 for files, Chapters 13 and 14 for relational databases, and Chapters 15 and 16 for object-oriented databases.
- **Graduate advanced data management course**. This course would follow a standard introduction to data management. The instructor could cover Chapters 1–3 and 8 quickly and then Chapters 5, 9, and 11–17 in more detail.

ACKNOWLEDGMENTS

We thank the many reviewers who took the time to read our manuscript and give us their thoughtful comments. The following persons reviewed all or part of the manuscript: Charlie Bachman, Brock Barkley, Grady Booch, Rick Cassidy, Michael Chonoles, Dave Curry, Fred Eddy, John Grosjean, Patricia Hawkins, Mike Hunt, Bill Huth, Chris Kelsey, Mary Loomis, John Putnam, Jim Rumbaugh, James Schardt, Hwa Shen, Lauren Slater, Rod Sprattling, and Barbara Zimmerman. The comments of Chris Kelsey and Jim Rumbaugh were particularly thorough and incisive.

Jim Keegan helped us by writing the first draft for several chapters from our outline. Nancy Talbert was our expert copy editor, improving the quality of expression and organization. Colleen O'Donnel-Nichols provided application content for the pharmacy exercises. Alan Apt and Laura Steele of Prentice Hall facilitated the production and distribution of this book. We are grateful to the Johnson Controls, Lockheed-Martin, and General Electric corporations for their financial support and the intellectual stimulation provided by various work assignments.

We acknowledge the seminal contributions of Mary Loomis and Jim Rumbaugh to the OMT methodology.

We especially acknowledge the contribution of Peter Dietz. He was the manager at GE who championed our first *Object-Oriented Modeling and Design* book. He sponsored our proposal to GE management and provided substantial company time for working on the book from his discretionary budget. Peter has continued to be an important influence in our careers since his move to Johnson Controls in 1993. Many of the ideas in this book were stimulated by the consulting work of Michael Blaha with Johnson Controls funded by Peter Dietz.

Finally, we thank our wives Jean and Judy for their patience and help during the four long years it has taken to write this book.

1

Introduction

First we discuss the purpose of object-oriented software engineering and the benefits of modeling. Then we explain object orientation's contribution to database applications and introduce the Object Modeling Technique (OMT) methodology. We explain the relationship of this book to our prior book [Rumbaugh-91] in the Bibliographic Notes at the end of the chapter.

1.1 WHY OBJECT ORIENTATION?

Object orientation is a strategy for organizing systems as collections of interacting objects that combine data and behavior. It applies to many technology areas, including hardware, programming languages, databases, user interfaces, and software engineering. An object-oriented approach to software development is superior to a procedural approach. For example, an object-oriented approach lets you organize database structures and program functionality around the same paradigm. Databases and programs can be developed together for ease of conceptualization, implementation, maintenance, and potential reuse.

The object-oriented philosophy creates a powerful synergy throughout the development life cycle by combining abstraction, encapsulation, and modularity. *Abstraction* lets you focus on essential aspects of an application while ignoring details. It also lets you preserve design freedom until later stages of development, when you can make more informed decisions. The ability to abstract is probably the most important skill required for object-oriented development, a skill that you can learn by studying this book and working the exercises. *Encapsulation* separates external specification from internal implementation. External specification must be addressed early in development; internal implementation can largely be deferred. *Modularity* promotes coherence, understandability, and symmetry by organizing a system into groups of closely related objects.

The ability to promote abstraction makes an object-oriented approach particularly appealing for software engineering. ***Software engineering*** is a systematic approach to software development that emphasizes thorough conceptual understanding prior to design and coding. Software is most malleable during the early stages of development; errors and oversights are easier to resolve during early stages of software development than once an application is cast in programming and database code. Thorough understanding of a problem can improve software reliability, increase flexibility, reduce development time, and reduce cost.

1.2 OBJECT-ORIENTED SOFTWARE ENGINEERING

You can perform object-oriented software engineering by constructing models. A ***model*** is an abstraction of some aspect of a problem; we express models with various kinds of diagrams. With an object-oriented approach you begin by describing relevant aspects of the real world and gradually elaborate and refine the description until you realize a working system. Thus models not only provide a basis for analyzing requirements but provide the genesis for design and implementation. There are additional advantages of models:

- **Reduced life cycle cost**. The clear documentation helps maintenance and allows more software reuse to occur—from code libraries, from related projects, and within a project. Code becomes smaller in size, more focused, more consistent, and easier to manipulate. You can use tools to generate code.
- **Better quality**. Models provide a language that enables deep thinking about a design. Developers can concentrate on the important issues for a system and defer attention to syntactic details. When a design is formally expressed and subjected to scrutiny by others, the resulting software is more likely to be robust and correct. A modeling language allows a more precise expression of ideas than does a natural language.
- **Faster time to market**. You can organize large projects into work units that can be assigned to different development teams. The clearly defined boundaries between portions of a system let new work occur with minimal impact on previous or parallel work. Software reuse also helps to shorten delivery time.
- **Communication**. Models bring important names and application concepts to the fore so they can be defined and understood by all parties. Models promote communication between developers and customers by separating deep conceptual issues from distracting implementation details.
- **Extensibility**. Software organized about an object-oriented theme parallels the real world and is flexible with respect to changes in requirements; to a large extent the software may be extended for new requirements without disrupting solutions to existing requirements. In contrast software that is decomposed into arbitrary functions (the procedural approach) is brittle and often difficult to evolve.
- **Traceability**. Software development is a seamless process. With object-oriented development the models derived from analysis of customer requirements are carried forward and permeate subsequent development steps. With tool support you can trace the impact of the requirements on the actual delivered system.

- **Better selection**. Models are useful even when software is purchased from a vendor rather than developed. Models can aid understanding of vendor software and enrich the basis for selecting the best product. You can choose the best product not only on the basis of functionality, user interface, and cost but also on the soundness of its supporting infrastructure.

Philosophically, we cannot think of any disadvantage of modeling. Modeling only consumes a modest amount of time; this is easily outweighed by the aforementioned advantages.

We are aware of one practical impediment to modeling—modeling can be difficult to learn. Many persons have difficulty with abstraction and the indirection of realizing an application through models. Our best advice is that developers should pore over this book and work the exercises to become proficient at modeling. You have to practice modeling to learn it. Developers can also benefit from university courses, commercial training, and expert consulting help as they experience modeling for their first few projects.

Do not confuse modeling with programming; object-oriented modeling is not the same as object-oriented programming. Object-oriented programming only concerns latter stages of the software life cycle. Object-oriented software engineering permeates the life cycle and even applies with implementation in a non-object-oriented language. The economic impact of object-oriented software engineering is much broader than that of object-oriented programming.

1.2.1 Benefits of OO Software Development to Database Applications

Object-oriented (OO) modeling is especially helpful for database applications. We give some specific advantages in this section. We assume that most readers know basic database terms, but for those who don't, we first define a few of the more important terms.

A ***database*** is a permanent, self-descriptive repository of data that is stored in one or more files. Self-description is what sets apart a database from ordinary files. A database contains the data structure or ***schema***—description of data—as well as the data. The premise of databases is that the data structure is expected to be relatively static while the actual data may rapidly evolve. Most database systems operate as a closed world; if a fact is not found in the database, the fact is assumed to be false.

A ***database management system*** (***DBMS***) is the software for managing access to a database. DBMSs are intended to provide generic functionality for a wide variety of applications. One of the broad objectives of object-oriented technology is to promote software reuse; for data-intensive applications DBMSs can substitute for much application code. You are achieving reuse when you can use generic DBMS code, rather than custom-written application code. A DBMS has one or more ***query languages***; a query language provides commands for reading and writing data interactively or via a program.

A ***relational database*** is a database in which the data is logically perceived as tables. A ***relational DBMS*** manages tables of data and associated structures that increase the functionality and performance of tables. ***Normal forms*** are guidelines for relational database design that increase the consistency of data. As tables satisfy higher normal forms, they are more likely to store correct data.

An ***object-oriented database*** can be regarded as a persistent store of objects created by an object-oriented programming language. With an ordinary programming language, objects cease to exist at program termination; with an object-oriented database, objects persist beyond the confines of program execution. An ***object-oriented DBMS*** manages the data, programming code, and associated structures that constitute an object-oriented database. In contrast to relational DBMSs, object-oriented DBMSs vary widely in their syntax and capabilities. The Object Data Management Group (ODMG) standards effort [Cattell-96] is gradually resolving this excessive variation and causing a convergence in OO-DBMS syntax and features.

Object-oriented modeling provides benefits for database applications in addition to the general benefits mentioned earlier.

- **One development paradigm**. Object-oriented models provide a uniform abstraction for the design of both programming code and database code. Object-oriented models map naturally to all major languages and the standard types of DBMSs. Unlike a procedural approach, you can apply the same paradigm throughout analysis, design, and implementation. You need not convert from one paradigm to another as software development proceeds.
- **Improved quality of data**. Rigorous modeling improves the quality of the data in a database. You can weave many constraints into the database structure. Modeling lets you clearly see nonstructural constraints, so you can enforce them with the programming code that accompanies the database. An object-oriented model naturally leads to a normalized relational database schema. (Chapter 13 presents mapping rules that you can apply to an object-oriented model to yield a normalized relational database schema.)
- **Better performance**. A well-conceived model simplifies database tuning. A database that crisply corresponds to the real world will simplify expression of queries and facilitate a rapid DBMS response.
- **Faster development**. Careful modeling is required to exploit DBMS capabilities fully. A clear and consistent database structure can simplify programming code. Furthermore, skillful developers can often substitute powerful DBMS commands for tedious, custom-written code.
- **Less debugging**. Flawed models cause errors in the corresponding applications. A poorly conceived database structure impairs the ability of the database to detect application errors. Database constraints can assure quality by providing a means to enforce integrity apart from application code.
- **Easier migration**. Software comes and goes as new systems are deployed and old systems become obsolete. However, data must carry forward and be preserved across systems. Sound models can survive system obsolescence and facilitate migration of legacy data.
- **Easier integration**. Integration of stand-alone systems can reduce human transcription of data and broaden the queries that can be answered. However, for integration to succeed, system architects must thoroughly understand the applications—purchased software, legacy systems, and newly developed software. Object-oriented models provide a uniform representation that facilitates understanding and integration.

1.3 THE OMT METHODOLOGY

There are many software development approaches that utilize object-oriented models, but not all of them are suitable for database applications. Some approaches lack sufficient abstraction and are tied too tightly to implementation details. Others are too programming oriented and fail to emphasize data structures and constraints adequately, which are particularly important to database applications.

Our approach, OMT, stresses the importance of models and uses models to achieve abstraction. We crisply differentiate the analysis stage, which focuses on the real world, from the design stage, which addresses the particular details of the computer resources. As such, the OMT methodology can apply to various implementation targets, including files, relational databases, and object-oriented databases. The OMT models are built around descriptions of data structure, behavior, and constraints. The OMT methodology addresses both transaction-processing systems (high volume, simple processing) and decision-support systems (a rich variety of unpredictable queries). The Bibliographic Notes at the end of this chapter comment on the relationship of this book to our prior book [Rumbaugh-91].

The OMT approach builds on past work. From the object-oriented programming community we have incorporated the notion of objects as coherent entities with identity, state, and behavior. Objects may be organized by their similarities and differences enabling the use of inheritance and polymorphism. From information modeling we have adopted the notion of entities that are connected by relationships. Information models tend to be *declarative* (stated as structural properties and constraints) rather than *imperative* (stated as a series of programming statements). We have also included ideas from software engineering, such as the notion of a formal process for software development with emphasis placed on early conceptual stages rather than later implementation details. Misunderstandings and flaws that are detected early are much easier and less costly to correct.

The OMT methodology emphasizes declarative specification of information. Declaration lets you cleanly capture requirements; imperative specification runs the risk of prematurely descending into design. Declarative specifications tend to be more concise and allow machine-based optimization. Many DBMSs provide direct implementation support for declarative specification.

1.3.1 OMT Development Process

A bonafide methodology consists of both process and concepts. A clear process must guide the practitioner through system development. The process is supported by underlying concepts and a notation for expressing the concepts. The OMT development process comprises the following sequence of steps. (The process preview in Chapter 6 expands the description of these steps.) Much iteration occurs within the steps and some iteration occurs across the steps.

- **Conceptualization**. Software development begins with business analysts or users conceiving an application and formulating tentative requirements. Conceptualization is often stimulated by business process reengineering (BPR)—critical examination of business processes and their impact on profitability. BPR identifies opportunities for new systems, specifies their purpose, and determines their scope. BPR is outside the scope of this book.

- **Analysis**. The requirements formed during conceptualization are scrutinized and rigorously restated by constructing real-world models. The goal of analysis is to specify *what* needs to be done without constraining *how* it is done.
- **System design**. The development team devises a high-level strategy—the system architecture—for solving the application problem. Policies must also be established that will serve as a default for the subsequent, more detailed portions of design.
- **Detailed design**. Real-world analysis models are augmented and transformed into a form amenable to computer implementation. We perform this shift in the models without descending into the particular details of the target database or programming language. We have renamed this step from *object design* in [Rumbaugh-91].
- **Implementation**. The design is translated into the actual programming language and database code. This step is straightforward, because much work has already been performed during analysis and design.
- **Maintenance**. The development documentation and the traceability from the models through the code facilitate subsequent maintenance. Treatment of maintenance, testing, and project management is outside the scope of this book.

The OMT methodology supports multiple development styles. You can use OMT as a waterfall approach performing the phases of analysis, design, and implementation in strict sequence for the entire system. More often, we adopt an iterative development strategy. First we develop the nucleus of a system—analyzing, designing, implementing, and dclivcring working code. Then we grow the scope of the system, adding properties and behavior to existing objects, as well as adding new kinds of objects.

Conceptualization and analysis are largely independent of the target platform. You must first understand an application before you can proceed, regardless of whether the application is being implemented with a programming language, database, or some other means. In contrast, design, implementation, and maintenance are heavily influenced by the target platform—in the case of this book, a database.

1.3.2 Entity-Based versus Attribute-Based Development

As we have said, the OMT approach focuses on objects that are coherent entities. This is in contrast to the attribute-based approach traditionally associated with relational database applications. We believe the distinction is purely philosophical, however, and not at all intrinsic to relational databases themselves.

In an entity-based approach, originated by [Chen-76], you note entities from the real world, describe them, and observe relationships among them. The OMT methodology adopts an entity-based approach. In contrast, with an attribute-based approach you list attributes that are meaningful to an application and organize them into groups.

In practice, entity-based development is clearly superior to attribute-based development. First, most applications have an order of magnitude fewer entities than attributes, so an entity-based approach is more tractable. Second, attribute-based development requires that you check normal forms—to ensure that attributes are combined in a reasonable manner so update anomalies are unlikely to occur. These checks can be tedious and confusing. With an entity-based approach attributes are introduced as part of describing real-world objects. To the extent that you directly describe objects and do not intermix different things, the anomalies addressed by normal forms do not arise. This real-world basis also makes an entity-based model more extensible, understandable, and tunable for fast performance. Finally, an entity-based approach is more likely to be compatible with the development approach for programming code and the overall system.

1.3.3 The Three OMT Models

The creator of a modeling language must confront the trade-off of expressiveness and understandability. A simple language is coherent but may fail to capture fully the nuances of a problem. A complex language permits more precise specification but may inhibit broad understanding. We have attempted to strike a balance between modeling power and simplicity with the OMT notation. As an example of the richness of the OMT notation, you can specify much programming behavior with simple navigation of an object model. Yet the OMT notation is understandable, as we have learned through experience with users on various projects.

As Figure 1.1 shows, the OMT methodology uses three complementary models to express aspects of a problem and its solution. Each model presents a perspective of the system, which by itself is incomplete. The models combine to describe a system fully. Each model contributes important information to the development process, but the models do not have equal importance for an individual application. Database applications, for example, tend to rely heavily on the object model.

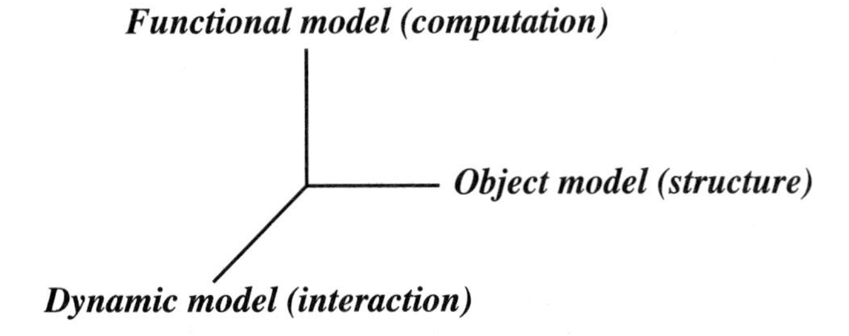

Figure 1.1 The three OMT models

We briefly describe the models here in preparation for more elaborate discussion in subsequent chapters.

- **Object model**. The object model characterizes the static structure of things (objects). It looks at structure in terms of groups of analogous objects (classes), their similarities and differences (generalization), and their important relationships with one another (associations). The object model defines the context for software development—the universe of discourse. An object model is expressed with one or more class diagrams. The object model is important for database applications because it concisely describes data structure and captures structural constraints.

- **Dynamic model**. The dynamic model describes temporal interactions between objects —various stimuli that occur and the response of objects to the stimuli. The objects in a class pass through states with similar responses to events. State diagrams express an aspect of behavior that is shared by the objects in a class. There is one state diagram for each class with significant dynamic behavior; the collection of interacting state diagrams constitutes the dynamic model.

  The typical database application is dominated by storage and retrieval, and the dynamic model is seldom important. (In contrast, the dynamic model is quite significant for many other kinds of applications, such as user interfaces and control applications.) Given that the focus of this book is on data management applications, we will omit treatment of the dynamic model. For a discussion of the dynamic model, see [Rumbaugh-91]. The full OMT methodology still includes the dynamic model, but we do not discuss the dynamic model in this book.

- **Functional model**. The functional model defines the computations that objects perform—how output values are computed from input values. The functional model specifies operations that arise within the object and dynamic models. We recommend several notations for expressing the functional model, though we emphasize pseudocode enhanced with the object navigation notation. (This is a departure from [Rumbaugh-91], which emphasizes data flow diagrams.) We believe it is beneficial to have more than one notation and to be able to mix them. The primary importance of the functional model for database applications is in showing navigation of the object model.

1.4 ORGANIZATION OF THIS BOOK

This chapter introduces basic concepts for object orientation, software engineering, and databases. The remainder of the book consists of four parts: concepts and notation, analysis and design, implementation, and large systems issues. The appendixes provide a glossary of terminology and summarize the functional modeling notation.

Part 1 (Chapters 2 through 5) describes OMT concepts and notation. We explain the object model in great detail because database applications tend to have rich data structure. We also cover the functional model, but omit the dynamic model. Our experience has been that data management applications seldom involve a significant dynamic model.

Part 2 (Chapters 6 through 11) presents the OMT analysis and design process. We introduce the financial portfolio manager case study and thread it throughout Parts 2 and 3.

Part 3 (Chapters 12 through 17) shows how to implement an OMT design for a variety of platforms: files, relational databases, and object-oriented databases (ObjectStore). We emphasize databases in this book, but also cover files because files are a reasonable option for managing data for many applications. We fully implement the portfolio manager case study for the MS-Access relational database. You can find the detailed code at our Web site (www.omtassociates.com).

Part 4 (Chapters 18 through 20) concludes the book with treatment of several large systems issues: distributed databases, integration of systems of applications, and reverse engineering of legacy databases.

This book is self-contained. However, the organization of the book deliberately parallels that of our past book [Rumbaugh-91], so you can see more clearly how database considerations affect the OMT development process.

All major chapters of this book contain exercises with indicated difficulty levels. The exercises bring out subtleties you may miss in the text. If you want to learn this material well, you should work the exercises.

1.5 FONT CONVENTIONS

We use several font conventions throughout this book. We begin most bullet items with a phrase in bold font to denote the theme. We italicize words that are modeling elements or for emphasis. We use bold, italic font to call attention to defined terms that are also listed in Appendix A.

1.6 CHAPTER SUMMARY

Object orientation is a strategy for organizing systems, especially hardware and software systems, as collections of interacting objects that combine data and behavior. Software engineering is a systematic approach to software development that emphasizes thorough conceptual understanding prior to design and coding. This book concerns object-oriented software engineering, the intersection of object orientation and software engineering. You can perform object-oriented software engineering by constructing models that abstract various aspects of a problem. You describe relevant aspects of the real world and gradually elaborate and refine the description until you realize a working system.

The OMT methodology is our approach for object-oriented software engineering. Any bonafide methodology consists of both process and concepts. A clear process must guide the practitioner through system development. The process is supported by underlying concepts and a notation for expressing the concepts. The OMT development process comprises the major steps of analysis, design, and implementation. During analysis the developer scrutinizes and rigorously restates requirements by constructing real-world models; the goal of analysis is to specify what needs to be done without constraining how it is done. During design the developer must shift focus from the real world toward the underlying computer resources. Implementation consists of relatively straightforward translation of the design into programming and database code.

The OMT methodology uses three complementary models that support the process. The object model characterizes the static structure of a system in terms of groups of analogous objects (classes), their similarities and differences (generalization), and their important relationships with one another (associations). The dynamic model describes temporal interactions between objects—various stimuli that occur and the response of objects to the stimuli. The functional model defines the computations that objects perform—how output values are computed from input values. Database applications seldom involve a significant dynamic model, so we omit coverage of the dynamic model from this book.

Figure 1.2 lists key concepts that we have discussed in this chapter.

abstraction	implementation
analysis	object model
conceptualization	object orientation
database	OMT methodology
database management system (DBMS)	schema
detailed design	software engineering
dynamic model	system design
functional model	

Figure 1.2 Key concepts for Chapter 1

BIBLIOGRAPHIC NOTES

The OMT object model is essentially an extended Entity-Relationship approach. [Chen-76] is the seminal paper on the Entity-Relationship approach to modeling. The OMT dynamic model is taken from [Harel-87] with some minor adjustments.

The general OMT methodology—including notation, concepts, and an analysis process that spans programming and database applications—is presented in an earlier book [Rumbaugh-91]. However, the treatment of databases, especially the design and implementation process, in [Rumbaugh-91] is incomplete. [Rumbaugh-96] extends the OMT methodology, mostly for programming applications.

This book is a sequel that supersedes [Rumbaugh-91] for data management applications. We have introduced additional advanced object modeling concepts and an entire chapter on metamodeling. We have also reworked the functional model. Nevertheless, our primary contribution lies in the areas of design, implementation, and large system issues. We have completely rewritten the design and implementation portions to address properly applications that are data intensive. Database applications require custom treatment because they are much different from programming applications. Database applications tend to have a much larger and more complex data structure and less rich behavior compared to programming applications. The exercises in this book complement those in [Rumbaugh-91].

Booch, Rumbaugh, and Jacobson have developed a new notation that unifies the Booch, OMT, and Objectory notations [UML-98]. They have made several changes in their notation that are pertinent to databases, and we have adopted them. Our emphasis in this book is on fleshing out the OMT process for data management applications, not on creating new notation. The UML authors are addressing programming applications; we are addressing database applications. The union of their efforts and our efforts constitutes the next generation of the OMT methodology.

In general, there are two major camps of object-oriented approaches to software development. The information-centric approaches start by building an enhanced Entity-Relationship model and then add behavior. OMT, Shlaer-Mellor [Shlaer-88], and Martin-Odell [Martin-92] are information-centric methodologies. In contrast, the behavior-centric approaches start by describing the behavior of a system and then develop an information model. Object-Oriented Software Engineering (OOSE) [Jacobson-92] and Requirements Driven Design (RDD) [Wirfs-Brock-90] illustrate the behavior-centric approach. Information-centric approaches tend to be declarative (organized about data structure), while behavior-centric approaches tend to be imperative (programming oriented).

We emphasize that software development is a seamless process; OMT models are gradually elaborated and optimized as the focus shifts from analysis to design to implementation. Some methodologists, such as [Shlaer-88], adopt an alternate viewpoint—shifting notations when proceeding from analysis to design. Our experience has been that a shift in notation is unnecessary, causes confusion, and can lose information.

REFERENCES

[Cattell-96] R Cattell, editor. *The Object Database Standard: ODMG-93, Release 1.2*. San Mateo, California: Morgan Kaufmann, 1996.

[Chen-76] PPS Chen. The Entity-Relationship model—toward a unified view of data. *ACM Transactions on Database Systems 1*, 1 (March 1976), 9–36.

[Harel-87] David Harel. Statecharts: a visual formalism for complex systems. *Science of Computer Programming 8* (1987), 231–274.

[Jacobson-92] Ivar Jacobson, Magnus Christerson, Patrik Jonsson, and Gunnar Overgaard. *Object-Oriented Software Engineering: A Use Case Driven Approach*. Reading, Massachusetts: Addison-Wesley, 1992.

[Martin-92] James Martin and James Odell. *Object-Oriented Analysis and Design*. Englewood Cliffs, New Jersey: Prentice Hall, 1992.

[Rumbaugh-91] J Rumbaugh, M Blaha, W Premerlani, F Eddy, and W Lorensen. *Object-Oriented Modeling and Design*. Englewood Cliffs, New Jersey: Prentice Hall, 1991.

[Rumbaugh-96] James Rumbaugh. *OMT Insights*. New York: SIGS Books, 1996.

[Shlaer-88] Sally Shlaer and Stephen J. Mellor. *Object-Oriented Systems Analysis: Modeling the World in Data*. Englewood Cliffs, New Jersey: Yourdon Press, 1988.

[UML-98] The following books are planned for the Unified Modeling Language:

Grady Booch, James Rumbaugh, and Ivar Jacobson. *UML User's Guide*. Reading, Massachusetts: Addison-Wesley.

James Rumbaugh, Ivar Jacobson, and Grady Booch. *UML Reference Manual*. Reading, Massachusetts: Addison-Wesley.

Ivar Jacobson, Grady Booch, and James Rumbaugh. *UML Process Book*. Reading, Massachusetts: Addison-Wesley.

[Wirfs-Brock-90] Rebecca Wirfs-Brock, Brian Wilkerson, and Lauren Wiener. *Designing Object-Oriented Software*. Englewood Cliffs, New Jersey: Prentice Hall, 1990.

PART 1: MODELING CONCEPTS

2

Basic Object Modeling

Part 1 presents the object and functional models. We define notation, explain the ideas embodied by the notation, and present many examples. We have made several changes in the original OMT notation to make it consistent with the work of Booch, Rumbaugh, and Jacobson [UML-98]. We summarize these changes at the end of each chapter.

The OMT notation spans the development lifecycle; the models that are developed during analysis carry forward to design and implementation. As software development proceeds, the developer elaborates and optimizes the OMT models. The emphasis in Part 1 is on analysis. Database applications require some additional notation for design and implementation; we present this additional notation in Parts 2 and 3.

We explore the object model in greater detail than the functional model and omit the dynamic model because database applications tend to be dominated by the object model. The dynamic model is seldom important for such applications, although it is integral to the OMT methodology for other purposes, such as user interfaces and control applications.

The object model describes a system's data structure, its objects and their relationships. Object models are built from classes that relate to each other through associations and generalizations. The object model defines the context for software development and provides the lattice to which behavior is attached. Chapter 2 describes object modeling concepts that frequently occur and that are sufficient for modeling many applications. Chapter 3 addresses additional aspects of object modeling that are required for advanced applications. Chapter 4 is very advanced and discusses object metamodeling.

The OMT approach places greater emphasis on data structure and constraints than do most other object-oriented methodologies. With the object model, there are often alternative ways to represent an application. (See Exercise 2.13.) It is important to exploit the richness of OMT modeling and capture data structure precisely. A thorough object model enforces many constraints and makes it easier to navigate among objects. The richness of the object model obviates much of the algorithmic code required by other methodologies.

2.1 OBJECT AND CLASS CONCEPTS

2.1.1 Objects

An ***object*** is a concept, abstraction, or thing that has meaning for an application. Objects often appear as proper nouns or specific references in problem descriptions or in discussions with users. Some objects have real-world counterparts (Albert Einstein and the General Electric company), while others are conceptual entities (simulation run 1234 and the formula for solving a quadratic equation). Still others (binary tree 634 and the array bound to variable *a*) are introduced for implementation reasons and have no correspondence to physical reality. The choice of objects depends on judgment and the nature of a problem; there can be many correct representations.

Each object exists and can be identified. For example, even though two persons may have the same name, they are separate. ***Identity*** is "that property of an object which distinguishes each object from all others" [Khoshafian-86]. During analysis, you can take for granted that objects have identity. During design, you must choose an approach for realizing identity such as memory addresses, assigned numbers, or combinations of attribute values.[*] This strong notion of identity is an integral and important part of object orientation.

Objects and their relationships are delineated in an instance diagram, such as that in Figure 2.1. We denote an object by a box with an object name in the top portion of the box followed by a colon and the class name. The object name and class name are both underlined. Our convention is to list the object name and class name in boldface. An optional second portion of the box may list each attribute name followed by an equal sign and the value. We do not list operations because they do not vary among objects of the same class. In the figure, object *aBinaryTree* belongs to the class *BinaryTree* and has no specified attribute values. Object *IAH* is in class *Airport* and has the values *IAH*, *Intercontinental*, and *Central*.

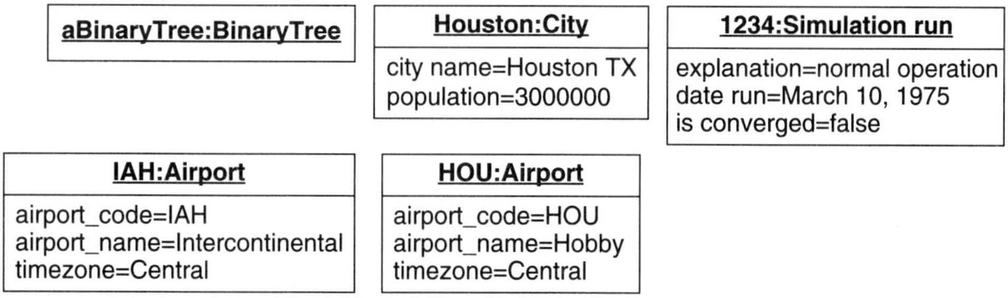

aBinaryTree:BinaryTree

Houston:City
city name=Houston TX
population=3000000

1234:Simulation run
explanation=normal operation
date run=March 10, 1975
is converged=false

IAH:Airport
airport_code=IAH
airport_name=Intercontinental
timezone=Central

HOU:Airport
airport_code=HOU
airport_name=Hobby
timezone=Central

Figure 2.1 Objects

[*] If you use an object-oriented DBMS, it will impose an approach to identity and save you much work. In contrast, relational DBMSs and files provide less support for identity. Parts 2 and 3 show how to deal with identity for relational DBMSs and files.

When you construct models you must exercise judgment in deciding which objects to show and which objects to ignore. An object represents only the *relevant* aspects of a problem. It is not helpful to model extraneous detail; the scope of a model is driven by what is needed for an application.

Figure 2.1 demonstrates three conventions for separating portions of a name: spaces, underscores, and intervening capital letters. For a particular problem, you should consistently use one of these. From now on, we will use the convention of mixed case.

2.1.2 Classes

An object is an ***instance***—or occurrence—of a class. A ***class*** is a description of a group of objects with similar properties (object attributes), common behavior (operations and state diagrams), similar relationships to other objects, and common semantics. Classes provide a mechanism for sharing across similar objects. Classes often appear as common nouns and noun phrases in problem descriptions or in discussions with users. Some classes have real-world counterparts (person and company) while others are conceptual entities (simulation run and equation). Still other classes (binary tree and array) are purely artifacts of implementation. The choice of classes depends on the nature and scope of an application and is a matter of judgment.

Classes and their relationships are delineated in a class diagram, such as that in Figure 2.2, which shows the classes that correspond to the objects in Figure 2.1. We denote a class by a box with the class name in the top portion of the box. An optional second portion of the box may list attributes. An optional third portion of the box may list operations. Our convention is to list the class name in boldface, capitalize the first letter of a class name, and use lowercase letters for the first letter of attribute and operation names. We use singular nouns for the names of classes. We center the class name and left-align the attributes and operations.

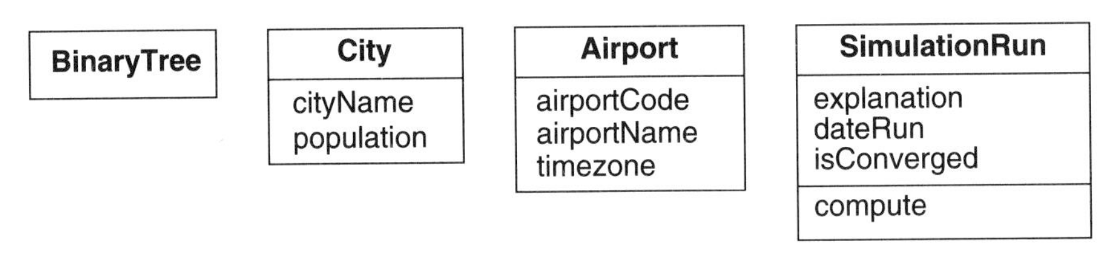

Figure 2.2 Classes

In Figure 2.2 class *BinaryTree* has no attributes and no operations specified. If *BinaryTree* were part of a large, multipage class diagram, it could have attributes and operations specified elsewhere, but they are not shown in Figure 2.2. Class *City* has attributes *cityName* and *population*. Class *Airport* has attributes *airportCode*, *airportName*, and *timezone*. We do not show operations for *City* and *Airport*. This does not imply a lack of operations; we just have not shown any. Class *SimulationRun* has attributes *explanation*, *dateRun*, and *isConverged* as well as operation *compute*.

Class diagrams and instance diagrams are the expression of the object model. As a class describes a group of objects, so too a class diagram describes a group of instance diagrams. Thus, a class diagram lets you abstract many individual instance diagrams to document data structure thoroughly. We will revisit class and instance diagrams after reviewing more object-modeling constructs.

The term "class" has been overloaded with several meanings in the literature. [Cattell-91] delineates three uses of the term "class."

- A class defines the ***intent***—the structure and behavior of objects of a particular type. The intent is the theme, essence, or semantic meaning of a class. For example, the intent for *SimulationRun* consists of the properties (*explanation*, *dateRun*, *isConverged*), behavior (*compute*), and relationships to other classes of objects (discussed in Section 2.2). Most object-oriented DBMSs and relational DBMSs adopt this definition of a class and it is the meaning we assign to "class" in the OMT methodology.
- A class defines a ***representative***—types represented by objects themselves. This meaning arises with the prototypical or delegation approach to object-oriented programming (such as the programming language *Self* [Ungar-87]). For example, you could define a new *SimulationRun* object by noting the deviations with respect to *SimulationRun 1234*.
- A class defines an ***extent***—the set of objects with a particular type. For example, the extent for *SimulationRun* is the set of all simulation run objects. In a relational database, the table combines the notions of intent and extent; a table name refers to both the structure of the table and the collection of instances stored in the table. Most object-oriented DBMSs do not automatically maintain the extent of a type.

2.1.3 Values and Object Attributes

A ***value*** is a piece of data. Values are found with enumerations and examples in problem documentation. An ***object attribute*** is a named property of a class that describes a value held by each object of the class. Object attributes are often discovered through adjectives or by abstracting typical values. The following analogy holds: Object is to class as value is to attribute. Object models are dominated by structural constructs, that is classes and relationships. Attributes are of lesser importance and serve to elaborate classes and relationships.

As the instance diagram in Figure 2.1 shows, values are listed in a second portion beneath the object name. The *IAH* object has values *IAH*, *Intercontinental*, and *Central*. Similarly, as the class diagram in Figure 2.2 shows, object attributes are listed in a second portion beneath the class name. The *Airport* class has attributes *airportCode*, *airportName*, and *timezone*.

Do not confuse values with objects. Objects have identity; values do not. Three persons with the name *Jim Smith* are three separate objects with their own identities. In contrast, all occurrences of the value "Jim Smith" (a string) are indistinguishable. Objects may be described with attributes and operations and participate in relationships; values merely exist.

Furthermore, a value should not refer to an object. The value of an attribute should be a simple primitive (string or number) or a collection of simple primitives. A reference to another object is properly modeled as a link, as we describe in Section 2.2.1.

When creating an analysis model, you should not include an attribute merely as an internal identifier for the instances of a class; object identifiers are implicit in analysis models. You should list only attributes that have meaning for the application. In contrast, during design you may show internal identifiers. For example, you may use internal identifiers to clarify the design of relational database tables.

In Figure 2.3, *airportID* is an inappropriate attribute for analysis because it has no intrinsic meaning. It merely refers to the identifier generated by an object-oriented DBMS or the sequence number that can be used to implement identity with a relational DBMS. In contrast, it can be quite proper to list *vehicleNumber* during analysis, because such a number can be found on the dashboard of a car.

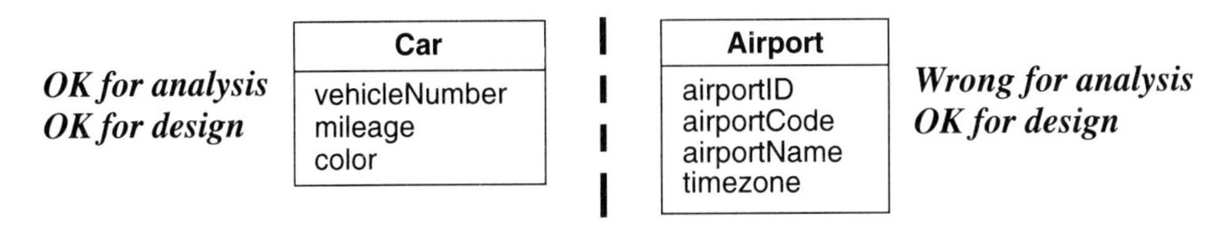

Figure 2.3 Do not show object identifiers in an analysis model

2.1.4 Operations and Methods

An ***operation*** is a function or procedure that may be applied to or by objects in a class. Each operation has a target object as an implicit argument. As Figure 2.2 shows, operations are listed in the third portion of the box. Thus, class *SimulationRun* has operation *compute* with an implicit target of a simulation run. Operation names may be followed by optional details such as an argument list and result type. Some operations are ***polymorphic***, that is applicable to many classes. A ***method*** is the implementation of an operation for a class.

For example, suppose class *Equipment* has an operation *calculateCost*. Operation *calculateCost* is polymorphic because there are different methods (formulas) for computing cost depending on the kind of equipment (tanks, pumps, conveyors, and so on). All these methods logically perform the same task but may be implemented by different pieces of code. Object-oriented software automatically chooses the appropriate method to use in computing an operation according to the class of the target object.

2.2 LINK AND ASSOCIATION CONCEPTS

2.2.1 Links and Associations

A ***link*** is a physical or conceptual connection between objects. Most links relate two objects, but some links relate three or more objects. An ***association*** is a description of a group of links with common structure and common semantics. Thus, a link is an instance of an association. The links of an association relate objects from the same classes and have similar properties (link attributes). An association describes a set of potential links in the same way a class describes a set of potential objects. Links and associations often appear as verbs in problem statements.

Figure 2.4 and Figure 2.5 illustrate notation for links and associations, respectively. A link is denoted by a line connecting related objects; a line may consist of several line segments. For example, both the IAH and HOU airports serve the city of Houston. An association connects related classes and is also denoted by a line (with possibly multiple line segments). For example, airports serve cities. (The two solid balls are multiplicity symbols, which we will discuss in the next section.) Our convention is to show link and association names in italics and to try to confine line segments to a rectilinear grid.

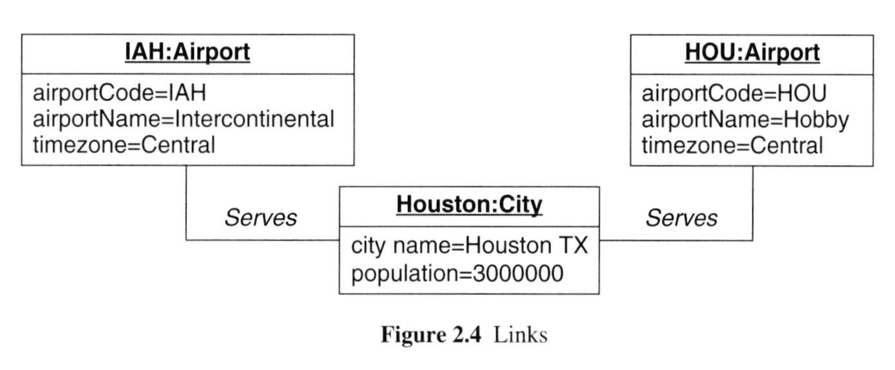

Figure 2.4 Links

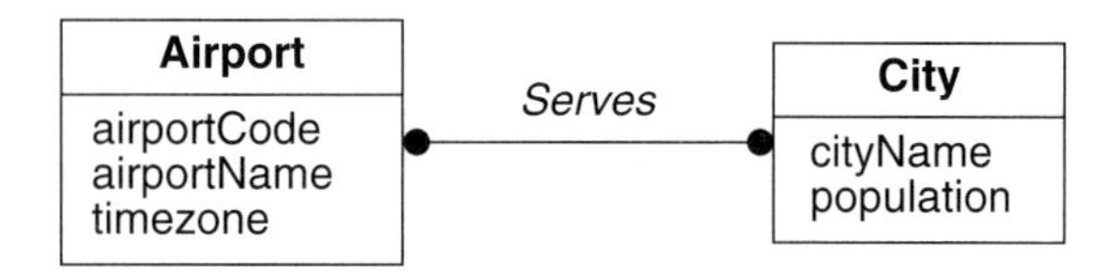

Figure 2.5 An association

The association name is optional, if the model is unambiguous. Ambiguity arises when there are multiple associations between the same classes (*person works for company* and *person owns stock in company*). When there are multiple associations, you must use association names or role names (Section 2.2.3) to resolve the ambiguity.

During analysis you may traverse links and associations in either direction, even though an association name may read in a single direction. For example, the association *an airport serves multiple cities* also implies that a city may be served by multiple airports.

Do not confuse links with objects. An object has inherent identity and exists in its own right. A link derives its identity from the related objects and cannot exist in isolation. A link can exist only if all objects to which it refers also exist.

Furthermore, do not confuse links with values. A value is a primitive such as a string or number. A link is a relationship between objects. During analysis, you should model all references to objects with links and recognize groups of similar links with associations. During design, you can choose to implement associations with pointers, foreign keys, or by some other means. Figure 2.6 illustrates two of several design models you could use to implement the *airport serves city* analysis model. You should delay the decision of how to implement associations until you complete analysis and understand a problem.

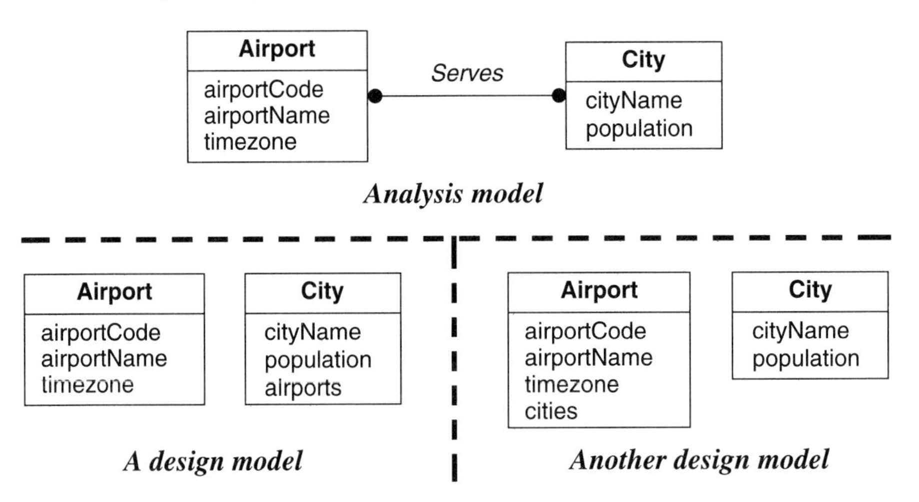

Figure 2.6 Use associations to model references to objects during analysis

Associations are an important and distinctive aspect of OMT modeling. Associations are taken for granted in the database community, where they have long been used with Entity-Relationship modeling. Most object-oriented DBMSs and the ODMG standard [Cattell-96] support simple associations. Most relational DBMSs provide referential integrity that also can support associations (see Chapter 14). However, associations have been overlooked by most object-oriented languages, as well as many object-oriented development methodologies. An association is not the same as a pointer. Proper use of associations will take some effort from those accustomed to programming and who may be less familiar with databases.

Figure 2.7 demonstrates the utility of associations with a simple model of information that might help a traveler choose a hotel. Each metropolitan area has a number of hotels, some of which are affiliated with a hotel chain. For example, the Hilton hotel in downtown Fort Wayne, Indiana belongs to the Hilton hotel chain. Most hotels accept a variety of credit

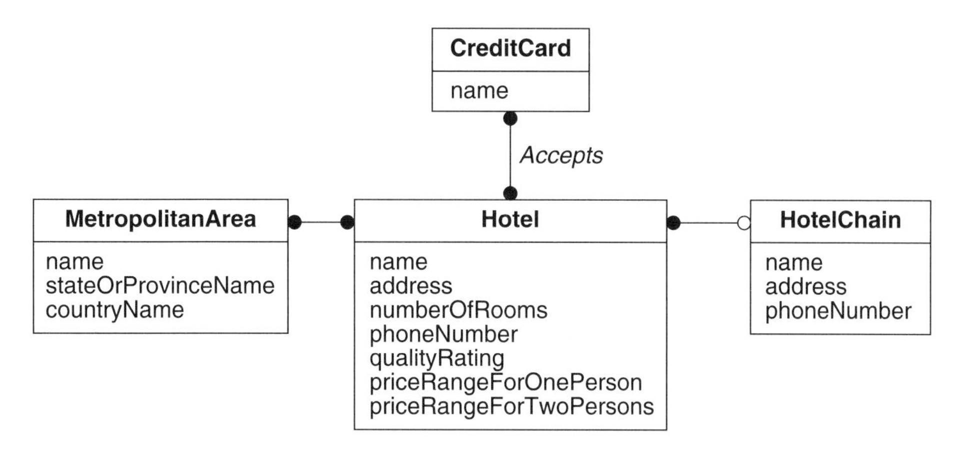

Figure 2.7 Associations in an object model for hotel selection

cards; the major credit cards are accepted by most hotels. (The solid and hollow balls are multiplicity symbols, which we will discuss in the next section.)

2.2.2 Multiplicity

Multiplicity specifies the number of instances of one class that may relate to a single instance of an associated class. Do not confuse "multiplicity" with "cardinality." Multiplicity is a *constraint* on the size of a collection; cardinality is the *count* of elements that are actually in a collection. Therefore, multiplicity is a constraint on the cardinality. Figure 2.8 summarizes multiplicity combinations. Solid balls denote "many" multiplicity, meaning zero or more. For example, a metropolitan area may have many hotels. A hollow ball denotes "zero or one" multiplicity. For example, a hotel may or may not belong to a hotel chain. The lack of a symbol at the end of an association line means exactly one. A driver's license belongs to a single person.

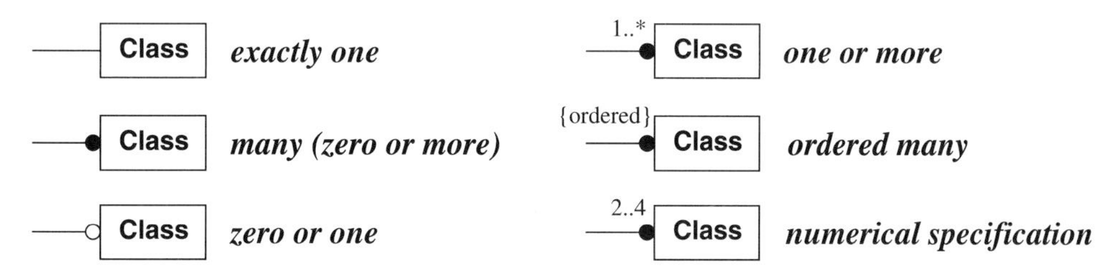

Figure 2.8 OMT notation for multiplicity

You may annotate a solid ball to indicate a more specific multiplicity, such as "1..*" or "2..4" multiplicity. You may also list several ranges such as "3..5,7..9,15..20" multiplicity. You may specify that the collection of objects for "many" multiplicity is ordered.

Figure 2.9 illustrates multiplicity with an excerpt of a model for airline flight reservations. A flight may be booked with multiple flight reservations. (Hopefully there will be numerous reservations or the plane will be empty!) The TW 250 flight from St. Louis to Milwaukee on October 23, 1995 is an example of a flight. The fare code for a flight reservation determines the fare that the airline will charge. For example, there are fare codes for first class, unrestricted coach, and weekend excursions. A trip reservation consists of multiple flight reservations that are ordered as a passenger proceeds from one city to the next. The airlines use a record locator to identify a trip reservation in their information systems. The passenger may note the record locator, especially for communicating with the airline about a complicated trip. A ticket may be issued for a trip, subject to the actual fare charged.

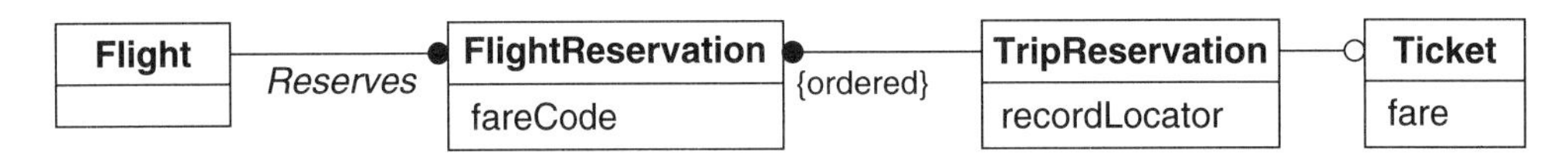

Figure 2.9 A simple object model for airline flight reservations

We do not show any attributes for *Flight* in Figure 2.9. A class need not have attributes to be included in an object model. A class may be significant because of its relationship to other classes, or a class may have attributes that have not been noted yet. We are going to elaborate the flight reservations example as we proceed in this chapter and will show attributes for *Flight* later.

A multiplicity of "many" specifies that an object may be associated with multiple objects. However, for each association there is at most one link between a given pair of objects. As Figure 2.10 and Figure 2.11 show, if you want two links between the same objects, you must have two associations.†

Figure 2.10 A pair of objects can be instantiated at most once per association

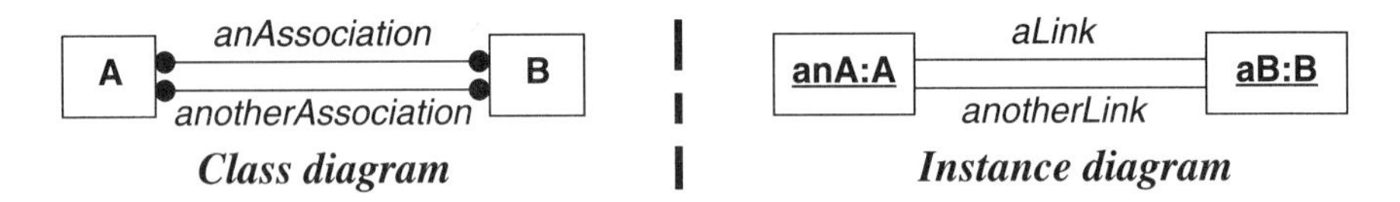

Figure 2.11 Use multiple associations to model multiple links between the same objects

† We thank Michael Chonoles for the suggestion to clarify this possible confusion.

Our notation for multiplicity follows that in [Rumbaugh-91], but you may use that in [UML-98] if you prefer. The notation in [UML-98] drops the solid and hollow balls and explicitly lists all multiplicities; "*" is a special symbol that denotes "many." Figure 2.12 restates Figure 2.9 to illustrate the UML notation. We believe the multiplicity notation in [Rumbaugh-91] is more readable than the UML multiplicity notation, especially for large models with many classes. Visual conciseness and brevity are important in presenting large models.

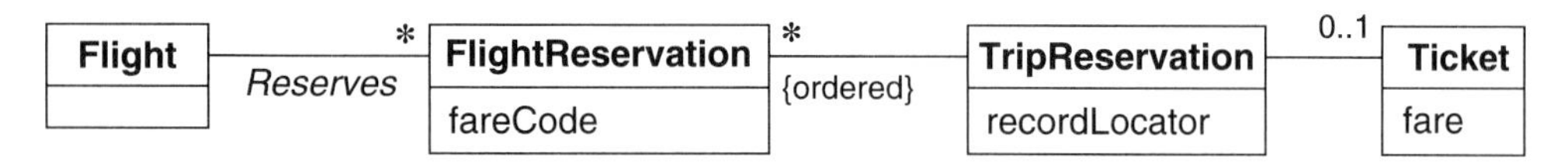

Figure 2.12 The flight reservations model using the multiplicity notation from [UML-98]

2.2.3 Roles

A ***role*** is one end of an association and may be assigned an explicit name. Role names often appear as nouns in problem descriptions. As Figure 2.13 shows, a role name is written next to the class-association intersection. When there is a single association between a pair of distinct classes (as in Figure 2.9), the names of the classes often serve as good role names, in which case you may omit the role names from the diagram.

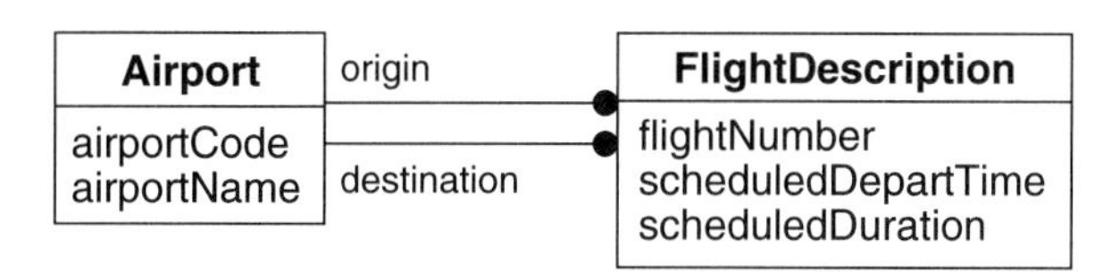

Figure 2.13 Roles in an object model for flight descriptions

FlightDescription in Figure 2.13 is different from *Flight* in Figure 2.9. A flight description refers to the published description of air travel between two airports. In contrast, a flight refers to the actual travel made by an airplane on a particular date. Thus, TW 250 is an example of a flight description; TW 250 on October 23, 1995 is an example of a flight.

Role names are especially convenient for traversing associations, because you can treat a role name as a pseudo attribute. In Figure 2.13 the notation *aFlightDescription.origin* refers to the airport at which a flight originates. (We use the convention of preceding a class name by "a" to refer to an object. Thus *aFlightDescription* refers to an object of the *FlightDescription* class.) Similarly, *aFlightDescription.destination* refers to the airport at which a flight terminates. Since a role is a pseudo attribute, the role name must not clash with any other attribute or role of the origination class. Thus, *FlightDescription* may not have any attributes or other target roles called *origin* or *destination*. Chapter 5 elaborates the notation for traversing object models.

Some modelers prefer to use directed association names, as Figure 2.14 shows. With directed association names you can more easily translate a model into natural language, but we consider this a minor benefit. Besides, a developer should never disseminate a "naked" model; there should always be definitions of key terms and a natural-language narrative to guide readers through the model and document the rationale of various modeling decisions. We strongly favor role names because they can be easily referenced when traversing an object model, unlike directed relationship names (see Chapter 5).

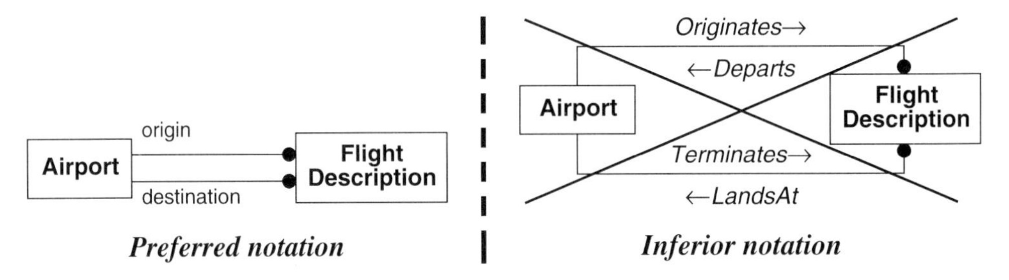

Figure 2.14 Use roles instead of directed association names

Role names are optional if a model is unambiguous. Ambiguity occurs when there are multiple associations between the same classes (Figure 2.13) or an association between objects of the same class (a ***reflexive association***, Figure 2.15). You can use association names or role names to resolve multiple associations between the same classes. You must use role names to clarify an association between objects of the same class.

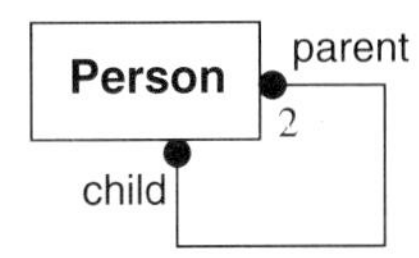

Figure 2.15 Roles can clarify an association between objects of the same class

Roles let you unify multiple references to the same class. When constructing class diagrams you should properly use roles and not introduce a separate class for each reference, as shown in Figure 2.16. In the wrong model, a person with a child is represented by two instances, one for the child role and one for the parent role. In the correct model, one person instance participates in two links, once as a parent and once as a child.

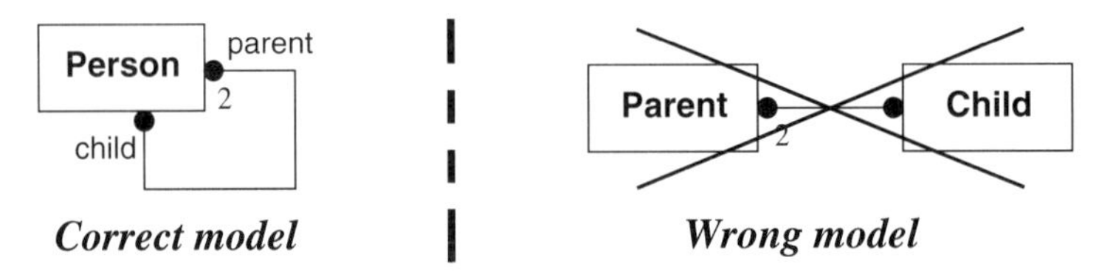

Figure 2.16 Use roles to model multiple references to the same class

2.2.4 Link Attributes

An object attribute is a named property of a class that describes a value held by each object of the class. Similarly, a ***link attribute*** is a named property of an association that describes a value held by each link of the association. A link attribute is a property of the association and should not be ascribed to an associated class. Link attributes are often discovered through adverbs in a problem statement or by abstracting known values.

A link attribute is denoted by a box attached to the association by a dashed line. The top portion of the box may be left blank or may contain the association name. One or more link attributes may appear in the second portion of the box. In Figure 2.17, a judge assigns a score to the efforts of a competitor (*Trial*) at an athletic meet. The *score* is a link attribute of the association between *Trial* and *Judge*.

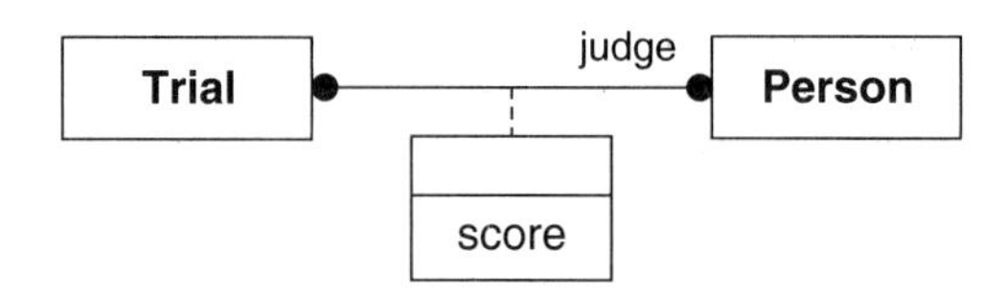

Figure 2.17 A link attribute is a property of the association

Many-to-many associations provide the most compelling rationale for link attributes. Such an attribute is unmistakably a property of the link and cannot be assigned to either object. In Figure 2.17, the score depends on both the trial and the judge. In contrast, for one-to-one and one-to-many associations, it is easier to confuse link attributes with object attributes. You will avoid errors in your applications if you construct your analysis models by directly describing objects and links. As a model evolves toward design and implementation, you can always restructure for more efficient execution.

2.2.5 Association Classes

An ***association class*** is an association whose links can participate in subsequent associations. An association class has characteristics of both an association and a class. Like the links of an association, the instances of an association class derive identity from instances of the constituent classes. Like a class, an association class can participate in associations.

An association class is denoted by a box attached to an association by a dashed line—the same notation as for a link attribute. This correspondence in notation is appropriate because a link attribute is really just a simple association class. Like an ordinary class, an association class may have attributes and operations. Figure 2.18 shows alternative models for authorization in a relational DBMS.

A user may own multiple tables. The owner of a table may authorize one or more other users to access the table; an authorized user may grant further permissions. For example, user *A* may own a table and authorize users *B* and *C* to access the table; for these associations user *A* is the *grantor* and users *B* and *C* are *grantees*. User *B* may, in turn, authorize user *D* to access the table.

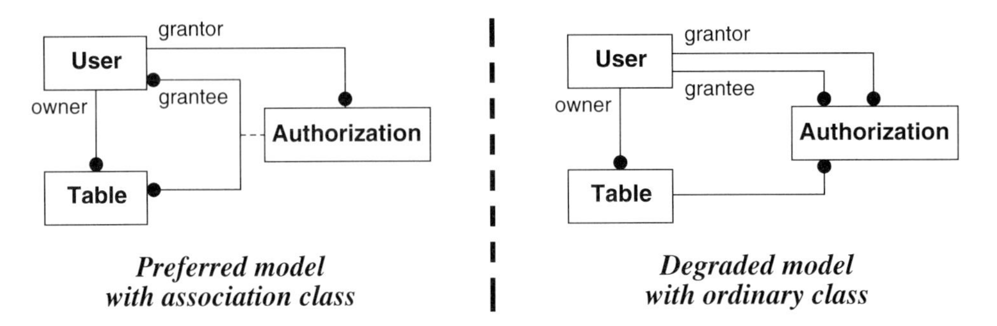

Figure 2.18 An association class lets you specify identity and navigation paths precisely

Association classes are an important facet of OMT modeling because they let you specify identity and navigation paths precisely. In Figure 2.18, the model with the association class clearly specifies that a table and a grantee imply an authorization and that a grantor may grant many such authorizations. The association class introduces a meaningful asymmetry to the model. In contrast, the degraded model in Figure 2.18 is symmetric and the normal path for traversing the model is not obvious. The degraded model has promoted authorization to a class with its own identity and has suppressed the fact that grantee and table directly relate.

In Figure 2.17, *Trial* is really an association class that describes a person competing in an athletic event. Figure 2.19 improves the model by using an association class.

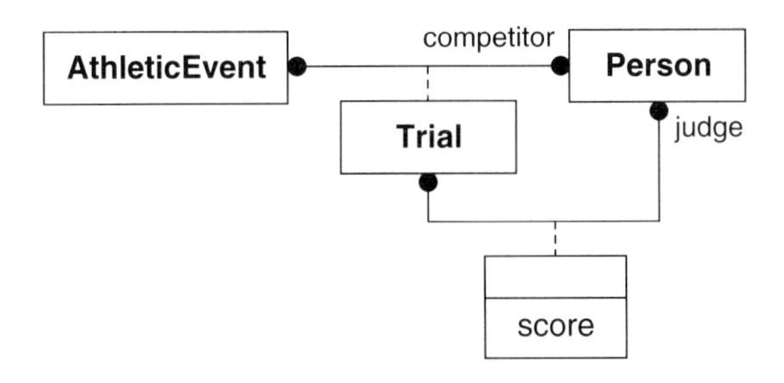

Figure 2.19 An association class for an athletic meet

2.2.6 Qualified Associations

A ***qualified association*** is an association in which the objects in a "many" role are partially or fully disambiguated by an attribute called the ***qualifier***. One-to-many and many-to-many associations may be qualified. A qualifier selects among the target objects, reducing the effective multiplicity, often from "many" to "one." Qualified associations with a target multiplicity of "one" or "zero-or-one" specify a precise path for finding the target object from the source object. As we will describe in Part 3, you can use a DBMS index to enforce a qualified association with a target multiplicity of "one."

Figure 2.20 illustrates the most common use of a qualifier—to yield a single target object. A bank has many accounts, with the combination of a bank and account number yielding a single account. Without a qualifier many objects are referenced; with the qualifier, each reference is to a specific object. In the "qualified" model, the second portion of the *Account* box is empty; this is because we have chosen not to list any attributes for *Account*.

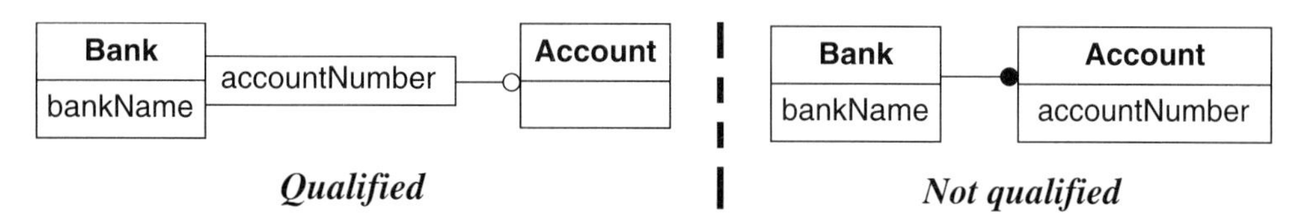

Figure 2.20 Qualification increases the precision of a model

Both models are acceptable, but the qualified model adds information. The qualified model adds a multiplicity constraint, that the combination of a bank and account number yields one account. The qualified model is actually more correct, because an account number really has no meaning unless the bank is also specified. The qualified model conveys the significance of *account number* in traversing the model, as methods will reflect. Chapter 3 presents more complex examples of qualified associations.

Figure 2.21 shows another example of a qualifier. A flight description pertains to many flights; the combination of a flight description and a departure date specifies a flight.

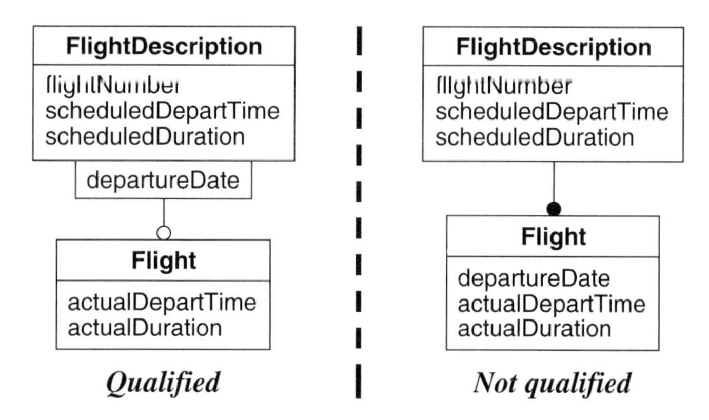

Figure 2.21 A simple qualified association for airline flights

The notation for a qualifier is a small box on the end of the association line near the source class. The qualifier box may grow out of any side (top, bottom, left, right) of the source class. The source class plus the qualifier yields the target class. In Figure 2.21, for example, *FlightDescription* plus *departureDate* yields *Flight*.

Qualification frequently arises through the use of names. Few names are globally unique; most names have meaning within some context. For example, an account number has meaning only when combined with the bank holding the account. Qualification can also arise through explicit ordering. For example, if the columns in a table are ordered, a table plus a column number yields a column.

2.3 GENERALIZATION CONCEPTS

2.3.1 Generalization

Generalization is the relationship between a class (the ***superclass***) and one or more variations of the class (the ***subclasses***). Generalization organizes classes by their similarities and differences, structuring the description of objects. The superclass holds common attributes, operations, state diagrams, and associations; the subclasses add specific attributes, operations, state diagrams, and associations. There can be multiple levels of generalization relationships. An instance of a subclass is simultaneously an instance (transitively) of all its superclasses.

Specialization provides another perspective on a system's structure. Specialization has the same meaning as generalization but takes a top-down perspective, starting with the superclass and splitting out variations. In contrast, generalization takes a bottom-up view, starting with the subclasses and abstracting commonality.

Simple generalization organizes classes into a hierarchy; each subclass has a single immediate superclass. (Chapter 3 discusses a more complex form of generalization relationship in which a subclass may have multiple immediate superclasses.) Figure 2.22 shows several kinds of financial instruments that could be managed within an investment portfolio. All financial instruments have a name and a current value that is expressed in a currency. Stocks have a quarterly dividend that is set by the board of directors. Bonds have a maturity date, maturity value, and either a fixed interest rate or a variable interest rate that is computed from a reference rate. There are various types of insurance, such as disability, life, and hazard. An annual payment may be required to maintain insurance coverage.

A large hollow arrowhead denotes generalization. The arrowhead points to the superclass. You may directly connect the superclass to each subclass, but we normally prefer to group subclasses as a tree. You may also show a discriminator next to a subclass tree.

The ***discriminator*** is an attribute that has one value for each subclass; the value indicates which subclass further describes an object. The discriminator is simply a name for the basis of generalization. Discriminators are superfluous for object-oriented DBMSs, but they can be useful for implementations that divide objects into multiple records. For example, with a relational DBMS you might implement each class with a separate table and store a record at each generalization level of an object. In Figure 2.22 *financialInstrumentType* and *bondType* are discriminators.

Generalization is an important construct for both implementation and conceptual modeling. Generalization facilitates type checking and reuse in object-oriented languages. Much of the literature emphasizes these programming uses of generalization. However, generalization has importance that transcends implementation. Generalization is also significant in the early stages of software engineering because it lets you call attention to patterns of similarities and differences. Often there are different ways to group classes; with generalization

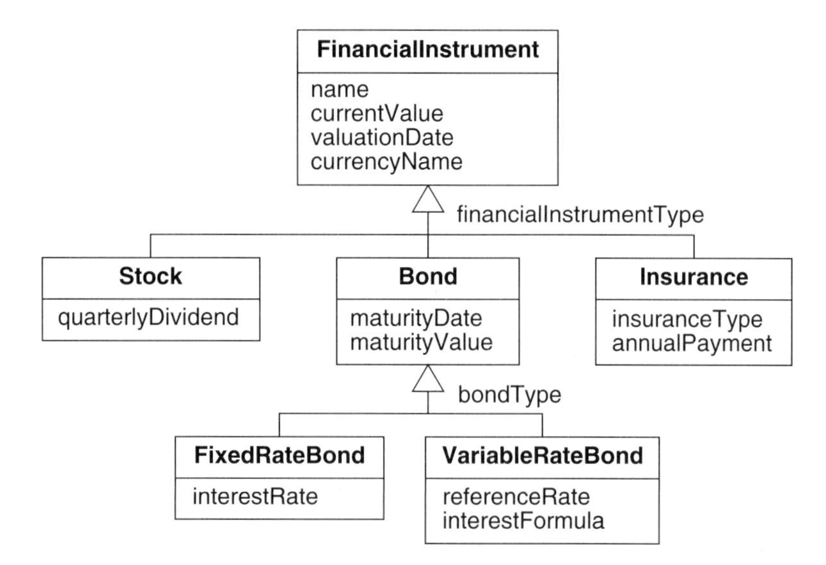

Figure 2.22 Generalization organizes classes by their similarities and differences

you make a statement about the relative importance of these different organizational bases. The improved depth of understanding leads to systems that are better conceived and better architected. The object-modeling fabric of classes, associations, and generalizations provides a rich, but concise, medium for thinking about applications.

2.3.2 Inheritance

Generalization is the structural relationship that permits the ***inheritance*** mechanism to occur. A subclass ***inherits*** the attributes, operations, state diagrams, and associations of its superclasses. Inherited properties can be reused from a superclass or overridden in the subclasses; new properties can be added to the subclasses. For example, in Figure 2.22 an object in the *FixedRateBond* class has seven attributes: *name*, *currentValue*, *valuationDate*, *currencyName*, *maturityDate*, *maturityValue*, and *interestRate*.

Simple generalization is really another term for ***single inheritance***, and the case in which a subclass may have multiple immediate superclasses is called ***multiple inheritance***. Chapter 3 describes multiple inheritance in more detail.

2.4 SAMPLE DIAGRAMS

2.4.1 Class Diagram

A ***class diagram*** is a graphical representation that describes objects and their relationships. Figure 2.23 presents a class diagram for an airline flight reservation system. We used portions of this diagram in earlier examples.

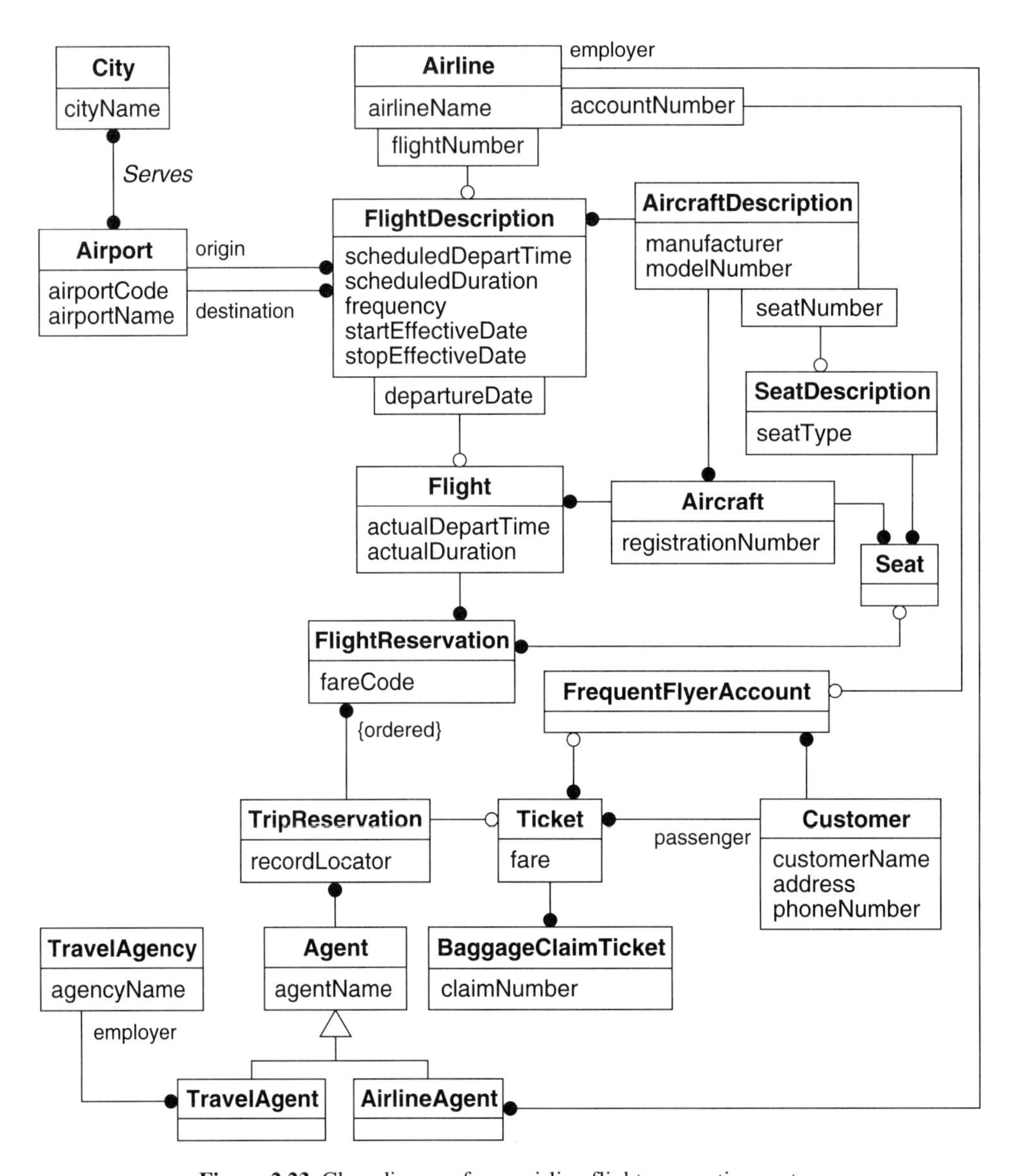

Figure 2.23 Class diagram for an airline flight reservation system

An airport serves many cities, and a city may have multiple airports. Airlines operate flights between airports. A flight description refers to the published description of air travel between two airports. In contrast, a flight refers to the actual travel made by an airplane on a particular date. The frequency indicates the days of the week for which the flight description applies. The start and stop effective dates bracket the time period for which the published flight description is in effect.

A sample flight description is TW 250 from St. Louis to Milwaukee. For the fall of 1995, TW 250 is scheduled to depart at 7:42 AM, has a scheduled duration of one hour eighteen minutes, is scheduled to be flown by a DC9 plane, and operates every day of the week except Sunday. A sample flight is TW 250 on October 23, 1995, which actually departed at about 7:45 AM and had an actual duration of one hour five minutes.

Each flight is normally booked by multiple reservations at various fare codes. The fare code governs the fare that is charged. Fare codes reflect multiple factors, such as the number of days of advance purchase, refundability, whether the passenger is staying over a Saturday night, and whether the seat is coach or first class (*seatType*). A passenger may be assigned a seat for a flight reservation. A seat may have multiple flight reservations, but hopefully no more than one for a flight!

A trip reservation consists of a sequence of flight reservations. For example, we could make a trip reservation for traveling from St. Louis to Milwaukee on TW 663 on October 20 and returning to St. Louis on TW 250 on October 23. The airlines use record locators to find a particular trip reservation quickly and unambiguously. Experienced passengers will often use these record locators for communicating with the airline. Trips are reserved by agents, who either work for an airline or a travel agency. A ticket may be issued for a passenger for a trip reservation, and a frequent flyer account may be noted.

We chose not to combine the *Customer* and *Agent* classes. In one sense, we were tempted to combine them; *Customer* and *Agent* could be regarded as roles of *Person*. However, there seems to be little commonality between agents and customers within a flight reservation system. Furthermore, we would like to allow some unusual customers, such as corporations, possibly animals, and inanimate objects. For example, a corporation might purchase an airplane seat to transport some delicate equipment.

For ease of layout, you may repeat class names within a class diagram, especially for large, multipage diagrams. However, each occurrence of a class name must refer to the same thing. For example, you cannot use *table* within the same diagram to refer to a relational database table and a piece of furniture. Attribute names (including roles as pseudo attributes) and method names may be globally unique, or unique within a class or association.

You should try to avoid intersecting association lines in your diagrams. For example, the right diagram in Figure 2.24 has three possible meanings: associations *AC* and *BD*, *AB* and *CD*, or *AD* and *BC*. Often you can adjust the layout of a diagram so there is no need for intersecting lines, as we did in Figure 2.23. Otherwise you can use a crossing symbol to avoid ambiguity.

By now we have presented enough object-modeling constructs, so that you should be able to read object diagrams. However, we have not yet shown you a process for constructing diagrams. This is not a problem in Part 1, because we have chosen problems with small models. However, for large problems you will need a systematic approach to constructing diagrams, and we present such a process in Chapter 8.

2.4.2 Instance Diagram

An ***instance diagram*** is a diagram that involves only objects, links, and values. Instance diagrams are useful for debugging and especially for understanding class diagrams.

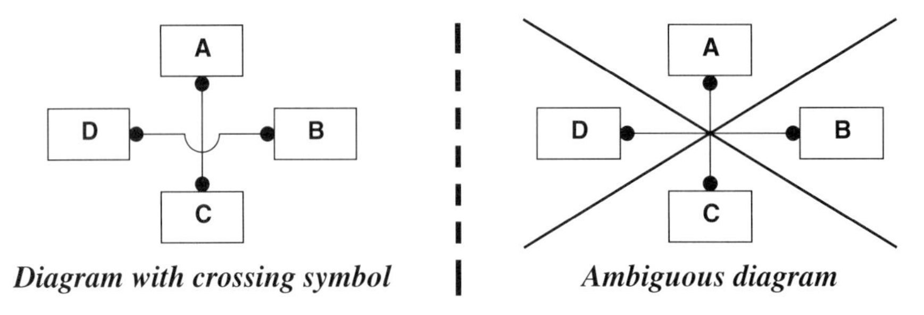

Figure 2.24 Intersecting association lines can be ambiguous

Often there are alternative ways to conceive a model, and it is not obvious which model is best. Instance diagrams can help you understand the capabilities and shortcomings of different representations. The most appropriate model then depends on the requirements of an application. As an example, consider the two class diagrams for directed graphs in Figure 2.25. (Figure E2.7 and Chapter 4 have additional models for directed graphs.)

A directed graph consists of a set of nodes and a set of edges. Each edge connects two nodes and has an arrow indicating the direction of flow. The left model describes directed graphs with at most one edge between nodes; edges are implicitly represented by the *parent-child* association between *Branch* and *Node*. By definition each node has one or more parents except root nodes. The right model is more powerful and can describe any directed graph.

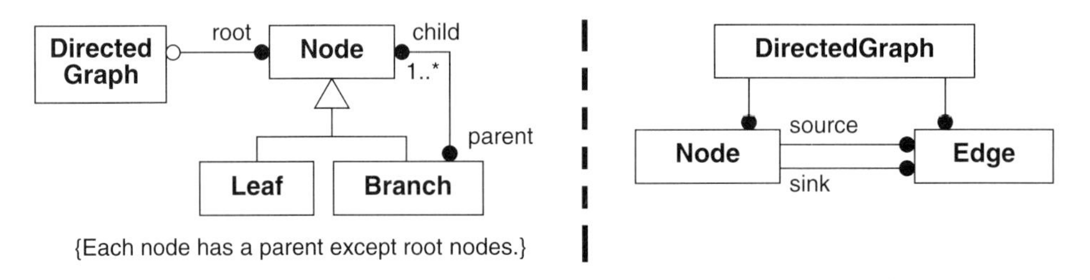

Figure 2.25 Class diagrams for directed graphs

Figure 2.26 shows a sample directed graph and the instance diagrams for the respective class diagrams. To reduce clutter we omit the links to the *DirectedGraph* object for the instance diagram for the right model. Note that the instance diagram for the left model directly corresponds to the graph. The graph consists of nodes that are connected by edges; similarly, the instance diagram consists of objects that are connected by links. In contrast, the instance diagram for the right model is more complex because it elevates both nodes and edges to the status of objects.

Figure 2.27 shows another directed graph. Again we omit the links to *DirectedGraph* for the instance diagram for the right model to reduce clutter. Such a graph cannot be represented by the left model; the left model cannot represent multiple edges between nodes.

Our point in presenting these diagrams is that both models are correct. If an application involves directed graphs with at most one edge between nodes, the left model is preferable

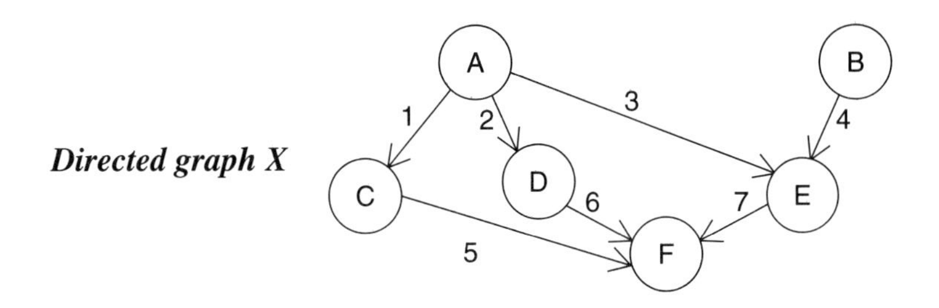

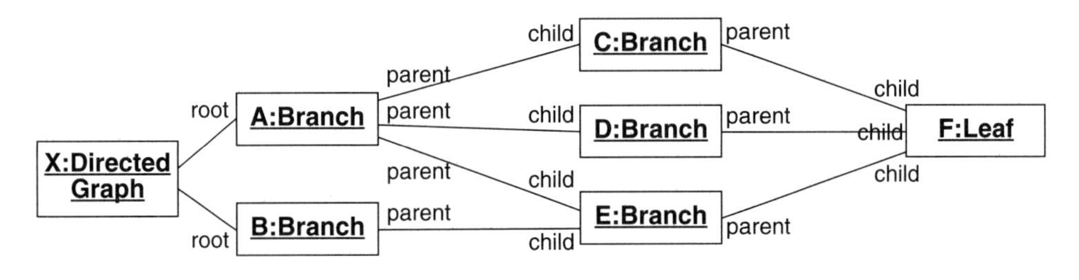

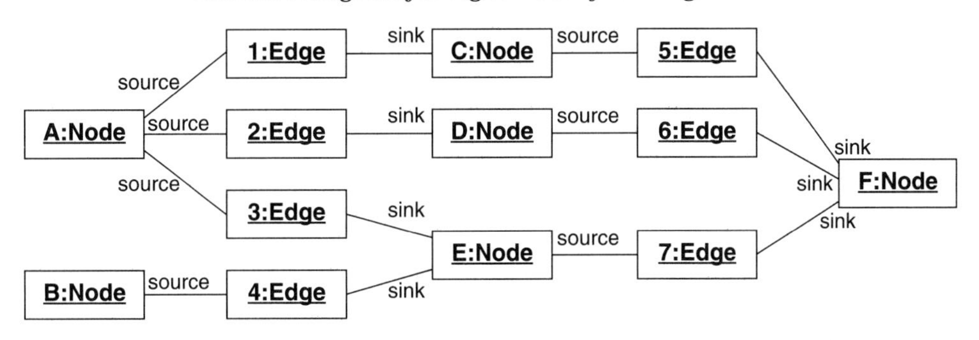

Figure 2.26 Instance diagrams for a directed graph

because it is simpler and structurally enforces the constraint that there not be multiple edges between nodes. On the other hand, the right model is more appropriate for applications that involve directed graphs in their more general form.

2.5 PRACTICAL TIPS

We have gleaned from our experience the following tips for constructing object models.

- **Scope of model**. Make sure you clearly understand the problem to be solved. The content of an object model is driven by relevance to an application. (Section 2.1.1)

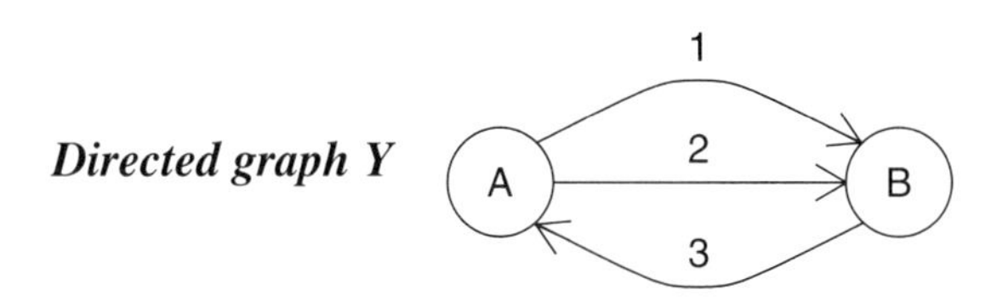

Instance diagram for left model

. . . cannot represent . . .

Instance diagram for right model

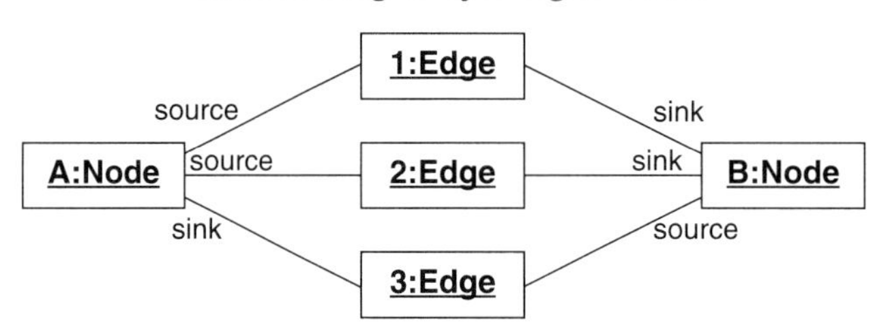

Figure 2.27 Instance diagrams for a directed graph

- **Simplicity**. Strive to keep your models simple. A simple model is easier to understand. Try to use a minimal number of classes that are clearly defined and not redundant. Be suspicious of classes that are difficult to define. You may need to reconsider such classes and restructure the model. (Section 2.4.2)
- **Layout of diagram**. Draw your class diagrams in a manner that elicits symmetry. Often there is a superstructure to a problem that lies outside the OMT notation that we have presented. Try to position important classes so that they are visually prominent on a diagram. Try to avoid crossing lines on diagrams.
- **Names**. Carefully choose names. Names carry powerful connotations. Names should be descriptive, crisp, and unambiguous. Choosing good names is a difficult aspect of object modeling. Persons often argue over names, but the arguments are worthwhile because they cause users and developers to think more deeply about a model and reach a consensus. You should use singular nouns for the names of classes.
- **Implementation of associations**. During analysis use associations to capture the true intent. During design you can choose to implement associations with pointers, foreign keys, or by some other means. (Section 2.2.1)
- **Multiplicity**. You should carefully review multiplicity decisions in your object models. Challenge roles with a multiplicity of exactly one. Often the object is optional and "zero-or-one" multiplicity is more appropriate. Sometimes "many" multiplicity may be needed. (Section 2.2.2)

- **Roles**. Be alert for multiple uses of the same class. Use roles to unify references to the same class. (Section 2.2.3)
- **Link attributes**. During analysis, do not collapse link attributes into one of the related classes. You should directly describe the objects and links in your models. During design and implementation, you can always combine information for more efficient execution. (Section 2.2.4)
- **Qualifiers**. Challenge roles with a multiplicity of "many." A qualifier can often improve the precision of an association and highlight important navigation paths. (Section 2.2.6)
- **Review of models**. Try to get others to review your models. Expect that your models will require revision. Object models require revision to clarify names, improve abstraction, repair errors, add information, and more accurately capture structural constraints. Nearly all of our models have required several revisions.
- **Instance diagrams**. Selectively use instance diagrams to explain subtleties of a class diagram. However, do not feel compelled to show a large number of instance diagrams, because they can be voluminous and are tedious to construct and read. (Section 2.4.2)

2.6 CHAPTER SUMMARY

The object model describes the data structure of a system, that is objects and their relationships. Object models are built from classes that relate to each other through associations and generalizations. An object is a concept, abstraction, or thing that can be individually identified and has meaning for an application. A class is a description of a group of objects with similar properties (object attributes), common behavior (operations and state diagrams), similar relationships to other objects, and common semantics.

A link is a physical or conceptual connection between objects and is an instance of an association. An association is a description of a group of links with common structure and common semantics. An association describes a set of potential links in the same way that a class describes a set of potential objects. Associations are an important and distinctive aspect of OMT modeling that most object-oriented DBMSs and relational DBMSs support.

Multiplicity specifies the number of instances of one class that may relate to a single instance of an associated class. A role is one end of an association. Each role may have an explicit name and a multiplicity. A link attribute is a named property of an association that describes a value held by each link of the association. An association class is an association whose links can participate in subsequent associations. A qualified association is an association in which the objects in a "many" role are partially or fully disambiguated by an attribute called the qualifier. The qualifier selects among the target objects, reducing the effective multiplicity, often from "many" to "one."

Generalization is the relationship between a class (called the superclass) and one or more variations of the class (called the subclasses). Generalization organizes classes by their similarities and differences, structuring the description of objects. Generalization is the structural relationship that permits the inheritance mechanism to occur. A subclass inherits the attributes, operations, state diagrams, and associations of its superclasses. Inherited properties can be reused from a superclass or overridden in the subclasses; new properties can be added to the subclasses.

This chapter has presented basic object modeling concepts—the concepts that frequently occur and are sufficient for modeling many applications. Figure 2.28 lists the key concepts for this chapter.

association	identity	multiplicity	qualifier
association class	inheritance	object	role
attribute	instance	object attribute	specialization
class	instance diagram	object model	subclass
class diagram	link	operation	superclass
discriminator	link attribute	polymorphism	value
generalization	method	qualified association	

Figure 2.28 Key concepts for Chapter 2

CHANGES TO OMT NOTATION

We have made the following changes to the notation in [Rumbaugh-91] for the sake of compatibility with the UML notation [UML-98]:

- **Object**. We changed the notation for an object to a box. The object name, a colon, and the class name in the top portion of the box are underlined.
- **Multiplicity**. For the situation where multiplicity is "many" and more restrictive than "0..*", we *overload* the solid ball with a textual specification. We show *both* the solid ball and the specification. This was not made clear in [Rumbaugh-91].
- **Link attribute**. A link attribute is denoted by a box attached to the association by a dashed line. A link attribute is a special case of an association class.
- **Association class**. An association class is also denoted by a box attached to the association by a dashed line.
- **Generalization** A large hollow arrowhead denotes generalization. The arrowhead points to the superclass. You may directly connect the superclass to each subclass, but we normally prefer to group subclasses as a tree. (This is our choice of style and is allowed by the UML.) You may show an optional discriminator next to a subclass tree.

We have made one further change to [Rumbaugh-91] not mentioned by [UML-98]:

- **Crossing symbol**. We use a crossing symbol to avoid ambiguity when lines in diagrams must cross.

BIBLIOGRAPHIC NOTES

Our presentation in this chapter is similar to that in [Rumbaugh-91]. Overall, we have seen little need to revise the object model. The current notation works well in practice and strikes a reasonable compromise between expressiveness and simplicity.

[Loomis-87] is the first publication about the OMT object-modeling notation. Loomis articulated the database viewpoint and Rumbaugh articulated the programming viewpoint in developing the initial object-modeling notation.

Pages 112–113 of [Wirfs-Brock-90] explain the proper use of inheritance in terms of Venn diagrams.The responsibilities of a subclass should completely encompass those of its superclasses. (Wirfs-Brock responsibilities are similar to OMT attributes and methods.) Subclass responsibilities cannot partially overlap superclass responsibilities; if this were to occur, then the subclass would not provide some superclass behavior, thus violating the meaning of inheritance. With inheritance every instance of a subclass must be an instance of all its superclasses.

The information model of [Shlaer-88] is similar to the OMT object model. Both notations emphasize structure and have been heavily influenced by database design techniques. A major difference is that the OMT object model applies to both analysis and design. During analysis OMT models are constructed in the conceptual realm; during design OMT models are elaborated and optimized for implementation needs. In contrast, the information model of [Shlaer-88] is built about relational database concepts. The heavy influence of relational database concepts is distracting during analysis, especially when a relational database is not the ultimate implementation medium.

REFERENCES

[Cattell-91] RGG Cattell. *Object Data Management: Object-Oriented and Extended Relational Database Systems*. Reading, Massachusetts: Addison-Wesley, 1991.

[Cattell-96] R Cattell, editor. *The Object Database Standard: ODMG-93, Release 1.2*. San Mateo, California: Morgan Kaufmann, 1996.

[Khoshafian-86] SN Khoshafian and GP Copeland. Object identity. *OOPSLA'86 as ACM SIGPLAN 21*, 11 (November 1986), 406–416.

[Loomis-87] Mary ES Loomis, Ashwin V Shah, and James E Rumbaugh. An object modeling technique for conceptual design. *European Conference on Object-Oriented Programming*, Paris, France, June 15–17, 1987, published as *Lecture Notes in Computer Science, 276,* Springer-Verlag.

[Rumbaugh-91] J Rumbaugh, M Blaha, W Premerlani, F Eddy, and W Lorensen. *Object-Oriented Modeling and Design*. Englewood Cliffs, New Jersey: Prentice Hall, 1991.

[Shlaer-88] Sally Shlaer and Stephen J Mellor. *Object-Oriented Systems Analysis: Modeling the World in Data*. Englewood Cliffs, New Jersey: Yourdon Press, 1988.

[UML-98] The following books are planned for the Unified Modeling Language:

Grady Booch, James Rumbaugh, and Ivar Jacobson. *UML User's Guide*. Reading, Massachusetts: Addison-Wesley.

James Rumbaugh, Ivar Jacobson, and Grady Booch. *UML Reference Manual*. Reading, Massachusetts: Addison-Wesley.

Ivar Jacobson, Grady Booch, and James Rumbaugh. *UML Process Book*. Reading, Massachusetts: Addison-Wesley.

[Ungar-87] David Ungar and Randall B Smith. SELF: The power of simplicity. *OOPSLA'87 as ACM SIGPLAN 22*, 12 (December 1987), 227–241.

[Wirfs-Brock-90] Rebecca Wirfs-Brock, Brian Wilkerson, and Lauren Wiener. *Designing Object-Oriented Software*. Englewood Cliffs, New Jersey: Prentice Hall, 1990.

EXERCISES

The number in parentheses next to each exercise indicates the difficulty, from 1 (easy) to 10 (very difficult).

2.1 (2) Explain the object model in Figure E2.1.

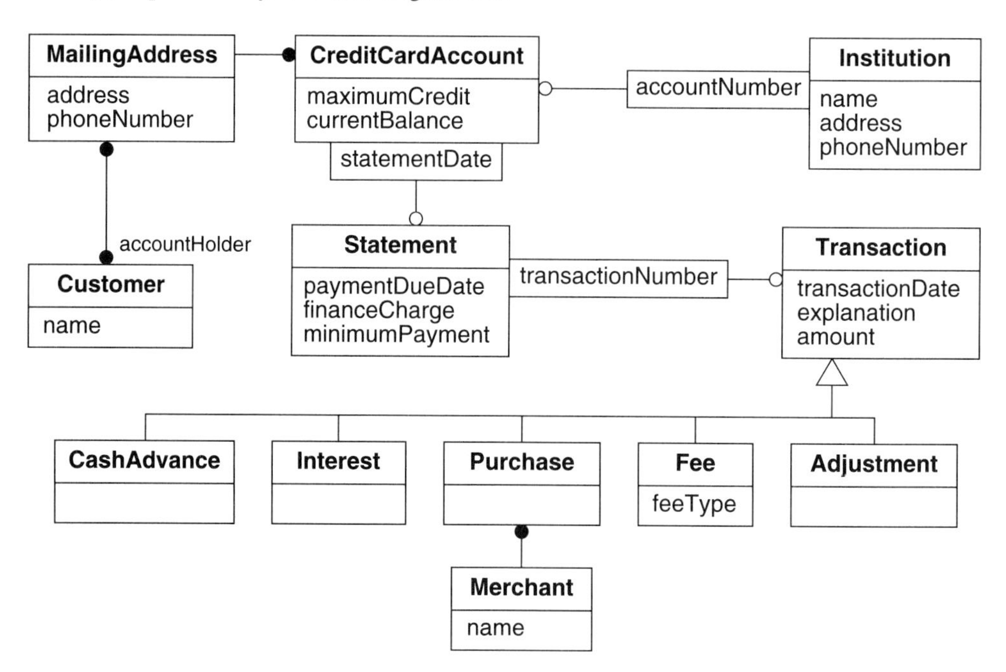

Figure E2.1 Object model for managing credit card accounts

2.2 (5) Add the following information to the object model of Figure 2.7:

a. A hotel may have a number of amenities, such as restaurant, lounge, fitness center, indoor pool, outdoor pool, golf, and tennis. Extend the model so that you can easily obtain the set of amenities for a given hotel and the set of hotels that offer one or more amenities.

b. Many hotels have variable pricing throughout the year; rates are low during off seasons and high during seasons of high demand. Extend the model so that it organizes pricing information by season.

2.3 (3) Construct an instance diagram for both models in Figure 2.18 with the following data. User *A* may own a table and authorize users *B* and *C* to access the table; user *B* may, in turn, authorize user *D* to access the table.

2.4 (4) Compare the object models in Figure E2.2. The left model represents *Subscription* as an association class; the right model treats *Subscription* as an ordinary class.

A person may have multiple magazine subscriptions. A magazine has multiple subscribers. For each subscription, it is important to track the date and amount of each payment as well as the current expiration date.

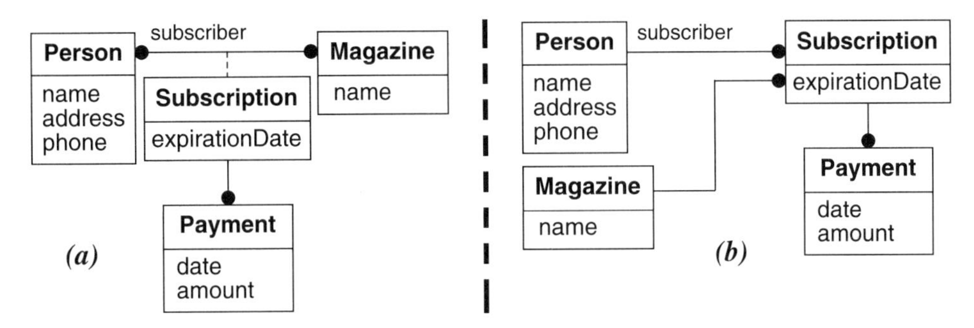

Figure E2.2 Object models for magazine subscriptions

2.5 (2) Refine the object model of Figure E2.3 by using generalization. The object model describes bibliographic citations such as those that appear in this book. A citation code is an abbreviation for referring to a paper or book. Position number refers to the ordering of the authors of a cited work.

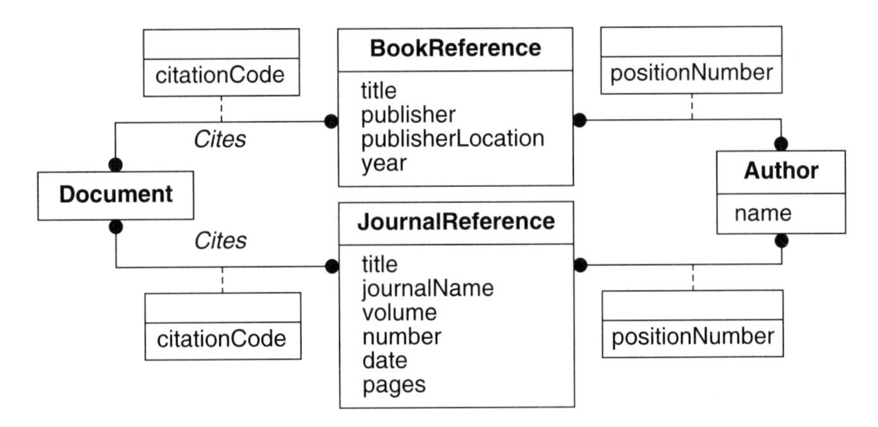

Figure E2.3 Object model for citations in a document

2.6 (5) Further improve the object model in Figure E2.3 through the use of qualifiers.

2.7 (4) Prepare an object model for personal automotive repair records. Include the following: date of repair, vehicle repaired, vehicle mileage, repair tasks, repair facility, cost, and comments. You should indicate tasks that fulfill warranty provisions. Hint: You should have the following classes: *Vehicle*, *RepairEvent*, *RepairTask*, and *VehicleRepairFacility*.

2.8 (7) Figure E2.4 shows a receipt for the purchase of a few items from a store. Prepare an object model that describes this information.

```
                   Acme food store
                   (314) 555-1234
                   198 State Street
                   Chesterfield, MO 63017

Description     Quantity     Price     Discount     Amount
-----------     --------     -----     --------     ------
milk                   2      3.50         0.00       7.00
beans                  5       .75         0.00       3.75
motor oil              1      2.00         0.00       2.00

Sub-total                12.75
Tax                       0.76
Total                    13.51
Amount on credit card    13.51
Change                    0.00

Customer name       Brian Smith
Credit card number  1234 5678 1234 5678
Expiration date     05/99
Approval number     AP654321

Customer signature

Receipt number      345127
Date                15 Sep 1996
Time                16:43
Employee            14
Register            8
```

Figure E2.4 A receipt for a purchase from a store

2.9 (6) Figure E2.5 is an excerpt of the certificate of title form for a motor vehicle. Prepare an object model that describes this information. Represent the title number as a qualifier.

2.10 (7) Figure E2.6 shows a typical frequent flyer account summary. Prepare an object model that describes this information.

```
Title number --- TE125023

Vehicle identification Number - KN4MW6335L1226731
Year - 89                    Make - Mazda 323 car
Cylinders - 6                Horsepower - 30
Previous state - NY          Mileage at time of transfer - 61515
Purchase date - 15 Mar 1992
Date issued - 10 Aug 1994

Owner - John and Mary Doe
        1561 Olive Street
        Scotia, NY 12309

First lien - First Bank
             1234 Main Street
             Schenectady, NY 12301

Second lien -
```

Figure E2.5 A simplified certificate of title for a motor vehicle

```
Frequent flyer account number: 123456789
Member since 06 Jun 1991

James Doe
1937 First Street
Creve Coeur, MO 63146

Activity       Prior     Credits   Remaining   Current   Current
as of:         balance   used      balance     Mileage   balance
-----------    -------   -------   ---------   -------   -------
17 Nov 1995    117,169         0   117,169       3,882   121,051

Date of   Airline   Flight   Service   Activity   Actual    Bonus
Travel    Partner   Number   Class                Mileage   Mileage
-------   -------   ------   -------   --------   -------   -------
20 Oct      TW       0663       Y      STL-MKE     1,000    600
23 Oct      TW       0250       Y      MKE-STL     1,000    600
21 Aug      MC                         hotel                500
17 Aug      SP                         phone                182
```

Figure E2.6 A typical frequent flyer account summary

2.11 (6) Prepare an object model for the program listings in a television guide. For movies you should include the network, date, time, movie name, description, quality rating, and release date. For other programs you should include the network, date, time, series name, episode name, and description.

2.12 (4) Construct an object model for movie information. Include the following: actor, actress, category, date, director, length, movie, producer, rating, synopsis, title, and writer.

2.13 (8) Discuss the relative merits of the object models for a directed graph that are shown in Figure E2.7. This exercise highlights the issue of alternative structures for an object model. A directed graph consists of a set of nodes and a set of edges. Each edge connects two nodes and has an arrow indicating direction of flow.

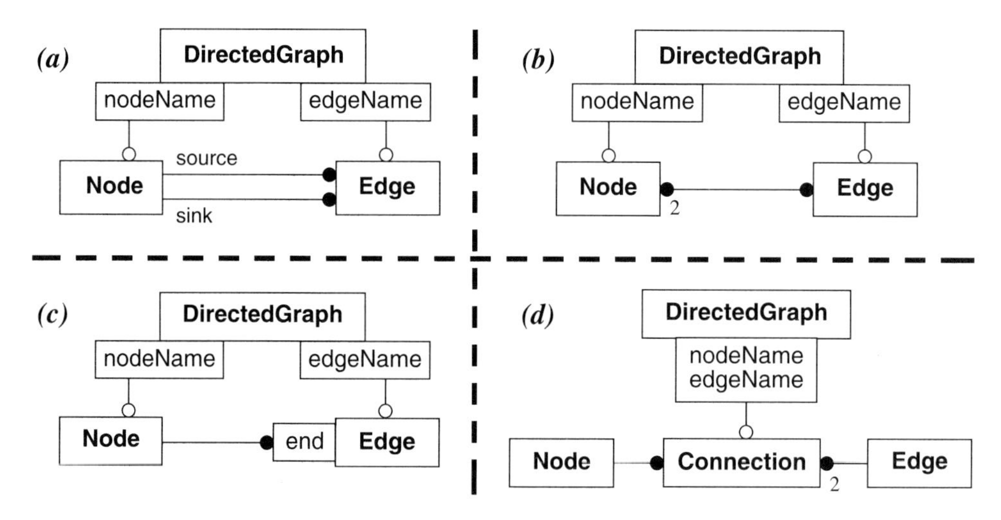

Figure E2.7 Alternative object models for a directed graph

2.14 (6) Construct instance diagrams for each of the four models in Figure E2.7 given the directed graph in Figure 2.27.

2.15 (7) Prepare an object model for checking accounts. Include the following classes: *Bank*, *CheckingAccount*, *Deposit*, *Check*, *InterestPayment*, *Fee*, *AccountOwner*, *Item*, and *Statement*.

A bank issues a checking account and assigns it a number. A checking account may have several account owners. An account owner may have multiple checking accounts. Deposits, checks, interest payments, and fees are all items that affect the balance for a checking account. There are different kinds of fees, including check printing fee, overdraft fee, and wire transfer fee. The bank issues periodic statements to the owners so that they can accurately monitor an account.

2.16 (7) Construct an object model for genealogical information. Include the following: names, sex, birth date, birth place, parents, marriages, children, death date, cause of death, and cemetery.

2.17 (7) Extend the genealogical object model to permit polygamy for males and females.

2.18 (7) Show two variations of the genealogical object model. One object model disallows remarriage between the same persons (as would occur with marriage, divorce, followed by marriage). The other object model permits remarriage between the same persons.

2.19 (6) Prepare an object model for words in a dictionary. Include the following: alternative spellings, antonyms, dictionary, grammar type (noun, verb, adjective, adverb), historical derivation, hyphenation, meanings, miscellaneous comments, prioritization by frequency of use, pronunciation, and synonyms. Some sample definitions are as follows (from *Webster's New World Dictionary*):

- **been** (bin; *also, chiefly Brit.*, bēn *&, esp. if unstressed*, ben), pp. of **be**.
- **kum·quat** (kum′kwot), *n.* [< Chin. *chin-chü*, golden orange], 1. a small, orange-colored, oval fruit, with a sour pulp and a sweet rind, used in preserves, 2. the tree that it grows on. Also sp. **cumquat**.
- **lac·y** (lās′i), *adj.* [-IER, -IEST], 1. of lace. 2. like lace; having a delicate open pattern. —**lac′i·ly**, *adv.* —**lac′i·ness**, *n.*
- **Span·ish** (span′ish), *adj.* of Spain, its people, their language, etc. *n.* 1. the Romance language of Spain and Spanish America. 2. the Spanish people.

2.20 (4) Prepare an instance diagram for the definition for *been* using your object model from the previous exercise.

3

Advanced Object Modeling

This chapter explains the advanced aspects of object modeling that you will need to model complex and large applications. It builds on the basic concepts in Chapter 2, so you should master the material there before reading this chapter. You can skip this chapter if you are just interested in getting a general understanding of object modeling.

3.1 OBJECT AND CLASS CONCEPTS

3.1.1 Instantiation

Instantiation is the relationship between an object and its class. The notation for instantiation is a dashed line from the instance to the class with an arrow pointing to the class; the dashed line is labeled with the legend *instance* enclosed by guillemets («»). Figure 3.1 shows this notation for *City* and its two instances *Bombay* and *Prague*. Making the instantiation relationship between classes and instances explicit in this way can be helpful in modeling complex problems and in giving examples.

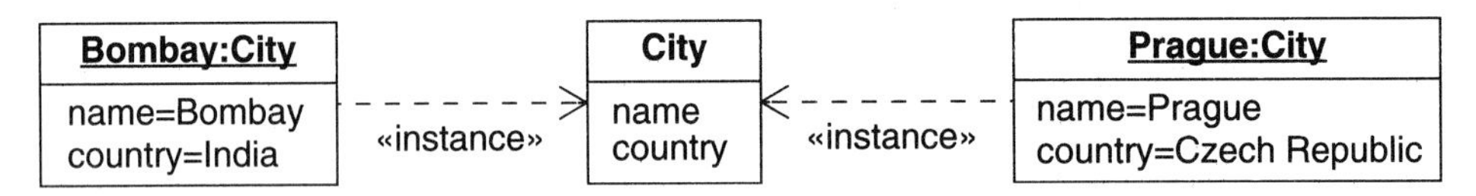

Figure 3.1 Instantiation relationships

3.1.2 Class Attributes and Operations

A ***class attribute*** is an attribute whose value is common to a group of objects in a class rather than peculiar to each instance. Class attributes can be used to store default or summary data for objects. A ***class operation*** is an operation on a class rather than on instances of the class. The most common kind of class operations are operations to create new class instances. You can denote class attributes and class operations with an underline. Our convention is to list them at the top of the attribute box and operation box, respectively.

In most applications class attributes can lead to an inferior model. We discourage the use of class attributes. Often you can improve your model by explicitly modeling groups and specifying scope. For example, the upper model in Figure 3.2 shows class attributes for a simple model of phone mail. Each message has an owner mailbox, date recorded, time recorded, priority, message contents, and a flag indicating if it has been received. A message may have a mailbox as the source or it may be from an external call. Each mailbox has a phone number, password, and recorded greeting. For the *PhoneMessage* class we can store the maximum duration for a message and the maximum days a message will be retained. For the *PhoneMailbox* class we can store the maximum number of messages that can be stored.

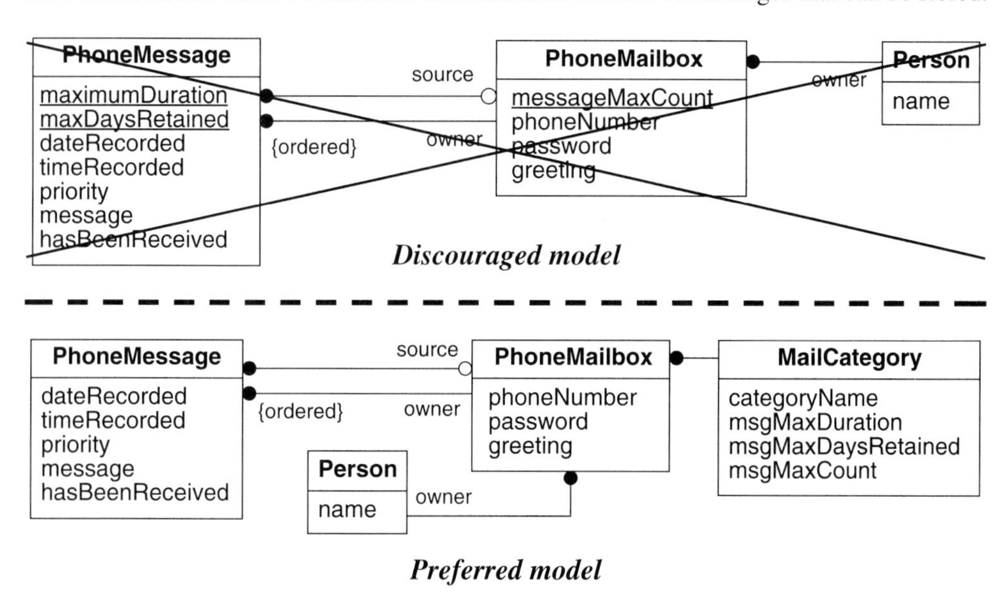

Figure 3.2 Instead of using class attributes, model groups explicitly

The upper model is inferior, however, because the maximum duration, maximum days retained, and maximum message count has a single value for the entire phone mail system. In the lower model these limits can vary for different kinds of users, yielding a phone mail system that is more flexible and extensible.

3.1.3 Attribute Multiplicity

Attribute multiplicity specifies the possible number of values for an attribute and is listed in brackets after the attribute name. You may specify a mandatory single value *[1]*, an optional single value *[0..1]*, an unbounded collection with a lower limit *[lowerLimit..*]*, or a collection with fixed limits *[lowerLimit..upperLimit]*. A lower limit of zero allows null values; a lower limit of one or more forbids null values. (***Null*** is a special value denoting that an attribute value is *unknown* or *not applicable*. See Chapter 9.) If you omit attribute multiplicity, an attribute is assumed to be single valued with nullability unspecified (*[0..1]* or *[1]*). In Figure 3.3 a person has one name, one or more addresses, zero or more phone numbers, and one birth date. Attribute multiplicity is similar to multiplicity for associations.

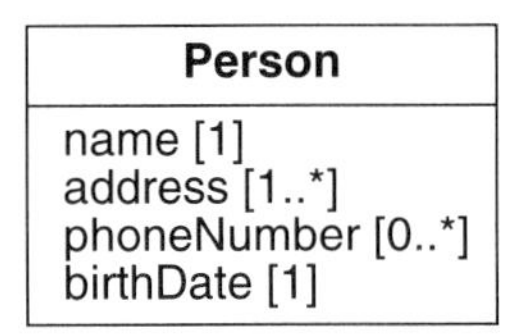

Figure 3.3 Attribute multiplicity

3.1.4 Candidate Keys for Classes

A ***candidate key*** for a class is a combination of one or more attributes that uniquely identifies objects within a class. (Section 3.2.3 discusses candidate keys for associations.) The collection of attributes in a candidate key must be minimal; no attribute can be discarded from the candidate key without destroying uniqueness. No attribute in a candidate key can be null. A given attribute may participate in multiple candidate keys.

For example, in Figure 3.4 *airportCode* and *airportName* are two candidate keys for *Airport*. The model specifies that each *airportCode* (such as IAH, HOU, STL, ALB) uniquely identifies an airport. Each *airportName* (such as Houston Intercontinental, Houston Hobby, Lambert St. Louis airport, and Albany NY airport) also uniquely identifies an airport.

Airport
airportCode {CK1} airportName {CK2}

Figure 3.4 Candidate keys for a class

We indicate a candidate key for a class with the notation *CKn* in braces next to the appropriate attributes. The *n* is a number that differentiates multiple candidate keys. For a multi-attribute candidate key, multiple attributes have the *CKn* designation with the same value of *n*.

Some readers may recognize the term "candidate key" from the database literature, but the notion of a candidate key is a logical construct, not an implementation construct. It is often helpful to be able to specify the constraint that one or more attributes taken together are unique. Relational database managers and most object-oriented database managers can readily enforce candidate keys.

3.1.5 Domains

A ***domain*** is the named set of possible values for an attribute. The notion of a domain is a fundamental concept in relational DBMS theory, but really has broader applicability as a modeling concept. As Figure 3.5 shows, an attribute name may be followed by a domain and default value. The domain is preceded by a colon; the default value is preceded by an equal sign. Some domains are infinite, such as the set of integers; others are finite. You can define a domain intensionally (by formula), extensionally (by explicitly listing occurrences), or in terms of another domain.

PhoneMessage
dateRecorded:Date timeRecorded:Time priority:PriorityType=NORMAL message:LongString hasBeenReceived:Boolean=FALSE

Figure 3.5 Assign a domain to an attribute rather than directly assign a data type

An ***enumeration domain*** is a domain that has a finite set of values. The values are often important to users, and you should carefully document them for your object models. For example, you would most likely implement *priorityType* in Figure 3.5 as an enumeration with values that could include *normal*, *urgent*, and *informational*.

A ***structured domain*** is a domain with important internal detail. You can use indentation to show the structure of domains at an arbitrary number of levels. Figure 3.6 shows two attributes that have a structured domain. An address consists of a street, city, state, mail code, and country. A birth date has a year, month, and day.

Person

name [1] : Name
address [1..*] : Address
    street
    city
    state
    mailCode
    country
phoneNumber [0..*] : PhoneNumber
birthDate [1] : Date
    year
    month
    day

Figure 3.6 Structured domains

During analysis you can ignore simple domains, but you should note enumerations and structured domains. During design you should elaborate your object model by assigning a domain to each attribute. During implementation you can then bind each domain to a data type and length.

Domains provide several benefits:

- **Consistent assignment of data types**. You can help ensure that attributes have uniform data types by first binding attributes to domains and then binding domains to data types.
- **Fewer decisions**. Because domains standardize the choices of data type and length, there are fewer implementation decisions.
- **Extensibility**. It is easier to change data types when they are not directly assigned.
- **Check on validity of operations**. Finally, you can use the semantic information in domains to check the appropriateness of certain operations. For example, it may not make sense to compare a name to an address.

Do not confuse a domain with a class. Figure 3.7 summarizes the differences between domains and classes. The objects of a class have identity, may be described by attributes, and may have rich operations. Classes may also be related by associations. In contrast, the values of a domain lack identity. For example, there can be many *Jim Smith* objects, but the value *normal* has only one occurrence. Most domain values have limited operations and are not described by attributes. During analysis we distinguish between domains and classes according to their semantic intent, even though some domains may be implemented as classes.

Classes	Domains
• A class describes objects.	• A domain describes values.
• Objects have identity.	• Values have no identity.
• Objects may be described by attributes.	• Most values are not described by attributes.
• Objects may have rich operations.	• Most values have limited operations.
• Classes may be related by associations.	• Domains do not have associations.

Figure 3.7 Classes and domains differ according to semantic intent

Do not confuse an enumeration domain with generalization. You should introduce generalization only when at least one subclass has significant attributes, operations, or associations that do not apply to the superclass. Do not introduce a generalization just because you have found an enumeration domain.

3.1.6 Secondary Aspects of Data

Occasionally you will encounter secondary aspects of attributes and classes that the OMT notation does not explicitly address [Blaha-93a]. This secondary data provides relevant information, but exists in a realm apart from the essence of an application. It is important to record secondary information without obscuring the focus of an application.

There are several kinds of secondary data for attribute values. Many scientific applications involve units of measure, such as inches, meters, seconds, and joules. Units of measure provide a context for values and imply conversion rules. For some numerical attributes you must specify accuracy—whether the values are exact, approximate, or have some standard deviation. You may wish to note the source of data—whether the values are obtained from persons, the literature, calculations, estimates, or some other source. You may require the time of the last update for each attribute value.

Figure 3.8 illustrates secondary data for a financial instrument. Because the value of a financial instrument is stated for some date and currency, *valuationDate* and *currencyName* are secondary data. This is one design approach for dealing with secondary data for attributes. Chapter 9 presents additional approaches.

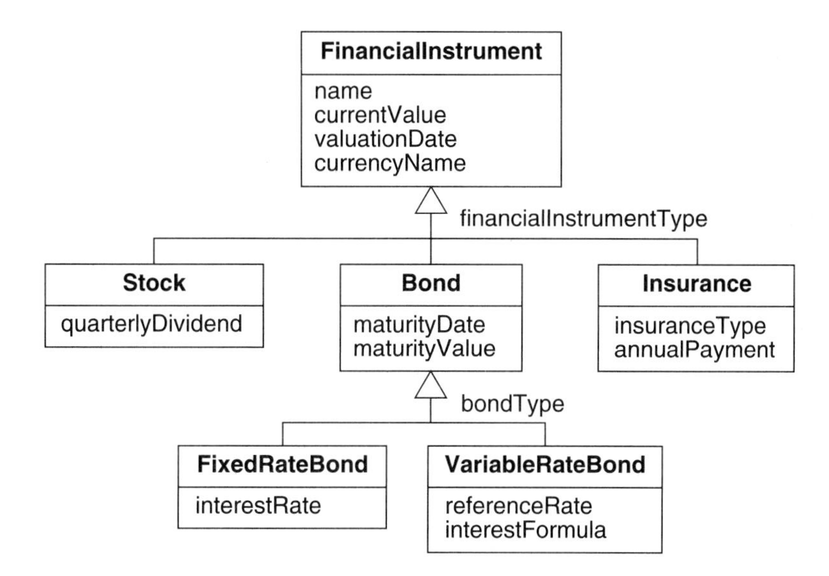

Figure 3.8 Secondary data for attribute values

Secondary data may also arise through policies and defaults that broadly apply to objects of various classes. Objects may have to be approved, logged, audited, distributed, and secured. They may also require multiple versions and may be persistent. For example, managers may need to approve critical data for some applications—the more important the data, the higher the level of approval. Permanent data must be logged to avoid accidental loss; transient objects may not be logged to speed processing. Updates to objects may necessitate an audit trail to protect against accidental and malicious damage. Some objects can be distributed over a network, while other objects may be limited to a single location. Objects may vary in their security level, such as none, unclassified, classified, and top secret. Some objects, such as alternative objects for an engineering design, may require versions. Other objects such as manufacturing records are not hypothetical and may not involve versions. Some objects may be persistent and require entry in the database, while other objects may be transient and need not exist beyond the confines of computer memory.

We have chosen not to augment the OMT notation for secondary data; too many variations are only occasionally required. We often use naming conventions to convey secondary data. Naming conventions are simple, orthogonal to notation, and enrich a model. The drawback is that naming conventions require discipline on the part of a modeler or a team of modelers. Comments are also helpful for documenting secondary data.

3.2 LINK AND ASSOCIATION CONCEPTS

3.2.1 Multiplicity

In Chapter 2, we introduced the notion of multiplicity. For database applications it is helpful to think in terms of minimum multiplicity and maximum multiplicity.

Minimum multiplicity is the lower limit on the possible number of related objects. Figure 3.9 shows several examples; the most common values are zero and one. We can implement a minimum multiplicity of zero by permitting null values and a minimum multiplicity of one by forbidding null values. A minimum multiplicity greater than one often requires special programming; fortunately such a minimum multiplicity seldom occurs.

OMT construct	*Minimum multiplicity*	*Maximum multiplicity*
—— Class	*1*	*1*
——● Class	*0*	*infinite*
——○ Class	*0*	*1*
1..* ——● Class	*1*	*infinite*
2..4 ——● Class	*2*	*4*

Figure 3.9 Examples of minimum and maximum multiplicity

A minimum multiplicity of one or more implies an existence dependency between objects. In our airline flight example (Figure 2.23) a flight reservation concerns one flight and a flight may be reserved by many flight reservations. *Flight* has a minimum multiplicity of one in this association. It does not make much sense to make a flight reservation unless you refer to a flight. Furthermore, if the airline cancels a flight, it must notify all passengers with a corresponding flight reservation. In contrast, *FlightReservation* has a minimum multiplicity of zero in this association. You can add a flight without regard for flight reservations. Similarly, the airline can cancel a flight reservation without affecting a flight.

Maximum multiplicity is the upper limit on the possible number of related objects. The most common values are one and infinite.

The choice of multiplicity depends on the application. For a multiplicity of exactly one, consider whether the target object is truly mandatory or possibly optional; use a hollow ball to indicate an optional target object. Furthermore, the target could be a set of objects rather than a single object; use a solid ball to allow a set of objects. The proper choice of multiplicity depends on application requirements. Be sure to review carefully whatever multiplicity decisions you make.

3.2.2 Ternary Associations

The ***degree of an association*** is the number of roles for each link. Associations may be binary, ternary, or higher degree. The vast majority are binary or qualified binary, and we described them in Chapter 2. Ternary associations occasionally occur, but we have rarely encountered an association of higher degree.

A ***ternary association*** is an association with three roles that cannot be restated as binary associations. The notation for a ternary association is a large diamond; each associated class connects to a vertex of the diamond with a line. In Figure 3.10 a professor teaches a listed course for a semester. The delivered course may use many textbooks; the same textbook may be used for multiple delivered courses. A ternary association may have link attributes or be treated as an association class, as in Figure 3.10.

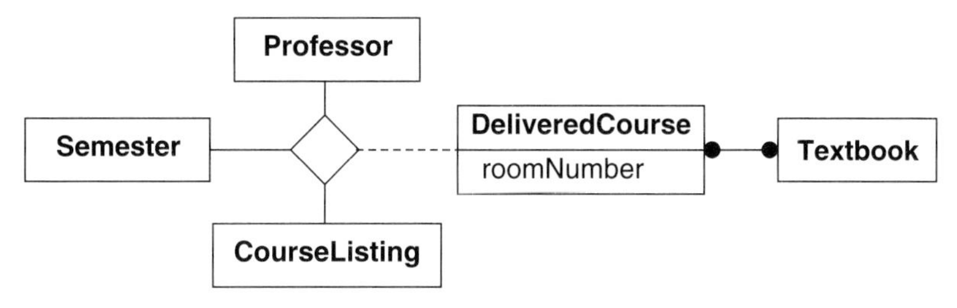

{Candidate key for ternary association = (semesterID, professorID, courseListingID)}

Figure 3.10 Ternary associations occasionally occur in models

3.2.3 Candidate Keys for Associations

Note that there are no balls next to the diamond or the classes of the ternary association in Figure 3.10. Although we could extend multiplicity notation to accommodate ternary associations, we prefer to use candidate keys to avoid confusion. A ***candidate key*** for an association is a combination of roles and qualifiers that uniquely identifies links within an association. Since the roles and qualifiers are implemented with attributes, we use the term "candidate key" for both classes and associations. The collection of roles and qualifiers in a candidate key must be minimal; no role or qualifier can be discarded from the candidate key without destroying uniqueness. Normally a ternary association has a single candidate key that is composed of roles from all three related classes. Occasionally you will encounter a ternary association with a candidate key that involves only two of the related classes.

The combination of *semesterID*, *professorID*, and *courseListingID* is a candidate key for *DeliveredCourse*. A professor may teach many courses in a semester and many semesters of the same course; a course may be taught by multiple professors. The confluence of semester, professor, and course listing is required to identify uniquely a delivered course.

We indicate a candidate key for an association with a comment in braces.

3.2.4 Exclusive-Or Associations

An ***exclusive-or association*** is a member of a group of associations that emanate from a class, called the ***source*** class. For each object in the source class exactly one exclusive-or association applies. An exclusive-or association relates the source class to a ***target*** class. An individual exclusive-or association is optional with regard to the target class, but the exclusive-or semantics requires that one target object be chosen for each source object. An exclusive-or association may belong to only one group.

Figure 3.11 shows an example in which *Index* is the source class and *Cluster* and *Table* are target classes. This example is an excerpt from the model for the *Oracle* relational DBMS. An index is associated with a table or a cluster, but not both, so a dashed line annotated by *or* cuts across the association lines close to the target classes. Interpreting the ball notation, a table may have zero or more indexes while a cluster has exactly one index. The alternative model using generalization is less precise and loses a multiplicity constraint: In the left model a cluster is associated with one index; in the right model a cluster can be associated with many indexes (via inheritance).

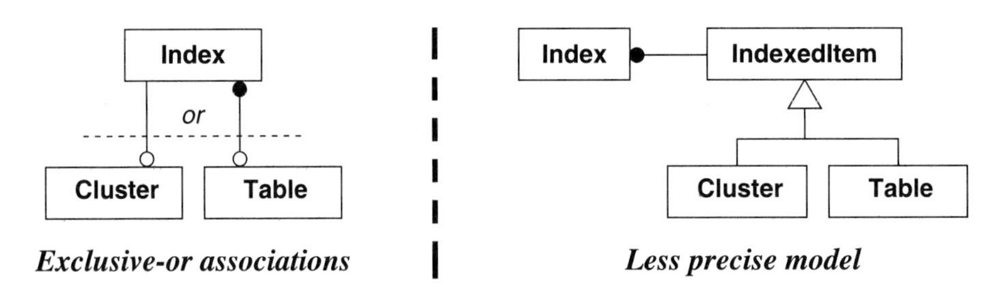

Figure 3.11 Exclusive-or associations can yield a more precise model

3.2.5 Qualified Associations

In Section 2.2.6, we introduced the notion of qualification and presented the most common situation, a single qualifier that reduces the maximum multiplicity of the target role from "many" to "one." Qualification does not affect the minimum multiplicity of an association. We now present more complex forms of qualification.

A qualifier selects among the objects in the target set and usually, but not always, reduces effective multiplicity from "many" to "one." In Figure 3.12 "many" multiplicity still remains after qualification. A company has many corporate officers, one president, and one treasurer but many directors and many vice presidents. Therefore, the combination of a company and office can yield many persons.

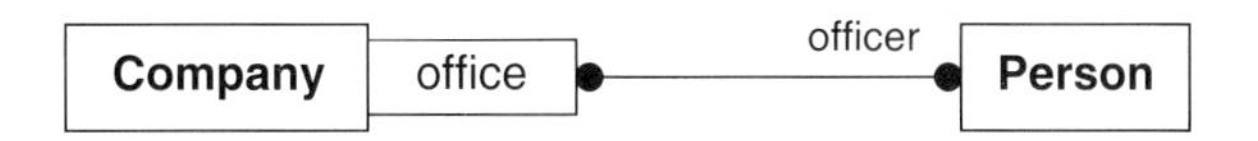

Figure 3.12 Qualification need not yield a target multiplicity of "one"

A ***qualification cascade*** is a series of consecutive qualified associations. Qualification cascades are encountered where an accumulation of qualifiers denotes increasingly specific objects. For example, in Figure 3.13 a city is identified by the combination of a country name, state name, and city name.

Figure 3.13 A qualification cascade denotes increasingly specific objects

A ***compound qualifier*** consists of two or more attributes that combine to refine the multiplicity of an association. The attributes that compose the compound qualifier are "anded" together. Figure 3.14 shows a compound qualifier. Both the node name and edge name are required to locate a connection for a directed graph.

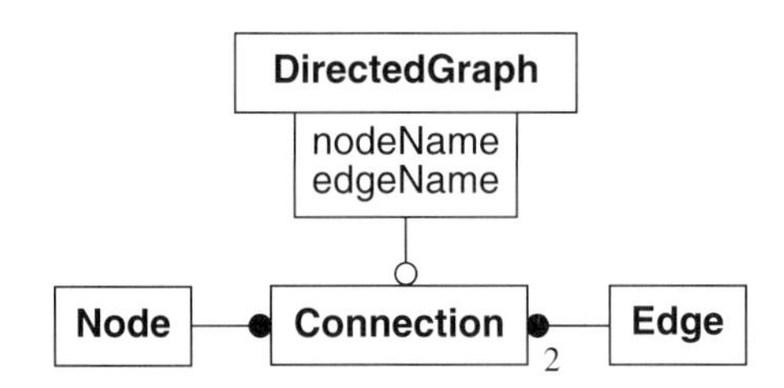

Figure 3.14 A compound qualifier refines the multiplicity of an association

3.3 AGGREGATION

Aggregation is a kind of association, between a whole, called the ***assembly***, and its parts, called the ***components*** [Blaha-93b]. Aggregation is often called the "a-part-of" or "parts-explosion" relationship and may be nested to an arbitrary number of levels. Aggregation bears the transitivity property: If *A* is part of *B* and *B* is part of *C*, then *A* is part of *C*. Aggregation is also antisymmetric: If *A* is part of *B*, then *B* is not part of *A*. Transitivity lets you compute the transitive closure of an assembly—that is, you can compute the components that directly and indirectly compose it. ***Transitive closure*** is a term from graph theory; the transitive closure of a node is the set of nodes that are reachable by some sequence of edges.

As Figure 3.15 shows, aggregation is drawn like an association with a small diamond added next to the assembly end. A book consists of front matter, multiple chapters, and back matter. Front matter, in turn, consists of a title page and a preface; back matter consists of multiple appendixes and an index.

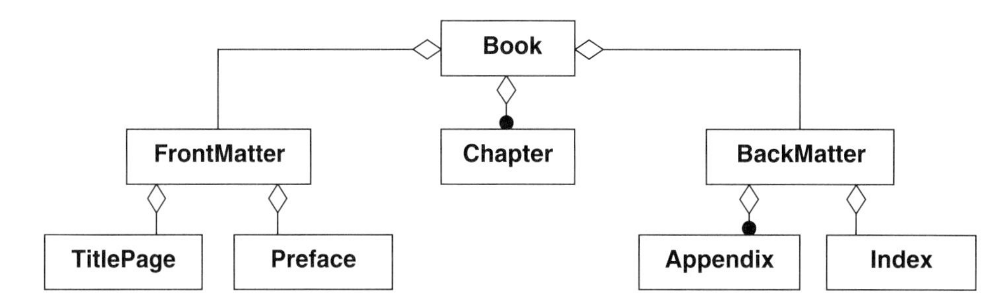

Figure 3.15 Aggregation is a kind of association with additional semantic properties

Having presented this example, we hasten to add that aggregation does not specify order. An assembly is a collection of components without any implied order. It just so happens that a book has a well-known and simple order of parts; the back matter follows the chapters, which follow the front mater. If the model had to capture component order, you would have to include comments, as we describe later in Section 3.7, or use a generic class, as described in Chapter 4.

Aggregation is often encountered with problems that involve bills-of-material. A ***bill-of-material*** is a report that lists each part on a separate line; the lines of the report are ordered by traversing the components in depth-first order starting from the root assembly. Each line may be indented according to its level in the hierarchy. Sibling parts (parts with the same parent) may be further ordered by some other criteria. Figure 3.16 shows a bill-of-material with two levels of parts. In practice, bills-of-material are often nested more deeply.

An aggregation relationship is essentially a binary association, a pairing between the assembly class and a component class. An assembly with many kinds of components corresponds to many aggregations. We define each individual pairing as an aggregation so that we can specify the multiplicity of each component within the assembly. This definition emphasizes that aggregation is a special form of association. An aggregation can be qualified, have roles, and have link attributes just like any other association.

For bill-of-material problems the distinction between association and aggregation is clear. However, for other applications it is not always obvious if an association should be modeled as an aggregation. To determine if an association is an aggregation, test whether the "is-part-of" property applies. The asymmetry and transitivity properties must also hold for aggregation. When in doubt about whether association or aggregation applies, the distinction is not important and you should just use ordinary association.

Bill-of-Material

Level	Part num	Name	Quantity
01	LM16G	Lawn mower	1
02	B16M	Blade	1
02	E1	Engine	1
02	W3	Wheel	4
02	D16	Deck	1

Figure 3.16 Aggregation often occurs with problems that involve bills-of-material

3.3.1 Physical versus Catalog Aggregation

It is important to distinguish between physical and catalog aggregation. ***Physical aggregation*** is an aggregation for which each component is dedicated to at most one assembly. ***Catalog aggregation*** is an aggregation for which components are reusable across multiple assemblies. As an example, consider physical cars (items with individual serial numbers) and car models (Ford Escort, Mazda 626). Customer service records refer to physical cars, while design documents describe car models. The parts explosion for a physical car involves physical aggregation, and the parts explosion for a car model involves catalog aggregation.

Figure 3.17 shows the canonical relationship between catalog aggregation and physical aggregation. A catalog part may describe multiple physical parts. Each catalog part and physical part may contain lesser parts. A catalog part may belong to multiple assemblies, but a physical part may belong to at most one assembly. (The text in braces is a constraint, which we describe later in Section 3.7.)

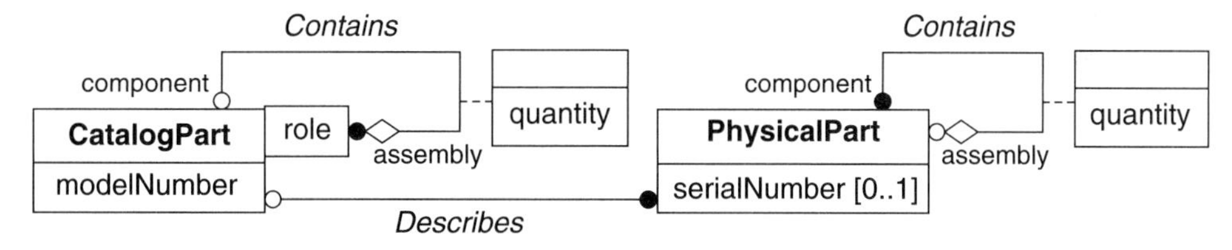

{The catalog parts and physical parts aggregations must be acyclic.}

Figure 3.17 Relationship between catalog aggregation and physical aggregation

Catalog parts may have a quantity specified within a context of usage. For example, two screws of a given type may be used for the windshield wiper assembly of a car and four screws of the same type used for the glove box assembly. A role, such as windshield wiper assembly and glove box assembly, may be specified to differentiate the various uses of a part. A series of roles provides a unique path for navigating a catalog-aggregation lattice.

A physical aggregation tree may also have quantities specified for parts. Parts with individual serial numbers always have a quantity of one, since each part must be individually noted. In contrast, other parts are interchangeable, such as nuts and bolts taken from a bin.

Interchangeable physical parts have identity in the real world, but the corresponding physical aggregation model may not preserve this identity.

The instances for physical aggregation form a collection of trees. Each part belongs to at most one assembly. The part at the root of the tree does not belong to any assembly, and all other parts within the tree belong to exactly one assembly.

In contrast, the instances for catalog aggregation form a directed acyclic graph. (The term "acyclic" means that you cannot start with a part and traverse some sequence of components and reach the starting part.) An assembly may have multiple components and a component may belong to multiple assemblies, but there is a strict sense of direction concerning which part is the assembly and which part is the component (antisymmetry).

The notation clearly indicates whether physical or catalog aggregation applies. With physical aggregation the assembly class has a multiplicity of "one" or "zero or one." With catalog aggregation the assembly class has a multiplicity of "many."

3.3.2 Extended Semantics for Physical Aggregation

Physical aggregation bears properties in addition to transitivity and antisymmetry.

- **Propagation of operations**. ***Propagation*** is the automatic application of some property to a network of objects when the property is applied to some starting object. With aggregation some operations of the assembly may propagate to the components with possible local modifications. For example, moving a window moves the title, pane, and border. For each operation and other propagated qualities, you may wish to specify the extent of propagation [Rumbaugh-88].
- **Propagation of default values**. Default values can also propagate. For example, the color of a car may propagate to the doors. It may be possible to override default values for specific instances. For example, the color of a door for a repaired car may not match the body.
- **Versioning**. A ***version*** is an alternative object relative to some base object. You can encounter versions with hypothetical situations, such as different possibilities for an engineering design. With aggregation, when a new version of a component is created, you may want to trigger automatically the creation of a new version of the assembly.
- **Composite identifiers**. The identifier of a component may or may not include the identifier of the assembly.
- **Physical clustering**. Aggregation provides a basis for physically clustering objects in contiguous areas of secondary storage for faster storage and retrieval. Components are often accessed in conjunction with an assembly. Composite identifiers make it easier to implement physical clustering.
- **Locking**. Many database managers use locking to facilitate concurrent, multiuser access to data. The database manager automatically acquires locks and resolves conflicts without any special user actions. Some database managers implement efficient locking for aggregate trees: A lock on the assembly implies a lock on all components. This is more efficient than placing a lock on each affected part.

3.3.3 Extended Semantics for Catalog Aggregation

Catalog aggregation has fewer properties than physical aggregation, but it is still important to recognize so that you do not confuse it with physical aggregation. For example, propagation is not helpful with catalog aggregation, because a component could have multiple, conflicting sources of information. Propagation across catalog aggregation is too specialized and unusual a topic for us to devise a general solution. Catalog aggregation still observes the basic properties of transitivity and antisymmetry.

With catalog aggregation a collection of components may imply an assembly. This situation is commonly encountered with structured part names in bills-of-material. Figure 3.18 shows a hypothetical object diagram for a lawn mower with two sample bills-of-material. The model number of a lawn mower is a structured name consisting of the prefix "LM" followed by two characters indicating the blade length followed by one character denoting a gas or electric engine. In this example, the blade length and engine type are sufficient to identify a lawn mower uniquely.

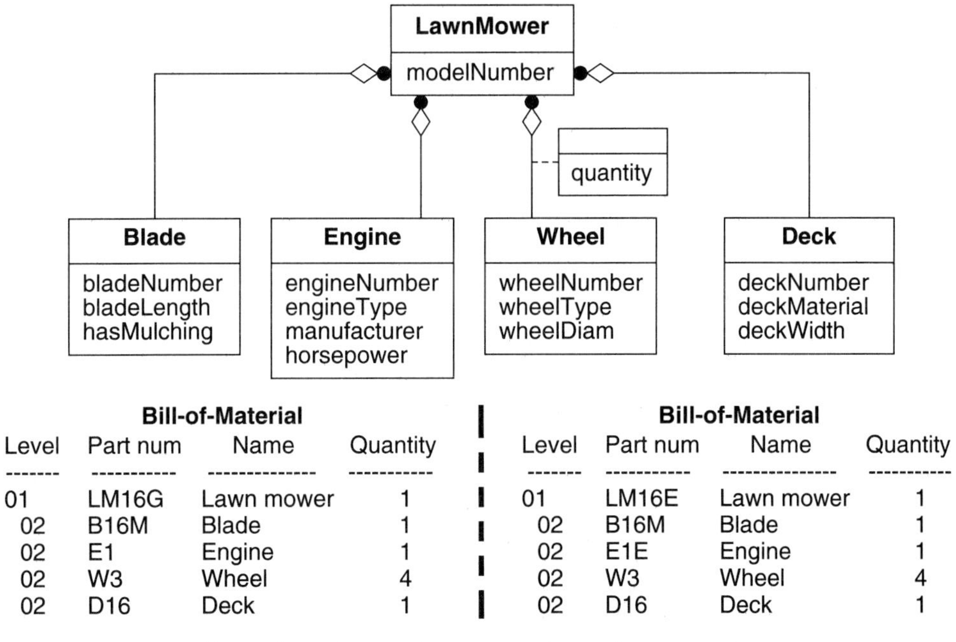

Bill-of-Material

Level	Part num	Name	Quantity
01	LM16G	Lawn mower	1
02	B16M	Blade	1
02	E1	Engine	1
02	W3	Wheel	4
02	D16	Deck	1

Bill-of-Material

Level	Part num	Name	Quantity
01	LM16E	Lawn mower	1
02	B16M	Blade	1
02	E1E	Engine	1
02	W3	Wheel	4
02	D16	Deck	1

Figure 3.18 Structured part names for catalog aggregation

Note that the object model does not capture the constraint that a collection of components implies one assembly. The object diagram in Figure 3.18 states that a lawn mower has "many" multiplicity with respect to each component. In other words, the engine design is useful for multiple lawn mower designs; a blade design applies to multiple lawn mower designs; and so on. You could add a comment to the object model if you wanted to note that some components taken together imply a single lawn mower design.

3.4 GENERALIZATION

3.4.1 Abstract and Concrete Classes

A ***concrete class*** is a class that can have direct instances. In Figure 3.19 *Stock*, *Bond*, and *Insurance* are concrete classes because they have direct instances. *FinancialInstrument* also is a concrete class because some *FinancialInstrument* occurrences (such as real estate) are not in the listed subclasses. The legend *concrete* below the class name indicates a concrete superclass.

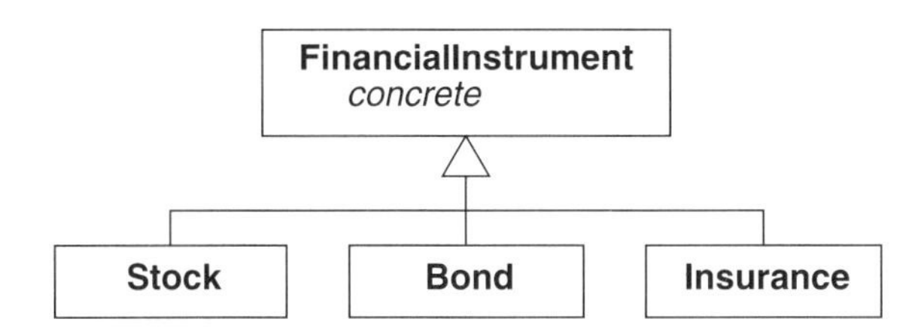

Figure 3.19 Concrete classes

An ***abstract class*** is a class that has no direct instances. The descendant classes can also be abstract, but the generalization hierarchy must ultimately terminate in subclasses with direct instances. In Figure 3.20 *Person* is an abstract class but the subclasses *Manager* and *IndividualContributor* are concrete. The legend *abstract* indicates an abstract superclass. You may define abstract operations for abstract classes. An ***abstract operation*** specifies the signature of an operation while deferring implementation to the subclasses. The ***signature*** of an operation specifies the argument types, the result type, exception conditions, and the semantics of the operation. The notation for an abstract operation is the legend *{abstract}* following the operation name.

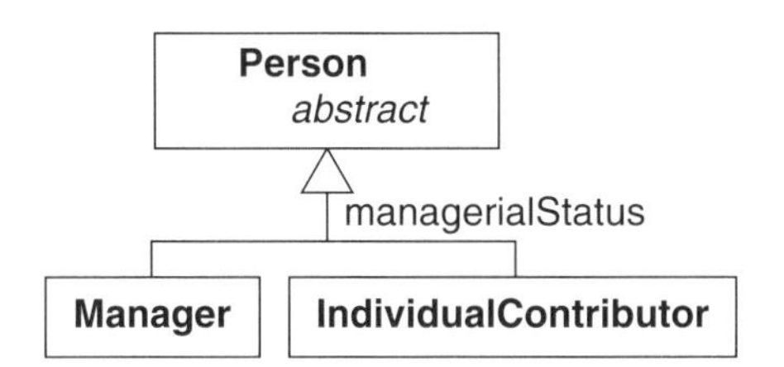

Figure 3.20 An abstract class

3.4.2 Generalization versus Other Object-Modeling Constructs

Figure 3.21 shows how generalization differs from association. Both generalization and association involve classes, but association describes the relationship between two or more instances, while generalization describes different aspects of a single instance.

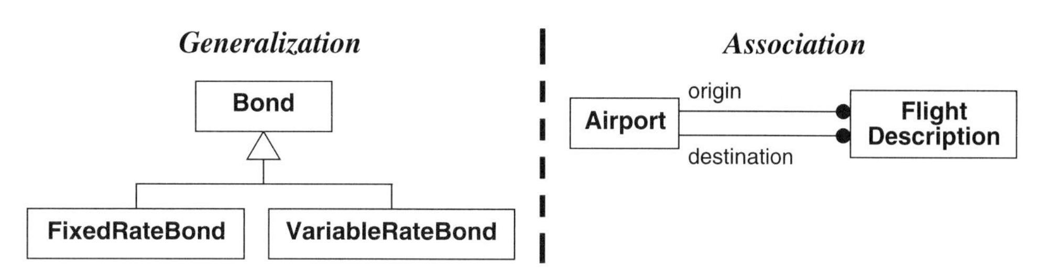

Figure 3.21 Generalization versus association

During analysis, associations are more important than generalization. Associations add information. Associations transcend class encapsulation boundaries and can have a broad impact on a system. In contrast, generalization eliminates duplications of shared properties (consolidates), but does not change the instances that conform to the model.

During design, generalization becomes more significant. Developers tend to discover data structure during analysis and behavior during design. Generalization provides a reuse mechanism for concisely expressing behavior and including code from class libraries. Judicious reuse reduces development time and substitutes carefully tested library code for error-prone application code.

Figure 3.22 shows how generalization differs from aggregation. Both generalization and aggregation give rise to trees through transitive closure, but generalization is the "or" relationship and aggregation is the "and" relationship. In the figure, a bond is a fixed-rate bond *or* a variable-rate bond. A book comprises front matter *and* many chapters *and* back matter. Generalization relates classes that describe different aspects of a single object. Aggregation relates distinct objects that compose an assembly.

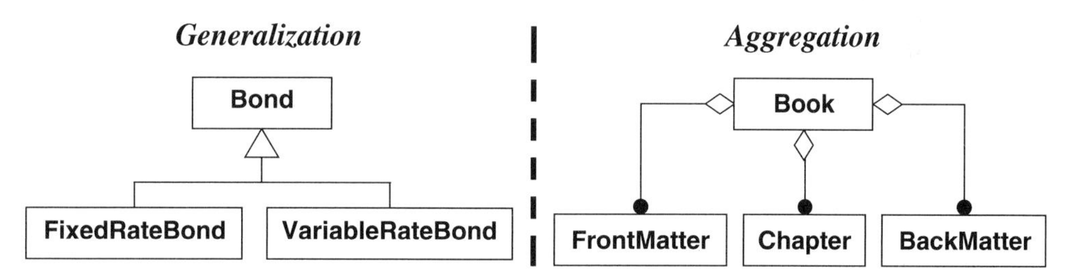

Figure 3.22 Generalization versus aggregation

Figure 3.23 shows how generalization differs from instantiation. Some modelers erroneously introduce an object as a subclass in a generalization. Generalization does not deal with individual objects; generalization relates *classes*—the superclass and the subclasses. Instantiation relates an instance to a class.

A common mistake is to confuse a subclass with a role. Figure 3.24 shows the difference between a subclass and a role. A subclass is a specialization of a class; a role is a usage of a class. By definition, a subclass pertains to only some superclass instances. There is no such constraint with a role; a role may refer to any or all instances. Figure 3.24 shows roles and

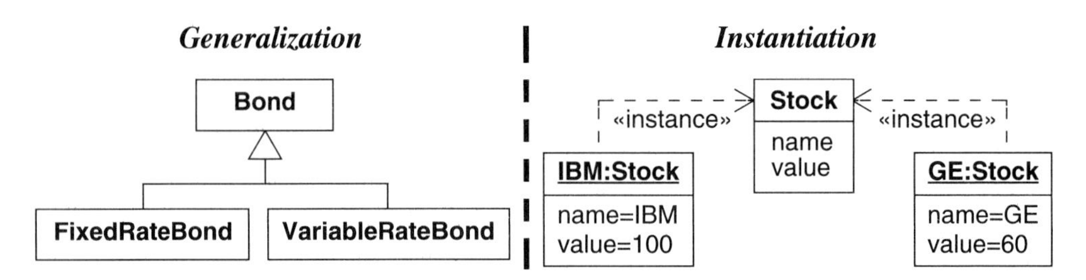

Figure 3.23 Generalization versus instantiation

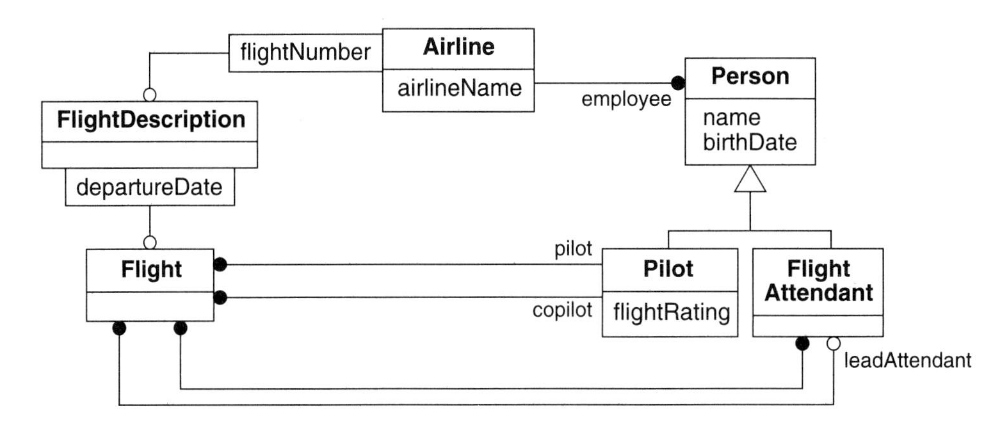

Figure 3.24 Subclass versus role

subclasses for the class *Person. Pilot* and *FlightAttendant* are modeled as subclasses, since they have different attributes and associations. In contrast, *employee* is merely a role for a *Person*; *pilot* and *copilot* are roles for *Pilot*; and *leadAttendant* is a role for *FlightAttendant.*

3.5 MULTIPLE INHERITANCE

Multiple inheritance permits a class to inherit attributes, operations, and associations from multiple superclasses, which, in turn, lets you mix information from two or more sources. Single inheritance organizes classes as a tree. Multiple inheritance organizes classes as a directed acyclic graph. Multiple inheritance brings greater modeling power but at the cost of greater complexity.

3.5.1 Multiple Inheritance from Different Discriminators

Multiple inheritance can arise through different bases (different discriminators) for specializing the same class. In Figure 3.25 a person may be specialized along the bases of manage-

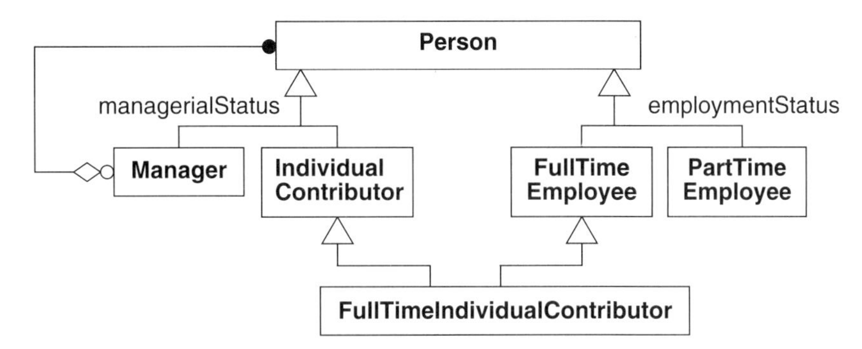

Figure 3.25 Multiple inheritance from different discriminators

rial status (manager or individual contributor) and employment status (fulltime or parttime). Whether a person is a manager or not is independent of employment status. Four subclasses are possible that combine managerial status and employment status. The figure shows one, *FullTimeIndividualContributor.*

3.5.2 Multiple Inheritance without a Common Ancestor

Multiple inheritance is possible, even when superclasses have no common ancestor. This often occurs when you mixin functionality from software libraries. When software libraries overlap or contradict, multiple inheritance becomes problematic.

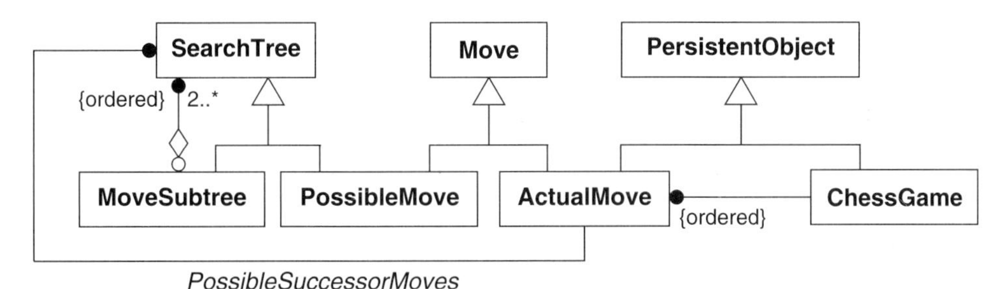

Figure 3.26 Multiple inheritance without a common ancestor

Figure 3.26 shows an excerpt of a model for a chess game. A chess program looks ahead from the current board position and examines multiple search trees to determine the most promising next move for actual play. You may wish to store the actual moves in a database for reconsideration at some later date or in order to replay a game from an intermediate point. In contrast, the exploration of the search space may be regarded as transient and unimportant to store.

In Figure 3.26 each *SearchTree* may be a *MoveSubtree* or a *PossibleMove*. Each *MoveSubtree*, in turn, is composed of lesser *SearchTree*s. Such a combination of object-modeling constructs can describe a tree of moves of arbitrary depth. Each *Move* may be a *Possible-Move* or an *ActualMove*. *PossibleMove* and *ActualMove* inherit common behavior from the *Move* superclass. *ActualMove* and *ChessGame* are the classes with the objects that we want to store permanently. In our model persistent objects must inherit from *PersistentObject*, as with the ODMG standard [Cattell-96]. In Figure 3.26 both *PossibleMove* and *ActualMove* use multiple inheritance.

3.5.3 Workarounds for Multiple Inheritance

Several workarounds are possible if you wish to avoid the complexity of multiple inheritance. Workarounds often make it easier to understand and implement a model.

- **Factoring**. Figure 3.27 avoids multiple inheritance by taking the cross product of the orthogonal bases of person (*managerialStatus*, *employmentStatus*) from Figure 3.25. The disadvantages are that you lose conceptual clarity and the reuse of similar code is more cumbersome.

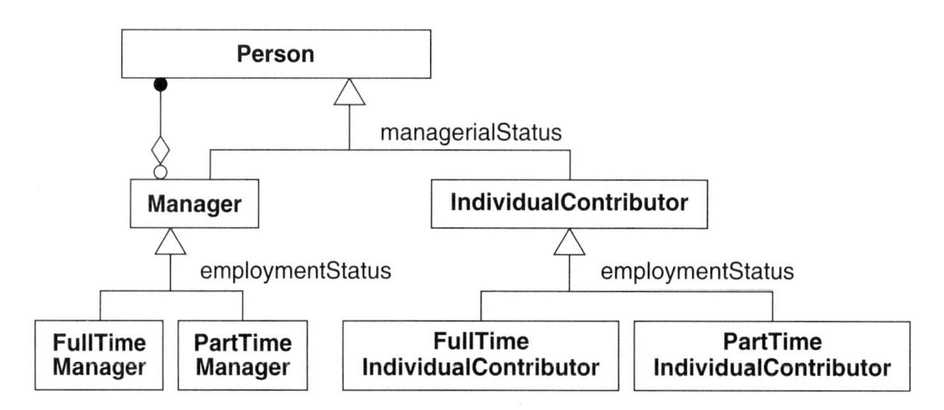

Figure 3.27 Multiple inheritance workaround: factoring

- **Fragmenting a subclass**. You may fragment a subclass into multiple classes—one for each superclass. For example, we could restate Figure 3.26 as Figure 3.28. The disadvantages are that objects are broken into multiple pieces and the added classes are often artificial and difficult to define.
- **Replacing generalization with associations**. You can replace a generalization with several exclusive-or associations. Figure 3.29 shows the result of doing this to the model in Figure 3.26. This option is most viable for a generalization with few subclasses. The disadvantages are the loss of identity and more cumbersome reuse of operations. Also, exclusive-or associations are not supported by most languages and database managers, so you may need to write application code to enforce exclusivity.

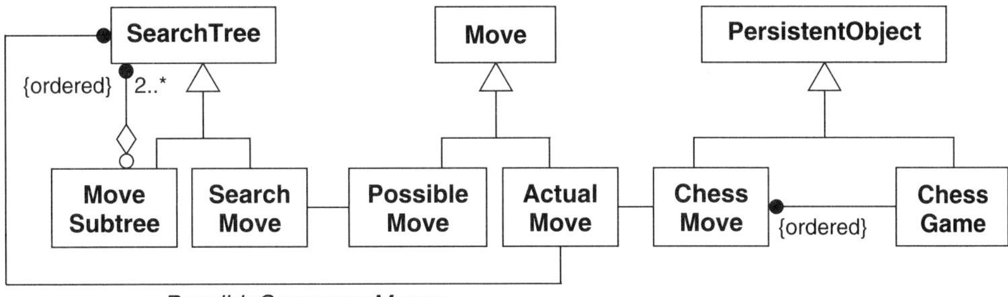

Figure 3.28 Multiple inheritance workaround: fragmenting a subclass

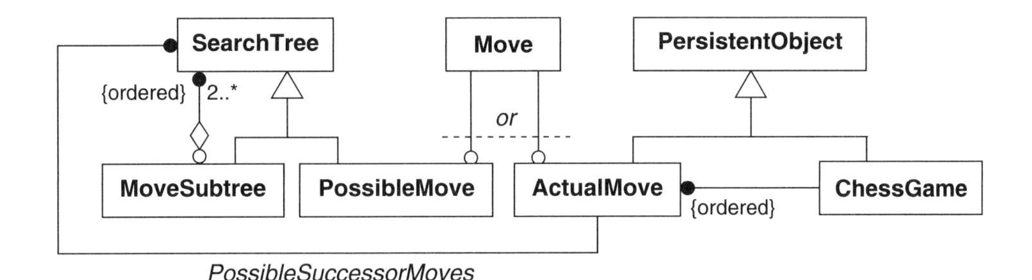

Figure 3.29 Multiple inheritance workaround: replacing generalization with exclusive-or associations

3.6 PACKAGES

You can fit an object model on a single page for many small and medium-sized problems. However, you will need to organize the presentation of large object models. A person cannot understand a large object model at a glance. Furthermore, it is difficult to get a sense of perspective about the relative importance of portions of a large model. You must partition a large model to allow comprehension.

A ***package*** is a group of elements (classes, associations, generalizations, and lesser packages) with a common theme. A package partitions a model, making it easier to understand and manage. Large applications may require several tiers of packages. Packages form a tree with increasing abstraction toward the root, which is the application, the top-level package. As Figure 3.30 shows, the notation for a package is a box with the addition of a tab. The purpose of the tab is to suggest the enclosed contents, like a tabbed folder.

There are various themes for forming packages: dominant classes, dominant relationships, major aspects of functionality, and symmetry. For example, many business systems

Figure 3.30 Notation for a package

have a *Customer* package or a *Part* package; *Customer* and *Part* are dominant classes that are important to the business of a corporation and appear in many applications. In an engineering application we used a dominant relationship, a large generalization for many kinds of equipment, to divide an object model into packages. Equipment was the focus of the model, and the attributes and relationships varied greatly across types of equipment. An object model of a compiler could be divided into packages for lexical analysis, parsing, semantic analysis, code generation, and optimization. Once some packages have been established, symmetry may suggest additional packages.

On the basis of our experience in creating packages, we can offer the following tips:

- **Carefully delineate each package's scope**. The precise boundaries of a package are a matter of judgment. Like other aspects of modeling, defining the scope of a package requires planning and organization. Make sure that class and association names are unique within each package, and use consistent names across packages as much as possible.
- **Make packages cohesive**. There should be fewer associations between classes that appear in different packages than between classes that appear in a single package. Classes may appear in multiple packages, helping to bind them, but ordinarily associations and generalizations should appear in a single package.
- **Define each class in a single package**. The defining package should show the class name, attributes, and possibly operations. Other packages that refer to a class can use a class icon, a box that contains only the class name. This convention makes it easier to read object diagrams because a class is most prominent in its defining package. It ensures that readers of the object model will not become distracted by possibly inconsistent definitions or be misled by forgetting a prior class definition. This convention also makes it easier to develop packages concurrently.

3.6.1 Logical Horizon

You can often use a class with a large logical horizon as the nucleus for a package. The ***logical horizon*** [Feldman-86] of a class is the set of classes reachable by one or more paths terminating in a combined multiplicity of "one" or "zero or one." A ***path*** is a sequence of consecutive associations and generalization levels. When computing the logical horizon, you may traverse a generalization hierarchy to obtain further information for a set of objects. You may not, however, traverse to sibling objects, such as by going up and then down the hierarchy. The logical horizons of various classes may, and often do, overlap. In Figure 2.23 on page 28 the logical horizon of *FlightDescription* is *Airport*, *Airline*, and *AircraftDescription*. The logical horizon of *Airport* is the empty set.

Figure 3.31 shows the computation of the logical horizon for *FlightReservation*.

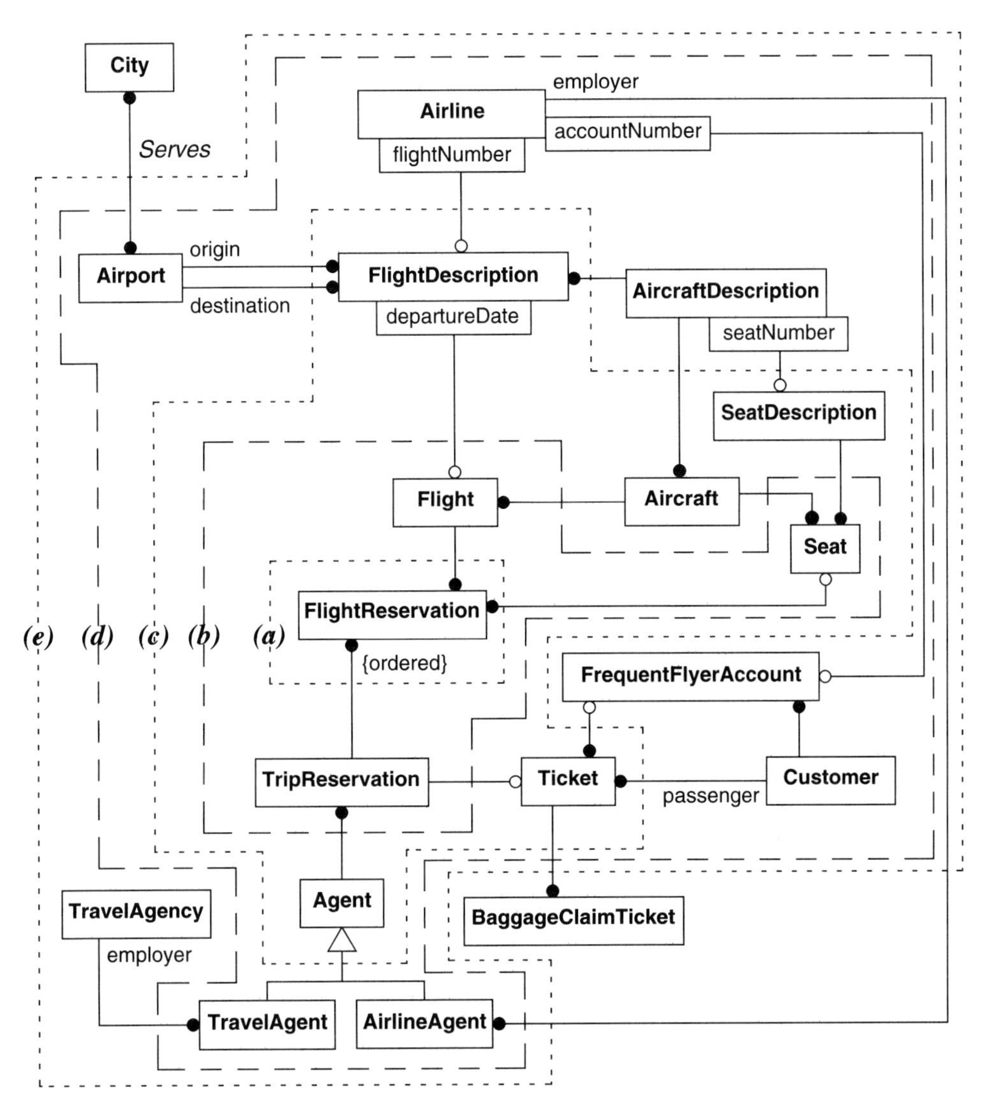

Figure 3.31 Computing the logical horizon for *FlightReservation*

- **(a)** We start with *FlightReservation.*
- **(b)** Each *FlightReservation* has a *Flight*, *Seat*, and *TripReservation.*
- **(c)** A *TripReservation* implies an *Agent* and a *Ticket*. A *Flight* implies a *FlightDescription* and an *Aircraft*. A *Seat* implies a *SeatDescription.*
- **(d)** A *Ticket* implies a *FrequentFlyerAccount* and a *Customer*; an *Agent* leads to *TravelAgent* and an *AirlineAgent* via generalization. The *FlightDescription* implies an *Airport*, *Airline*, and *AircraftDescription*.

- **(e)** A *TravelAgent* has a *TravelAgency* as an *employer*. Thus the logical horizon of *FlightReservation* includes every class in the diagram except *City* and *BaggageClaimTicket*.

When computing the logical horizon, you should disregard any qualifiers and treat the associations as if they were unqualified. The purpose of the logical horizon is to compute the objects that can be inferred from some starting object.

3.6.2 Example of Packages

Figure 3.32 shows a model for an airline information system with packages organized on a functional basis. (We are elaborating the model presented in Figure 2.23.) The reservations package records customer booking of airline travel. Flight operations deals with the actual logistics of planes arriving and departing. The aircraft information package stores seating layout and manufacturing data. Travel awards tracks bonus free travel for each customer; a person may submit a frequent flyer account number at the time of a reservation, but does not receive credits until after taking the flight. Baggage handling involves managing bags in conjunction with flights and accommodating errant pieces of luggage. Crew scheduling involves scheduling to staff flight needs. The subsequent diagrams elaborate all packages except *CrewScheduling*.

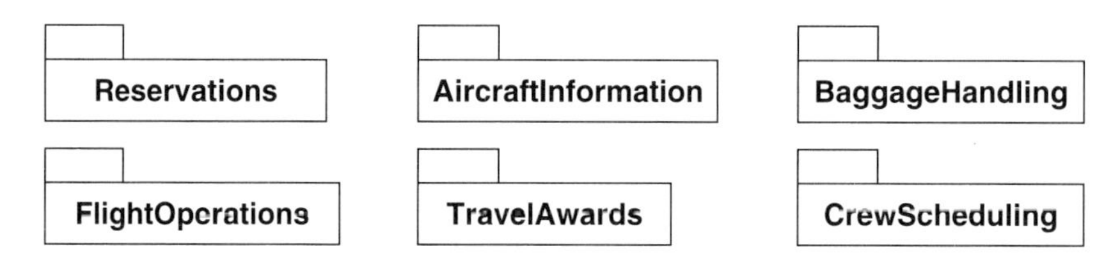

Figure 3.32 A partial high-level object model for an airline information system

Figure 3.33 describes the *Reservations* package. A trip reservation consists of a sequence of flight reservations, where each flight reservation refers to a specific flight. Sometimes another flight is substituted for a booked flight because of equipment problems, weather delays, or customer preference. The passenger may reserve a seat for each flight. A trip reservation is made on some date; the passenger must purchase a ticket within a certain number of days or the reservation becomes void. The airlines use record locators to find a particular trip reservation quickly and unambiguously. A trip is reserved by an agent, who either works for an airline or a travel agency. The frequent flyer account may be noted for a passenger. Although the structure of the model does not show it, the owner of the frequent flyer account must be the same as the passenger. We directly associate *TripReservation* with *FrequentFlyerAccount* and *Customer*, because a customer can make a reservation and specify a frequent flyer account before a ticket is even issued. Multiple payments may be made for a trip, such as two credit-card charges. Payment may also be made by cash or check.

Figure 3.34 describes the *FlightOperations* package. An airport serves many cities, and a city may have multiple airports. Airlines operate flights between airports. A flight descrip-

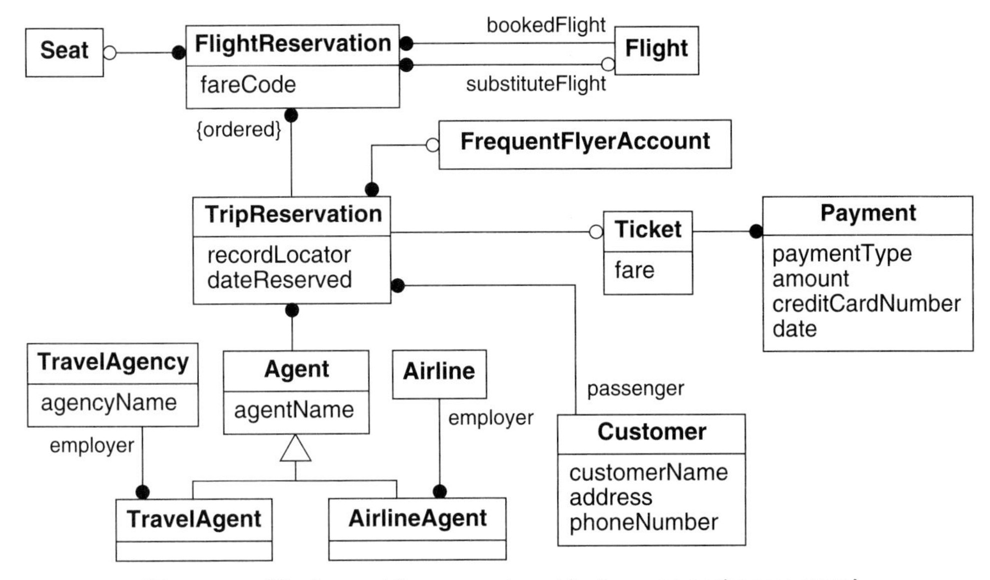

Figure 3.33 An object model for the *Reservations* package

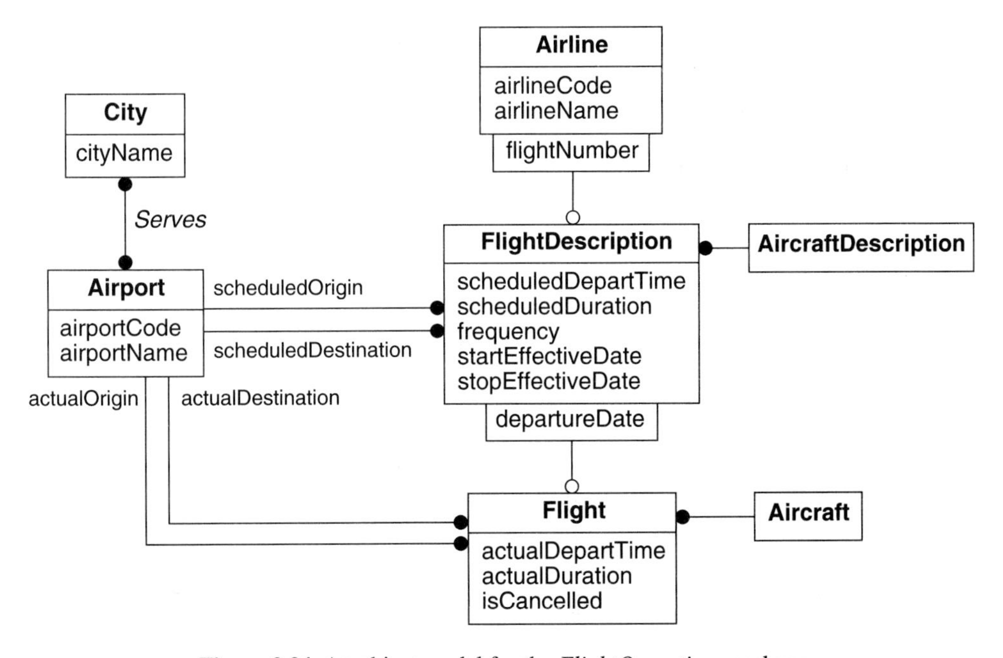

Figure 3.34 An object model for the *FlightOperations* package

tion refers to the published description of air travel between two airports. In contrast, a flight refers to the actual travel made by an airplane on a particular date. The frequency indicates the days of the week for which the flight description applies. The start and stop effectivity dates bracket the time period for which the published flight description is in effect. The actual origin, destination, departure time, and duration of a flight can vary because of weather and equipment problems.

Figure 3.35 presents a simple model of the *AircraftInformation* package. Each aircraft model has a manufacturer, model number, and specific numbering for seats. The seat type may be first class, business, or coach. Each individual aircraft has a registration number and refers to an aircraft model.

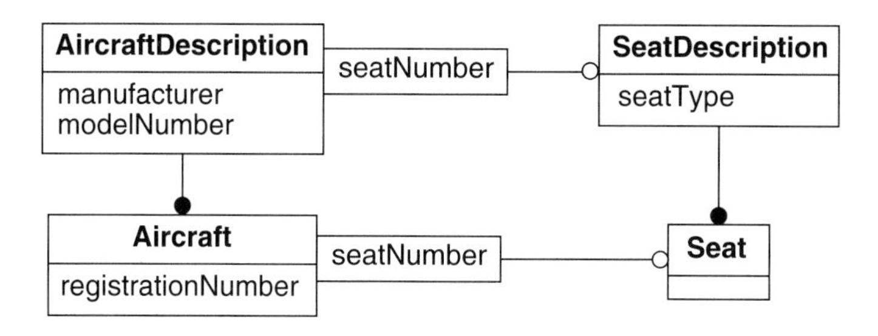

Figure 3.35 An object model for the *AircraftInformation* package

Figure 3.36 describes the *TravelAwards* package. A customer may have multiple frequent flyer accounts. Airlines identify each account with an account number. An account may receive numerous frequent flyer credits. Some frequent flyer credits pertain to flights; others (indicated by *creditType*) concern adjustments, redemption, long distance mileage, credit card mileage, hotel stays, car rental, and other kinds of inducements to patronize a business.

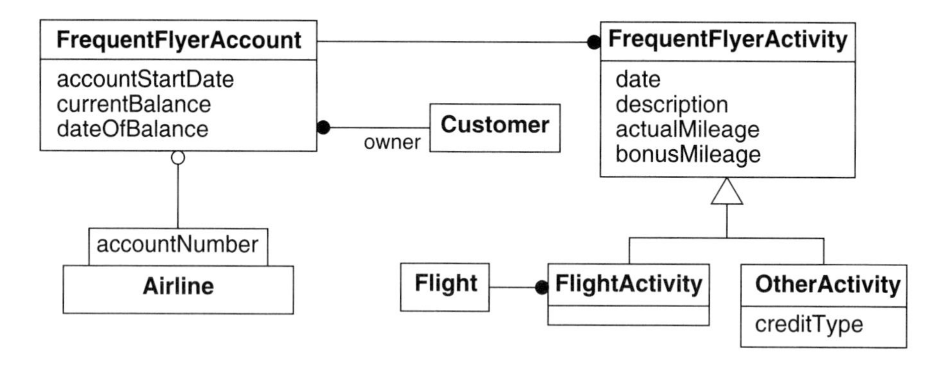

Figure 3.36 An object model for the *TravelAwards* package

Figure 3.37 describes the *BaggageHandling* package. A customer may check multiple bags for a trip and receives a claim ticket for each bag. Sometimes a bag is lost, damaged, or delayed, in which case the customer completes a baggage complaint form for each problem bag.

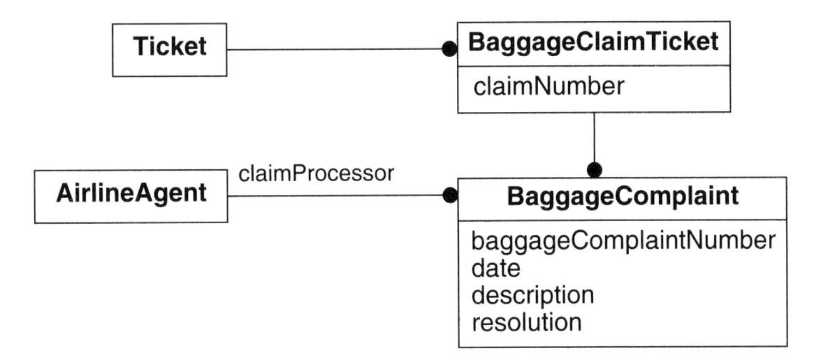

Figure 3.37 An object model for the *BaggageHandling* package

3.7 DERIVED DATA AND CONSTRAINTS

Derived data is data that can be completely determined from other data. Classes, attributes, and associations can all be derived. The underlying data can, in turn, be base data or further derived. Do not confuse our use of the term "derived" with the C++ derived class. A C++ derived class refers to the subclass of a generalization; it has nothing to do with OMT's meaning of derived data.

As a rule, you should not show derived data during analysis unless the data appears in the problem description. During design you can add derived data to improve efficiency and ease implementation. During implementation you can compute derived data on demand from constituent data (lazy evaluation) or precompute and cache it (eager evaluation). Derived data that is precomputed must be marked as invalid or recomputed if constituent data is changed.

The notation for derived data is a slash preceding the name of the attribute, class, association, or role. Figure 3.38 shows an example of a derived attribute for airline flight descriptions. Exercise 3.8 illustrates derived associations.

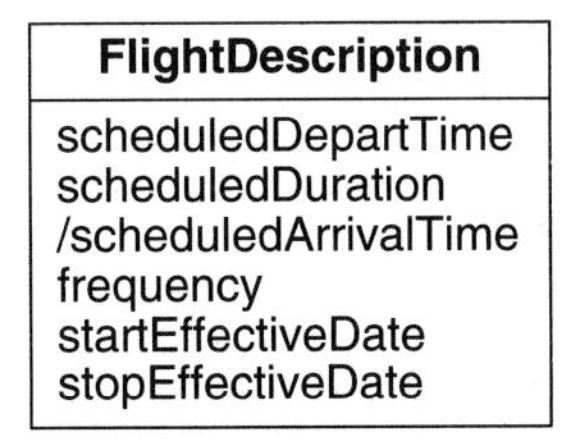

Figure 3.38 A derived attribute

A ***constraint*** is a functional relationship between modeling constructs such as classes, attributes, and associations. A constraint restricts the values of data. You may place simple constraints in the object model. You should specify complex constraints in the functional model.

A "good" model should capture many constraints with its very structure. In fact, the ability of a model to express important constraints is one measure of the quality of a model. (See Exercise 2.13.) Most object models require several iterations to strike a proper balance between rigor, simplicity, and elegance. However, sometimes it is not practical to express all important constraints with the structure of a model. For example, in Figure 3.33 we found it difficult to express structurally that the owner of the frequent flyer account must be the same as the passenger. In Figure 3.17 we specified that the catalog parts and physical parts aggregations must be acyclic.

Constraints are denoted by text in braces ("{" and "}"). The text of a constraint should clearly indicate the affected data. Similarly, comments are also delimited by braces. We often use comments to document the rationale for subtle modeling decisions and convey important enumerations.

Sometimes it is useful to draw a dotted arrow between classes or associations to indicate the scope of a constraint. For example, in Figure 3.39 a table has many columns; the primary key columns are a subset of the overall columns.

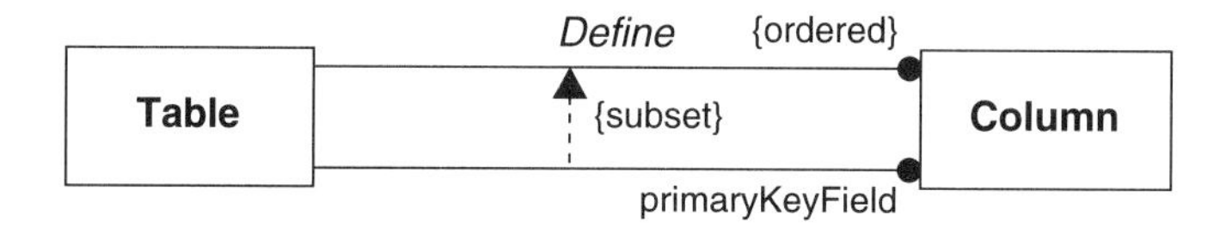

Figure 3.39 A subset constraint between associations

3.8 ADVANCED PRACTICAL TIPS

The following list summarizes the practical aspects of the object-modeling constructs described in this chapter.

- **Class attributes**. Try to avoid class attributes in your models. Often you can restructure a model, both avoiding class attributes and making the model more precise. (Section 3.1.2)
- **Domains**. Specify domains for attributes instead of data types. Domains promote uniform assignment of data types and convey additional semantic information. (Section 3.1.5)
- **Enumeration domains**. Do not create unnecessary generalizations for attributes of enumeration domain. Only specialize a class when the subclasses have distinct attributes, operations, or associations. (Section 3.1.5)

- ***N*-ary associations**. Try to avoid general ternary and *n*-ary associations. Most of these can be decomposed into binary associations, with possible qualifiers and link attributes. (Section 3.2.2)
- **Aggregation**. Consider aggregation when the "is-part-of" relationship holds. An aggregation must satisfy the transitivity and antisymmetry properties. Be careful not to confuse physical and catalog aggregation. (Section 3.3)
- **Roles**. Do not confuse classes with roles. A role is a use of a class in an association; a class may assume various roles. Do not introduce multiple classes in a model when there really is just one class with multiple roles. It is a good practice to label a class with its intrinsic name rather than a role name. (Section 3.4.2)
- **Multiple inheritance**. Try to avoid multiple inheritance during analysis because it is often confusing. Multiple inheritance is more helpful during design because of the need to mixin orthogonal aspects of objects. (Section 3.5)
- **Large models**. Organize large models so that the reader can understand portions of the model at a time, rather than the whole model at once. Packages are useful for organizing large models. (Section 3.6)
- **Constraints**. You may be able to restructure an object model to improve clarity and capture additional constraints. Use comments to express constraints that are awkward to represent with object-model structure. Also add comments to document modeling rationale and important enumeration values. (Section 3.7)

3.9 CHAPTER SUMMARY

Classes may have attributes and operations whose value is common to a group of objects. We advise that you restructure your object models to minimize use of class attributes. Attribute multiplicity specifies whether an attribute may be single or multivalued and whether an attribute is optional or mandatory. A domain is the set of possible values for an attribute. During design you should assign a domain to each attribute, instead of just directly assigning a data type.

The degree of an association is the number of distinct roles for each link. The vast majority of associations are binary or qualified binary. Ternary associations occasionally occur, but we have rarely encountered an association of higher degree.

A qualification cascade is a series of consecutive qualified associations. Qualification cascades often occur where an accumulation of qualifiers denotes increasingly specific objects.

Aggregation is a kind of association in which a whole, the assembly, is composed of parts, the components. Aggregation is often called the "a-part-of" or "parts-explosion" relationship and may be nested to an arbitrary number of levels. Aggregation bears the transitivity and antisymmetry properties. Do not confuse physical and catalog aggregation. With physical aggregation each component is dedicated to at most one assembly. With catalog aggregation components are reusable across multiple assemblies.

Generalization superclasses may or may not have direct instances. A concrete class can have direct instances; an abstract class has no direct instances. Multiple inheritance permits a class to inherit attributes, operations, and associations from multiple superclasses. Multiple inheritance brings greater modeling power but at the cost of greater conceptual and implementation complexity.

You will need multiple pages of diagrams to express object models for large problems. Large object models can be organized and made tractable with packages. Packages partition an object model into groups of tightly connected classes, associations, and generalizations.

Figure 3.40 lists the key concepts for this chapter.

abstract class	derived association	multiple inheritance
aggregation	derived attribute	package
association degree	derived class	path
attribute multiplicity	domain	physical aggregation
candidate key	enumeration domain	qualification cascade
catalog aggregation	exclusive-or association	secondary data
class attribute	instantiation	signature
class operation	logical horizon	structured domain
concrete class	maximum multiplicity	ternary association
constraint	minimum multiplicity	

Figure 3.40 Key concepts for Chapter 3

CHANGES TO OMT NOTATION

This chapter has introduced the following changes to the notation in [Rumbaugh-91] for compatibility with the UML notation [UML-98]. These changes are in addition to those from Chapter 2.

- **Instantiation.** The notation for instantiation is a dashed line from the instance to the class with an arrow pointing to the class; the dashed line is labeled with the legend *instance* enclosed by guillemets («»).
- **Class attribute and class operation.** You can indicate class attributes and class operations with an underline.
- **Exclusive-or association.** You can use a dashed line annotated by the legend "or" to group exclusive-or associations.
- **Compound qualifier.** You may use more than one attribute as the qualifier for an association.
- **Abstract and concrete classes.** You can indicate these by placing the legend *abstract* or *concrete* below the class name.
- **Package.** The notation for a package is a box with the addition of a tab.

- **Derived data**. Classes, attributes, and associations can all be derived. A slash in front of a name denotes derived data.

We have made some further minor notation extensions of our own.

- **Attribute multiplicity**. You can specify the number of values for an attribute within brackets after the attribute name. You may specify a mandatory single value *[1]*, an optional single value *[0..1]*, an unbounded collection with a lower limit *[lowerLimit..*]*, or a collection with fixed limits *[lowerLimit..upperLimit]*.
- **Candidate key**. You can specify a candidate key for a class with the notation *{CKn}* next to the participating attributes.
- **Structured domain**. You can use indentation to show the structure of domains.

BIBLIOGRAPHIC NOTES

Chapter 4 of [Booch-94] presents an insightful treatment of inheritance in the context of the broader classification literature.

This chapter contains much new material that complements our earlier book [Rumbaugh-91]. Our most significant improvements are in the areas of domains, secondary data, and aggregation.

REFERENCES

[Blaha-93a] Michael Blaha. Secondary aspects of modeling. *Journal of Object-Oriented Programming 6*, 1 (March 1993), 15–18.

[Blaha-93b] Michael Blaha. Aggregation of Parts of Parts of Parts. *Journal of Object-Oriented Programming 6*, 5 (September 1993), 14–20.

[Booch-94] Grady Booch. *Object-Oriented Analysis and Design with Applications*. Reading, Massachusetts: Benjamin/Cummings, 1994.

[Cattell-96] RGG Cattell, editor. *The Object Database Standard: ODMG-93, Release 1.2*. San Francisco, California: Morgan-Kaufmann, 1996.

[Feldman-86] P Feldman and D Miller. Entity model clustering: Structuring a data model by abstraction. *Computer Journal 29*, 4 (August 1986), 348–360.

[Rumbaugh-88] James Rumbaugh. Controlling propagation of operations using attributes on relations. *OOPSLA'88 as ACM SIGPLAN 23*, 11 (November 1988), 285–296.

[Rumbaugh-91] J Rumbaugh, M Blaha, W Premerlani, F Eddy, and W Lorensen. *Object-Oriented Modeling and Design*. Englewood Cliffs, New Jersey: Prentice Hall, 1991.

[UML-98] The following books are planned for the Unified Modeling Language:
Grady Booch, James Rumbaugh, and Ivar Jacobson. *UML User's Guide*. Reading, Massachusetts: Addison-Wesley.
James Rumbaugh, Ivar Jacobson, and Grady Booch. *UML Reference Manual*. Reading, Massachusetts: Addison-Wesley.
Ivar Jacobson, Grady Booch, and James Rumbaugh. *UML Process Book*. Reading, Massachusetts: Addison-Wesley.

EXERCISES

3.1 (2) Add domains to the object model in Figure E2.1.

3.2 (3) Figure 3.16 is an example of catalog aggregation. Construct an instance diagram for Figure 3.17 using the instances in Figure 3.16. Assume that the *role* shown in Figure 3.17 is *normal.*

3.3 (3) The object model in Figure E3.1 describes a reporting hierarchy within a company. Change the object model to accommodate matrix management (where a person may report to more than one manager).

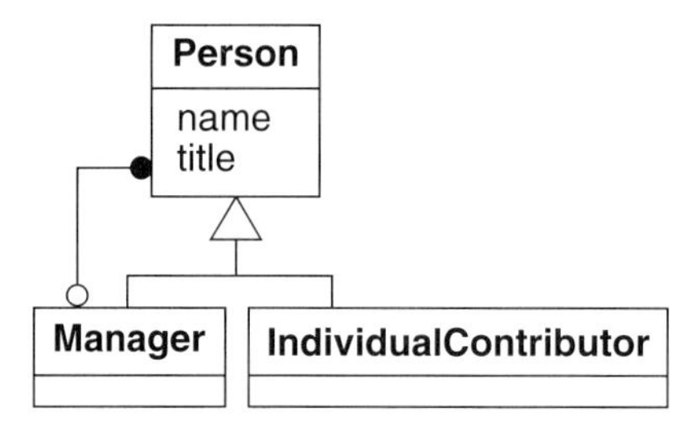

Figure E3.1 An object model for the management hierarchy in a corporation

3.4 (7) Extend the object model in Figure E3.1 to track the evolution of the reporting hierarchy over time. The reporting hierarchy changes as persons join and leave a company. The hierarchy also changes due to promotions and demotions.

3.5 (4) What is the logical horizon of *Statement* in Figure E2.1?

3.6 (4) What is the logical horizon of *City* in Figure 3.34?

3.7 (5) What is the logical horizon of *OtherActivity* in Figure 3.36? You should also consider relationships for the other packages in Section 3.6.2.

3.8 (7) Add the following derived information to extend your answer to Exercise 2.16: age, grandparent, ancestor, descendant, aunt, uncle, sibling, and cousin.

3.9 (9) Construct an object model that describes the baseball statistics listed in Figure E3.2. Use any baseball knowledge you may have. Even if you are unfamiliar with the game of baseball, this exercise is still useful. For legacy applications, it is not uncommon to be given examples of data structure with little explanation. We have chosen data that illustrates most multiplicity combinations that would be found in a more comprehensive set of data. Note that the St. Louis Browns moved at the end of the 1953 season and became the Baltimore Orioles; the Philadelphia Athletics moved at the end of the 1954 season and became the Kansas City Athletics.

3.10 (9) Prepare an object model for the game of hearts. A game of hearts typically involves four players. The objective of the game is to play a series of hands and score the fewest points. There are two phases to each hand: exchanging cards and then playing cards.

The following cycle must be observed for exchanging cards. For the first hand each player passes three cards to the player on the left (and receives three cards from the player on the right). For the second hand each player passes three cards to the right. Each player passes three cards

Player Statistics—Batting

Year	League	City	Team name	Player	Field position	Bat pos.	At bat	HR	RBI	BA
1953	American	St. Louis	Browns	Vern Stephens	3B	R	165	4	17	.321
1953	American	Chicago	White Sox	Vern Stephens	3B,SS	R	129	1	14	.186
1953	American	St. Louis	Browns	Bob Elliott	3B	R	160	5	29	.250
1953	American	Chicago	White Sox	Bob Elliott	3B,OF	R	208	4	32	.260
1953	American	Phil.	Athletics	Dave Philley	OF,3B	B	620	9	59	.303
1953	American	St. Louis	Browns	Don Larsen	P,OF	R	81	3	10	.284
1955	American	Detroit	Tigers	Al Kaline	OF	R	588	27	102	.340
1955	American	Detroit	Tigers	Fred Hatfield	2B,3B,SS	L	413	8	33	.232

Player Statistics—Pitching

Year	League	City	Team name	Player	Pitch pos.	IP	Win	Loss	Save	ERA
1953	American	St. Louis	Browns	Don Larsen	R	193	7	12	2	4.15
1953	American	St. Louis	Browns	Bobo Holloman	R	65	3	7	0	5.26
1953	American	St. Louis	Browns	Satchel Paige	R	117	3	9	11	3.54
1953	American	Phil.	Athletics	Alex Kellner	L	202	11	12	0	3.92

Team Statistics

Year	League	City	Team name	Win	Loss	Manager	Save	ERA	AB	HR	RBI	BA
1953	Amer.	St. L	Browns	54	100	Marty Marion	24	4.48	5264	112	522	.249
1953	Amer.	Phil.	Athletics	59	95	Jimmy Dykes	11	4.67	5455	116	588	.256
1953	Amer.	Det.	Tigers	60	94	Fred Hutchinson	16	5.25	5553	108	660	.266
1955	Amer.	Balt.	Orioles	57	97	Paul Richards	22	4.21	5257	54	503	.240
1955	Amer.	KC	Athletics	63	91	Lou Boudreau	23	5.35	5335	121	593	.261
1955	Amer.	Det.	Tigers	79	75	Bucky Harris	12	3.79	5283	130	724	.266
1953	Natl.	Phil.	Phillies	83	71	Steve O'Neill	15	3.80	5290	115	657	.265
1953	Natl.	St. L	Cardinals	83	71	Eddie Stanky	36	4.23	5397	140	722	.273
1953	Natl.	NY	Giants	70	84	Leo Durocher	20	4.25	5362	176	739	.271
1955	Natl.	NY	Giants	80	74	Leo Durocher	14	3.77	5288	169	643	.260
1955	Natl.	Phil.	Phillies	77	77	Mayo Smith	21	3.93	5092	132	631	.255
1955	Natl.	St. L	Cardinals	17	19	Eddie Stanky	15	4.56	5266	143	608	.261
1955	Natl.	St. L	Cardinals	51	67	Harry Walker						

Figure E3.2 Sample baseball statistics

across for the third hand. No cards are passed for the fourth hand. The fifth hand starts the passing cycle over again with the left player. Each player chooses the cards for passing. A good player will assess a hand and try to pass cards that will reduce his or her own likelihood of scoring points and increase that for the receiving player.

The card-playing portion of a hand consists of 13 tricks. The player with the two of clubs leads the first trick and play continues clockwise. The player who plays the largest card of the lead suit "wins" the trick and adds the cards in the trick to his or her pile for scoring at the end of the hand. The winner of the trick also leads the next trick. The sequence of cards in a suit from largest to smallest is ace, king, queen, jack, and ten down to two. Each card is played exactly once in a game.

If possible, a player must play the same suit as the lead card on a trick. Furthermore, on the first trick, a player cannot play a card that scores points (see next paragraph), unless that is the only possible play. For subsequent tricks, a player without the lead suit can play any card. A player cannot lead hearts until they have been broken (a heart has been thrown off-suit on a preceding trick) or only hearts remain in the player's hand.

At the end of a hand, each player's pile is scored. The queen of spades is 13 points; each heart counts one point; all other cards are zero points. Ordinarily, a player's game score is incremented by the number of points in that player's pile. The exception is a shoot, when one player takes the queen of spades and all the hearts. The player accomplishing a shoot receives zero points, and all opponents receive 26 points. The game ends when one or more players have a game total of at least 100 points.

4

Object Metamodeling

In this chapter we discuss object metamodeling, which you will need for some very advanced applications. You should be proficient with the material in Chapters 2 and 3 before continuing with this chapter. Those doing an initial reading can skip this chapter.

4.1 METADATA AND METAMODELS

Data is information about application concepts and relationships; a ***model*** is an abstraction of data. Similarly, ***metadata*** is information about a model and a ***metamodel*** is an abstraction of metadata. The relationship of metadata to data is relative, and you can continue the abstraction process to create models at successively higher levels, as Figure 4.1 shows. The lowest level of abstraction is data—information that directly relates to an application. Data is abstracted into classes, relationships, events, states, and operations that can, in turn, be abstracted into metaconstructs.

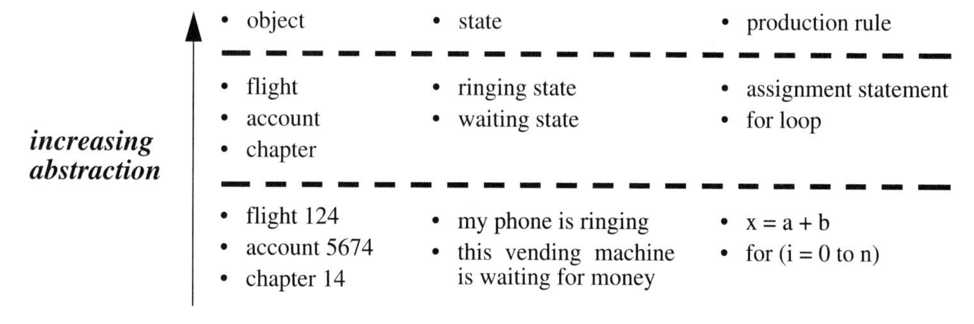

Figure 4.1 Levels of abstraction

Metamodels arise for some complex applications, such as DBMSs and model editors. DBMSs facilitate a wide variety of applications that manage persistent data. The data dictionary of a DBMS can span many applications and thus must deal with metadata. A model editor, such as an object model editor, can also handle many applications. Since a model editor is an application in its own right, you can describe it with a model, in this case a metamodel. Parts catalogs, blueprints, and dictionaries are other examples of metadata.

Figure 4.2 shows a progression of data and models in order of increasing abstraction. The employee data and model could arise for an application like payroll processing. The metadata and metamodel are an excerpt from an object model editor that could be used to construct the application model. Of course, this is not the only possible metamodel. The contents of a metamodel, as for any other application, depend on the problem requirements and the skill and style of the modeler.

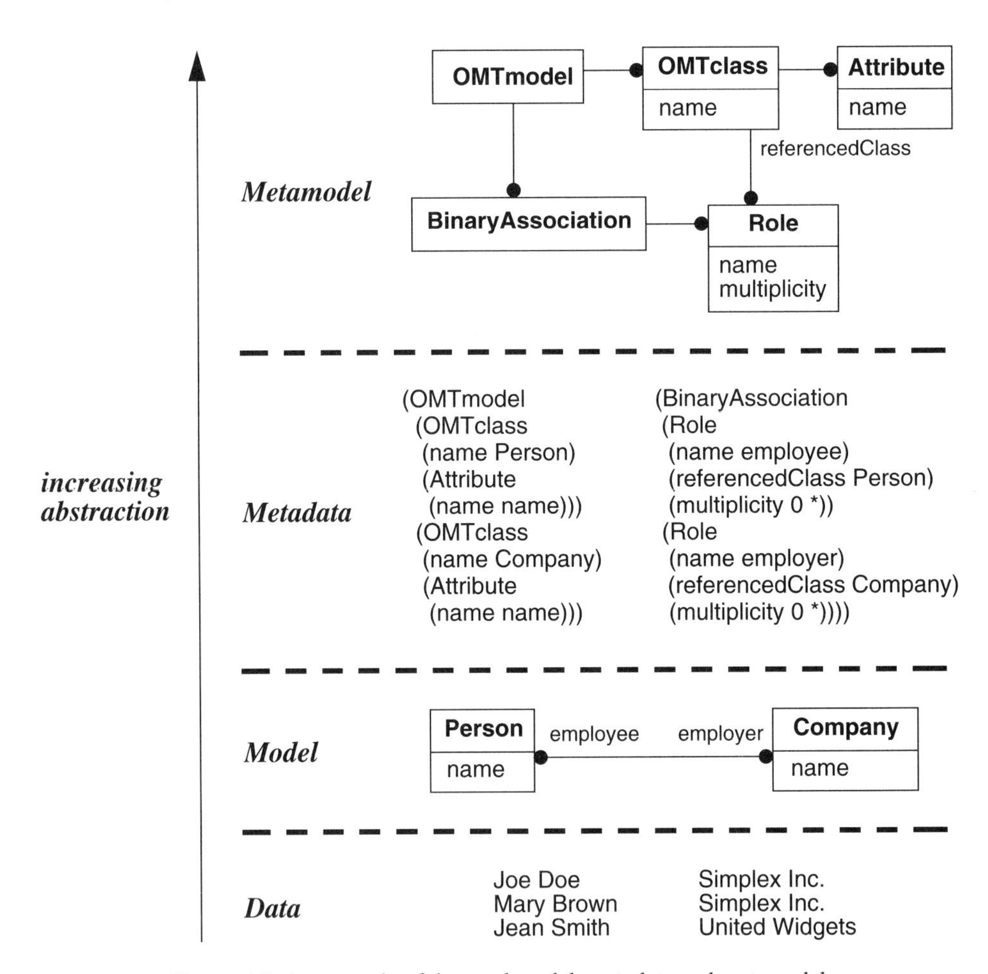

Figure 4.2 An example of data and model, metadata and metamodel

There are many reasons for using metamodels. Metamodels promote reuse by making software serve a category of applications. An application can be conceived broadly to satisfy multiple needs. Expert system shells, DBMSs, software development tools, and the UNIX tools LEX and YACC are examples of software that intrinsically involve metamodeling. Metamodels also isolate software from changes in an application model, in effect detaching the software from the details of an application.

However, metamodels have some disadvantages. They are complex, making them confusing to understand and difficult to debug. A developer must experience several simpler applications before taking the intellectual leap of abstraction that metamodels require. Also, metamodel-driven software may perform more slowly than custom software. For example, a DBMS provides general data management routines. You could perform faster data management by custom coding the data access routines for an application. However, the trade-off for the performance overhead of a DBMS is the savings in development time and effort.

Languages and DBMSs vary in their accessibility for metadata. Some languages, like Lisp and Smalltalk, let metadata be inspected and altered by programs at run-time. In contrast, languages like Ada, C, C++, and Eiffel deal with metadata at compile-time but do not make the metadata explicitly available at run-time. Relational DBMSs incorporate a data dictionary that describes the structure of all tables, attributes, indexes, and other system objects, including the data dictionary tables. Programmers can read and write to the data dictionary at run-time. Object-oriented DBMSs also make metadata available at run-time, but access can be complicated by programming language restrictions.

4.1.1 Generic Classes

A ***generic class*** is a class that combines data and metadata. In Figure 4.3 and Figure 4.4 we have replaced application classes with generic classes. In Figure 4.3 *WindowParameter* is a generic class that pairs the name of a window default attribute with its corresponding value. The generic *DocumentComponent* class in Figure 4.4 eliminates the *Page*, *Paragraph*, and *Line* classes, capturing the distinction with the *documentType* attribute.

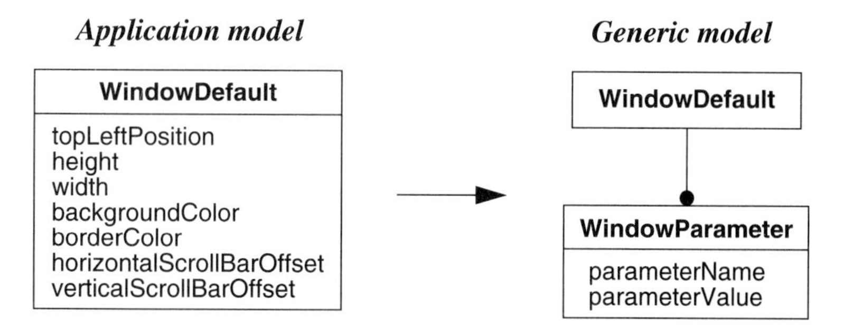

Figure 4.3 Making attributes generic: Application model versus generic model

Generic classes have several advantages. They can often simplify a model; a few generic classes and attributes can substitute for many application classes and attributes. Generic

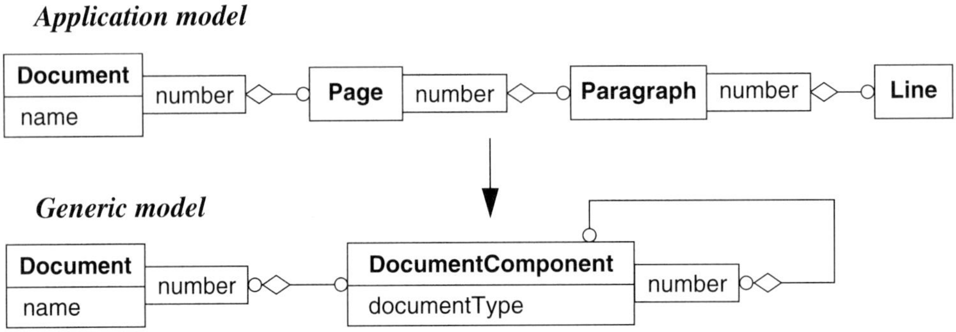

Application model

Document | name — number — Page — number — Paragraph — number — Line

Generic model

Document | name — number — DocumentComponent | documentType — number

Figure 4.4 Making classes generic: Application model versus generic model

classes let you accommodate classes and attributes that may not be known as the application is written. Application software need only conform to the loose generic model rather than to a more rigid application model. For example, adding a window default parameter to a generic model would not change the application code, but if you changed an application model, you would have to modify and recompile the application code. Generic classes let you define objects at run-time, even in environments with compile-time type checking.

Generic classes also have their disadvantages. The metadata complicates programming. With an application model, you represent the various application attributes directly as programming variables. With a generic model, you must generate application code with a preprocessor at compile-time or query a class at run-time to determine the attributes and their values. Method code is more difficult to write for generic data structures than for application data structures. Finally, generic classes can slow performance because of the extra processing required to bind a variable to a value.

In summary, generic classes provide a valuable modeling technique that is occasionally helpful. The increased complexity and performance overhead can be offset by greater brevity and extensibility.

4.1.2 Reification

Reification is the promotion of something that is not an object into an object [Rumbaugh-92]. Reification is a helpful technique for meta applications because it lets you shift the level of abstraction. On occasion it is useful to promote attributes, methods, constraints, and control information into objects so you can describe and manipulate them as data.

In Figure 4.5 we have promoted the *substanceName* attribute to a class to capture the "many-to-many" relationship between *Substance* and *SubstanceName*. Each chemical substance may be referenced by multiple aliases. For example, propylene may be referred to as *propylene* and C_3H_6. Also, an alias may pertain to multiple chemical substances. Various mixtures of ethylene glycol and automotive additives may have the alias of *antifreeze*. Exercise 4.8 also illustrates reification.

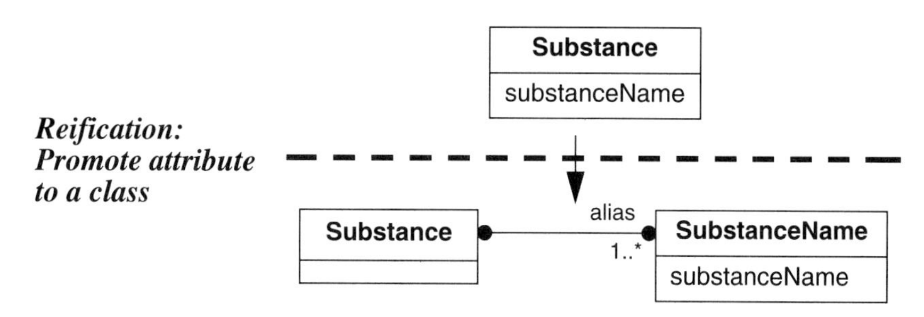

Figure 4.5 Reification: Promotion of an attribute to a class

4.1.3 OMT Object Metamodel

Figure 4.6 presents an abridged OMT object metamodel. Object models embody three major constructs: class, association, and generalization. Our metamodel recognizes the similarities between classes and associations with the *Element* superclass. Both classes and associations have names, deal with instances, may participate in associations, and have candidate keys. The difference is that objects in a class have intrinsic identity, while the identity of links in an association is derived from the related elements. You should try to maintain the uniqueness of element names within an object model.

An element is described by zero or more instance attributes. Each attribute has a name. An attribute also has attribute multiplicity and a domain once a model is fully specified during design. As noted in the model, the name of an instance attribute is unique within the scope of a class or association but need not be globally unique. Enumeration domains have multiple enumeration values. A domain may be structured and consist of lesser domains.

An element may assume different roles in relationships with other elements. Role names may not collide with attribute names for a given element. (See Section 2.2.3.) The metamodel includes qualified associations because a role may involve qualifiers. Each element may have multiple candidate keys. Each candidate key consists of one or more roles, qualifiers, and instance attributes.

Generalization and association are the primary relationships in OMT object modeling. Both generalization and association involve two or more classes. The difference is that association describes the relationship between two or more instances; generalization describes different aspects of a single instance. Aggregation is a special kind of association and consequently also describes a relationship between instances.

Figure 4.6 supports multiple inheritance because a class can participate in multiple generalizations as a superclass and multiple generalizations as a subclass. The *abstractOrConcrete* attribute indicates if every object must be further described by a subclass. (See Section 3.4.1.) Each generalization may have a discriminator. The discriminator is an enumeration attribute that indicates which subclass further describes a given superclass object.

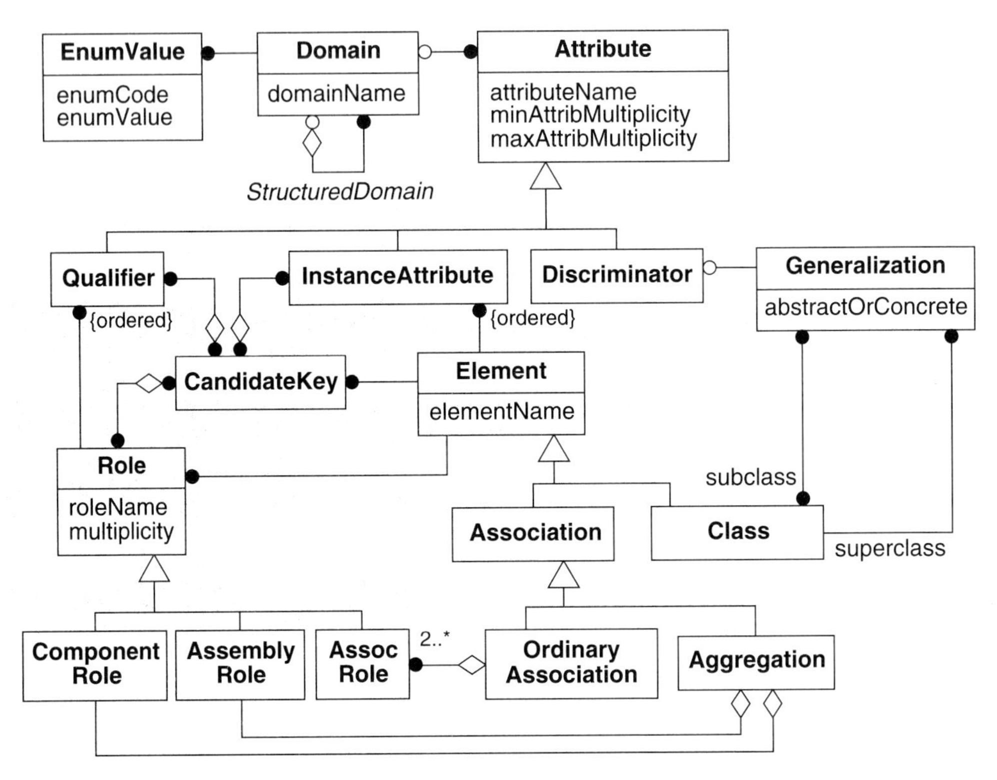

Figure 4.6 An abridged OMT object metamodel

4.2 FRAMEWORKS

A ***framework*** [Johnson-88] is a skeletal structure of a program that must be elaborated to build a complete application. This elaboration often consists of specializing abstract classes with behavior specific to an individual application. A framework may be accompanied by a class library so that the user can perform much of the specialization by choosing the appropriate subclasses rather than programming subclass behavior from scratch. Frameworks consist of more than just the classes involved and include a paradigm for flow of control and shared invariants. Frameworks tend to be specific to a category of applications; framework class libraries are often application specific and not suitable for general use.

[Blaha-90] describes a framework that generates bills-of-material from functional requirements. The bill-of-material generator has no class libraries. Instead it consists of generic code that is customized with an object model and decision table definitions for the desired application. The framework includes a declarative language to compensate for the limitations of decision tables. Flow of control is implicit in the decision tables and declarative programming code. The bill-of-material generator is intended for applications that configure a mass-produced mechanical assembly from a modest number of constituent parts that mix and match in many ways. Thus the bill-of-material generator would be suitable for lawn mowers (mass production) but would not apply to the space shuttle (few produced, too many kinds of component parts).

The EXODUS extensible database manager [Carey-86] is another example of a framework. The goal of EXODUS is to "facilitate the fast development of high-performance, application-specific database systems." An increasing variety of applications require database services and the EXODUS developers believe a single DBMS will be unable to satisfy the diverse requirements. The EXODUS framework includes generic code, class libraries, and a programming language specially designed for writing database system software. Application database managers with either a relational flavor or an object-oriented flavor can be developed with EXODUS.

Frameworks are important not only for their ability to organize large systems, but also for their contribution to reuse. A framework supports reuse at a larger granularity than classes, promoting reuse of designs instead of code. Frameworks can be built on other frameworks, yielding a hierarchy of abstractions.

However, frameworks are difficult to build and understand. They often involve metamodels and generic classes, both of which are intrinsically complex. You should experience several applications before attempting a framework. Tangible applications are easier to grasp than the abstractions of a framework. With a framework it is difficult to foresee all requirements, exceptions, and deviations; the inevitable peculiarities of applications stress a framework.

4.3 PATTERNS

A ***pattern*** is an excerpt of an object model with one or more parameters as placeholders for classes and associations. You incorporate a pattern into a model by substituting specific classes and associations for the parameters, that is by instantiating the pattern.

Most of the literature, including recent OOPSLA conference workshops and an Internet pattern user group, has been concerned with *design* patterns [Gamma-95] and *implementation* patterns, called "idioms" in [Coplien-92]. In this section we focus on analysis patterns. Analysis patterns are more relevant to databases, while design and implementation patterns are more relevant to programming. Some frequently encountered analysis patterns are tree, graph, item description, and homomorphism. We describe each of these in detail later.

There are many benefits to using patterns in model construction:

- **Enriched OMT language**. Patterns provide a higher level of building blocks for models than the base primitives of class, association, and generalization. Patterns are prototypical modeling fragments that distill some of the knowledge of experts.

- **Improved documentation**. Patterns capture some of the model's semantic intent. A proficient modeler will not only show classes and relationships, but will try to present them using canonical forms.
- **Basis for transformations**. Patterns let you apply a general transformation to a specific model. Chapter 10 explains transformations.

You cannot model an application by just combining patterns. Typically you will use only a few patterns in practice, but they often embody key ideas that are the crux of applications.

Do not confuse a pattern with a framework. A pattern is typically a small number of classes and relationships, while a framework is much broader in scope. Furthermore, a pattern refers only to a single model, which in this book is the object model. (Dynamic model patterns and functional model patterns are beyond the scope of this book.) In contrast, a framework has rich behavior and must be described with all three OMT models.

A pattern is different from an Eiffel generic or C++ template, each of which is a single class. A pattern, on the other hand, consists of several related classes and can have multiple parameters.

4.3.1 Trees

Figure 4.7 presents the complex object-modeling pattern for trees. (We denote all pattern parameters with angle brackets.) Each branch has many child nodes; each node has a parent, except the root node. Exactly one node is designated the root of the tree. You need not include the root association in an instantiation of the pattern.

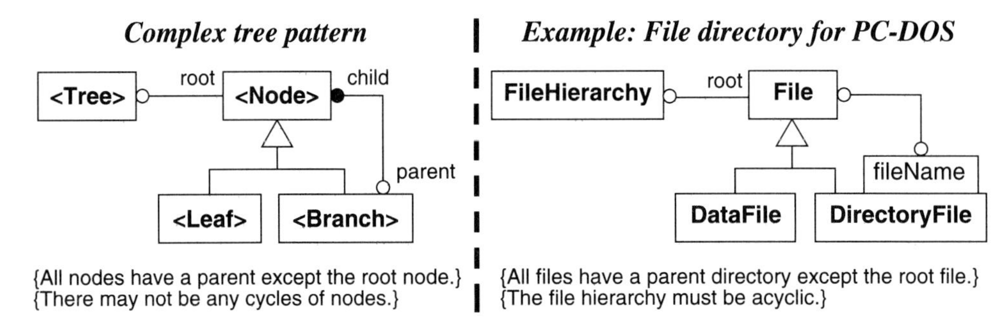

Figure 4.7 Object-modeling pattern: Complex tree

The right model in Figure 4.7 demonstrates the complex tree pattern. A file may be a data file or a directory file. The combination of a directory and a file name yields a specific file. All files belong to a single directory except the root file (as is true for PC-DOS). Directories contain multiple files, some or all of which may be subdirectories. Directories can be nested to an arbitrary depth, with data files and empty directories terminating the recursion.

We assert that the tree cannot have any cycles. (Otherwise, there is nothing in the model to stop *node A* from having *node B* as a parent and *node B* from having *node A* as a parent.)

You can describe the associations in the pattern with link attributes as well as qualifiers, and construct trees with aggregation instead of association, since aggregation is a special kind of association.

The left model in Figure 4.8 simplifies the tree pattern by not distinguishing between leaves and branches. The right model is an example of the simple tree pattern. A manager has many subordinates. Each subordinate reports to at most one manager: The CEO reports to no manager, and all others report to one manager. The management hierarchy can be arbitrarily deep.

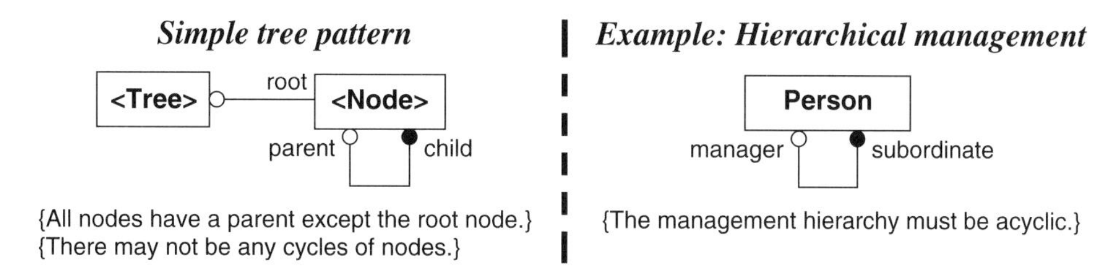

Figure 4.8 Object-modeling pattern: Simple tree

The complex tree pattern is most appropriate when the behavior of branch nodes differs from the behavior of leaf nodes. For example, the command *dir directoryFileName* elicits a different response from that of *dir dataFileName*. The complex form is also preferred when attributes and associations apply only to branch nodes or only to leaf nodes.

The simple form of a tree suffices when tree decomposition is merely a matter of data structure and there are no significant differences between branch and leaf nodes. Both forms of the tree pattern often occur in practice.

4.3.2 Directed Graphs

Graphs arise for applications with important topology or connectivity. We present graph patterns for both directed and undirected graphs (Section 4.3.3). A directed graph is a set of nodes and edges, where an edge originates at a source node and terminates in a sink node. A directed graph can have nodes with any number of edges.

Figure 4.9 shows the complex form of the directed graph pattern. We arrived at the figure by relaxing the multiplicity constraints for the parent and root nodes in Figure 4.7. Each branch has many child nodes; each node may have multiple parents.

The right model in Figure 4.9 illustrates the pattern. With the UNIX operating system a file may belong to multiple directories via symbolic links. A file may have a different name in each directory where it is referenced. For directories of files we require a directed acyclic graph, so we have added a constraint to the model.

You can fully traverse a directed graph by starting at the root nodes and successively navigating from parent nodes to child nodes. The associations in the patterns may have link attributes and qualifiers, as well as be aggregations. In general, a directed graph may have

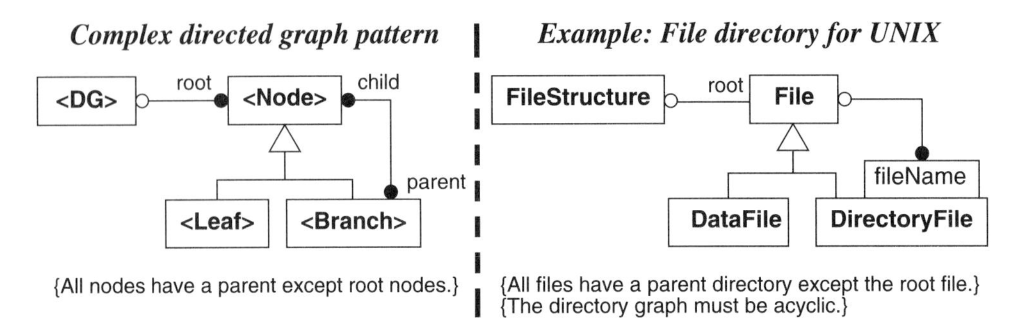

Figure 4.9 Object-modeling pattern: Complex directed graph

cycles. However, cycles are often disallowed for instantiations of the directed graph pattern. You need not include the root association in a pattern instantiation.

The left model in Figure 4.10 shows the simple form of the directed graph pattern that does not distinguish between leaves and branches. The right model is an example of the pattern. A manager has many subordinates, and a subordinate can report to more than one manager. We add the constraint that the management graph have no cycles because a manager would not normally report to a subordinate.

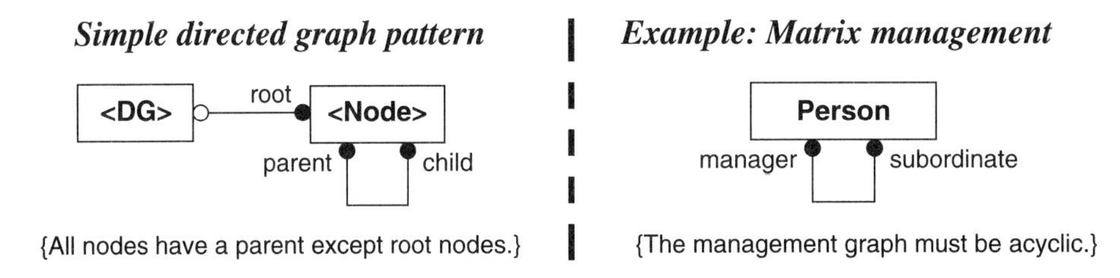

Figure 4.10 Object-modeling pattern: Simple directed graph

The left model in Figure 4.11 shows a third pattern for directed graphs for which we have reified edges. Nodes and the edges that connect them can both bear information. The right model illustrates the pattern. Engineers simulate chemical plants and calculate the performance of equipment and the chemical composition in the piping that connects equipment. A process simulation involves many unit operations (representing equipment) and streams (representing the piping). The chemicals in each stream flow from a source unit operation to a sink unit operation. Every stream has one source and one sink, because we treat the net input and output from the plant as special unit operations.

The complex directed graph pattern is appropriate when you must differentiate the behavior of branch nodes from leaf nodes and do not wish to call attention to the edges themselves. The simple directed graph pattern is useful if you need only record data structure. The reified directed graph pattern is useful when nodes and edges are both important semantic constructs.

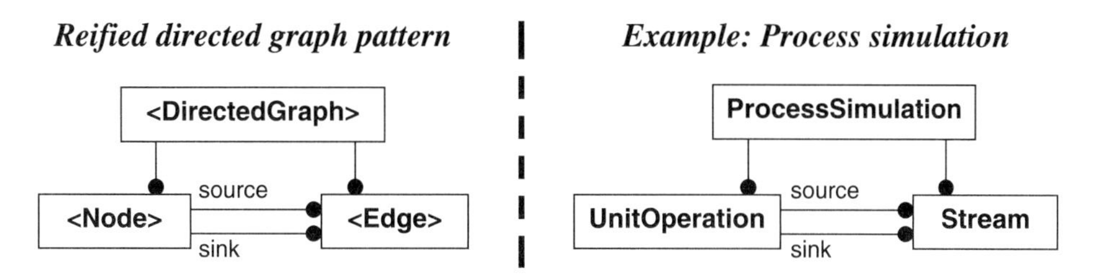

Figure 4.11 Object-modeling pattern: Reified directed graph

4.3.3 Undirected Graphs

An undirected graph is also a set of nodes and edges, but in contrast to a directed graph an edge merely connects two nodes. Given the symmetry of an undirected graph, there is no counterpart to the complex pattern in Figure 4.9.

The left model in Figure 4.12 shows the undirected graph pattern that is the analog to Figure 4.10. An undirected graph consists of a set of root nodes. All nodes, including root nodes, may connect to multiple other nodes. With an undirected graph there is no sense of direction, just the notion that nodes are connected. The right model is an example of a simple undirected graph. A contract may be associated with other contracts. Contracts can be related in many ways—such as in terms of legal consequences, complementary business, or underlying technology.

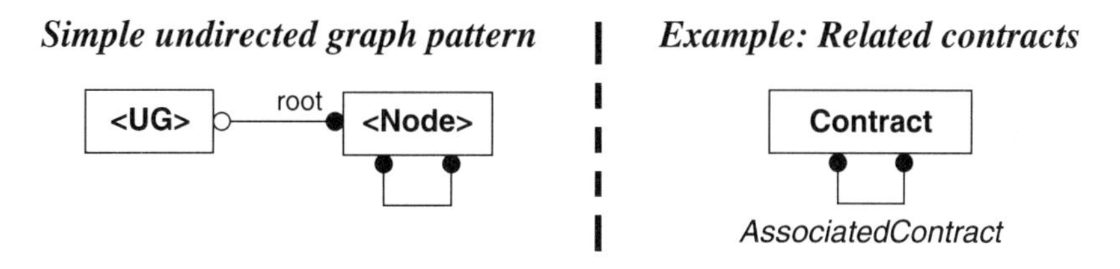

Figure 4.12 Object-modeling pattern: Simple undirected graph

Figure 4.13 shows another pattern for undirected graphs that reifies the edges. As the right model shows, we can also use this pattern to model contract information. For example, we might want to describe the relationship between contracts and record the kind of relationship, date of occurrence, and commentary.

You should use the simple undirected graph pattern if you merely need to record that nodes are related. The reified pattern allows you to store information about the relationship.

4.3.4 Item Descriptions

The item description pattern [Coad-92] in Figure 4.14 associates an item with its description. The association can be an aggregation, and link attributes and qualifiers can be involved.

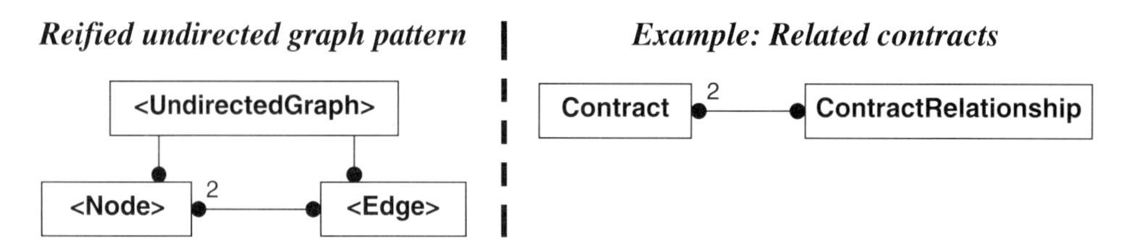

Figure 4.13 Object-modeling pattern: Reified undirected graph

This pattern involves data (the item) and metadata (the item description), and typically appears when a model concerns both an item and its description. The item description pattern frequently occurs, but is visually less striking than other patterns. You must exercise judgment in deciding when one class is in some sense an instance of another.

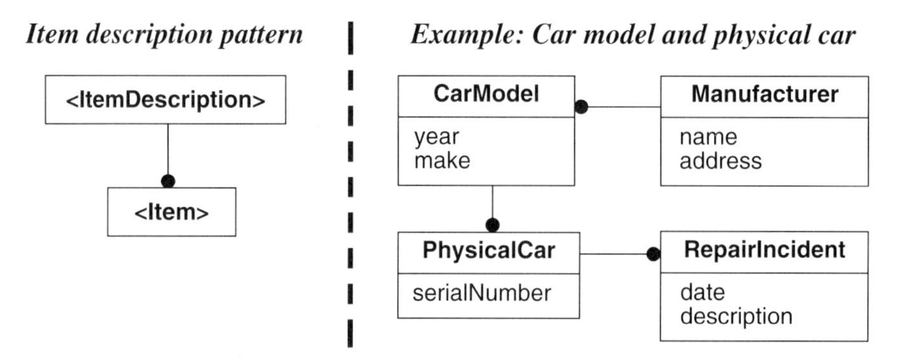

Figure 4.14 Object-modeling pattern: Item description

The right model in Figure 4.14 illustrates the pattern. *CarModel* corresponds to *ItemDescription* and *PhysicalCar* corresponds to *Item*. Physical cars are items with individual serial numbers. In contrast, car models refer to the year and make, such as *1986 Ford Escort* and *1989 Mazda 626*. Customer repair records refer to physical cars, while design documents describe car models. In Chapter 3, Figure 3.17 provides another example of the item description pattern.

The item description pattern is useful if you cannot fully describe the data as software is being developed. The item description pattern lets you enter data and the description of data at run-time, as opposed to the more typical approach of defining data structure at compile-time.

4.3.5 Homomorphisms

The homomorphism pattern [Rumbaugh-91] involves two item description patterns that are themselves related, as Figure 4.15 shows. A homomorphism is essentially an analogy: The description of item 1 is to item 1 as the description of item 2 is to item 2. You need not include the association between association classes in a pattern instantiation.

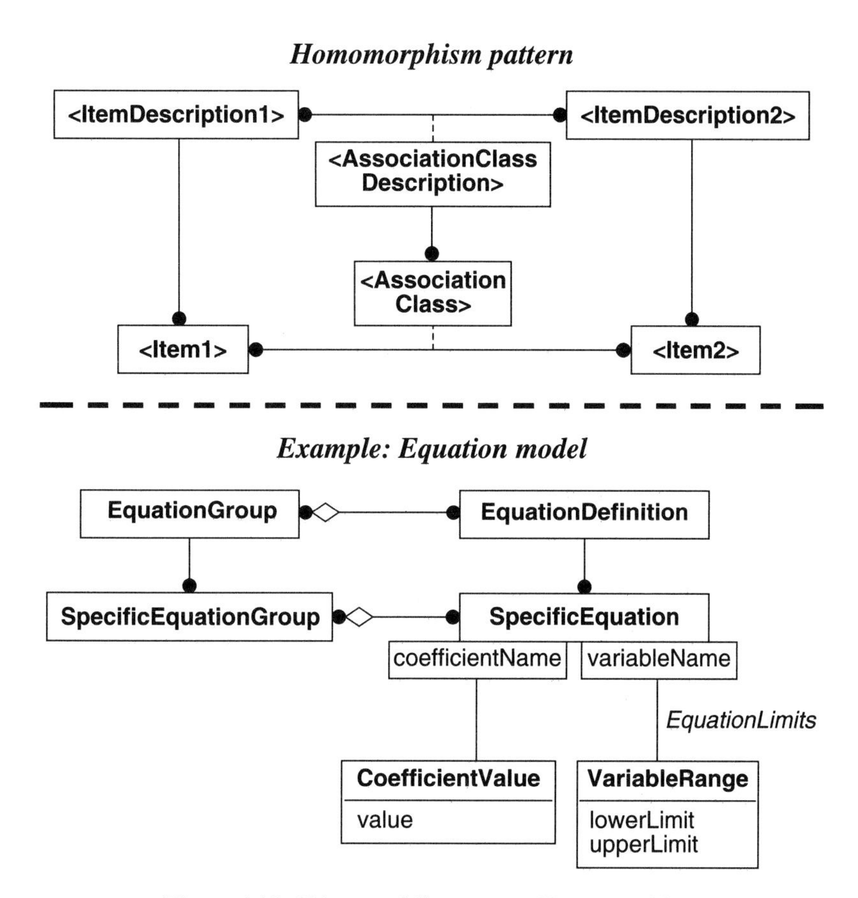

Figure 4.15 Object-modeling pattern: Homomorphism

The bottom model in Figure 4.15 illustrates the homomorphism pattern. An equation definition has symbolic variables and symbolic coefficients. A value must be supplied for each coefficient when an equation definition is instantiated, yielding a specific equation. You can then evaluate a specific equation by substituting a value for each independent variable and computing the dependent variable. The validity of a specific equation may be limited to some ranges of variables.

An equation group is a collection of related equation definitions. The notion of an equation group is particularly useful when several equations are taken together and must be solved by iteration. A specific equation group is a collection of specific equations corresponding to an equation group. The relationship between equation group and specific equation group is homomorphic to the relationship between equation definition and specific equation.

Our need for an equation model arose in the context of chemical engineering applications. Many chemical engineering applications must compute physical properties (such as density and vapor pressure) for chemical substances (such as water, ethanol, and gasoline).

Engineers often need detailed physical properties when simulating the performance of chemical processing equipment.

In Chapter 3, Figure 3.35 is another example of a homomorphism pattern. An *Aircraft-Description* is to an *Aircraft* as a *SeatDescription* is to a *Seat*.

4.4 CHAPTER SUMMARY

Data is information about application concepts and relationships; a model is an abstraction of data. Similarly, metadata is information about a model, and a metamodel is an abstraction of metadata. The relationship of metadata to data is relative, and you can continue the abstraction process to create models at successively higher levels. Metamodels arise for frameworks and some complex applications.

A pattern is an excerpt of an object model with one or more parameters as placeholders for classes and associations. You incorporate a pattern into a model by substituting specific classes and associations for the parameters, that is by instantiating the pattern. We have presented object-modeling patterns for trees, directed graphs, undirected graphs, item descriptions, and homomorphisms.

Figure 4.16 lists the key concepts for this chapter.

directed graph pattern	item description pattern	reification
framework	metadata	tree pattern
generic class	metamodel	undirected graph pattern
homomorphism pattern	pattern	

Figure 4.16 Key concepts for Chapter 4

BIBLIOGRAPHIC NOTES

In this chapter, we introduced one extension to the OMT object-modeling notation from [Rumbaugh-91]. We use angle brackets to denote parameters for patterns.

[Coad-92] and [Coad-95] also discuss analysis patterns, but our patterns are more abstract and mathematical. Our patterns essentially extend the OMT language. For example, our tree pattern is at the same level of abstraction as class, association, and generalization. In contrast, Coad's patterns embody more application knowledge. These comments are not intended as criticism of Coad's work; we are just trying to compare his work to ours.

[Buschmann-96] distinguishes among architectural patterns, design patterns, and idioms. An architectural pattern occurs across subsystems. In contrast, a design pattern occurs within a subsystem but is independent of the implementation paradigm. An idiom is a low-level pattern that is language specific. Out treatment of patterns is essentially at the same level of abstraction as their architectural patterns, but we only discuss data structure. [Buschmann-96] has a much more thorough presentation and also covers behavior.

Charlie Bachman provided us with many slides documenting his approach to modeling graphs. We believe our approaches to be quite similar, and his comments clarified our thinking.

REFERENCES

[Blaha-90] MR Blaha, WJ Premerlani, AR Bender, RM Salemme, MM Kornfein, and CK Harkins. Bill-of-material configuration generation. *Sixth International Conference on Data Engineering*. February 5–9, 1990, Los Angeles, California, 237–244.

[Buschmann-96] Frank Buschmann, Regine Meunier, Hans Rohnert, Peter Sommerlad, and Michael Stal. *Pattern-Oriented Software Architecture: A System of Patterns*. Chichester, United Kingdom: Wiley, 1996.

[Carey-86] M Carey, D DeWitt, D Frank, G Graefe, J Richardson, E Shekita, and M Muralikrishna. The Architecture of the EXODUS Extensible DBMS. *Proceedings of the International Workshop on Object-Oriented Database Systems*. September 1986, Pacific Grove, California, 52–65.

[Coad-92] Peter Coad. Object-oriented patterns. *Communications ACM 35*, 9 (September 1992), 152–159.

[Coad-95] Peter Coad, David North, and Mark Mayfield. *Object Models: Strategies, Patterns, and Applications*. Englewood Cliffs, New Jersey: Yourdon Press, 1995.

[Coplien-92] James O. Coplien. *Advanced C++ Programming Styles and Idioms*. Reading, Massachusetts: Addison-Wesley, 1992.

[Gamma-95] Erich Gamma, Richard Helm, Ralph Johnson, and John Vlissides. *Design patterns: Elements of Reusable Object-Oriented Software*. Reading, Massachusetts: Addison-Wesley, 1995.

[Johnson-88] Ralph E. Johnson and Brian Foote. Designing reusable classes. *Journal of Object-Oriented Programming 1*, 3 (June/July 1988), 22–35.

[Rumbaugh-91] J Rumbaugh, M Blaha, W Premerlani, F Eddy, and W Lorensen. *Object-Oriented Modeling and Design*. Englewood Cliffs, New Jersey: Prentice Hall, 1991.

[Rumbaugh-92] James Rumbaugh. Let there be objects: a short guide to reification. *Journal of Object-Oriented Programming 5*, 7 (November-December 1992), 9–14.

EXERCISES

4.1 (4) Given the instance diagram in Figure E4.1, prepare an instance diagram for the generic model in Figure 4.3.

wd01:WindowDefault
topLeftPosition = (0,0) height = 100 width = 234 backgroundColor = white borderColor = dark grey horizontalScrollBarOffset = null verticalScrollBarOffset = null

Figure E4.1 An application object for window defaults

4.2 (3) Given the instance diagram in Figure E4.2, prepare an instance diagram for the generic model in Figure 4.4.

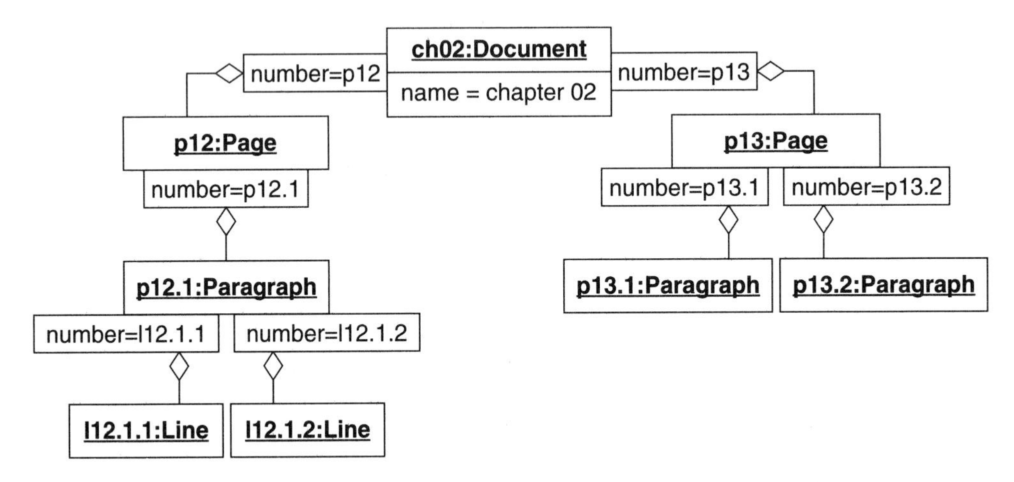

Figure E4.2 An application instance diagram for document composition

4.3 (4) What is the logical horizon of *Generalization* in Figure 4.6? (In Chapter 3, Section 3.6.1 discusses the logical horizon.)

4.4 (6) Construct an instance diagram of Figure 4.6 for the object model in Figure E4.3.

Figure E4.3 A sample object model

4.5 (7) Construct an instance diagram of Figure 4.6 for the object model in Figure E4.4.

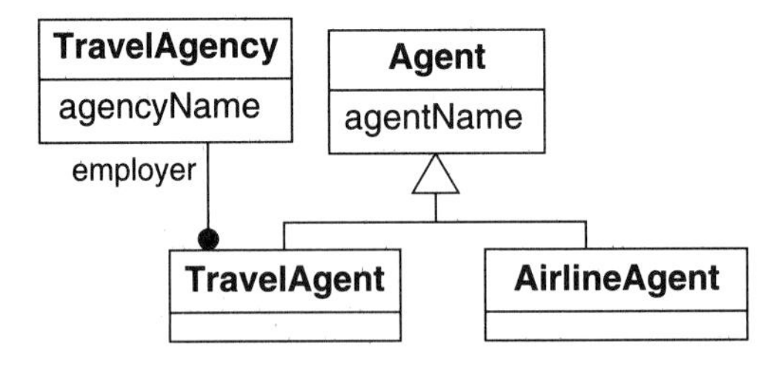

Figure E4.4 A sample object model

4.6 (8) Construct an instance diagram of Figure 4.6 for the object model in Figure E4.5.

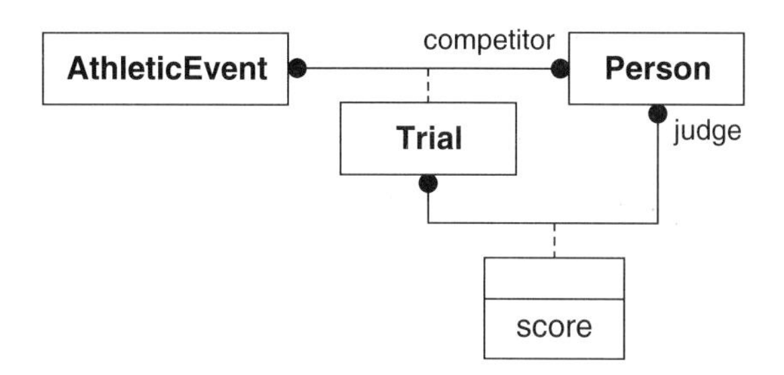

Figure E4.5 A sample object model

4.7 (8) Extend the metamodel of Figure 4.6 to include exclusive-or associations.

4.8 (6) Prepare a simple object model, sufficient for representing recipes. Use the recipe in Figure E4.6 as a basis. This exercise is an example of reification. In one sense the tasks of a recipe could be operations; in another sense they could be data in an object model.

Lasagna

2.5 tbsp. salad oil for browning
1 cup minced onion
1 clove garlic
1 lb. ground beef
2 tsp. salt
3.5 cups whole tomatoes (large can)
1 tsp. oregano
.5 box lasagna noodles
1 lb. ricotta cheese
1 cup grated mozzarella cheese
.5 cup parmesan cheese
2 cans tomato paste

Cook onion, clove garlic, ground beef, 1 tsp. salt in salad oil until meat is browned. Add tomatoes, tomato paste, 1 tsp salt, oregano, and simmer, covered, 1 hour until thick. Cook noodles 15 minutes in water until tender. Drain and blanch. Butter 12×8 inch pan and place in layers of noodles, sauce, mozzarella, ricotta cheese, and parmesan. Bake at 350 degrees for 45 to 60 minutes.

Figure E4.6 A simple recipe

4.9 (7) Extend your object model of recipes to handle alternate ingredients. For example, some lasagna recipes allow cottage cheese to be substituted for ricotta cheese.

4.10 (5) A grocery store has many cashier stations for checking out groceries; various employees staff the cashier stations. The cashiers process a customer by scanning the selected grocery items and entering auxiliary information into the cash register.

Figure E4.7 presents alternative models for a grocery store checkout. The upper model reifies the assignment of a person to a cashier station. The lower model treats assignment as an association class. Discuss the differences between the two models. Which model do you prefer and why?

4.11 (5) Prepare an object model for data from scientific experiments. Include the following classes: *Experiment*, *Variable*, and *VariableValue*. An experiment involves many dependent and independent variables. A scientist sets independent variables and observes dependent variables.

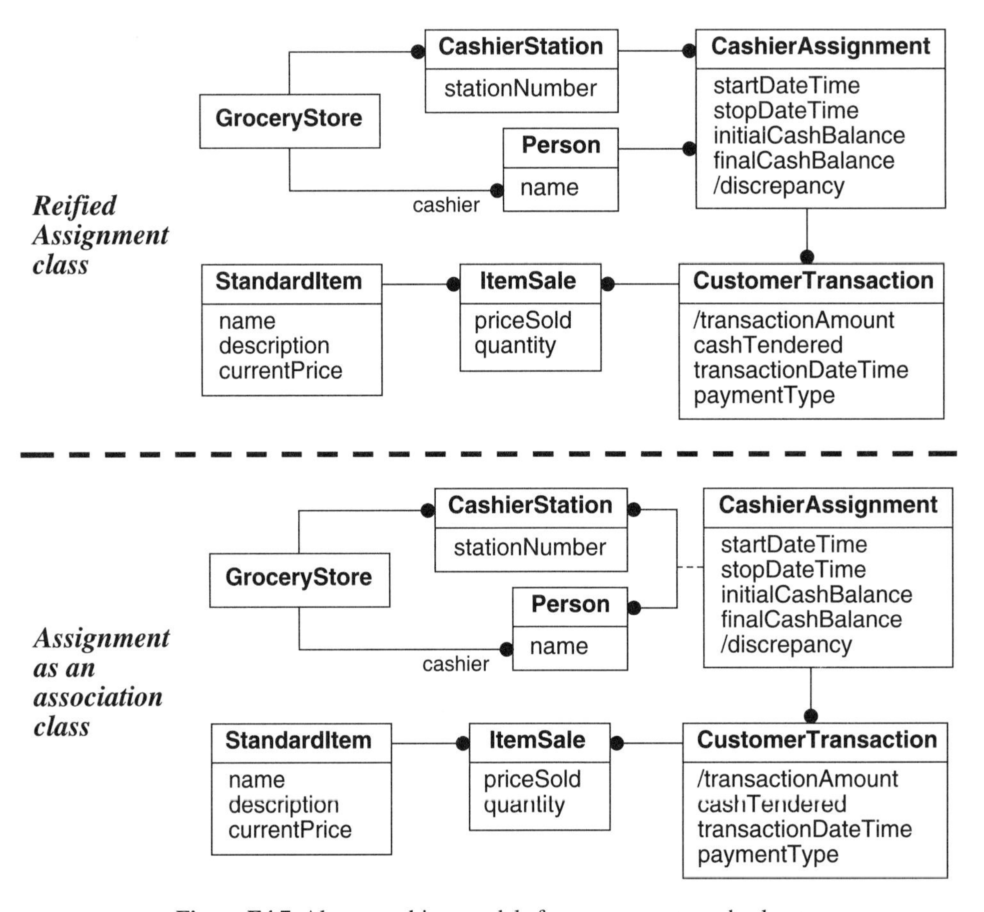

Figure E4.7 Alternate object models for a grocery store checkout

Your model should also include the unit of measure and estimated accuracy for each variable value.

4.12 (8) Prepare an object model for the data in Figure E4.8 from the 1990 U.S. census. You should include generic classes so that you can extend the model to include additional census data.

4.13 (7) Figure E4.9 shows an abridged Backus-Naur form (BNF) grammar for equations. Literal values are quoted; all other names are tokens in the grammar. Vertical bars separate alternate acceptable values. Informal definitions are in uppercase. Prepare an object model that corresponds to the grammar. (Chapter 12 describes BNF grammars in more detail.)

4.14 (3) List all the patterns and their occurrences in the airline information model from Chapter 3 (Figure 3.33–Figure 3.37).

United States Urban and Rural and Size of Place	1990 population				1980 population			
	Number of places	Total population	Pct of total population	Pct distribution	Number of places	Total population	Pct of total population	Pct distribution
Total	23,435	248,709,873	100.0		22,529	226,542,199	100.0	
Urban	9,421	187,053,487	75.2	100.0	8,765	167,050,992	73.7	100.0
Inside urbanized area	5,483	158,258,878	63.6	84.6	4,938	139,170,683	61.4	83.3
Central place	549	78,847,406	31.7	42.2	431	67,035,302	29.6	40.1
1,000,000 or more	8	19,952,631	8.0	10.7	6	17,530,248	7.7	10.5
500,000 to 999,999	15	10,107,184	4.1	5.4	16	10,834,121	4.8	6.5
250,000 to 499,999	41	14,585,006	5.9	7.8	33	11,900,309	5.3	7.1
100,000 to 249,999	97	14,602,452	5.9	7.8	82	12,295,543	5.4	7.4
50,000 to 99,999	177	12,274,504	4.9	6.6	125	8,649,031	3.8	5.2
Less than 50,000	211	7,325,629	2.9	3.9	169	5,826,050	2.6	3.5
Urban fringe	4,934	79,411,472	31.9	42.5	4,507	72,135,381	31.8	43.2
100,000 or more	39	5,100,382	2.1	2.7	36	4,976,800	2.2	3.0
50,000 to 99,999	178	11,752,941	4.7	6.3	165	11,137,456	4.9	6.7
25,000 to 49,999	443	15,118,958	6.1	8.1	424	14,539,729	6.4	8.7
10,000 to 24,999	1,167	18,482,502	7.4	9.9	1,087	17,232,991	7.6	10.3
5,000 to 9,999	1,200	8,679,826	3.5	4.6	1,108	7,900,939	3.5	4.7
2,500 to 4,999	996	3,641,246	1.5	1.9	671	2,424,502	1.1	1.5
2,000 to 2,499	162	362,540	.1	.2	183	406,475	.2	.2
1,500 to 1,999	159	276,809	.1	.1	210	367,921	.2	.2
1,000 to 1,499	192	240,177	.1	.1	238	297,473	.1	.2
Less than 1,000	398	199,377	.1	.1	385	188,377	.1	.1
Other urban		15,556,714	6.3	8.3		12,662,718	5.6	7.6
Outside urbanized area	3,938	28,794,609	11.6	15.4	3,827	27,880,309	12.3	16.7
25,000 or more	121	3,917,665	1.6	2.1	118	3,773,752	1.7	2.3
10,000 to 24,999	652	9,907,357	4.0	5.3	642	9,708,035	4.3	5.8
5,000 to 9,999	1,135	7,909,614	3.2	4.2	1,073	7,455,198	3.3	4.5
2,500 to 4,999	2,030	7,059,973	2.8	3.8	1,994	6,943,324	3.1	4.2
Rural	14,014	61,656,386	24.8	100.0	13,764	59,494,813	26.3	100.0
2,000 to 2,499	931	2,074,977	.8	3.4	918	2,048,678	.9	3.4
1,500 to 1,999	1,378	2,381,156	1.0	3.9	1,318	2,280,677	1.0	3.8
1,000 to 1,499	2,115	2,594,725	1.0	4.2	2,198	2,708,485	1.2	4.6
Less than 1,000	9,590	3,801,051	1.5	6.2	9,330	3,863,470	1.7	6.5
Other rural		50,804,477	20.4	82.4		48,593,503	21.4	81.7

Figure E4.8 Data from Table 6 of the 1990 U.S. census

United States Urban and Rural and Size of Place	1990 housing units			1990 land area			
	Total housing units	Pct of total housing units	Pct dis-tribu-tion	Square kilometers	Square miles	Pct of total land area	Pct dis-tribu-tion
Total	102,263,678	100.0		9,158,960.4	3,536,278.1	100.0	
Urban	76,212,052	74.5	100.0	226,303.8	87,376.0	2.5	100.0
Inside urbanized area	64,201,132	62.8	84.2	158,027.6	61,014.5	1.7	69.8
Central place	33,030,250	32.3	43.3	61,503.5	23,746.5	.7	27.2
1,000,000 or more	8,133,674	8.0	10.7	6,329.6	2,443.9	.1	2.8
500,000 to 999,999	4,214,279	4.1	5.5	6,890.7	2,660.5	.1	3.0
250,000 to 499,999	6,351,594	6.2	8.3	12,137.9	4,686.5	.1	5.4
100,000 to 249,999	6,135,006	6.0	8.0	13,370.0	5,162.2	.1	5.9
50,000 to 99,999	5,111,887	5.0	6.7	13,374.9	5,164.1	.1	5.9
Less than 50,000	3,083,810	3.0	4.0	9,400.4	3,629.5	.1	4.2
Urban fringe	31,170,882	30.5	40.9	96,524.1	37,268.0	1.1	42.7
100,000 or more	1,973,834	1.9	2.6	3,700.0	1,428.6	0	1.6
50,000 to 99,999	4,541,303	4.4	6.0	8,136.5	3,141.5	.1	3.6
25,000 to 49,999	5,904,257	5.8	7.7	13,674.8	5,279.9	.1	6.0
10,000 to 24,999	7,271,797	7.1	9.5	21,088.6	8,142.3	.2	9.3
5,000 to 9,999	3,478,491	3.4	4.6	12,974.9	5,009.6	.1	5.7
2,500 to 4,999	1,489,043	1.5	2.0	6,970.8	2,691.4	.1	3.1
2,000 to 2,499	165,671	.2	.2	725.8	280.2	0	.3
1,500 to 1,999	123,159	.1	.2	675.2	260.7	0	.3
1,000 to 1,499	107,868	.1	.1	532.5	205.6	0	.2
Less than 1,000	84,614	.1	.1	642.8	248.2	0	.3
Other urban	6,030,845	5.9	7.9	27,401.9	10,579.9	.3	12.1
Outside urbanized area	12,010,920	11.7	15.8	68,276.2	26,361.5	.7	30.2
25,000 or more	1,592,895	1.6	2.1	6,185.8	2,388.3	.1	2.7
10,000 to 24,999	4,051,675	4.0	5.3	17,717.1	6,840.6	.2	7.8
5,000 to 9,999	3,322,499	3.2	4.4	19,978.3	7,713.6	.2	8.8
2,500 to 4,999	3,043,851	3.0	4.0	24,395.0	9,418.9	.3	10.8
Rural	26,051 626	25.5	100.0	8,932,656.6	3,448,902.2	97.5	100.0
2,000 to 2,499	901,912	.9	3.5	9,951.6	3,842.3	.1	.1
1,500 to 1,999	1,057,343	1.0	4.1	11,615.8	4,484.8	.1	.1
1,000 to 1,499	1,138,151	1.1	4.4	14,006.5	5,407.9	.2	.2
Less than 1,000	1,740,468	1.7	6.7	50,087.9	19,339.0	.5	.6
Other rural	21,213,752	20.7	81.4	8,846,994.8	3,415,828.1	96.6	99.0

Figure E4.8 (continued) Data from Table 6 of the 1990 U.S. census

```
EquationDefinition :  Expression '=' Expression ;
Expression         :  UnaryExpression
                   |  BinaryExpression
                   |  Constant
                   |  FunctionInvocation
                   |  StdConstantRef
                   |  VariableRef
                   |  CoefficientRef ;
UnaryExpression    :  UnaryOperator Expression ;
UnaryOperator      :  '+'
                   |  '-' ;
BinaryExpression   :  Expression BinaryOperator Expression ;
BinaryOperator     :  '+'
                   |  '-'
                   |  '*'
                   |  '/'
                   |  '**'
                   |  'DIV'
                   |  'MOD' ;
Constant           :  A REAL OR INTEGER NUMBER
FunctionInvocation :  StandardFunction '(' Arguments ')'
                   |  StandardFunction '()' ;
StandardFunction   :  'ln'
                   |  'log10'
                   |  'exp'
                   |  'sinh'
                   |  'cosh'
                   |  AND OTHERS
Arguments          :  Expression ',' Arguments
                   |  Expression ;
StdConstantRef     :  'e'
                   |  'pi'
                   |  AND OTHERS
VariableRef        :  A NAME
CoefficientRef     :  A NAME
```

Figure E4.9 An abridged BNF grammar for equations

5

Functional Modeling

In the last three chapters, we have described the OMT object model. The object model captures the static data structure of a system. It looks at structure in terms of groups of analogous objects (classes), their similarities and differences (generalization), and their important relationships with one another (associations).

In this chapter we now discuss the functional model. The functional model specifies operations from both the object and dynamic models. As such, it defines the computations that objects perform. Operations arise in the object model from queries, updates, derived entities, and constraints. We do not discuss operations for the dynamic model, because the dynamic model is seldom important for database applications.

A major advantage of the OMT methodology is that much functionality is already implicit in the object model. You can satisfy many requests for information by simply traversing the object model. As a consequence, for database applications you will normally need to specify only a small number of methods.

We advocate a variety of paradigms for expressing the functional model, with the paradigm depending on the application. For database applications we most often use pseudocode enhanced with a notation for navigating object models. For that reason, in this chapter we focus on enhanced pseudocode and only briefly describe the other paradigms. Chapter 8 provides guidance for choosing between the paradigms.

5.1 PSEUDOCODE

Pseudocode is an informal language that provides sequence, conditionality, and iteration. It is similar to many programming and database languages and can be readily translated to them. The remainder of this section summarizes our conventions for pseudocode, but you can use any common programming or mathematical construct.

- **Sequence**. The listed order of the statements implies sequence. We separate successive statements with a semicolon.

- **Conditionality**. The syntax for conditionality is as follows, with literal keywords in bold font. The *condition* must evaluate to true or false. The *statement* denotes a single statement or multiple statements. The *else if* clause is optional and can be repeated multiple times. The *else* clause is also optional. The indentation emphasizes the scope of each block.

```
if condition then
   statement
else if condition then
   statement
else
   statement
end if
```

- **Iteration through a collection**. You can sequentially loop through a collection, making each element available for processing. The *statement* denotes a single statement or multiple statements.

```
for each anElement in aCollection
   statement
end for each
```

- **Iteration through a fixed loop**. This construct causes one or more statements to be processed a fixed number of times.

```
for loopCounter := loopMinimum to loopMaximum
   increment by loopIncrement
   statement
end for
```

- **Method signature**. A method is the implementation of an operation for a class. We specify method signatures as shown in the following code. You need not show the parentheses for a method with zero arguments. A function has the *returns domain* clause; a procedure does not have this clause. (Section 3.1.5 explains the importance of assigning domains instead of directly assigning data types.)

```
className::operationName (arguments) returns domain
```

- **Method invocation**. We denote method invocation with *objectName#methodName(arguments)*. This notation is a bit awkward, but we want to avoid confusion with programming languages and the low-level distinction between by-value and by-reference invocation (such as the "." and "->" operators with C++).

- **Function return**. The *return* keyword terminates execution of a method and returns the value following the keyword.

- **Local variables**. We try to keep our methods simple and use few local variables. We implicitly define local variables and have no explicit declaration.

- **Implicit method argument**. By convention in object-oriented languages, a method has an implicit argument of the object that is executing the method. We use the keyword *self* to denote the implicit argument. Smalltalk uses the keyword *self*; C++ uses the keyword *this*.

5.2 PSEUDOCODE WITH THE OBJECT NAVIGATION NOTATION

We augment pseudocode with a notation for navigating object models that we call the *Object Navigation Notation* (*ONN*). The ONN mitigates a disadvantage of straight pseudocode; the procedurality of pseudocode can encourage premature descent into design. Instead the ONN is a declarative notation; you specify what you want, not how you implement it. Furthermore, you can readily implement the ONN with many database languages (as we will show in Part 3). The ONN has several other advantages:

- **Tightens coupling between the functional and object models**. The object model not only describes data structure, but also expresses the potential for computation. You can satisfy many requests for information by simply traversing the object model. The ONN lets us integrate the three models more tightly than accomplished in [Rumbaugh-91a].
- **Increases productivity**. You can specify methods more quickly and with fewer errors than you can when using pseudocode alone.
- **Facilitates checking**. You can check the correctness of an object model by constructing hypothetical queries. You should be able to traverse from a source object to a target object via a meaningful path and obtain the correct multiplicity. Otherwise, you must revise the object model.
- **Fosters object-oriented thinking**. Perhaps the biggest advantage of the ONN is that it gets you accustomed to *thinking* in terms of object models. The object-oriented paradigm not only lets you concisely specify data structure, but it also lets you think abstractly about computation. Once you have internalized this idea, it is quite natural to read an object model and directly write many methods and database queries.

5.3 ONN CONSTRUCTS

In this section we present basic ONN constructs. In the next section we describe ONN's power in combining these constructs. In explaining the various constructs, we use *aClassName* to refer to a single object of *className*. Most of our examples are based on the airline flight reservation model in Figure 2.23. We do not explicitly mention aggregation, but aggregation has the same behavior as association. Appendix B provides a BNF grammar for the ONN.

5.3.1 Traversal of Simple Binary Association

The *dot* operator traverses an association to find objects. You can specify the traversed association with either a target role name or a source role name.

- **objectOrSet.targetRole**. This construct traverses an association *to* a role—the target role. The target role may be a role name or, when there is no ambiguity, a class name. You can start the traversal with a single object or a set of objects. In the latter case, the ONN construct has the following meaning: (1) Iterate for each object in *objectOrSet*, (2) for each object find the associated object(s), and (3) return the union of the associated object(s).

In Figure 5.1, for example, the expression *aCity.Airport* denotes the airports that serve a city. We start with a *City* object and traverse the association to find the airports that serve a city. The expression *aTravelAgency.TravelAgent* denotes the travel agents that work for a travel agency. The expression *aTravelAgent.employer* denotes the employer for a travel agent. We can also find the employer for a travel agent with *aTravelAgent.TravelAgency*, since there is no ambiguity. (There is a single association between *TravelAgency* and *TravelAgent*.) All these examples start with a single object, but when we combine ONN constructs in Section 5.4 we will encounter traversals that start with sets of objects.

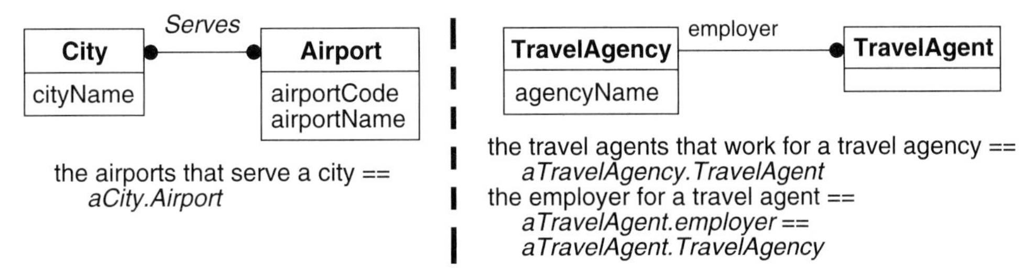

Figure 5.1 ONN examples: Traversal of simple binary associations, using the target role

- **objectOrSet.~sourceRole**. A dot combined with a tilde traverses a binary association *from* a role—the source role. The source role may be a role name or, when there is no ambiguity, a class name. The tilde is significant because it lets you access the associated object(s) when there is no target role name and the target class name is ambiguous. You can start the traversal with a single object or a set of objects.

  In Figure 5.2, for example, the expression *aCity.~City* denotes the airports that serve a city. The expression *aTravelAgency.~employer* denotes all the travel agents that work for a travel agency. These first examples read more naturally using a target role (Figure 5.1), but we show navigation with the source role for comparison. In contrast, we cannot traverse to *FlightDescription* without using the source role (or adding target role names to the object model). The expressions *anAirport.~origin* and *anAirport.~destination* denote the flight descriptions that originate and arrive at an airport.

In the first bullet (and much of the remainder of this section) we explain traversal from a set of objects by presenting an algorithm. You should be careful not to confuse this explanation with the construct's abstract meaning. You could implement traversal from a set of objects procedurally, as we explained. Or you could write a query with a relational or object-oriented DBMS, as we discuss in Part 3. Our point is that the ONN is an analysis notation; you specify what is desired and choose from multiple approaches to realize your specification.

From now on we will use *role* to designate *targetRole* or *~sourceRole*. Thus *objectOrSet.role* denotes *objectOrSet.targetRole* and *objectOrSet.~sourceRole*.

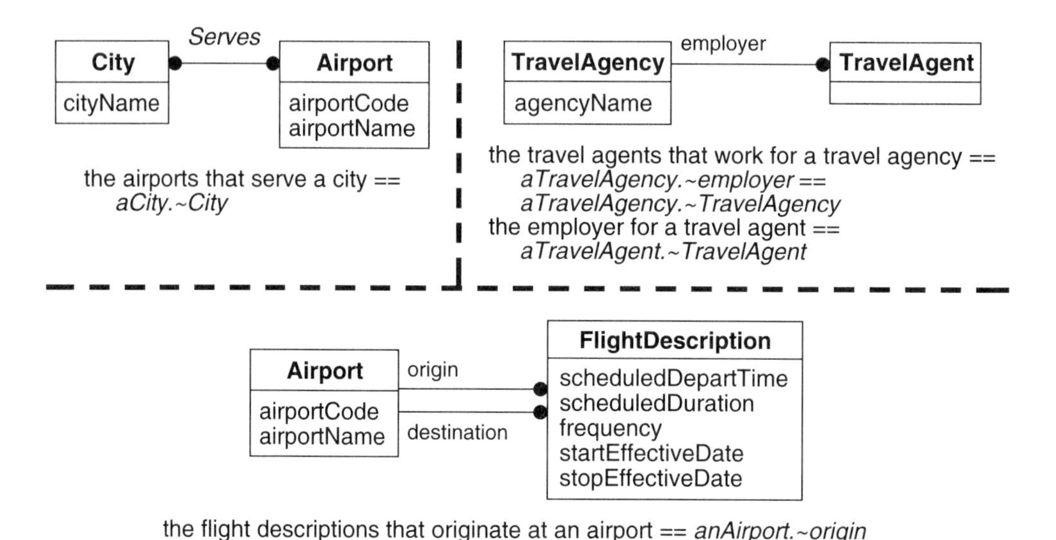

Figure 5.2 ONN examples: Traversal of simple binary associations, using the source role

5.3.2 Traversal of Qualified Association

You can also use the *dot* operator to traverse a qualified association. The construct in the first bullet ignores the qualifier and traverses only the underlying association.

- **objectOrSet.role**. You can always traverse a qualified association without specifying the qualifier. Such an expression normally yields a set of objects. In Figure 5.3, for example, the expression *anAirline.FlightDescription* denotes the multiple flight descriptions for an airline.

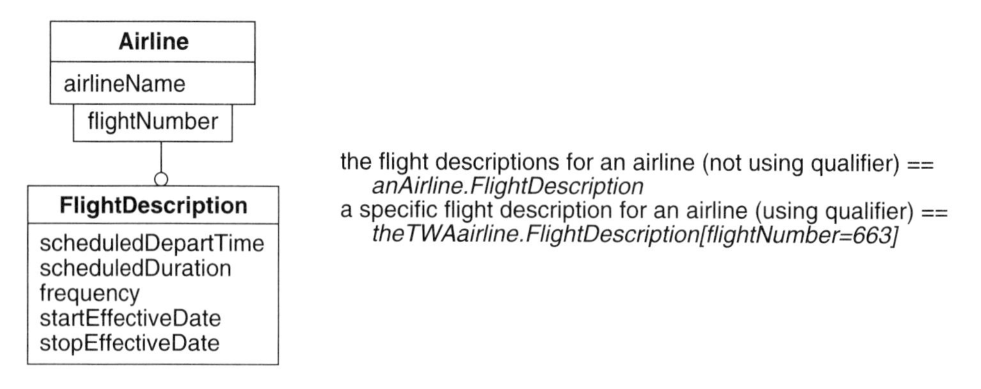

Figure 5.3 ONN examples: Traversal of a qualified association

Or you can be more precise and use a qualifier to select objects.

- **objectOrSet.role[qualifier=value]**. In Chapter 2 we explained that qualifiers deepen the structure of an object model. Qualifiers let you specify multiplicity more precisely; you can specify that an airline and flight number yield one flight description, for example, rather than merely noting that an airline has many flight descriptions.

  Qualifiers not only improve the structure of an object model, but also make it easier to navigate the model. You can specify a smaller and more precise set of associated objects than you can without the qualifier. Most qualified associations yield a single object. For example, the expression *theTWAairline.FlightDescription[flightNumber=663]* yields a single flight description object. Without a qualifier you could not reference a particular flight description object and would need to write additional pseudocode.

5.3.3 Traversal of Generalization

The *colon* operator traverses a generalization. The generalization may be a hierarchy (single inheritance) or a lattice (multiple inheritance).

- **objectOrSet:superclass**. You can traverse up a generalization from a subclass to a superclass, which is called upcasting. You can upcast either a single object or a set of objects by implicit iteration.

  In Figure 5.4, for example, the expression *aTravelAgent:Agent* starts with the *TravelAgent* subclass and navigates to the *Agent* superclass. This example is rather trivial, but upcasting can be useful when the subclass object is computed from some other expression, as we will see in Section 5.4.

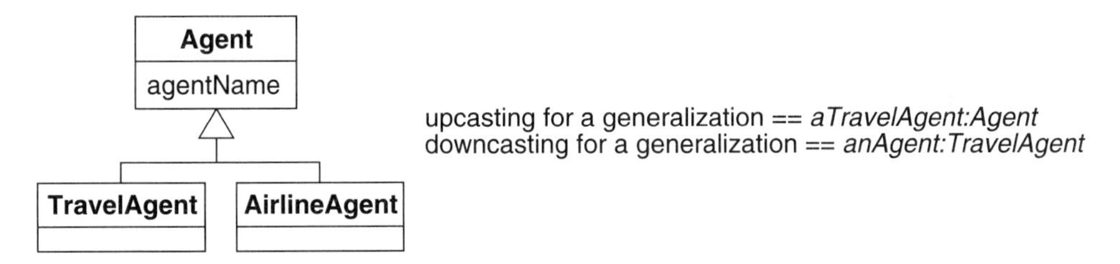

Figure 5.4 ONN examples: Traversal of generalization

- **objectOrSet:subclass**. You can also traverse down a generalization from a superclass to a subclass, which is called downcasting. You can downcast either a single object or a set of objects by implicit iteration. The ONN does not indicate whether traversal is up or down; the direction of traversal is determined from the context of the model.

  You should apply downcasting with caution, because downcasting to a particular subclass is not possible for some superclass instances. Thus downcasting yields a set of objects with cardinality less than or equal to the cardinality of the original set. If *anAgent* happens to be an *AirlineAgent*, the expression *anAgent:TravelAgent* yields the value of null. If *someAgents* happens to be a set that contains two *TravelAgents* and three *AirlineAgents*, the expression *someAgents:TravelAgent* yields a set with two objects.

5.3.4 Traversal from Link to Object

The third use of the *dot* operator is traversing from a link to an object. The link may be for a binary, ternary, or *n*-ary association.

- **linkOrSet.role**. You can traverse from a link to a related object. By definition a link implies one object for each role. So a single link will always yield a single object for a given role. A set of links will yield a set of related objects for a given role; the cardinality of the set of objects is less than or equal to the cardinality of the set of links. (The same object may participate in multiple links.)

  Traversal from link to object is often useful with association classes. You can traverse a model to find association links and then use the links to find constituent objects.

  You can always find an object by traversing a link to a target role. For binary associations you can also traverse a link by specifying a source role. You cannot specify a source role for a ternary or *n*-ary association, because a source role is ambiguous and does not fully imply an object. For example, the expression *HobbyAirportForHouston.City* and *HobbyAirportForHouston.~Airport* both clearly yield the same object (*Houston*). For Figure 3.10 (most of which is repeated as the second example in Figure 5.5) the expression *aDeliveredCourse.Professor* evaluates to the professor who teaches a course. In contrast, the expression *aDeliveredCourse.~Semester* is ambiguous and we cannot tell whether the intent is to refer to a professor or to a course listing.

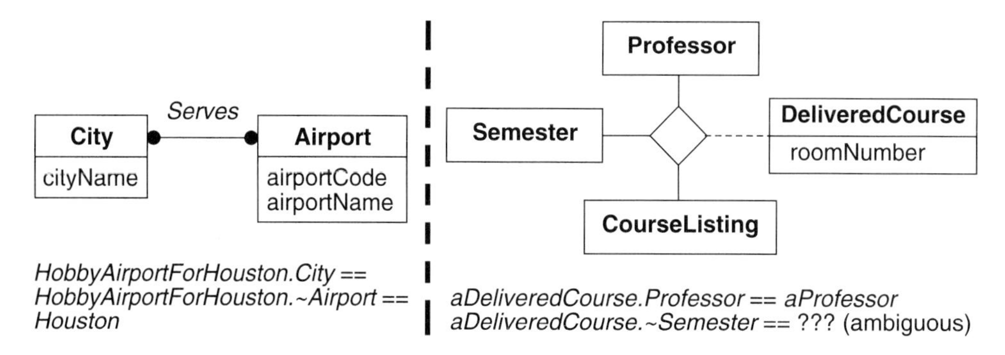

Figure 5.5 ONN examples: Traversal from link to object

5.3.5 Traversal from Object to Link

There are two ways you can traverse from objects to find links:

- **{role1=objectOrSet1, role2=objectOrSet2, . . . , associationName}**. This construct finds links by using multiple roles. You can specify two objects for a binary association, and the construct yields one link if the objects are related and null otherwise. If you provide sets for one or more roles, the construct yields the links found for all possible combinations of objects from the sets.

You can specify at most two roles for a binary association, at most three roles for a ternary association, and so on. It is acceptable to specify fewer roles than the degree of the association; the expression just evaluates to the links that have the specified roles. For binary associations you can specify either target or source roles; for ternary and *n*-ary associations, you must specify target roles. The association name is required only if there is ambiguity. You can specify the roles in any order.

Traversing from object to link can help you traverse association classes or find link attributes. For example, for the left model in Figure 2.18 (repeated in Figure 5.6) the expression *{grantee=aUser, Table=aTable}* yields an authorization. We can then use the authorization to find the user who was the grantor.

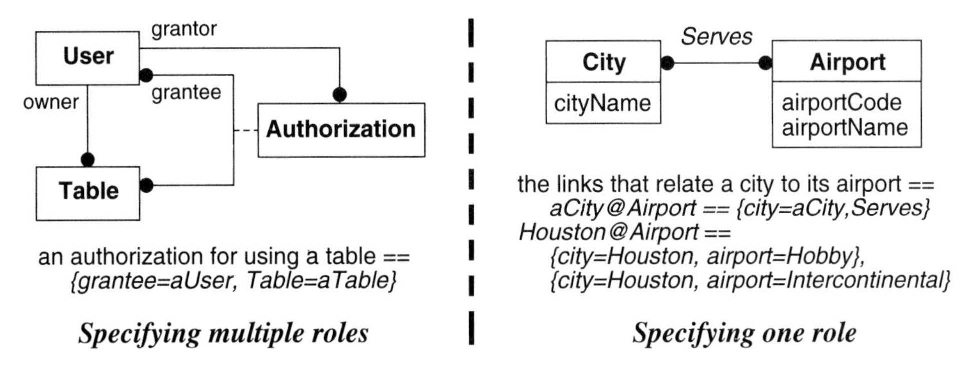

Figure 5.6 ONN examples: Traversal from object to link

- **objectOrSet@role**. You can also retrieve links by specifying only one role and using the *at sign* operator. For example, *aCity@Airport* yields the links that relate a city to its airports. Thus *Houston@Airport* would yield the links *{city=Houston, airport=Hobby}* and *{city=Houston, airport=Intercontinental}*. This is an optional construct; you can always restate an *at sign* expression by using the brace notation in the previous bullet. We include the *at sign* construct for convenience.

5.3.6 Filtering

You can specify a general expression that winnows a set of objects:

- **objectOrSet[filter]**. The filter can be any expression that evaluates to true or false. In Figure 5.7 we find the flight descriptions that originate at an airport and are effective in the first quarter of 1997. The predicate *qualifier = value* in Section 5.3.2 is a special kind of filter. You can use multiple qualifiers ANDed together to specify traversal for a compound qualifier (see Section 3.2.5).

Similarly, you can specify a general expression that winnows a set of links:

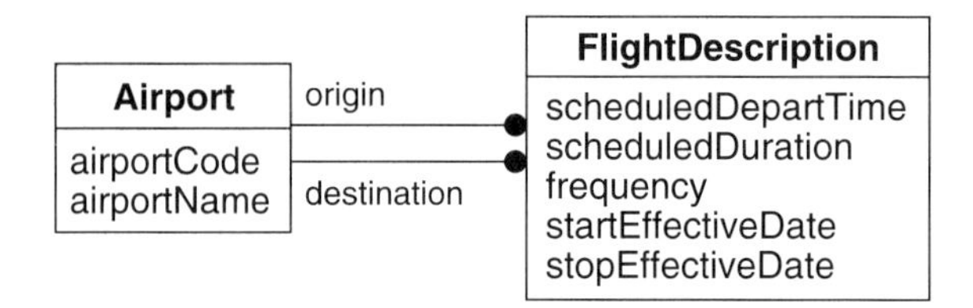

someFlightDescriptions:= anAirport.~origin
theFlightDescriptions1Q97:= someFlightDescriptions[startEffectiveDate<=1-Jan-1997
AND stopEffectiveDate >= 31-Mar-1997]

Figure 5.7 ONN example: Filtering a set of objects

- **linkOrSet[filter]**. The filter can be any expression that evaluates to true or false.

5.3.7 Traversal from Object to Value

You can use the *dot* operator to find attribute values for an object:

- **objectOrSet.attribute**. You can traverse from an object to an attribute value. Similarly, you can access an attribute for each object in a set, returning a set of attribute values. In Figure 5.8, for example, *theHobbyAirport.airportCode* evaluates to *HOU* and *theIntercontinentalAirport.airportCode* evaluates to *IAH*.

Figure 5.8 ONN example: Traversal from object to value

5.3.8 Traversal from Link to Value

You can also use the *dot* operator to find attribute values for a link:

- **linkOrSet.attribute**. You can traverse from a link to an attribute value. Similarly, you can access an attribute for each link in a set, returning a set of attribute values. In Figure 5.9 (based on Figure 2.19), for example, the expression *aJudgedTrial.score* yields the score that a judge assigns to the efforts of a competitor at an athletic event.

  You can also use the *dot* operator to access a qualifier value for a link. It is appropriate to access a qualifier via a link rather than a related object, because a qualifier is a property of an association and not the related classes. In Figure 5.9 you can find the link to an airline for a frequent flyer account and then use the link to find the account number.

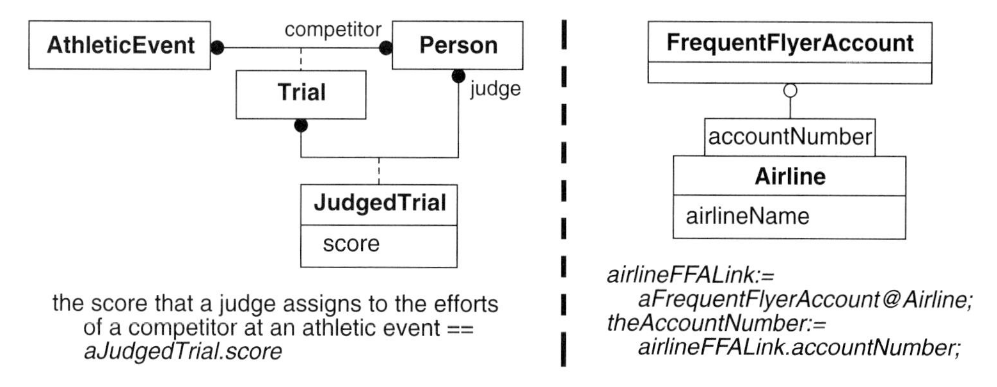

Figure 5.9 ONN examples: Traversal from link to value

5.3.9 Summary

Table 5.1 summarizes the primitive ONN constructs. The token *role* denotes a target role or a source role. (Remember that a source role is always preceded by a tilde.) Traversal from a set has the following meaning: (1) Iterate for each object or link in the set, (2) for each object or link find the associated object(s), link(s), or value(s) and (3) return the union of the associated object(s), link(s), or value(s).

5.4 COMBINING ONN CONSTRUCTS

The real power of the ONN (as with any other language) comes from combining primitive constructs into expressions. You may substitute an expression that evaluates to an object or a set of objects anywhere *objectOrSet* appears. For example, you may evaluate an expression yielding an object or an object set and then traverse an association. Similarly, you may substitute an expression that evaluates to a link or a set of links anywhere *linkOrSet* appears. An expression may contain lesser ONN expressions, so you can build expressions of arbitrary complexity. Most of our examples in this section are based on Figure 2.23.

5.4.1 Examples of ONN Expressions

The following examples illustrate the use of ONN expressions:

- **theStLouisAirport.~origin.destination**. This expression finds all the airports serviced from the St. Louis airport via a nonstop flight. This expression traverses two associations. We start with a single object, the St. Louis airport, and find all the flight descriptions that originate at St. Louis. Because the association has the multiplicity of one *origin Airport* to many *FlightDescriptions*, we can expect to find multiple *FlightDescriptions*. So the first traversal starts with an object and yields an object set.

Concept	ONN construct	Meaning
Traverse binary association	objectOrSet.role	Traverse the association denoted by role.
Traverse qualified association	objectOrSet.role [qualifier=value]	Traverse the association and return the object(s) that correspond to the qualifier value.
Traverse generalization	objectOrSet:superclass	Traverse up a generalization from the subclass to a superclass.
	objectOrSet:subclass	Traverse down a generalization from the superclass to a subclass. *Warning*: You should apply downcasting with caution, because downcasting to a particular subclass is not possible for some superclass instances.
Traverse from link to object	linkOrSet.role	Traverse from the link(s) to the object(s) denoted by role.
Traverse from object to link	{role1=objectOrSet1, role2=objectOrSet2, . . . , associationName}	Find the link(s) for the specified objects. You can specify at most two roles for a binary association, three roles for a ternary association, and so on. The association name is only required if there is ambiguity.
	objectOrSet@role	Traverse from the object(s) to the link(s) denoted by role.
Filter objects	objectOrSet[filter]	Restrict the objects in a set using a boolean expression.
Filter links	linkOrSet[filter]	Restrict the links in a set using a boolean expression.
Traverse from object to value	objectOrSet.attribute	Find the attribute value(s) for object(s).
Traverse from link to value	linkOrSet.attribute	Find the attribute value(s) for link(s).

Table 5.1 Summary of the primitive constructs in the Object Navigation Notation

The second traversal starts with an object set and yields an object set. For each flight description we find the destination airport. We collect the destination airports and return the union as the answer.

- **theStLouisAirport.~origin.destination.~origin.destination**. This expression finds all the airports serviced from the St. Louis airport via a one-stop flight. The result of the expression includes the St. Louis airport, because the St. Louis airport is reachable from itself via a one-stop flight.

- **aFrequentFlyerAccount@Airline.accountNumber**. This expression retrieves the account number for a frequent flyer account and combines the two separate statements used in the second example in Figure 5.9.
- **aFlight.FlightDescription.AircraftDescription.modelNumber**. This expression finds the model of aircraft being used for a flight.
- **aTripReservation.FlightReservation.Flight.FlightDescription.Airline**. This expression specifies the airlines providing transport for a trip.

5.4.2 Examples of ONN Combined with Pseudocode

We were able to express the previous examples by just using the ONN. We will now define some methods that combine pseudocode with ONN expressions. We will intuitively assign ownership of the methods to classes. Chapter 10 provides guidelines for determining ownership.

- **Airport::findZeroOneStops**. This method finds all airports serviced from a given airport via a nonstop or one-stop flight. The "+" denotes the mathematical operator of set union.

```
Airport::findZeroOneStops returns set of Airport
   return self.~origin.destination +
      self.~origin.destination.~origin.destination;
```

- **TripReservation::hasOnlyAisleSeats**. This method determines if aisle seats have been assigned for every flight reservation within a trip reservation. You can extend the object model by adding *aisleCenterOrWindow* to *SeatDescription*, as we do in this example.

```
TripReservation::hasOnlyAisleSeats returns boolean
   for each aFlightReservation in self.FlightReservation
      aSeat := aFlightReservation.Seat;
      if aSeat is NULL then return FALSE;
      else if aSeat.SeatDescription.aisleCenterOrWindow <>
         "aisle" then return FALSE;
      end if
   end for each
   return TRUE;
```

- **Airline::calcFractionLate(month,year)**. This method computes the fraction of flights that were more than 15 minutes late for a given month and airline. We assume we have functions *getMonth(date) returns month* and *getYear(date) returns year*.

```
Airline::calcFractionLate (month,year) returns number
   totalFlights := 0;
   lateFlights := 0;
   for each aFlight in self.FlightDescription.Flight
      aDepartureDate :=
         aFlight@FlightDescription.departureDate;
      if getMonth(aDepartureDate) == month AND
```

```
            getYear(aDepartureDate) == year then
            totalFlights ++;
            aScheduledDepartTime :=
               aFlight.FlightDescription.scheduledDepartTime;
            if aFlight.actualDepartTime >
               aScheduledDepartTime + 15 then lateFlights ++;
            end if
         end if
      end for each
      if totalFlights <= 0 then return ERROR;
      else return lateFlights / totalFlights;
      end if
```

- **TravelAgency::calcMonthlySales(month,year)**. This method computes the total ticket sales for a given month for a travel agency. We will assume we have functions *getMonth(date) returns month* and *getYear(date) returns year*. The object model in Figure 2.23 lacks the date of sale, so we add attribute *datePaid* to the *TripReservation* class.

```
TravelAgency::calcMonthlySales (month,year) returns money
   totalSales := 0;
   for each aTripReservation in
      self.TravelAgent:Agent.TripReservation
      if getMonth(aTripReservation.datePaid) == month AND
         getYear(aTripReservation.datePaid) == year then
         totalSales += aTripReservation.Ticket.fare;
      end if
   end for each
   return totalSales;
```

- **TripReservation::setFrequentFlyerAccount (aFrequentFlyerAccount)**. This method updates the frequent flyer account link for a trip reservation. This time we must revise the object model to fix an error; in Figure 5.10 we associate *TripReservation* with *FrequentFlyerAccount* and *Customer*. These changes are necessary because a customer can make a reservation and specify a frequent flyer account before a ticket is issued. It is common to make modest revisions to a model as you formulate methods.

```
TripReservation::setFrequentFlyerAccount
   (aFrequentFlyerAccount)
   if self.Customer.FrequentFlyerAccount contains
      aFrequentFlyerAccount then
      self.FrequentFlyerAccount := aFrequentFlyerAccount;
   else ERROR
   end if
```

5.5 ADDITIONAL ONN PROPERTIES

So far we have explained the primitive ONN constructs and presented some examples. Now we will discuss additional properties of the ONN.

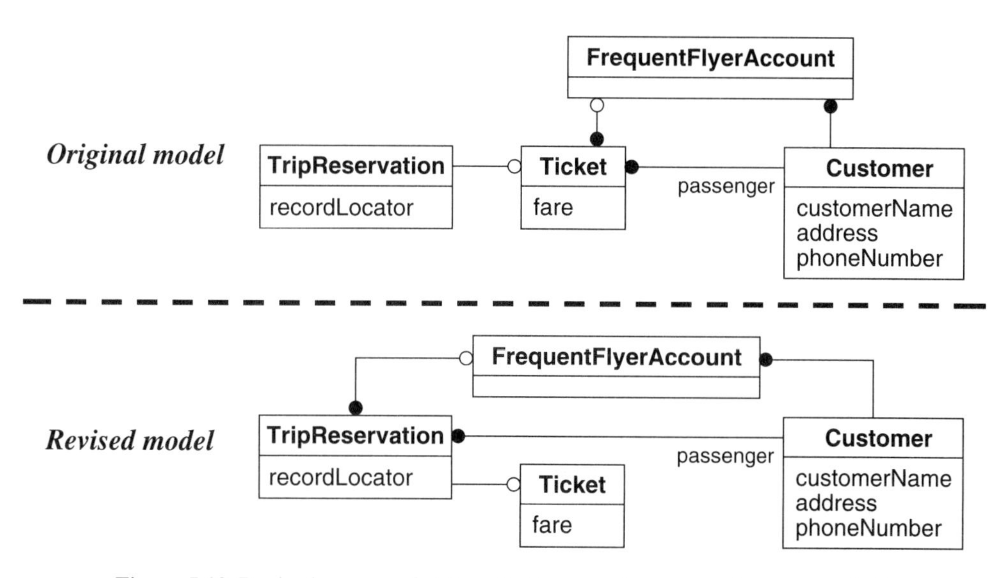

Figure 5.10 Revised excerpt of airline flight reservation system from Figure 2.23

- **Mathematical properties**. ONN expressions are not commutative. For example, we can evaluate *theStLouisAirport.~origin*; in contrast, *~origin.theStLouisAirport* is undefined. (Appendix B contains a BNF grammar for the ONN.) Furthermore, ONN expressions are not associative. We can evaluate *(theStLouisAirport.~origin).destination*; in contrast, *theStLouisAirport.(~origin. destination)* is undefined. However, ONN expressions do observe closure (aside from traversal to a value that terminates an expression). The argument(s) to an ONN expression are sets of objects and links; the result of an ONN expression is a set of objects or links. This closure is an important property because it lets you build expressions using primitive constructs in flexible ways.

- **Nulls**. At this point we would like to be more precise with the arguments and results of expressions. In Table 5.1 for *objectOrSet*, you may provide an object, a set of objects, or null. The ONN accepts either an object or a set of objects because a single object often initiates an ONN expression or is computed by an ONN expression. It is easier to deal with a single object than a set with one object. Similarly, in Table 5.1 for *linkOrSet*, you may provide a link, a set of links, or null.

  All ONN constructs evaluate to null if they have a null argument. The ONN allows nulls because you can encounter them in traversing associations. For example, if a flight reservation does not have an assigned seat, *aFlightReservation.Seat* evaluates to null. A null can also result from downcasting, because superclass instance(s) may not belong to the specified subclass.

- **Attribute access**. The ONN directly accesses attributes. In contrast, object-oriented programming books emphasize that you should only indirectly access attributes via read and write methods. There is no contradiction between these different points of view. Directly reading and writing attributes in an *analysis* model does not violate the *design* principle of encapsulation. During analysis you should acquire a broad understanding of a problem. During design you can then devise the best way to access information. You may choose to implement an ONN expression with a relational DBMS or OO-DBMS query and violate encapsulation. Or you can decompose an ONN expression into a series of methods that carefully encapsulate access to objects. In Chapter 10 we discuss the trade-off between encapsulation and query optimization.

- **Qualifiers and association classes**. In Chapters 2, 3, and 4 we emphasized the importance of qualifiers and association classes. These constructs not only let you increase the precision of an object model but also help you navigate the model. Qualifiers that yield a multiplicity of "one" or "zero or one" let you traverse a model and find individual objects. In contrast, for notations that lack qualifiers you can find only ambiguous sets of objects, not *the* desired object. For example, in Figure 2.23 given an airline and a flight number, you can find at most one flight description. If we omitted the *flightNumber* qualifier and merely stated that an *Airline* has many *FlightDescriptions*, it would not be clear how we should find an individual flight description.

  Association classes also let you traverse to a specific object. In Figure 5.9 given an *AthleticEvent* and a *competitor*, we can find a *Trial*. Given a *Trial* and a *judge* we can find the assigned *score*.

5.6 OTHER PARADIGMS

As we mentioned at the beginning of this chapter, we normally prefer to specify functionality by using pseudocode combined with the ONN. This section presents other paradigms. We first discuss data flow diagrams. Although they are not helpful for database applications, we discuss them for continuity with [Rumbaugh-91a] and explain why they are inappropriate. Then we discuss decision tables and mathematical equations that are occasionally helpful.

5.6.1 Data Flow Diagrams

A ***data flow diagram*** is a directed graph whose nodes are objects that generate data, objects that consume data, and operations on data; the arcs are data flows that convey data between the objects and operations. Data flows imply functional dependencies between operations. You can expand each operation in a data flow diagram into a lower-level diagram, leading to a hierarchy of data flow diagrams. The expansion eventually terminates with operations that are easy to explain and understand. [Rumbaugh-91a] presents a detailed explanation of data flow diagrams.

Data flow diagrams provide an effective paradigm for some kinds of applications. For example, they are helpful for business process re-engineering because the focus is on processes—documenting existing processes and inventing new processes. Data flow diagrams let you decompose complex processes and convey your understanding to other persons.

However, we do not find data flow diagrams useful for database applications. In fact, the use of data flow diagrams, as in the structured analysis/structured design techniques of the past, is the antithesis of a proper approach to database applications. You cannot reasonably design a database by emphasizing process and discovering data as a side effect. The objects that can be discovered via data flow diagrams are fragmented, disjointed, and usually incomplete. Instead we strongly advocate that you organize software about real-world objects. A database is organized about data, so your analysis techniques should also be organized about data. Data tends to be more stable and less vulnerable to change than the way in which the data happens to be used.

5.6.2 Decision Tables

Decision tables [Montalbano-74] provide a useful representation for discrete logic. In a ***decision table*** the rows are individual rules and the columns are the attributes (antecedents and consequences) that are the subjects of the rules. A decision table is an unordered set of rules: Input1 through input*n* implies output1 through output*m*. The inputs for a decision table rule are implicitly ANDed together, and multiple rows express OR combinations of inputs.

Figure 5.11 shows a simple object model and a corresponding decision table. There could be many decision tables for the object model, but we only show one. The combination of the deck width and engine type for a lawn mower determines the proper deck part number. Thus a deck width of *16* and engine type of *gas* implies the deck with part number *D16*. A deck width of *16* and engine type of *electricity* also implies the deck part *D16*.

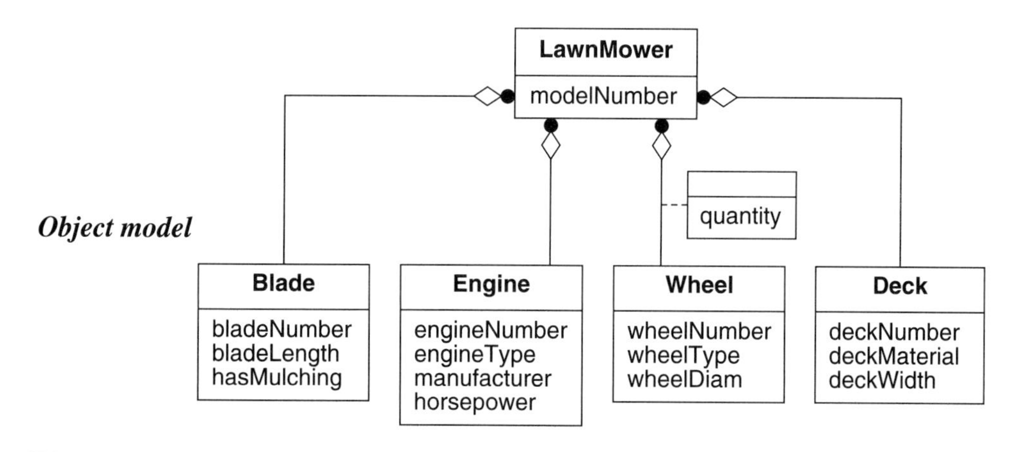

Functional model

deckNumber	deckWidth	engineType
D16	16	gas
D16	16	electricity
D18	18	gas
D20	20	electricity
D20S	20	gas
D22S	22	gas

Figure 5.11 Decision table for choosing a mechanical part

There are many minor variations in notation for decision tables. In Figure 5.11 we list outputs (one output in this case) to the left and inputs to the right. The double vertical line separates outputs from inputs. Attributes that are the subjects of the rules are shown at the top delimited by a horizontal line.

Decision tables provide a declarative representation for functionality because the means for searching the decision table is unspecified. A decision table does not specify how to compute an answer. Declarative representation is highly desirable for analysis and is often practical as a basis for implementation. Decision tables are especially useful for systems that learn from experience, because rules can be added without altering programming code or data structure.

There are subtleties to consider with decision tables. For example, several interpretations are possible when multiple rows have the same combination of inputs:

- **Arbitrary answer**. A solution may be chosen at random or based on some optimization criteria. For example, multiple parts may satisfy the functional needs for a manufacturing application. An engineer may select the least expensive part or the part that is more likely to be in stock.
- **Conflict or contradiction**. The decision table may be missing one or more of the input attributes needed to make a decision. These attributes would lead to additional columns that must be filled for each row.
- **Set of answers**. The proper answer may be the union of the answers found for rows with the same inputs.

Similarly, alternative interpretations are possible when a decision table lacks a combination of inputs. All logic must be explicit in a decision table (closed-world assumption); you cannot conjecture for missing combinations of inputs.

- **Missing rule**. The decision table may be missing information. If so, you can add one or more rules.
- **Missing output**. The combination of inputs may be reasonable and simply lack a corresponding output. You can add a rule with a null output to differentiate missing rules and missing outputs.
- **Illogical combination of inputs**. The search against the decision table may have involved a nonsensical combination of inputs, in which case there is no answer to the query as stated.

Various extensions can increase the expressiveness of decision tables. Strictly speaking, all decision table inputs and outputs should be boolean predicates: The output predicates are asserted to be true when the input predicates are true. However, you can use various conventions to abbreviate inputs and outputs. A common convention is to assume equality predicates, in which case you can simply list the target values of the predicates as inputs and outputs. Frequently, the predicate values are of enumeration type and you can list rules for all possible values.

You can also devise a convention for an attribute that is irrelevant for some combination of other inputs. Such an attribute is a "wildcard" or "don't care" for a decision table row. For example, in Figure 5.11 if there are only two possible values, engine type is irrelevant for a deck width of *16*; you could replace the first two rows with one row where the engine type is set to a wildcard value. The use of a wildcard is more concise than listing a rule for each possible input. The resulting abstraction renders rules less fragmented and easier to understand. Wildcards may cause overlapping combinations for different inputs, leading to partial redundancy of the rules.

Decision tables are inflexible in terms of the kinds of rules they can express. They naturally represent the case in which something is selected by a set of precise conditions (that is, equality predicates ANDed together). More complex expressions, such as nested combinations of ANDs and ORs, mathematical formulas, and inequality predicates, do not easily fit the decision table paradigm.

5.6.3 Mathematical Equations

Mathematical equations provide an appropriate medium for expressing functionality for some applications. Equations can be useful for both analysis and design. During analysis, equations declaratively describe what is needed. During design, equations can be directly solved if suitable software is available. You should organize large numbers of equations into groups that are meaningful for the application. Such groups are easier to understand and make it less likely that you will accidentally overspecify or underspecify a system.

For some applications you may be able to specify functionality with linear simultaneous equations. These are usually straightforward to solve. Matrix solvers are readily available and can meaningfully explain numerical difficulties that arise in computation. Linear programming techniques use optimization criteria to solve underspecified systems of linear equations. For example, the simplex algorithm is often used to determine the product mix (gasoline, diesel fuel, kerosene, and others) of an oil refinery.

Nonlinear simultaneous equations also arise. For example, engineers sometimes use nonlinear equations to describe the behavior of chemical processing plants. Advanced mathematical techniques are required to solve nonlinear simultaneous equations. Such equation solvers can be limited by convergence difficulties for ill-behaved and large sets of equations.

5.7 PRACTICAL TIPS

On the basis of our experience, we offer the following tips for constructing the functional model:

- **Consider multiple paradigms**. Consider alternative paradigms for functional modeling. For a given problem you should choose the paradigm or combination of paradigms that is most natural and effective.
- **Avoid procedural decomposition**. Do *not* model a system by constructing data flow diagrams and adding object models as an afterthought. (This is the traditional structured analysis/structured design approach to system development.) Instead first construct an object model. Then build the dynamic and functional models as needed for your appli-

cation. The premise of the object-oriented approach is that software will be more robust, extensible, and maintainable if software is organized about things rather than about functionality.

- **Use pseudocode carefully**. We normally use pseudocode enhanced with the Object Navigation Notation to express the functional model. But do not regard pseudocode as an excuse to descend into premature design.
- **Do not write obscure pseudocode**. Pseudocode is most helpful when methods are brief (no more than one-half page) and statements average one to two lines.
- **Avoid ambiguous decision tables**. If you use decision tables, establish conventions to resolve any ambiguity and apparent conflicts.
- **Document your functional models**. When you document your functional models, others can understand them and you can understand them at a later date. Note side effects, computational complexity, and semantic intent as needed for each method. The methods in this chapter are relatively simple and did not require much explanation. Some of our methods for the case study in Part 2 will require more explanation.

5.8 CHAPTER SUMMARY

The functional model is important because it lets you realize the potential for computation that is latent in the object model. We have presented alternative notations (enhanced pseudocode, decision tables, and equations) for the functional model that are helpful for database applications. You can use them separately or together.

Most often, we use enhanced pseudocode for expressing the functional model. Pseudocode is an informal language that provides sequence, conditionality, and iteration. We augment pseudocode with a notation for navigating object models that we call the *Object Navigation Notation* (*ONN*). The ONN mitigates a disadvantage of straight pseudocode; the procedurality of pseudocode can encourage premature descent into design. Instead the ONN is a declarative notation; you specify what you want, not how you implement it. The ONN takes advantage of the precision of the object model in navigating to specific objects. Qualifiers and association classes are especially important for navigation.

Decision tables provide a useful representation for discrete logic. A decision table is a table in which the rows are individual rules and the columns are the attributes (antecedents and consequences) that are the subjects of the rules. All logic must be explicit in a decision table; no conjecture can be made for a combination of inputs that are missing. Decision tables provide a declarative representation that is highly desirable for analysis and often practical for implementation.

Mathematical equations provide a suitable medium for expressing functionality for some applications. Equations can serve both analysis and design. During analysis, equations declaratively describe what is needed. During design, equations can be directly solved, if suitable software is available.

Figure 5.12 lists the key concepts for this chapter.

combining ONN constructs	Object Navigation Notation
decision table	pseudocode
functional model	traversal of an object model
mathematical equations	

Figure 5.12 Key concepts for Chapter 5

BIBLIOGRAPHIC NOTES

[Rumbaugh-91a] presents data flow diagrams as the only notation for the functional model. This recommendation stemmed from the prominence of data flow diagrams in traditional structured analysis/structured design. Since then, we have reconsidered and now advocate a variety of paradigms for expressing the functional model.

[Blaha-90] provides an example of a large database application that extensively uses decision tables. The paper describes a generic framework for analyzing designs of mass-produced mechanical products to determine configuration rules. Figure 5.11 shows a very simple example of configuration rules. The framework has a large object model and a large functional model consisting of decision tables and equations.

The IDEF1X notation [Bruce-92] for relational database design incorporates some concepts related to navigation. IDEF1X distinguishes between independent entities and dependent entities. An ***independent entity*** is an entity that can be distinguished with intrinsic attributes. In contrast, a ***dependent entity*** derives identity from other entities (in database jargon, has a foreign key as part of its primary key). The cascade of dependent entities that occurs in IDEF1X models is similar to the ONN. However, IDEF1X uses traversal only to establish keys; the ONN uses it to express functionality.

[Blaha-94] discusses semantic implications of the object model for identifying objects. We have improved the ONN since this column. [Rumbaugh-91b] also includes an early form of the ONN in the answers to some of the exercises.

[Feldman-86] discusses another related topic, the logical horizon. The *logical horizon* of a class is the set of classes reachable by one or more paths terminating in a combined multiplicity of "one" or "zero or one." A path is a sequence of consecutive associations and generalization levels. The logical horizon is a powerful idea that has improved our understanding of object models. Chapter 3 discusses the logical horizon in detail.

You can augment the functional model with preconditions, postconditions, and invariants as described in [Meyer-88]. A ***precondition*** is a boolean expression that holds upon entry to a method. A ***postcondition*** is a boolean expression that holds upon exit from a method. An ***invariant*** is a boolean expression that applies for an object across all its methods. Preconditions, postconditions, and invariants increase the likelihood of software correctness. Preconditions let you define the conditions under which a call to a method is legitimate. Postconditions define the conditions that the method must ensure on return.

REFERENCES

[Blaha-90] MR Blaha, WJ Premerlani, AR Bender, RM Salemme, MM Kornfein, and CK Harkins. Bill-of-material configuration generation. *Sixth International Conference on Data Engineering*, February 5–9, 1990, Los Angeles, California, 237–244.

[Blaha-94] Michael Blaha. Finding objects in object diagrams. *Journal of Object-Oriented Programming 6*, 9 (February 1994).

[Bruce-92] Thomas A. Bruce. *Designing Quality Databases with IDEF1X Information Models*. New York: Dorset House, 1992.

[Feldman-86] D Feldman and D Miller. Entity model clustering: structuring a data model by abstraction. *Computer Journal 29*, 4 (1986), 348–360.

[Meyer-88] Bertrand Meyer. *Object-Oriented Software Construction*. Hertfordshire, England: Prentice Hall International, 1988.

[Montalbano-74] Michael Montalbano. *Decision Tables*. Palo Alto, California: Science Research Associates, 1974.

[Rumbaugh-91a] J Rumbaugh, M Blaha, W Premerlani, F Eddy, and W Lorensen. *Object-Oriented Modeling and Design*. Englewood Cliffs, New Jersey: Prentice Hall, 1991.

[Rumbaugh-91b] J Rumbaugh, M Blaha, W Premerlani, F Eddy, and W Lorensen. *Solutions Manual for Object-Oriented Modeling and Design*. Englewood Cliffs, New Jersey: Prentice Hall, 1991.

EXERCISES

5.1 (2) For Figure 5.6 write an ONN expression that finds the grantor for *aTable* and grantee *aUser*.

5.2 (1) For Figure 2.23 write an ONN expression that finds the *seatType* for a seat.

5.3 The following exercises are based on Figure 2.23 and the revisions in Figure 5.10. Write methods using pseudocode enhanced with the ONN to specify the following queries. Clearly state any assumptions you make.

a. (2) Find the airlines that serve an airport.

b. (3) Determine how many pieces of baggage are supposed to be on a flight.

c. (4) Find the seat number(s) assigned to a passenger on a flight. A passenger normally is assigned at most one seat, but on occasion may reserve multiple seats for a flight.

d. (4) Compute the number of passengers that are served by an airport on a given date. You can ignore the complexities of long distance travel, in which a flight may begin on one day and end on the next day. (Hint: If a person flies through an airport several times in the same day, you can count the person only once.)

5.4 (7) Compute the flight mileage that should be credited to a frequent flyer account each month using the object model of Figure 2.23 and the revisions in Figure 5.10. You will need to extend the object model for base and bonus miles.

For example, TWA awards a minimum of 750 miles for each flight. Gold and red card holders receive a minimum of 1000 miles per flight. Gold card holders receive a 25% bonus for any flight. Red card holders receive a 50% bonus for any flight.

For this exercise, do not consider the effect *fareCode* may have on credited miles. (Some airlines award more credits for first-class travel than coach travel.) You will also need to record whether each flight reservation is actually used. (Airlines do not award credit for paid tickets, only for travel that is actually taken.)

a. Extend the object model for this new information.

b. Using pseudocode and the ONN, write a method to compute the flight mileage earned each month. (Hint: The method should have arguments of month and year.) Assume you have functions *getMonth(date) returns month* and *getYear(date) returns year.*

5.5 Chapter 3 revises the airline flight reservation model from Chapter 2. The remainder of this exercise lists some navigation expressions for Figure 2.23. Rewrite these expressions using Figure 3.33 through Figure 3.37.

a. (1) theStLouisAirport.~origin.destination

b. (1) aTripReservation.FlightReservation.Flight.FlightDescription.Airline

5.6 (5) Chapter 3 also extends the airline flight reservation model. Find the flights for an airline in a specified month that were diverted from their scheduled destination to some other destination. Write a method using pseudocode enhanced with the ONN. Assume you have functions *getMonth(date) returns month* and *getYear(date) returns year.*

5.7 (6) Using pseudocode and the ONN, write the following methods for the directed graph model in Figure E2.7a.

a. Find all edges leaving a node.

b. Find all edges with the same node as source and sink, that is, the direct cycles in the graph.

c. Find all nodes that are reachable from a source node, that is, the transitive closure.

5.8 (8) Repeat Exercise 5.7 for Figure E2.7b. Assume link attribute *sourceOrSink* has been added to the model. Note that the queries are more difficult to express with model (b) than model (a).

5.9 (7) Repeat Exercise 5.7 for Figure E2.7c. Note that the queries are more difficult to express with model (c) than model (a). However, the queries are easier for model (c) than model (b).

5.10 (8) Repeat Exercise 5.7 for Figure E2.7d. Assume class *Connection* has an attribute *connectionType* with values *source* and *sink*. Note that the queries are more difficult to express with model (d) than models (a) and (c). The queries for model (d) are comparable to those for (b).

5.11 Write ONN expressions or methods with enhanced pseudocode for the following queries, which refer to the abridged OMT metamodel in Figure 4.6.

a. (2) Find all the direct superclasses for a subclass.

b. (2) Find all the direct subclasses for a superclass.

c. (5) Find all the components for an assembly.

d. (5) Find all the assemblies that include a component.

e. (5) Find all the attributes that describe the objects of a class. (Compute the transitive closure of inherited attributes for a class.) You need traverse only up the inheritance lattice.

PART 2: ANALYSIS AND DESIGN PROCESS

6

Process Preview

We have concluded our presentation of concepts and notation for the two OMT models of interest to database applications. We now shift our focus to the process that uses these models. Part 2 addresses the front portion of software development—conceptualization, analysis, and design. Part 3 shows how to implement a design for various targets—files, relational databases, and object-oriented databases.

Each chapter in this part addresses a stage in the OMT software development process. In this chapter, we give you a sneak preview of the process steps:

1. **Conceptualization**. (Chapter 7) Conceive an application and formulate tentative requirements.
2. **Analysis**. (Chapter 8) Expand your understanding of the requirements by constructing models of the real world. The goal of analysis is to specify *what* needs to be done, not *how* it is done, which is the goal of design. You must understand a problem before attempting a solution. The same OMT analysis process applies uniformly to both programming and database applications though the relative importance of the three models varies.

   During analysis consider any and all information sources. We normally prefer to assimilate existing written documentation before interviewing domain experts. We have found that interviews are more efficient and effective if we begin with a tentative understanding rather than a clean sheet of paper.
3. **System design**. (Chapter 9) Devise a high-level strategy—the system architecture—for solving the application problem. Also establish policies to guide the subsequent detailed design. The design and implementation of databases differs significantly from that for programming code. It is at this phase that our specific treatment for data management applications diverges from the general process in [Rumbaugh-91].

   First choose a data management approach. Some applications require formal DBMS services; for other applications, files or memory are adequate solutions. Then if you are going to use a DBMS, choose the specific DBMS paradigm, normally a relational DBMS or an object-oriented DBMS. You can adopt various strategies for coordinating programming code with DBMS code.

4. **Detailed design**. (Chapter 10) Augment and transform the real-world analysis models into a form amenable to computer implementation. We have renamed this phase from *object design* in [Rumbaugh-91]. During detailed design you selectively override the general policies from system design. Detailed design deals with decisions that are platform independent.

   Our most important contribution to detailed design is in object model transformations. Transformations let you shift the object model from the analysis focus on the real world to the design focus on the computer resources. We list some transformations in Chapter 10 and apply them to our case study example.

   During detailed design we typically convert the functional model into methods on classes and choose algorithms. The exception occurs when we have a decision table interpreter, equation solver, or some other mechanism that lets us use the functional model directly.

In *implementation* (Part 3), you translate the design into the actual database schema and programming code. You can use tools to automate much of this phase.

The OMT software development process is seamless. You gradually elaborate and optimize the OMT models as your focus shifts from analysis to design to implementation. Throughout the development process the same concepts and notation apply; the only difference is the shift in perspective from an emphasis on describing the real world to a focus on computer resources.

The OMT methodology supports multiple development styles. You can use OMT as a waterfall approach performing the phases of analysis, design, and implementation in strict sequence for the entire system. More often, however, we adopt an iterative development strategy. First we develop the nucleus of a system—analyzing, designing, implementing, and delivering working code. Then we grow the scope of the system, adding properties and behavior to existing objects, as well as adding new kinds of objects. For any development style, it is common to complete analysis nominally and move forward to design, only to uncover some flaws that should have been resolved during analysis. It is much easier to redress such mistakes when dealing with models rather than code.

As much as possible, your analysis models should be ***declarative*** (stated as structural properties and constraints) rather than ***imperative*** (stated as a series of programming statements). Declarative specifications tend to be more concise and avoid premature descent into design. During design your models will tend to become more imperative. Occasionally you can carry declarative models directly into design, such as if you are using a decision table interpreter or equation solver.

Practical application of the OMT methodology to real problems normally requires the use of CASE (computer-aided software engineering) tools. CASE tools can enforce style conventions, validate syntax, and maintain consistency across models. You should definitely use CASE tools for large, multideveloper projects. Sophisticated CASE tools can generate code from models, maintain traceability from analysis through design and implementation, and manage the progress of a project.

REFERENCE

[Rumbaugh-91] J Rumbaugh, M Blaha, W Premerlani, F Eddy, and W Lorensen. *Object-Oriented Modeling and Design*. Englewood Cliffs, New Jersey: Prentice Hall, 1991.

7

Conceptualization

Conceptualization deals with the genesis of an application. Initially some person thinks of an idea for an application and prepares a statement of intent. This chapter introduces the portfolio manager case study, which threads throughout this part and Part 3. We chose the portfolio manager problem because it is representative of many data management problems. Also, the portfolio manager has interesting design content that will be helpful for illustrating important points in later chapters.

7.1 OVERVIEW

The input to conceptualization is an idea and rationale for a new system. The output is a statement of specific requirements that are then studied during the subsequent analysis phase. During conceptualization you should ask yourself high-level questions to refine and elaborate the initial concept. Ask yourself the basic questions: who, what, where, when, why, and how.

- **Who is the application for?** You should clearly understand which persons and organizations are the sponsors for the new system. Two of the most important kinds of sponsors are the financial sponsors and end users.

  The financial sponsors are important because they are paying for the new system. They expect the project to be on schedule and within budget. You should get the financial sponsors to agree to some measure of success. You need to know when the system is complete and meets their expectations.

  The users are also sponsors, but in another sense. The users will ultimately determine the success of the new system by an increase (or decrease) in their productivity or effectiveness. Users can help you if they are receptive and provide critical comments. They can improve your system by telling you what is missing and what could be improved. In general, users will not consider new software unless they have a compelling interest—either personal or business. You should try to help them find a vested interest in your project so you can obtain their buy-in.

- **What problems will it solve?** You must clearly bound the size of the effort for the new system and establish its scope. You should determine which features will be in the new system and which features will not. You must reach various kinds of users in different organizations with their own viewpoints and political motivations. You must not only decide which features are appropriate, but you must also obtain the agreement of influential persons.
- **Where will it be used?** At this early stage, it is helpful to get a general idea of where the new system might be used. You should determine if the new system is mission-critical software for the organization, experimental software, or a new capability that you can deploy without disrupting the workflow. You should have a rough idea about how the new system will complement the existing systems. It is important to know if the software will be used locally or will be distributed via a network.
- **When is it needed?** Two aspects of time are important. The first is the feasible time, the time in which the system can be developed within the constraints of cost and available resources. The other is the required time, when the system is needed to meet business goals. You must make sure that the timing expectations driven by technical feasibility are consistent with the timing the organization requires. If there is a disconnect, you must initiate a dialogue between technologists and business experts to reach a solution.
- **Why is it needed?** You may need to prepare a business case for the new system if someone has not already done so. The business case contains the financial justification for the new system, including the cost, tangible benefits, intangible benefits, risk, and major alternatives. You should consider the cost of development, purchase, and deployment. You must be sure that you clearly understand the business motivation for the new system. The business case will give you insight into what sponsors expect, roughly indicate the scope, and may even provide information for seeding your models.
- **How will it work?** You should brainstorm about the feasibility of the problem. For large systems you should consider the merits of different architectures. The purpose of this speculation is not to choose a solution, but to increase confidence that the problem can be solved reasonably. You might need some prototyping and experimentation. You can often salvage code, data, or ideas from obsolete systems. Often the best decision is to purchase software, if you can find software that satisfies most of the requirements.

7.2 THE PORTFOLIO MANAGER CASE STUDY

Figure 7.1 lists our original system concept for the portfolio manager. We ask high-level questions to refine and elaborate the initial concept.

- **Who is the application for?** We are developing the portfolio manager software for our own use. The case study is a bit artificial in that we are serving multiple roles—sponsor,

We would like to have a program that helps us assess the performance of an investment portfolio over time. We want to record investment decisions, track net worth, and compute return-on-investment (ROI).

Figure 7.1 System concept for portfolio manager

developer, and end user. Different persons assume these roles for most industrial applications.

- **What problems will it solve?** We desire a simple system that tracks personal net worth and summarizes investment history as a guide for future investment decisions. We want to be able to study investment performance by grouping financial assets in different ways. The software should focus on stock and bond investments, but we should be able to add other kinds of financial investments as future extensions. The system need not handle complex issues, such as estates and trusts, and accounting details, such as double entry ledgers. It also need not generate paperwork for tax returns.

  The system should not require entry of the full price and dividend history for a stock. We need enter only the data required for a computation. For example, if we want to compute the value of a portfolio at year-end 1985, the system must store a price for that date for all assets in the portfolio. To compute the ROI of a portfolio for 1994 and 1995, we must supply complete data for the interval, including interest and dividend payments.

- **Where will it be used?** We anticipate using the portfolio manager at home to analyze personal investments. For simplification, we will not explicitly address interfaces to other applications, such as spreadsheets and tax software.

- **When is it needed?** We want to develop a useful program with a few person-months of effort. This is consistent with the rapid prototyping approach to software development: Quickly develop some functionality, test the embryonic system, and then extend the software. We don't expect efficiency to be an issue; the most difficult calculation is ROI, which we must solve through iteration.

- **Why is it needed?** The commercial financial software we have encountered does not allow us to compute portfolio value at various points in time or recursively structure portfolios in terms of lesser portfolios. There is little risk to our development of the portfolio manager, because we have computed ROI for past applications and understand the algorithms. (Of course, to some extent our rationalization for constructing the portfolio manager is contrived; we want to choose some reasonable and tractable problem to illustrate the methodological concepts in this book.)

- **How will it work?** The software must decouple program functionality from the user interface. This may require one model for financial concepts and another for the user interface. We only briefly discuss development of the user interface (Chapter 14) because it is outside the scope of this book.

The system should let us browse the database directly (such as with interactive DBMS commands), but discourage direct database update. All updates should occur via the user interface.

We will implement portions of the portfolio manager with files, a relational DBMS, and an object-oriented DBMS for illustrative purposes. It would be acceptable just to store the data in files. However, we would prefer to store all the data in a formal database to facilitate ad hoc queries and extensibility.

The next chapter shows the resulting problem statement for the portfolio manager.

7.3 CHAPTER SUMMARY

Conceptualization is the first phase of the OMT process for software development. Conceptualization begins with a person thinking of an idea for a new system to provide new functionality or simplify delivery of existing functionality. The output of conceptualization is a statement of specific requirements that serve as grist for the subsequent phase of analysis.

BIBLIOGRAPHIC NOTE

James Rumbaugh first sensitized us to the need for a conceptualization stage that precedes analysis.

8

Analysis

The purpose of analysis is to understand the problem requirements thoroughly and devise a model of the real world. During analysis the developer describes *what* must be done, not *how* it should be done. You must understand a problem before attempting a solution. The models constructed during analysis provide the basis for subsequent design and implementation. The OMT analysis process applies to various implementation targets, including databases and programming code.

Generally, you need all three OMT models—object, dynamic, and functional. However, a particular application may be unbalanced and not fully exercise all three models. In fact, most database applications lack a significant dynamic model. The case study that we introduced in the preceding chapter (portfolio manager) has a rich and robust object model, a weak dynamic model, and a modest functional model.

8.1 OVERVIEW

Various sources of information can serve as inputs to analysis. Often you will have a formal statement that describes your problem. In this book we use a concise problem statement; in commercial settings you may encounter RFPs (requests for proposal) and RFQs (requests for quotation). Sometimes domain experts will provide scenarios, storyboards, and use cases for a new system. (A ***use case*** [Jacobson-92] is a theme for interacting with a system. See Section 8.6.) You may also have access to problem experts for interviews and advice. On fortunate occasions, you yourself may be a source of problem expertise. Replacements for existing systems can benefit from prior artifacts: manuals, data entry forms, and other sources of grist for reverse engineering (see Chapter 20). We aggressively seek all existing information at the start of a project.

The outputs from analysis are three kinds of models that capture crucial aspects of a system. The object model describes objects and their relationships. The dynamic model characterizes temporal interactions of objects and their responses to events. The functional model defines operations and constraints. The data dictionary serves as a detailed repository for the three models, documenting names, concepts, definitions, examples, and rationale. You should also prepare a textual narrative to guide the reader through the models.

We have included use cases as part of the OMT methodology. (See Section 8.6.) But we have given use cases less prominence than some readers might expect. Use cases are important because they specify requirements for interaction in a manner that is understandable to both the software developer and the business client. But use cases are only one source of information for system development. We desire to exploit *all* information sources—including statements of requirements, existing documentation, and reverse engineering. Furthermore, our experience has been that artifacts for reverse engineering are normally available for database applications, as well as statements of requirements. We encounter use cases less often.

8.2 PROBLEM STATEMENT

Always begin analysis with a written problem statement. The statement need not be rigorous or complete, but you should at least have the motivation for a new system in writing. The problem statement should focus on the minimal functionality needed to make a useful system. You can always add functionality later. Carefully delimit the scope of a new system from other applications with which it will interact. Keep in mind that the problem statement is only a starting point and you will not fully know the true requirements until after analysis.

Beware of problem statement pitfalls. Problem statements are often ambiguous. Natural language is expressive, but it is also vague and subject to multiple interpretations. Problem statements often contain errors. When working with natural language, most persons have difficulty grasping the full scope of a problem and recognizing inconsistencies. Problem statements may be underspecified and lack important information. Similarly, a problem statement may be overspecified and contain details that would be better left to design. You must be careful not to accept every thought and every word literally; instead apply reasoned judgment and act upon perceived intent.

Figure 8.1 is the problem statement for the portfolio manager case study (described in Chapter 7). The examples (they are real) help clarify the requirements, but do not convey additional requirements.

8.3 OBJECT MODEL

As we stated earlier, the object model is the most stable part of a system. The object model defines the universe of discourse for the other two models and provides the lattice to which behavior is attached. Objects are the things that interact and perform computations.

Within the object model the class-association structure is critical, and you should develop this first. Only then should you add inheritance—so that you do not prematurely bias the model. You will discover operations as a byproduct of dynamic and functional modeling.

Object modeling begins with an analysis of the problem statement and has the following steps. (The numbers in brackets refer to sections that explain these steps in detail.)

Develop software for managing investment portfolios. The following capabilities must be provided:

- Permit a mixture of assets and liabilities, including bonds, stocks, options, commodities, mutual funds, precious metals, collectibles, cash, insurance, real estate, and loans. A financial instrument is an asset, liability, or portfolio.
- Support hierarchies of portfolios. Portfolios contain financial instruments.
- Record dates of purchase and sale of financial instruments and their price. Allow for stock splits and issuance of different kinds of securities.

    Example: In 1991 KV Pharmaceuticals issued one share of class A stock for each existing share. Existing shares were designated class B stock. Class A stock has a higher dividend; class B stock is supervoting. (Supervoting stock has additional votes per share.)

    Example: In 1993 the Chile Fund issued one warrant for each share of stock. Seven warrants entitled the holder to purchase two shares of stock at $26 per share. The warrants expired about one month after their issuance. The warrants traded on the New York Stock Exchange as well as the primary stock.
- Track value of assets over time.
- Handle dividends, both stock and cash. Handle interest payments.

    Example: In 1992 the Mexico Fund issued both cash dividends and stock dividends.
- Compute ROI.
- Handle currency conversions with possible fees.
- The developer should look at broker statements, stock exchange listings, existing financial programs, and other sources of information in order to discover relevant classes, attributes, and relationships for this problem.

Figure 8.1 Problem statement for the portfolio manager

- Discover classes by listing tentative classes and then rejecting spurious classes. [8.3.1–8.3.2]
- Discover associations by listing tentative associations, rejecting spurious associations, and refining the remaining associations. [8.3.3–8.3.5]
- Add tentative attributes of objects and links and then eliminate the spurious attributes. [8.3.6–8.3.7]
- Use generalization to note similarities and differences. [8.3.8]
- Test access paths. [8.3.9]
- Iterate and refine the model. Consider shifting the level of abstraction. [8.3.10–8.3.13]
- Organize the object model. [8.3.14]

8.3.1 Listing Tentative Classes

You should list tentative classes according to the following considerations:

- **Nouns and noun phrases**. Look for nouns and noun phrases in the problem statement. Figure 8.2 highlights the first occurrence. (We have omitted the examples shown in Figure 8.1.)

> Develop **software** for managing **investment portfolios**. The following **capabilities** must be provided:
>
> - Permit a **mixture** of **assets** and **liabilities**, including **bonds**, **stocks**, **options**, **commodities**, **mutual funds**, **precious metals**, **collectibles**, **cash**, **insurance**, **real estate**, and **loans**. A **financial instrument** is an asset, liability, or **portfolio**.
> - Support **hierarchies** of portfolios. Portfolios contain financial instruments.
> - Record **dates** of **purchase** and **sale** of financial instruments and their **price**. Allow for **stock splits** and **issuance** of different **kinds** of **securities**.
> - Track **value** of assets over **time**.
> - Handle **dividends**, both stock and cash. Handle **interest payments**.
> - Compute **ROI**.
> - Handle **currency conversions** with possible **fees**.
> - The **developer** should look at **broker statements**, **stock exchange listings**, existing **financial programs**, and other **sources** of **information** in order to discover relevant **classes**, **attributes**, and **relationships** for this **problem**.

Figure 8.2 Problem statement with highlighted nouns and noun phrases

- **Passive voice**. Watch for missing subjects in passive sentences. In the problem statement, the second sentence in the first paragraph has passive voice with an implicit subject of the "software developer." There is no other passive voice because the problem statement is worded as a directive to the software developer. Many problem statements are phrased this way.

- **Implicit classes**. Some classes are implicit or taken from general knowledge.

  The advisory statement at the end of the problem statement "the developer should look at. . ." accentuates some other sources of information. You must consider all available sources of information, as we described in Section 8.1.

  We add the following classes based on general knowledge: *FinancialInstitution* (the holder of portfolios) and *Person* (the owner of a portfolio).

- **Kinds of classes**. Model both physical entities (cash, precious metal) and conceptual entities (financial instrument, portfolio).

- **Reuse**. When possible, reuse class definitions from previous development efforts. We are developing the portfolio manager as an example and have no related software with which to achieve reuse. However, a colleague may develop software for forecasting stock market prices. We would expect to share some class definitions between the portfolio manager and the forecast software.

- **Naming**. Carefully choose class names. Crisp names promote understanding and concurrence as to the precise scope of the software. For example, we decided to represent option with the class name *StockOption* for clarity. Don't be surprised by fervent arguments between proponents of alternative names. You should try to use vocabulary from the application and avoid inventing terms that confuse application experts.
- **Missing classes**. Missing classes eventually become apparent as a gap in access paths or a missing target or argument of an operation.

8.3.2 Eliminating Spurious Classes

After listing the classes, you should review them to eliminate unnecessary and incorrect ones. Use the following criteria. (The last bullet in Figure 8.2 is really just admonishing the developer and has no description of requirements. Thus we disregard it in the following list.)

- **Redundant classes**. Do not keep multiple classes that express the same information. Choose the most descriptive and precise name and systematically use this name. For the portfolio manager, *InvestmentPortfolio*, *Mixture*, and *Portfolio* are redundant; we retain *Portfolio*. We eliminate *Security* and keep *Asset*. We discard *CurrencyConversion* as being redundant with *Purchase* and *Sale*.

  We decide to combine the notions of *Asset* and *Liability*. A liability really is an asset with a negative quantity. Then a positive quantity of stock is owned stock; a negative quantity is borrowed stock. This combination yields a nice symmetry; any asset can be bought or sold, owned or borrowed. Because we are combining *Asset* and *Liability*, we should also combine *Bond* and *Loan*. A bond is money owned; a loan is money borrowed.
- **Irrelevant or vague classes**. Every class should be clearly defined and necessary. This requires some judgment because the relevance of a class is dictated by the purpose of an application. For the portfolio manager, the following tentative classes are vague and should be discarded: *Software*, *Capabilities*, *Issuance*, and *Kinds*.
- **Attributes**. Restate as attributes tentative classes that are used only as values. If the independent existence of a property is important, make it a class and not an attribute. We should model *date*, *price*, and *value* as attributes for the portfolio manager. *Fee* also is an attribute and a consequence of purchases and sales.
- **Operations**. If an operation is manipulated as an entity, it is a class. Classes describe objects that have identity, may form relationships to other objects, are described with attributes, and bear functionality. If an operation fails these tests, eliminate it as a class.

  ROI is an operation to be provided by the portfolio manager software. The computation of return on investment will be complex, involving discounted cash flow to reflect the temporal value of money.
- **Roles**. The name of a class should reflect its intrinsic nature and not a role that it plays in an association. We clarify our portfolio manager model by adding roles. *FinancialInstitution* assumes the role of *accountHolder*. *Person* assumes the role of *accountOwner*. Various roles for persons often arise in applications.

- **Derived classes**. During analysis, avoid derived classes unless they appear in the application description. You should try to defer derived classes until design. The portfolio manager problem statement mentions stock dividends. For now we will treat each stock dividend as a combination of a cash dividend and a purchase.
- **Implementation information**. During analysis, avoid implementation details and concentrate on essential information. All classes must make sense for the application; avoid implementation constructs, such as linked lists and trees.

  We discard *Hierarchy*. Various implementation approaches can all yield the effect of hierarchies of portfolios. The purpose of analysis is to capture the essence of a problem, not a solution approach. We cannot emphasize this point enough. We have encountered programming books that would stress reuse from a *Tree* class in the class library at this point in the problem understanding. This is shortsighted; resist the urge to commit prematurely to an implementation approach until you fully grasp the problem. The key observation during analysis is that the portfolio manager involves recursion and the approach for realizing recursion is not yet decided.

  The problem statement refers to time; the composition and value of a portfolio change over time. During analysis, it is not apparent how we will handle temporal information in the final system. We will postpone the issue of temporal data until the design phase (Chapter 9). Our analysis model will deal only with the current value of portfolios.

We know little about commodities and have arbitrarily decided to eliminate them from further consideration. However, we will make sure the architecture readily extends to additional investment types, such as commodities. Such a revision to the problem statement is common. In most cases, the problem statement is not an absolute statement of requirements. Requestors can make mistakes and may reconsider.

The following valid classes remain at this stage in our analysis: *FinancialInstitution*, *Person*, *Asset*, *Bond*, *Stock*, *StockOption*, *MutualFund*, *PreciousMetal*, *Collectible*, *Cash*, *Insurance*, *RealEstate*, *FinancialInstrument*, *Portfolio*, *Purchase*, *Sale*, *Dividend*, and *InterestPayment*. We have observed proper OMT protocol for class names, restating plural nouns in the singular form.

For now we retain an additional class that we will revisit later in analysis. The notion of a *StockSplit* is relevant, but we must broaden the concept to accommodate the examples in Figure 8.1. (We also defer consideration of time until the system design phase.)

8.3.3 Listing Tentative Associations

You should list tentative associations found using the following guidelines:

- **Verb and prepositional phrases**. Look for verb phrases and prepositional phrases in the problem statement. These phrases follow the pattern *noun verb noun* or *noun preposition noun*. Figure 8.3 highlights candidate phrases.

  Note that although it has a noun verb noun construction, *a financial instrument is an asset, liability, or portfolio* is not a candidate association. Rather it is an indicator of generalization (an *is-a* phrase; see Section 8.3.8).

Develop **software for managing investment portfolios**. The following capabilities must be provided:

- Permit a **mixture of assets and liabilities**, including bonds, stocks, options, commodities, mutual funds, precious metals, collectibles, cash, insurance, real estate, and loans. A financial instrument is an asset, liability, or portfolio.
- Support **hierarchies of portfolios**. **Portfolios contain financial instruments**.
- Record dates of **purchase and sale of financial instruments** and their price. Allow for stock splits and **issuance of different kinds of securities**.
- Track **value of assets** over time.
- Handle dividends, both stock and cash. Handle interest payments.
- Compute ROI.
- Handle **currency conversions with possible fees**.

Figure 8.3 Problem statement with highlighted verb and prepositional phrases

- **Dependencies**. A dependency between two or more classes may be an association. The problem statement for the portfolio manager expresses dependencies rather vaguely. For example, the system must record *purchase and sale of financial instruments*. Many problems have more explicit statements of dependency, such as *person works for company* and *flights arrive and depart from airports*. Regardless, note any dependencies you can glean from the problem statement, because they represent possible associations.
- **Implicit associations**. Some associations are implicit. For example, *a financial institution holds portfolios* and *persons own portfolios*.
- **General knowledge**. Other associations are taken from general knowledge. For example, *a bond has interest payments*, *a stock has dividends*, and *a stock option has an underlying stock*.
- **Kinds of associations**. Model various kinds of associations, including physical location (*next to*, *contained in*), directed actions (*drives*), communication (*talks to*), ownership (*has*, *part of*), or satisfaction of some condition (*works for*, *married to*, *manages*). For example, *portfolios contain financial instruments*.
- **Avoid pointers**. Do not bury object references as pointers or pseudo attributes. There are different ways to implement associations, but these are considerations for design, not analysis. For example, *Stock* is not an attribute of *Dividend*; it is a class with which *Dividend* is associated.

Figure 8.4 summarizes possible associations for further consideration.

8.3.4 Eliminating Spurious Associations

The next task is to eliminate unnecessary and incorrect associations using the following criteria:

Verbs and prepositional phrases:
- software for managing investment portfolios
- mixture of assets and liabilities
- hierarchy of portfolios
- portfolios contain financial instruments
- purchase and sale of financial instruments
- issuance of different kinds of securities
- value of assets
- currency conversions with possible fees

Implicit verb phrases:
- a financial institution holds portfolios
- persons own portfolios

Knowledge of problem:
- a bond has interest payments
- a stock has dividends
- a stock option has an underlying stock

Figure 8.4 Tentative associations for the portfolio manager

- **Associations between eliminated classes**. If you have eliminated one of the classes in an association, you must eliminate the association or restate it in terms of other classes. For the portfolio manager we can eliminate the following candidate associations: *software for managing investment portfolios*, *hierarchy of portfolios*, *value of assets*, and *currency conversions with possible fees*.
- **Irrelevant or implementation associations**. Eliminate any associations that are outside the problem domain or deal with implementation constructs. For example, we do not want to be induced to make premature implementation decisions by the phrase *hierarchy of portfolios*.
- **Actions**. An association should describe a structural property of the application, not a transient event. The portfolio manager has no actions that are mislabeled as associations. We retain the clause *purchase and sale of financial instruments*. In one sense purchase and sale are events that occur, but we need to record descriptions of these events. We must track date of purchase or sale, price, and profit or loss.
- **Ternary associations**. You can decompose most associations between three or more classes into binary associations or phrase them as qualified associations. We can restate the phrase *mixture of assets and liabilities* as *mixture of assets* (because we have discarded liabilities as a class). This is a rather trivial decomposition. You will often encounter phrases that follow the pattern *noun verb noun preposition noun* (such as *person works-for division of company*). Decompose them into their binary constituents.

- **Derived associations**. Omit derived associations or explicitly mark them. None of the tentative associations for the portfolio manager appear to be derived. Later in our analysis we will encounter derived information. For example, portfolios have values and profits or losses that are systematically computed from assets and lesser portfolios.

We again defer consideration of complex situations, such as stock splits. Thus we postpone analysis of the following clause: *issuance of different kinds of securities*. Figure 8.5 shows the associations remaining at this stage of analysis. (We have substituted preferred class names for synonyms.)

- portfolio of assets
- portfolios contain financial instruments
- purchase of financial instruments
- sale of financial instruments
- a financial institution holds portfolios
- persons own portfolios
- a bond has interest payments
- a stock has dividends
- a stock option has an underlying stock

Figure 8.5 Remaining associations for the portfolio manager

8.3.5 Refining Associations

The last activity in discovering associations is to refine those that remain. You should further specify the semantics of associations using the following criteria:

- **Premature implementation**. Be careful to represent the logical intent, not an implementation approach. We have seen some developers of relational database applications become confused by associations. These developers try to express their thoughts directly with tables and bury associations as pointers. In reality, there are many ways to implement associations. By thinking in terms of tables, these developers violate the precepts of software engineering—limiting their conceptual ability, obscuring communication with others, and prematurely making detailed decisions before a broad understanding is in place.
- **Misnamed associations**. Choose names carefully. Don't say how or why a situation came about, say what it is. Names are important to understanding, and you should choose them with great care. You may omit association names when there is no ambiguity. Restate the awkward phrase *portfolio of assets* as *portfolio contains assets*.
- **Role names**. Add role names that appear in the problem statement. Also consider adding role names to distinguish multiple associations between classes. Role names are especially important for reflexive associations (see Section 2.2.3). In the association be-

tween *Stock* and *StockOption*, we can refer to *Stock* with the role of *underlyingSecurity*. Each person who owns a portfolio assumes the role of *accountOwner*; financial institutions assume the role of *accountHolder*.

- **Qualifiers**. Use qualifiers where possible. Qualifiers often arise from the need to specify a name within some context; most names are not globally unique. You will occasionally encounter qualification cascades (see Section 3.2.5), in which an accumulation of names denotes an object. Qualifiers are not apparent at this step of the portfolio manager analysis. They will appear later.

- **Multiplicity**. Specify multiplicity for the associations in your model, but don't dwell on it. The multiplicity often changes as your model evolves. You should always challenge multiplicity values of "exactly one." Usually the multiplicity is "many" (zero or more) or "zero or one." Similarly, you should challenge multiplicity values of "many." Sometimes you can restate a multiplicity of "many" with a qualifier or with a designation of *ordered*.

  We make the following multiplicity adjustments based on our understanding of the problem. We allow assets to belong to multiple portfolios. A portfolio may have multiple owners (such as a husband and wife) and need not be held by an institution (maybe the portfolio is hypothetical and exists only to facilitate ROI calculations). A stock may have many stock options.

- **Association classes**. You may have a class that would be better stated as an association class. An association class can participate in associations, just like an ordinary class. The difference is that an association class derives identity from its constituent classes, while a class has intrinsic identity (see Section 2.2.5). The portfolio manager does not have any association classes.

- **Missing associations**. Add missing associations. You can often discover these by testing access paths in your object model, especially during construction of the functional model.

- **Aggregations**. Sometimes it is difficult to distinguish aggregation from association. Do not worry if this occurs; for these cases the distinction is probably unimportant. When in doubt, use ordinary association. For bill-of-material problems, you should carefully consider whether to use physical aggregation or catalog aggregation (see Section 3.3.1).

  The word *contain* in the association *portfolios contain financial instruments* is highly suggestive of aggregation. Similarly, note the phrase *portfolio contains assets* (see the "Misnamed associations" bullet).

Figure 8.6 presents our initial class diagram for the portfolio manager.

8.3.6 Listing Tentative Attributes of Objects and Links

Once you have established the basic structure of your object model, you should elaborate it by describing the classes and associations. List tentative attributes according to the following criteria:

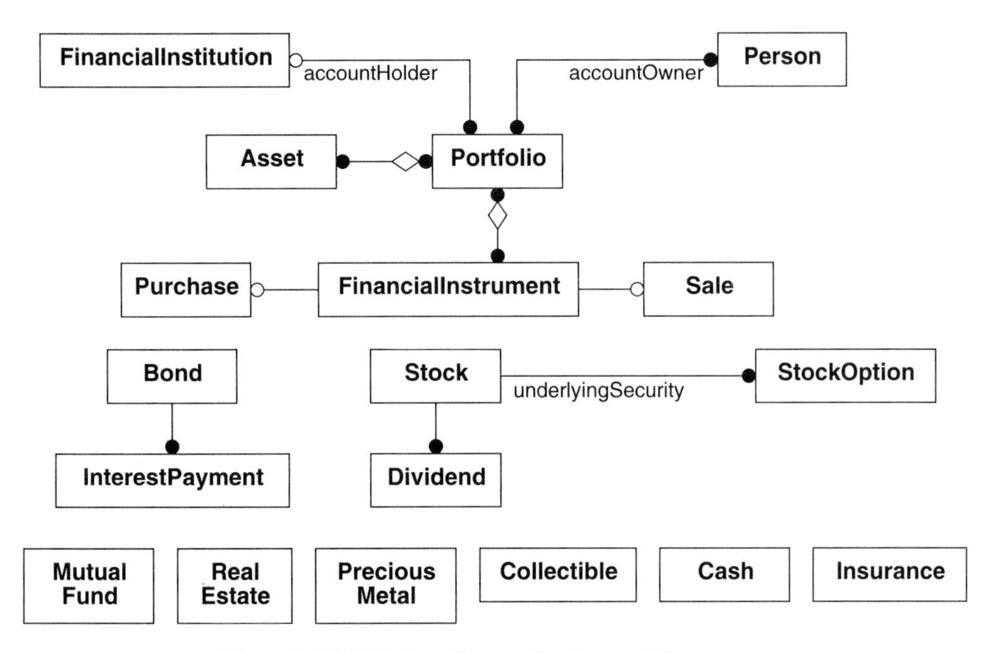

Figure 8.6 Initial class diagram for the portfolio manager

- **Possessive phrases and enumerations.** Look for nouns followed by possessive phrases in the problem statement, such as *the population of the city* or *the password of the user*. Adjectives often represent specific enumerated values, such as *red*, *green*, or *blue*. Figure 8.7 highlights candidate phrases in the problem statement.

> Develop software for managing investment portfolios. The following capabilities must be provided:
>
> - Permit a mixture of assets and liabilities, including bonds, stocks, options, commodities, mutual funds, precious metals, collectibles, cash, insurance, real estate, and loans. A financial instrument is an asset, liability, or portfolio.
> - Support hierarchies of portfolios. Portfolios contain financial instruments.
> - Record **dates of purchase and sale** of financial instruments and **their price**. Allow for stock splits and issuance of different kinds of securities.
> - Track **value of assets** over time.
> - Handle dividends, both stock and cash. Handle interest payments.
> - Compute ROI.
> - Handle currency conversions with possible fees.

Figure 8.7 Problem statement with highlighted attribute phrases

- **Missing attributes**. Unlike classes and associations, attributes are less likely to be described fully in the problem statement. You must draw on your knowledge of the application and expert advice to find them. Fortunately, attributes seldom affect the basic structure of a problem. For the portfolio manager, we know that there is preferred stock and common stock, so we add the *isPreferred* attribute to *Stock*.
- **Limit discovery of attributes**. Do not carry discovery of attributes to excess. Consider only attributes that directly relate to the application being developed.
- **Avoid pointers**. Attributes should not refer to objects; use associations to describe dependencies between objects.

8.3.7 Eliminating Spurious Attributes

You should eliminate unnecessary and incorrect attributes using the following criteria:

- **Objects**. If the independent existence of an entity is important, rather than just its value, it is an object. Scrutinize the classes and tentative attributes to make sure you have not mislabeled any attributes or classes.
- **Qualifiers**. If the value of an attribute depends on a particular context, consider restating the attribute as a qualifier. Names are often better modeled as qualifiers rather than object attributes. Qualifiers are not yet apparent for the portfolio manager problem.
- **Identifiers**. Object-oriented languages and object-oriented databases incorporate the notion of an object identifier for unambiguously referencing an object. Do not list these object identifiers, because they are implicit in object models. List only attributes that exist in the application.
- **Link attributes**. If a property depends on the presence of a link, it is an attribute of the link, not of a related object. Link attributes are obvious for many-to-many associations; you cannot ascribe them to either class because of the multiplicity. Link attributes also arise for one-to-many and one-to-one associations. Assign each attribute to the class or association it describes most directly and naturally.

  For the portfolio manager *quantity* is a link attribute of the association between *Asset* and *Portfolio*. We could also allow for unequal ownership of portfolios. The association between *Person* and *Portfolio* would then have the link attribute *fractionalOwnership*. We will not add this complication.
- **Fine detail**. Omit minor attributes that are not important for the application.
- **Derived attributes**. Omit or clearly label derived attributes.
- **Discordant attributes**. An attribute that seems completely different from and unrelated to all other attributes may indicate a class that should be split into two distinct classes. A class should be simple and coherent. If you encounter many discordant attributes, your analysis is likely flawed. By its very nature an object-oriented approach is centered about things and descriptions of things, so awkward combinations should seldom arise.

- **Boolean attributes**. Reconsider all boolean attributes. Often (but not always) you can broaden a boolean attribute and restate it as an enumeration [Coad-95].

  For the portfolio manager we will restate the *isPreferred* attribute of *Stock* more broadly as *stockType* with the values *common*, *preferred*, *closed-end*, and *other*.

After applying these criteria to the portfolio manager model, we get the diagram in Figure 8.8. We continue to defer the handling of stock splits. For simplicity, we assume that stocks have a single ticker symbol used on all stock exchanges. Attribute *portfolioType* characterizes portfolios for tax purposes; portfolio types of *ordinary*, *IRA*, and *401K* would be appropriate for the United States.

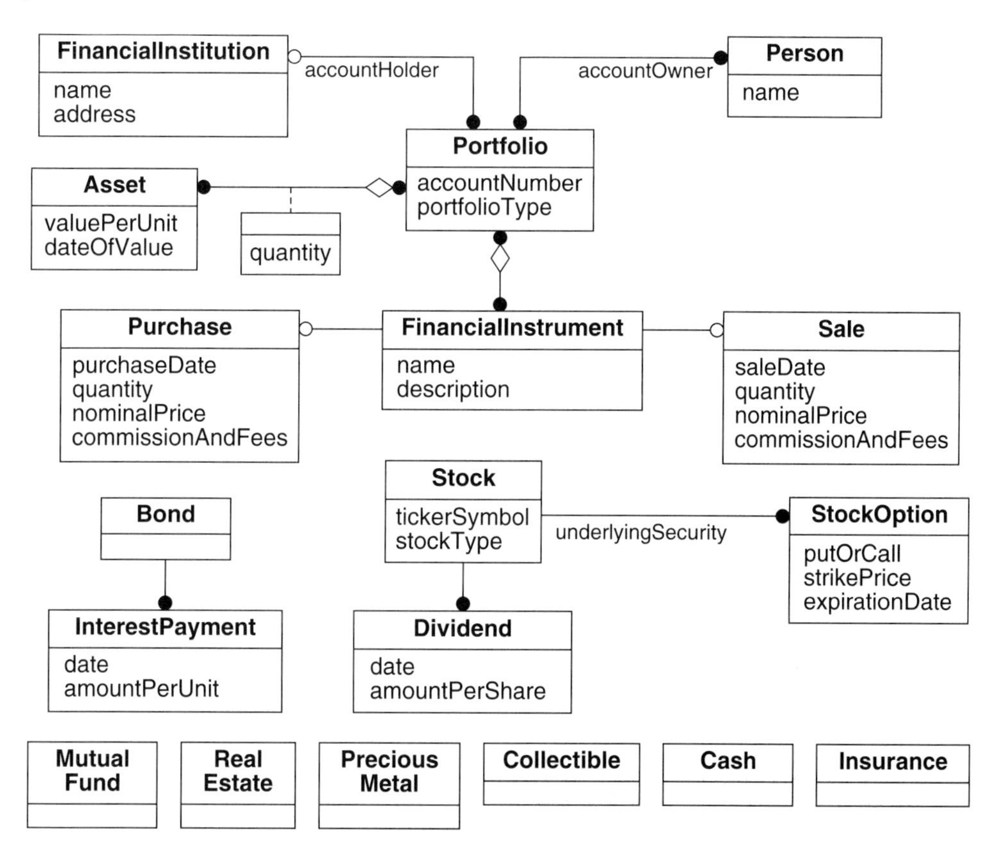

Figure 8.8 Class diagram with attributes for the portfolio manager

8.3.8 Using Generalization to Note Similarities and Differences

The next step in object modeling is to organize classes using inheritance. You can add inheritance in one of two directions: top down or bottom up.

- **Top-down generalization**. Specialize existing classes into subclasses. Sometimes there is an obvious need to partition. Look for lists of special subcases in the problem statement. Also look for noun phrases composed of various adjectives on a class name. Often you can think of additional special cases. Unify any redundant attributes that now become apparent. You should avoid creating unnecessary generalizations for enumeration attributes. Specialize a class only when the subclasses have distinct attributes, associations, methods, and/or semantic behavior.

  The first bullet of the portfolio manager problem statement (see Figure 8.1) lists several forms of assets (the "including" clause). If these examples were of minor importance and were not described by attributes, associations, or operations, they would merely be enumeration values. However, different kinds of investments have distinctive behavior. Thus we will treat *Bond*, *Stock*, *StockOption*, *MutualFund*, *PreciousMetal*, *Collectible*, *Cash*, *Insurance*, and *RealEstate* as subclasses of *Asset*. (Recall that in Section 8.3.2 we discarded *Commodity* as a minor refinement of the problem statement. We also unified the notions of *Bond* and *Loan*.)

  The second sentence in the first bullet states that *a financial instrument is an asset, liability, or portfolio*. The *is-a* phrase is an obvious indicator of generalization. (Once again there are only two subclasses because we have combined *Asset* and *Liability*.)

- **Bottom-up generalization**. Find classes with similar attributes, associations, or operations. Create an abstract class to share common information. You may have to alter the definitions a bit; this is acceptable as long as generalization semantically applies. Symmetry will suggest missing classes.

  In Figure 8.8 a bond may have many interest payments. However, a mutual fund comprised of bonds also may pay interest. Furthermore, both a stock and a mutual fund of stocks may pay dividends. We decide to split *MutualFund* into *BondFund* and *StockFund*. Then we generalize *Bond* and *BondFund* into *BondAsset*, which pays interest. We also generalize *Stock* and *StockFund* into *StockAsset*, which pays dividends.

  In principle, a fund could combine bonds and stocks. However, most funds are more focused to suit investor taste. We could always add mixed funds to the portfolio manager in the future if needed. This process of defining the precise boundary for a system often occurs in practice. The software requestor and the developer must decide what is within the scope of a system and what is not. A major virtue of object-oriented modeling is that the scope is made clearly visible. You can readily extend a carefully modeled and architected system.

  We do not show the composition of a stock fund in terms of its underlying stocks. Similarly, we do not show the composition of a bond fund. You should only model information that is important to an application. Stock funds and bond funds are assets and therefore have a direct market value. We do not need the internal structure of stock funds and bond funds for computing portfolio value and ROI.

  At this point we notice that *Stock* and *StockFund* have the same attributes, associations, methods, and behavior. Hence there is no need to show the subclasses; we can merely show *StockAsset* in the object model. Similarly, the distinction between *Bond* and *BondFund* is inconsequential and we can simply show *BondAsset*.

We could also choose to regard stock option as a stock asset, and we considered this at first. However, stock options do not pay dividends, and we decided that the similarity between stock and stock fund was much more significant than the commonality with stock option. We avoid this further generalization.

- **Adjust placement of associations**. As you introduce inheritance, you may need to adjust the targets of some associations. Move associations up or down the inheritance hierarchy as semantically appropriate. Symmetry may suggest additional attributes to distinguish among subclasses more clearly. You may need to decompose large classes so you can find similar components.

  We define an asset as something that has value and can be *directly* bought or sold. Thus a portfolio is not an asset; a portfolio of financial instruments cannot be directly bought and sold. Therefore, *Purchase* and *Sale* do not relate to a *FinancialInstrument* but to an *Asset*. Also, we shift the aggregation from *Asset* and *Portfolio* to *FinancialInstrument* and *Portfolio*; a portfolio contains financial instruments that can be assets or lesser portfolios.

- **Multiple inheritance**. You may use multiple inheritance to increase sharing, but only if necessary, because it increases conceptual complexity. Use of multiple inheritance is more appropriate during design, especially if you mixin functionality from a class library. (See Section 3.5.2.)

Figure 8.9 shows the portfolio manager object model after adding inheritance.

8.3.9 Testing Access Paths

You cannot fully verify access paths until you have constructed the functional model. Nevertheless, it is worthwhile to peruse the object model quickly for obvious omissions. Later, during construction of the functional model, we will check more thoroughly.

Trace access paths through the object model diagram to see if they yield sensible results. Where a unique value is expected, is there a path yielding a unique result? For multiplicity "many," is there a way to pick out unique values when needed? You may need to introduce some qualifiers. Are there useful questions you cannot answer?

The *accountNumber* really is a qualifier on the association between *Portfolio* and *FinancialInstitution*. The financial institution assigns the account number; an account number is meaningful only within the context of a financial institution. Account numbers uniquely identify a portfolio for a financial institution. *InterestPayment* and *Dividend* are qualified by *date*.

It is perfectly acceptable to have classes that are "disconnected" from other classes. This usually occurs when the relationship between a disconnected class and the remainder of the model is diffuse. The portfolio manager as presented here has no such disconnected classes. However, a prior draft had an *InflationIndex* class for adjusting ROI calculations for inflation. *InflationIndex* was stand-alone and had no relationships to other classes. (We cut the mention of inflation to simplify the case study.)

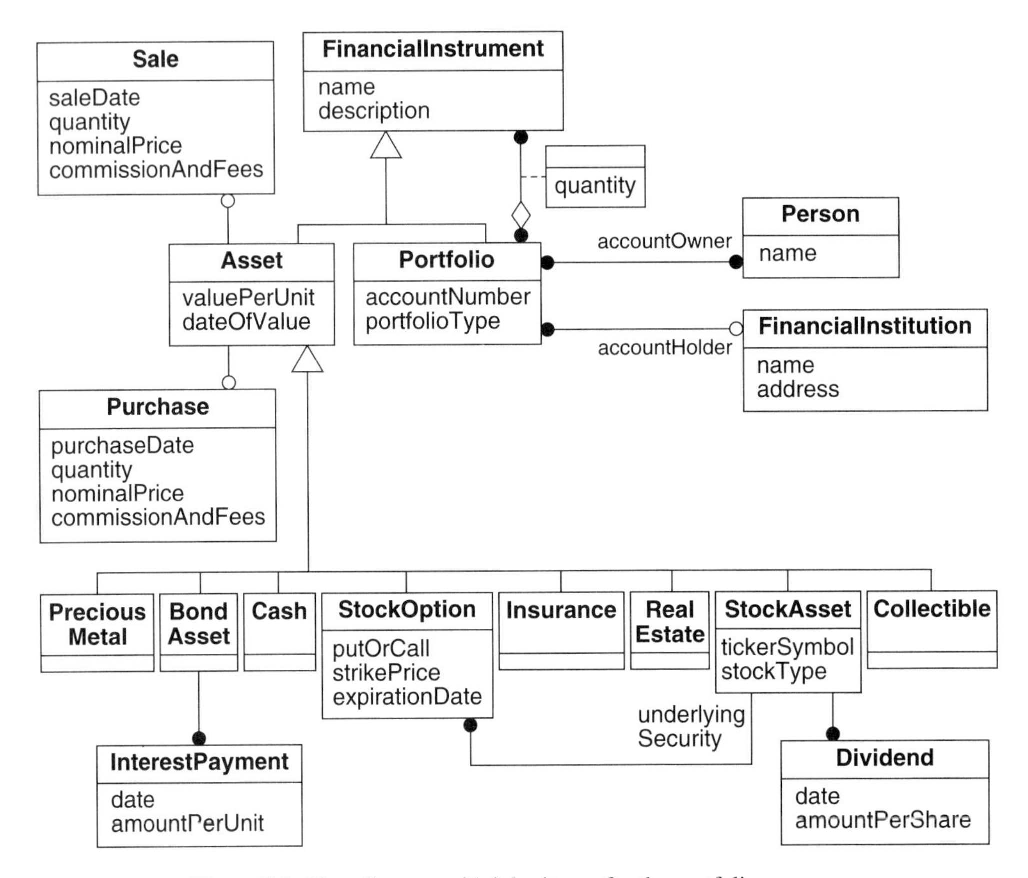

Figure 8.9 Class diagram with inheritance for the portfolio manager

8.3.10 Iterating and Refining the Model

A model is rarely correct after a single pass; our models are seldom correct on the first try. Most object models take several passes, interleaved with refinement of the dynamic and functional models. Actively seek the advice and criticism of others. You can improve a model by reconciling different perspectives.

We have found that revision often improves the ability of a model structurally to enforce application constraints. Qualifiers and association classes are especially important because they let you specify identity and navigation paths precisely. However, do not carry the enforcement of constraints to excess and distort your model. Elegance is also an important goal.

Look for patterns (see Section 4.3) as you refine your object model. You should be alert to patterns for several reasons. First, if you recognize a pattern, you can ensure that you use it correctly and fully understand the implications. If you draw your models so that a pattern can be easily recognized, other persons will better understand your models. In Figure 8.9 the relationship between *FinancialInstrument*, *Asset*, and *Portfolio* instantiates the complex pattern for a directed graph (Figure 4.9).

You may wish to rework your object model to eliminate multiple inheritance. Multiple inheritance may not be worth the nuisance of explaining to your reviewers. Substituting a workaround for multiple inheritance may simplify the explanation of your model. In general, we are reluctant to use multiple inheritance in our analysis models; we are much more likely to introduce multiple inheritance during design for mixins, such as inheriting persistence from the class library of an object-oriented DBMS.

As you refine your object model, it is worthwhile to revisit names. Reviewers can often suggest better names with crisper meaning. Be wary of names that seem fine but have deleterious connotations within an organization. Reviewers will become confused by such names and draw erroneous conclusions, so you must revise them.

Next we address the requirements we have postponed since Section 8.3.2, building on the nucleus of our object model.

8.3.11 Shifting the Level of Abstraction

So far in the OMT analysis process, we have taken the problem statement quite literally. We have regarded nouns and verbs in the problem description as direct analogs of classes and associations. This is a good way to begin analysis, but it does not always suffice. Sometimes you must shift the level of abstraction to solve a problem. On occasion it is helpful to introduce more abstraction with metamodels, generic classes, and reification. (See Chapter 4.)

For example, we have developed several bill-of-material applications. (See Section 3.3.) For several reasons, it is often difficult to model precisely the parts explosion. Our understanding of the problem domain is incomplete. Furthermore, different organizations have different opinions about the appropriate decomposition. The manufacturing staff believe a bill-of-material should be organized by the sequence of manufacturing steps. In contrast, engineers prefer to organize a bill-of-material on the basis of functionality. And there are additional viewpoints. We have found that the preferred organization varies from company to company and can change over time.

For some applications our solution has been to elevate the abstraction of the model and shift toward a generic model of parts (Figure 3.17) rather than a tangible model that directly corresponds to the application (Figure 3.18). Then our bill-of-material software straddles the argument about how best to decompose a bill-of-material. At run-time our software stores both the part definition and part instantiation, and we can change the structure of the bill-of-material without rewriting our software.

The bill-of-material example highlights another important philosophical point. Developers should try to decouple software from business decisions. Software should enable strategic business decisions, not dictate business decisions. Business leaders should be able to make decisions without inhibition from software systems. For our bill-of-material applications we have straddled the business decision of how best to organize the bill-of-material. We wrote software in a manner that is neutral with regard to the eventual decision, and it can tolerate various and possibly changing decisions.

Our first revision of the object model for the portfolio manager is motivated by a shift in abstraction.

8.3.12 First Revision: Introduce the Notion of a Transaction

We need to represent more abstractly the notion of purchase and sale. *Purchase* and *Sale* have similar attributes and differ only with regard to which party receives the proceeds. We introduce the notion of a transaction that may be a purchase or a sale of an asset. We must specify the portfolio that is the source of the assets that are sold or the portfolio that is the recipient of the assets that are purchased. Since *Purchase* and *Sale* have the same attributes, associations, and behavior, we need not represent them as subclasses and can instead add an enumeration attribute to *Transaction* indicating purchase or sale.

This revision corrects a flaw in our earlier model (Figure 8.9). Previously we indicated that each asset has at most one sale and at most one purchase. Then it becomes awkward to represent the situation in which someone purchases 500 shares of a stock and then sells 200 shares on one date and 300 shares on another date. More abstractly representing an asset as having many transactions resolves this problem.

Next we broaden our definition of a transaction. Each transaction may have a type of *purchase*, *sale*, *deposit*, *withdrawal*, *barter*, or *journal*. A *purchase* occurs when a person gives cash and receives an asset. A *sale* occurs when a person gives an asset and receives cash. A *deposit* is the simple addition of an asset to a portfolio. A *withdrawal* is the simple removal of an asset from a portfolio. *Barter* accommodates complex situations, such as stock splits and the issuance of different kinds of securities. A *journal* entry occurs when there is a discrepancy in the portfolio manager records that we (as users) wish to adjust manually. In reality, we can deduce *transactionTypes* other than *journal* from the links to *Asset*; however, we retain *transactionType* for convenience.

The *Transaction* class reifies trading actions that occur against a portfolio. We reify transactions so that we can record past financial activity. We must store both *transactionDate* (when the transaction occurs) and *recordDate* (when the transaction is recorded). The notion of a transaction is a general concept that arises with many other kinds of applications. This increases our confidence that the abstraction is correct.

A consequence of recording transactions is that the assets in a portfolio are derived. A person must define a portfolio, supplying a name, description, and portfolio type. A person must also specify which portfolios are contained in other portfolios. However, software can reprocess the log of transactions to determine the assets in a portfolio. The transaction is the *only* mechanism for changing the assets in a portfolio. We must now replace the association between *FinancialInstrument* and *Portfolio* by two associations—from *Portfolio* to *Asset* and *Portfolio* to *Portfolio*. Figure 8.10 illustrates that the association from *Portfolio* to *Asset* is derived; the association from *Portfolio* to *Portfolio* is not derived. We simplify the model by restricting portfolio composition to a collection of trees; a portfolio has at most one parent.

We can treat currency exchanges (such as exchanging U.S. dollars for Deutschemarks) as *purchase* or *sale* transactions involving two cash assets. For example, we can treat a sale of Deutschemarks as a transaction in which Deutschemarks are the asset (cash asset) that is given and U.S. dollars are received. We treat expired options as a *sale* with receipt of zero cash. We model exercised options as a *barter*.

In Figure 8.10 we define *quantity* as positive if an asset is received in a transaction and negative if an asset is given. The *Asset* and *Portfolio* classes appear twice in the diagram for ease of layout. Our normal convention is to show the attributes (and possibly operations) for a class in one location where the class is "defined" and show the class icon (attributes and operations suppressed) for all other references to the class.

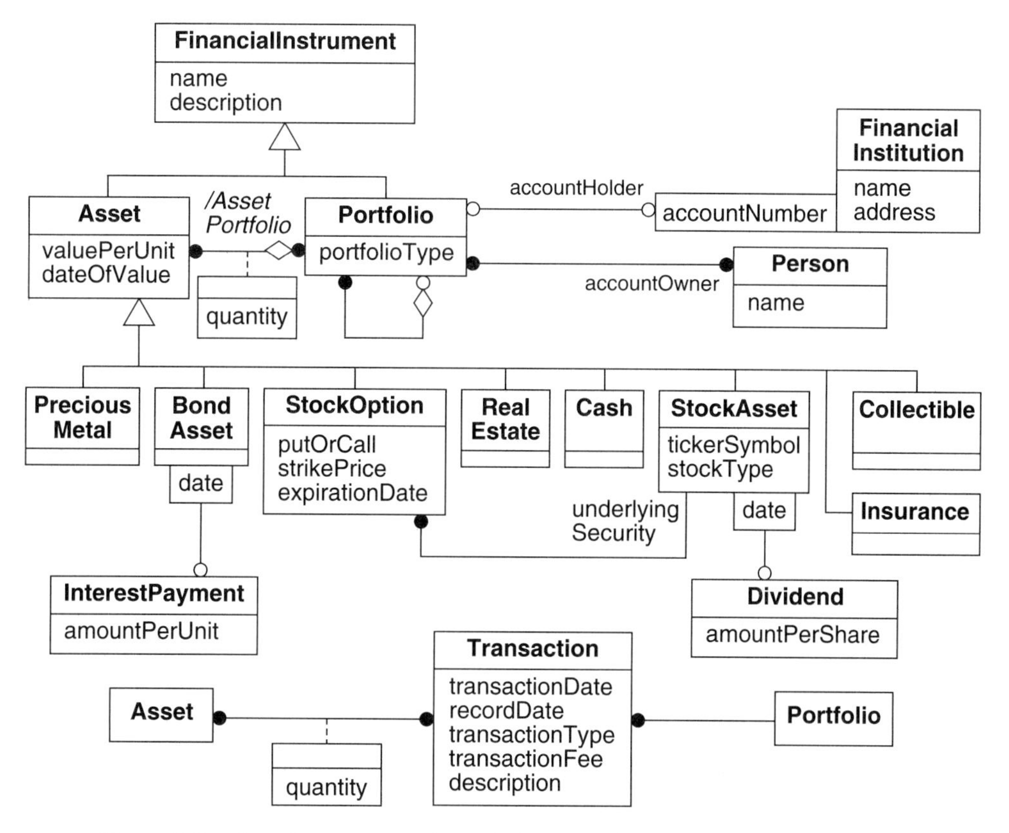

Figure 8.10 Class diagram revised to include transactions for the portfolio manager

8.3.13 Second Iteration: Analyze Existing Forms and Refine Transaction

After inspecting some financial statements, we found an oversight in the object model. Many bonds have a single return of principal; a bond fund may have periodic partial returns of principal. Consequently, we would add *ReturnOfPrincipal* to the object model.

The inspection also prompted us to extend our definition of a transaction to include *dividend*, *interest*, and *return of principal*. This extension eliminates several classes and lets us directly handle stock dividends. We no longer need to treat a stock dividend as a cash dividend and a purchase (see the "Derived class" bullet in Section 8.3.2). The change clarifies the processing of dividends, interest, and return of principal; we intend that a transaction be the only means of changing the assets in a portfolio. We add an association to the *Transaction* class denoting the *assetReferenced*. For a dividend, the *assetReferenced* is a stock. For interest and return of principal, the *assetReferenced* is a bond. For all other transaction types, *assetReferenced* is null.

Figure 8.11 shows the final result of object modeling.

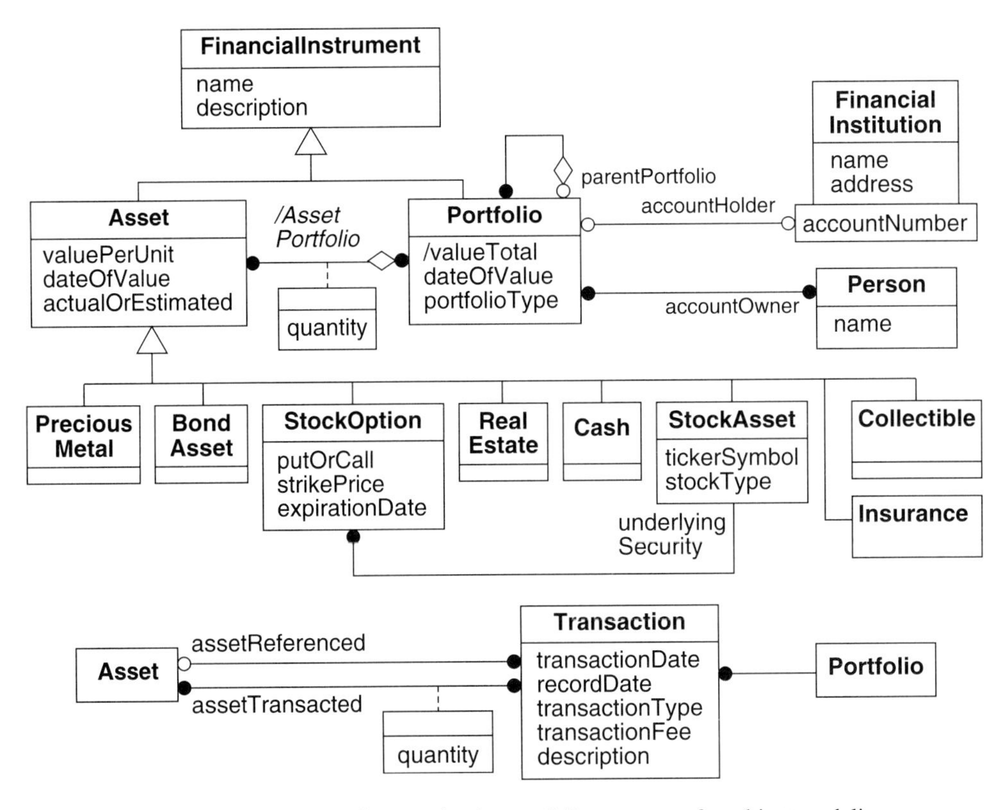

Figure 8.11 Final class diagram for the portfolio manager after object modeling

In practice, model building is not as rigidly ordered as we have shown. As you become proficient, you can perform the modeling steps in an alternative order. For example, for an experienced modeler the generalization of precious metals, bonds, cash, stocks, and so forth into an asset is apparent from the initial reading of the problem statement. Also, the generalization of *Asset* and *Portfolio* into *FinancialInstrument* is readily noted. You may infuse such obvious top-down generalization early in the analysis process. However, you should defer the search for similarities and differences to achieve reuse (bottom-up generalization) until you have determined associations and attributes. This will help you avoid choosing a basis for generalization prematurely, which can lead to an inferior model.

8.3.14 Organizing the Object Model

The last step in object modeling is to organize your model. Flat class diagrams suffice for small and medium-sized problems. However, it is difficult to understand large applications (applications with more than five pages of class diagrams) all at once, and you should organize them into packages. (See Section 3.6.) Large applications may require several tiers as you decompose packages; this cascade of packages forms a hierarchy.

Each package should encompass a logical group of classes (with corresponding associations and generalizations). Each package should have a coherent theme; look for dominant classes, dominant relationships, and major aspects of functionality. A class or association class may appear in multiple packages, but ordinarily we show each association and generalization only once. By convention, we show the attributes (and possibly operations) for a class where the class is defined and show the class icon (attributes and operations suppressed) for all other references to a class.

The portfolio manager, as described, is too small to require packages. We have had to simplify the problem for presentation. A more complete portfolio manager could have a package for each type of asset. For example, a stock package might include information for forecasting future stock market prices. A real estate package might track depreciation, capital improvements, rental income, taxes, and cash flow.

It is also worth devoting some effort to the aesthetics of your diagrams. Place important concepts so they are immediately noticed. For the portfolio manager, we made *FinancialInstrument*, *Asset*, *Portfolio*, and *Transaction* visually prominent. The layout of your diagrams should elicit symmetry. For example, we try to place subclasses at the same level in the portfolio manager diagram. Minimize crossings and bends for association and generalization lines. Try to place the generalization superclass above its subclasses.

8.4 THE DATA DICTIONARY

The data dictionary defines all modeling entities (classes, associations, attributes, operations, domains, and enumeration values) and explains the rationale for key modeling decisions. The data dictionary highlights important assumptions and restrictions. You should not define by example, but it is helpful to give examples of the definition. You should also avoid circular definitions.

As you prepare a data dictionary, it is a good time to check for adherence to naming standards. Naming standards can help coordinate the work of multiple persons and projects. You may want to use a prefix for classes and associations indicating their primary package. Sometimes we make attributes globally unique by giving them a prefix of their class or association name. Also, you may want to standardize suffixes for different kinds of data. For example, in the portfolio manager, we incorporate the phrase *date* in the names of attributes that are dates. All enumerations incorporate the phrase *type* or are named by the enumeration values separated by *or*.

You should also enforce conventions for abbreviations. Otherwise you will have trouble searching the data dictionary and your models. For example, you may require that the phrase *description* either be fully spelled or abbreviated as *desc*.

We now present definitions for the portfolio manager, organized about classes.

- *Asset*—something that has value and can be *directly* bought or sold. An asset with a negative quantity in a portfolio is a liability. Assets vary in their liquidity, that is the period of time typically required to achieve a purchase or sale. Assets may belong to multiple portfolios. We record the value of each asset, the date of valuation, and all transactions by which an asset is acquired or divested. Asset values may be actual or estimated. The portfolio manager supports the following types of assets: precious metal, bond, stock, stock option, real estate, cash, collectible, and insurance.
- *BondAsset*—a right to receive periodic interest payments and one or more returns of principal. The value of the principal is fixed on the redemption date(s), unlike stocks, whose value varies according to corporate fortunes and the economy. Companies must pay interest on bonds before issuing dividends on stock. Thus corporate bond holders receive a more secure income stream in return for lesser participation in corporate fortunes. The term "bond" encompasses annuities, municipal bonds, federal bonds, corporate bonds, zero coupon bonds, certificates of deposit, and savings bonds. A loan is a bond that is held short (money owed).
- *Cash*—currency or coinage, such as U.S. dollars, Deutschemarks, or Japanese yen. The name of the financial instrument (via inheritance) denotes the country of issue.
- *Collectible*—an item of artistic or aesthetic merit, such as paintings, coins, stamps, and antiques. Many collectibles have value that can be measured with purchases and sales in the marketplace.
- *FinancialInstitution*—a business that holds portfolio accounts. Banks, insurance companies, and stock brokerages are examples of financial institutions.
- *FinancialInstrument*—an asset or a portfolio.
- *Insurance*—a guarantee of payment for a specified situation (usually a crisis or disaster) in return for periodic payment of premiums. For the portfolio manager, we treat insurance as a financial instrument only if it has a surrender value. For example, life insurance is a financial instrument if the insured pays more than the current cost of benefits and accrues a residual value.
- *Person*—the owner of portfolios. We exclude other kinds of persons, such as stock brokers and fund managers, who are peripheral to the portfolio manager.
- *Portfolio*—a collection of financial instruments that may evolve over time through transactions. A portfolio consists of financial instruments which can be assets and lesser portfolios. A portfolio may be owned by multiple persons and may be held by some financial institution. We permit various portfolio types (such as normal, 401K, and IRA). We compute ROI and record the value of portfolios.

- *PreciousMetal*—a metal that trades at a great premium compared with base metals. For example, gold, silver, and platinum are commonly regarded as precious metals, while copper, lead, and aluminum are considered base metals. Precious metals derive their high values from high demand relative to supply. Precious metals complement a portfolio because they are liquid and resistant to inflated supply (unlike currency, which governments can print at whim). Precious metals have the drawback of not generating cash flow.

- *RealEstate*—land and/or buildings. Real estate incurs taxes and investment for capital improvement and returns rent or the benefits of personal use. This case study superficially considers the impact of a home on net worth. We would need to enhance the portfolio manager to handle investment property.

- *StockAsset*—an ownership stake in a company. Holders of stock benefit from possible dividends and anticipated increases in corporate net worth reflected in higher future stock prices. For most kinds of stock, the maximum possible loss is limited to the amount of investment. Stocks trade on various exchanges throughout the world that establish current value. For simplicity, we assume that stocks have a single ticker symbol across all exchanges. There are different types of stock (such as common, preferred, closed-end stock funds, and special classes) with various combinations of advantages and restrictions. A stock may have corresponding stock options.

- *StockOption*—a call or put option. A call (put) option is a right to purchase (sell) a stock on or before an expiration date at a strike price. The holder of a stock option has the choice of exercising the option or waiting; this choice ends on the expiration date. The seller of a stock option is subject to the whims of the option holders.

- *Transaction*—a trade of one collection of assets for another collection of assets. Each transaction may have a type of *purchase*, *sale*, *deposit*, *withdrawal*, *barter*, *journal*, *dividend*, *interest*, and *return of principal*. A *purchase* occurs when a person gives cash and receives an asset. A *sale* occurs when a person gives an asset and receives cash. Purchase and sale transactions often involve fees. A *deposit* is the simple addition of an asset to a portfolio. A *withdrawal* is the simple removal of an asset from a portfolio. *Barter* accommodates complex situations, such as stock splits, stock option exercises, and issuance of different kinds of securities. A *journal* entry occurs when there is a discrepancy in the portfolio manager records that the user wishes to adjust manually.

  A *dividend* is a periodic payment of cash or stock to the holder of a stock. *Interest* is the periodic payment of cash to the holder of a bond. *Return of principal* is the payment of the principal (original cash tendered) for a bond. Bonds may return principal in one final payment or in multiple payments throughout the life of the bond.

  In reality, we can deduce most *transactionType* values from the links to *Asset*; however, we retain *transactionType* for convenience. Note that the *Transaction* class reifies trading actions that occur against a portfolio. The transaction is the *only* mechanism for changing the assets in a portfolio. We reify transactions so we can recall past financial activity. We note two kinds of dates for transactions: *transactionDate* (the date when the transaction occurs) and *recordDate* (the date when the transaction is recorded).

8.5 DYNAMIC MODEL

As we described earlier, the dynamic model characterizes the temporal interactions of objects and their responses to events. The first step in dynamic modeling is to construct a state diagram for each class with important temporal behavior—that is, a class with multiple states or multiple responses to stimuli. Other classes have degenerate state diagrams, and you need not show them. Some temporally important classes are ***active*** and control an application, such as the controller for a user interface; active objects determine the sequence of states for other objects. The remaining temporally important classes are ***passive*** and impose no control on other classes; passive objects have an interesting life history or constraints that depend on the current state. As an example of a passive state diagram, consider an order entry system; an order passes through well-defined states, and corporate policies may need to be enforced on a change in state. The collection of state diagrams constitutes the dynamic model.

The portfolio manager case study has a trivial dynamic model. The user interface is the only active class, and there are no passive classes. All the other classes of the portfolio manager problem have no change of state. Database applications in general tend to have simple dynamic models.

In principle, we could construct a state diagram for the user interface of the portfolio manager. But such a state diagram would be of little help. For many data management applications, you can build a user interface with a form and report builder, essentially a user interface framework. We used such a framework with our MS-Access implementation (see Chapter 14). The developers of a user interface framework have essentially already constructed a dynamic model for us. Further exploration of the user interface is outside the scope of this book.

If you are interested in learning more about the dynamic model, [Rumbaugh-91] gives a detailed explanation.

8.6 FUNCTIONAL MODEL

The functional model defines the operations that objects perform and to which they are subjected. Operations arise in the object model from queries, updates, derived data, and constraints. Operations arise in the dynamic model from events, guard conditions, actions, and activities. During analysis you should establish that a reasonable algorithm exists and try to defer the details of the algorithm until design.

We use multiple paradigms to express the functional model, including decision tables, mathematical equations, and enhanced pseudocode. The choice of appropriate paradigm depends on the particular application.

You should perform the following steps during functional modeling:

- Specify use cases. [8.6.1]
- Choose one or more functional modeling paradigms. [8.6.2]
- Specify operations from the object model. [8.6.3]

- Specify operations from the dynamic model. [8.6.4]
- Verify the entire analysis model. [8.6.5]

8.6.1 Specifying Use Cases

Once you have a sound object model, you should specify use cases. A ***use case*** [Jacobson-92] is a theme for interacting with a system. You should think of the various external entities that interact with the system and the different ways they each use the system. For the portfolio manager we can only think of two external entities: the user of the system and financial institutions that provide financial data and handle transactions. Use cases let you concisely list different kinds of behavior. Figure 8.12 lists use cases for the portfolio manager. It is by no means complete; you can always devise new ways to explore a database.

- Add, delete, and update a person.
- Add, delete, and update a financial institution.
- Add, delete, and update a financial instrument.
- Add and delete account owners.
- Add and delete a portfolio-to-portfolio link (adjust portfolio composition).
- Add, delete, and update a transaction.
- Add, delete, and update asset prices for various dates.
- Find the account number for a portfolio.
- Find all dividends for a stock.
- Find all interest payments and returns of principal for a bond.
- Find all transactions for an asset other than dividend, interest, and return of principal.
- Find all transactions for a portfolio.
- Calculate the total transaction fees incurred for a portfolio for an interval of time.
- Calculate the assets in a portfolio on a given date.
- Find the stock options in a portfolio that will expire one month from now.
- Calculate the value of a portfolio on a date.
- Calculate the ROI for a portfolio for an interval of time.

Figure 8.12 Use cases for the portfolio manager

Many developers prefer to let the specification of use cases drive the discovery of classes and associations during construction of the object model. This is a reasonable variation of the OMT process. In a sense this is what we actually did in our construction of the object model. The portfolio manager requirements embody the rudiments of use cases.

But our expression of use cases early in analysis would have been raw, because our understanding of the relevant objects and their relationships was incomplete. For example, at the beginning of analysis we were not aware of the importance of transactions. We discovered transactions as we worked through the process of constructing the object model. If you introduce use cases early in analysis, you should regard them as tentative and subject to revision, similar to the way in which our object model evolved.

We have found that programming professionals tend to favor early consideration of use cases. Most database professionals would rather deal directly with object models, probably because of prior exposure to entity-relationship modeling. We have found that customers can intelligently review both use cases and object models, as long as there is thorough documentation.

8.6.2 Choosing Functional Modeling Paradigms

Chapter 5 described various functional modeling paradigms. We now provide advice for choosing the most appropriate paradigms for a data management application.

- **Data flow diagrams**. Data flow diagrams have the advantage of being well understood. Many software developers are familiar with them from the structured analysis/structured design approaches of the past. Furthermore, process decomposition is intuitive to many persons, especially persons accustomed to programming.

  However, data flow diagrams have substantial disadvantages. First, they are impaired as an analysis notation. The very nature of process decomposition is an artifact of design, the kind of artifact we are trying to avoid during analysis. A careful developer can construct data flow diagrams with the intent of describing what is desired and not a solution approach. Nevertheless, at best such a specification is verbose and contains irrelevant information. At worst, other persons may regard a data flow diagram as a literal specification. It is just not helpful to take a paradigm that is organized about process and use it to design data structure.

  We recommend that you avoid data flow diagrams for data management applications.

- **Decision tables**. Decision tables have the virtue of simplicity. Persons can quickly grasp the notion of using records in a table to specify simple if-then rules. Furthermore, the declarative paradigm reinforces the purpose of analysis—specifying what functionality is desired and not how to realize it. Decision tables can also be practical for design, with the support of an efficient execution engine.

  But decision tables are not a panacea. They can be verbose; you may require many rules to specify simple functionality. Furthermore, decision tables are inflexible and cannot easily represent complex expressions, such as nested combinations of ANDs and ORs, mathematical formulas, and inequality predicates. Decision tables can have complex subtleties.

  You can use decision tables to represent simple discrete logic exhaustively. The number of rules must be modest if provided by a person. It is practical to use a much larger rule set for systems in which the software learns and itself acquires the rules. Decision tables are attractive when you can directly use the analysis specification for implementation.

- **Mathematical equations**. Mathematical equations are a declarative paradigm that is rigorous. With a set of mathematical equations, there is no presumption of a solution approach, although many solution approaches may be possible and viable.

  The disadvantages of mathematical equations become apparent if you have a large system, however. Large systems of equations can be difficult to understand and can involve esoteric numerical effects when you solve them. To mitigate this complexity, you must carefully group and document large systems of equations.

  Mathematical equations are highly desirable for systems with less than 20 equations. You can occasionally use mathematical equations for large systems. Mathematical equations are practical for direct implementation only when robust equation solvers are available.

- **Enhanced pseudocode**. Enhanced pseudocode refers to the combination of pseudocode with the Object Navigation Notation. Pseudocode is straightforward to read and write and does not require any special tool support. Furthermore, the pseudocode paradigm is intuitive to developers who are accustomed to programming. The ONN provides a powerful declarative representation that complements pseudocode. It tightly couples to the object model, simplifying ultimate implementation.

  However, pseudocode can be dangerous when used by undisciplined developers. It is all too easy to slip into premature design and implementation. Programmers who are not accustomed to declarative expression can have trouble learning the ONN.

  Despite its flaws, enhanced pseudocode is usually our functional modeling notation of choice. We use the other functional modeling paradigms only in special situations. Our use of pseudocode is informal. The purpose of the pseudocode is to help us understand the meaning of an operation and to document this meaning for later implementation. We find the detail of enhanced pseudocode to be a useful intermediate level of complexity, before we deal with the full complexities of programming.

8.6.3 Specifying Operations from the Object Model

The first seven use cases in Figure 8.12 are straightforward. All involve the creation and deletion of objects and links, as well as the update of their attribute values and links. We use enhanced pseudocode to specify the remaining use cases. We intersperse excerpts of the object model so that you can more easily follow the pseudocode.

- **Portfolio::findAccountNumber**. (See Figure 8.13.) Find the account number for a portfolio. This is a simple retrieval given a portfolio. Note that the account number can be null.

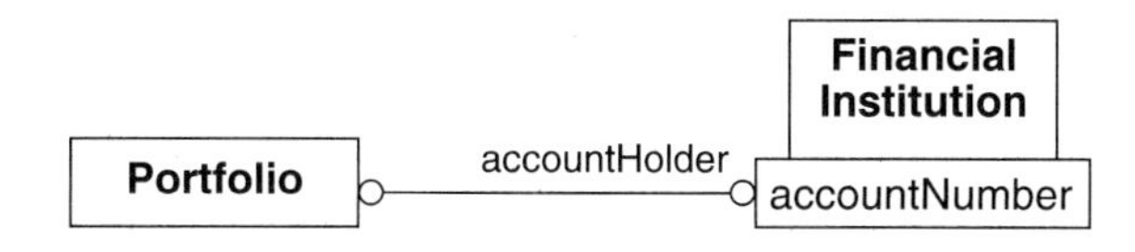

Figure 8.13 Excerpt of the final class diagram for the portfolio manager

```
Portfolio::findAccountNumber returns accountNumber
   return self@FinancialInstitution.accountNumber;
```

- **StockAsset::findDividends**. (See Figure 8.14.) Find all dividends for a stock. We find all the transactions that reference a stock and return the transactions with a transaction type of *dividend*. Given the definition of *assetReferenced*, we should not find any other transaction types, but we check for a possible error.

```
StockAsset::findDividends returns set of Transaction
   answer := emptySet;
   for each aTransaction in self:Asset.~assetReferenced
      if aTransaction.transactionType == "dividend"
         then answer += aTransaction;
      else ERROR;
      end if
   end for each
   return answer;
```

Figure 8.14 Excerpt of the final class diagram for the portfolio manager

- **BondAsset::findInterestPayments**. (See Figure 8.14.) Find all interest payments for a bond. We find all the transactions that reference a bond and return the transactions with a transaction type of *interest payment*. We also check for illegal transaction types.

```
BondAsset::findInterestPayments returns set of Transaction
   answer := emptySet;
   for each aTransaction in self:Asset.~assetReferenced
      if aTransaction.transactionType == "interest payment"
         then answer += aTransaction;
      else if aTransaction.transactionType ==
         "return of principal" then ;
      else ERROR;
      end if
   end for each
   return answer;
```

- **BondAsset::findReturnsOfPrincipal**. (See Figure 8.14.) Find all returns of principal for a bond. We find all the transactions that reference a bond and return the transactions with a transaction type of *return of principal*. We also check for illegal transaction types.

```
BondAsset::findReturnsOfPrincipal returns set of Transaction
   answer := emptySet;
   for each aTransaction in self:Asset.~assetReferenced
```

```
        if aTransaction.transactionType ==
            "return of principal" then answer += aTransaction;
        else if aTransaction.transactionType ==
            "interest payment" then ;
        else ERROR;
        end if
    end for each
    return answer;
```

- **Asset::findTradingTransactions**. (See Figure 8.15.) Find all transactions for an asset other than dividend, interest, and return of principal. Starting with an asset, we traverse the association denoted by *assetTransacted* to find the relevant transactions.

```
Asset::findTradingTransactions returns set of Transaction
    answer := emptySet;
    for each aTransaction in self.~assetTransacted
        if aTransaction.transactionType not in
            {"dividend", "interest", "return of principal"}
            then answer += aTransaction;
        end if
    end for each
    return answer;
```

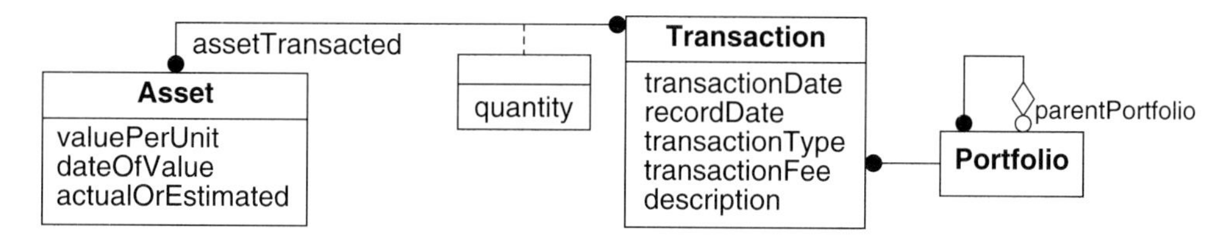

Figure 8.15 Excerpt of the final class diagram for the portfolio manager

- **Portfolio::findTransactions**. (See Figure 8.15.) Find all transactions for a portfolio.

```
Portfolio::findTransactions returns set of Transaction
    return self.Transaction;
```

- **Portfolio::computeFees (startDate, stopDate)**. (See Figure 8.15.) Calculate the total transaction fees incurred for a portfolio for an interval of time. We find all transactions for a portfolio that lie within the time interval and sum the fees.

```
Portfolio::computeFees (startDate, stopDate) returns money
    answer := 0;
    for each aTransaction in self.Transaction
        if aTransaction.transactionDate >= startDate AND
            aTransaction.transactionDate <= stopDate then
            answer += aTransaction.transactionFee;
        end if
    end for each
    return answer;
```

- **Portfolio::computeDirectAssets (date)**. (See Figure 8.15.) This is an intermediate operation for the use case "calculate the assets in a portfolio on a given date." Calculate the assets that are directly held by a portfolio on a given date. This operation computes the derived association between *Asset* and *Portfolio*. First we accumulate the transactions that have been posted to the given portfolio by the given date. Then we sum the quantity transacted for each asset and return assets with a nonzero quantity (assets that still remain in a portfolio). The operation returns a set with the assets and their quantity. Note: A bag is a collection like a set, but a bag allows duplicates.

```
Portfolio::computeDirectAssets (date)
   returns set of {Asset, number} /* return a set of pairs */

   /* first collect the assets traded and their quantity */
   aBag := emptyBag;
   for each aTransaction in self.Transaction
      if aTransaction.transactionDate <= date then
         for each anAsset in aTransaction.assetTransacted
            quantity := {assetTransacted=anAsset,
               Transaction=aTransaction}.quantity;
            aBag += {Asset=anAsset,number=quantity};
         end for each
      end if
   end for each

   /* now sum the quantity for each asset */
   answer := emptySet;
   for each groupOfPairs in aBag, group by Asset
      quantity := sum (groupOfPairs.number);
      anAsset := groupOfPairs.Asset;
      if quantity <> 0 then
         answer += {Asset=anAsset,number=quantity};
      end if
   end for each
   return answer;
```

- **Portfolio::computeAllAssets (date)**. (See Figure 8.15.) Calculate all assets in a portfolio on a given date, that is the assets that are directly held by a portfolio, as well as the assets that are indirectly held by descendant portfolios. We traverse the portfolio hierarchy and accumulate the assets. We specify this operation using recursion, but the operation could be implemented other ways. (Chapter 14 shows an alternate implementation.)

```
Portfolio::computeAllAssets (date)
   returns set of {Asset, number}

   /* first collect the assets traded and their quantity */
   aBag := self#computeDirectAssets(date);
   for each aPortfolio in self.~parentPortfolio
```

```
        aBag += aPortfolio#computeAllAssets(date);
    end for each

    /* now sum the quantity for each asset */
    answer := emptySet;
    for each groupOfPairs in aBag, group by Asset
        quantity := sum (groupOfPairs.number);
        anAsset := groupOfPairs.Asset;
        if quantity <> 0 then
            answer += {Asset=anAsset,number=quantity};
        end if
    end for each
    return answer;
```

- **Portfolio::findExpiringOptions**. (See Figure 8.15.) Find the stock options in a portfolio (directly and indirectly held assets) that will expire one month from now.

```
Portfolio::findExpiringOptions returns set of StockOptions
    aSet := self#computeAllAssets (now);
    answer := emptySet;
    for each aPair in aSet
        anExpirationDate :=
            aPair.Asset:StockOption.expirationDate;
        if anExpirationDate is null then ;
            /* the expiration date is unspecified  */
            /* or the asset is not a stock option. */
        else if anExpirationDate <= now + 1-month then
            answer += aPair.Asset:StockOption;
        end if
    end for each
    return answer;
```

- **Portfolio::computeValue(date)**. (See Figure 8.15.) Calculate the value of a portfolio on a date. For now, we assume that all assets in a portfolio (directly and indirectly held assets) have a value for the given date. We will revisit this assumption in Chapter 9 when we deal with temporal data.

```
Portfolio::computeValue (date) returns money
    answer :=0;
    aSet := self#computeAllAssets (date);
    for each aPair in aSet
        anAsset := aPair.Asset;
        quantity := aPair.number;
        answer += quantity * anAsset.valuePerUnit;
    end for each
    return answer;
```

- **Portfolio::computeROI (startDate, stopDate)**. Calculate the ROI for a portfolio for an interval of time. The ROI is the effective interest rate that, when applied to the invested funds, will yield the portfolio value on the stop date. We will not write pseudocode for this operation. It is difficult to specify ROI more precisely without descending into extensive algorithmic details, which we wish to postpone until design.

8.6.4 Specifying Operations from the Dynamic Model

The portfolio manager does not have a dynamic model, so we have no further operations to specify.

8.6.5 Verifying the Entire Analysis Model

We can detect no flaws in the models for the portfolio manager at this time. If you wish, you can collect the operations from the dynamic and functional model and add them to the object model now.

8.7 LESSONS LEARNED

We began the portfolio manager case study by rapidly jotting down requirements. (We had been thinking about the problem for a while.) Then we quickly developed an analysis object model. At this point we paused to reconsider our model and discovered two serious flaws (which we did not show in our earlier presentation). These flaws would have been difficult to repair if they had remained undetected until implementation. A benefit of thorough modeling and software engineering is early discovery of oversights and misconceptions.

In our initial analysis model we poorly conceived transactions. We had separate classes for purchases, sales, deposits, withdrawals, and "asset exchanges" (for unusual situations, such as stock splits). As we tried to explain the model, we encountered redundancy; for example we could also treat a purchase as an asset exchange. The notion of a transaction abstracts and subsumes all these separate concepts, providing flexibility and consistency for complex situations. The *transactionType* attribute is the only vestige of the original model. You can often improve models through insightful use of abstraction. You will tend to discover abstractions when broadening the scope of models or resolving inconsistencies. You should be wary of models with multiple ways of representing the same information.

Our second mistake was the improper handling of versioning. We forgot that there are two means by which a portfolio changes value over time: The underlying assets change value and portfolio composition changes via transactions. Our initial analysis object model inadvertently included a design of versions that would record the changes to asset values over time. Then we suddenly realized that our model did not address the structural change of a portfolio over time; the constituent financial instruments in a portfolio can also change. We had fallen into the trap of designing an approach to versions, instead of properly deferring the decision of how to realize versions until design. The analysis model should treat versioning as implicit and postpone the choice of a solution approach. Careless handling of versioning is particularly troublesome for the portfolio manager, since the best approach to versioning depends on the implementation platform. Many OO-DBMSs have intrinsic semantic support for versions; files and relational DBMSs do not. Figure 8.16 shows the initial flawed model and the correction. We did not show this mistake in our presentation of analysis in the previous sections of the chapter.

Excerpt from initial flawed analysis object model . . .

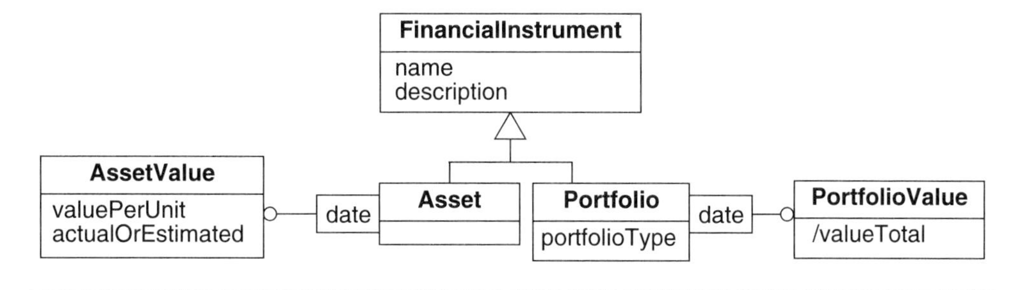

Excerpt from proper analysis object model . . .

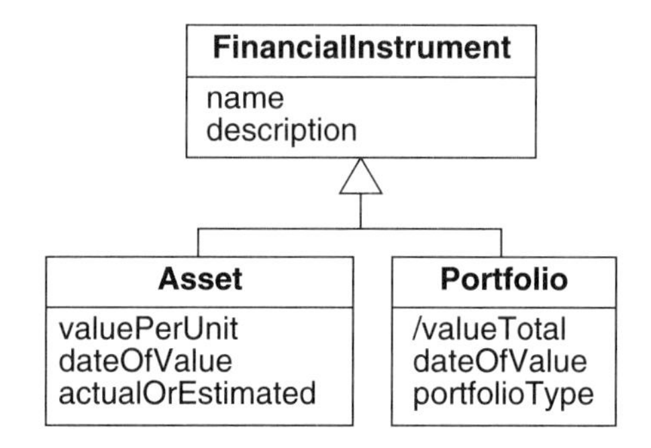

Figure 8.16 Correction of error in analysis class diagram: handling of versions

8.8 CHAPTER SUMMARY

The purpose of analysis is to construct models of the real world. The analysis models deepen the understanding of problem requirements and serve as a basis for subsequent design and implementation. During analysis you should describe *what* must be done, not *how* it should be done. You must understand a problem before attempting a solution. The OMT analysis process applies to various implementation targets, including databases and programming code.

You can synthesize the analysis models from various information sources: a problem statement, use cases, storyboards, advice from problem experts, and artifacts from prior systems. You should take care to exploit *all* information sources; many developers tend to overlook information sources. The outputs of analysis are three models—the object, dynamic, and functional models—which capture the essence of a system. The data dictionary is the repository of definitions and rationale for modeling entities. You should also write a textual narrative to guide reading of the models.

Begin analysis with a written problem statement, flawed as it may be. We construct the object model first, before the dynamic and functional models, because the object model describes the most stable portion of a system and is less affected by implementation details. Within the object model the class-association structure is critical and you should develop this first. Only then should you add inheritance so that you do not prematurely bias the model. You will discover operations as a byproduct of the dynamic and functional models.

Construct a state diagram for each class with important temporal behavior, that is a class with multiple states or multiple responses to stimuli. The collection of state diagrams constitutes the dynamic model. Most data management applications have a simple dynamic model.

The functional model specifies operations from the object and dynamic models. We use multiple paradigms to express the functional model, including decision tables, mathematical equations, and enhanced pseudocode. The choice of appropriate paradigm depends on the particular application, but most often we use enhanced pseudocode. Use cases can help you conceive important operations for specification with the functional model.

We present a basic process you should use until you become proficient. Expert analysts often perturb the sequence of the steps to fit their particular problem or personal preferences better. We present the steps of the OMT development process in a linear order. However, actual development need not be linear. You may use various development styles, such as waterfall and rapid prototyping. You may introduce use cases earlier in analysis than we show here.

Figure 8.17 lists the key concepts for this chapter.

analysis	data dictionary	levels of abstraction
analysis model	finding associations	organization of models
building the dynamic model	finding attributes	requirements
building the functional model	finding classes	testing access paths
building the object model	finding generalizations	use case

Figure 8.17 Key concepts for Chapter 8

BIBLIOGRAPHIC NOTES

Our treatment of analysis has changed little from [Rumbaugh-91]. The analysis process is a strength of our earlier book. [Abbott-83] was our original source for the idea of seeding a model with the nouns, verbs, and phrases from a problem statement. We have made the following major changes to the analysis process from [Rumbaugh-91]:

- **Levels of abstraction**. In the process for the object model we now explicitly consider a shift in the level of abstraction. This is a difficult step for persons just learning modeling. In contrast, proficient modelers readily shift levels of abstraction.
- **Use cases**. We have incorporated the notion of a use case as advocated by [Jacobson-92]. Use cases have an intuitive appeal and have clarified our thinking.
- **Multiparadigm functional model**. We do not use data flow diagrams for data management applications. In [Rumbaugh-91] we were trying to synthesize past experience with data flow diagrams in structured analysis with object-oriented technology. Data flow diagrams did not work as well as we had hoped. We have been successfully applying the multiparadigm functional model, with emphasis on enhanced pseudocode, in our industrial work.

Fusion [Coleman-94] is a second-generation methodology that builds on our first OMT book and other methodologies. However, Fusion is a programming methodology and does not address database applications. As with every methodology, Fusion has its strengths and weaknesses. We especially like the emphasis Fusion places on preconditions and postconditions. Fusion also benefits from its broad basis in assimilating ideas from other methodologies. But this strength is also a weakness: The methodology is more a *fusion* of ideas rather than a *synthesis*. For example, Fusion has a seam when shifting from analysis to design; information that is entered during analysis must be converted to a different form during design.

[Batini-92] has an excellent analysis of information in forms (pages 92–101) that complements our presentation in this chapter. They distinguish four parts of a form. The *certificating* part (such as signatures and stamps) validates a form and is of little interest during analysis. In contrast, the *extensional* part is highly relevant and consists of the user entries for the form—the data that would be stored in a database. The *intensional* part consists of the labels on a form and corresponds to the schema of a database. The *descriptive* part contains instructions and rules and provides context for understanding a form.

REFERENCES

[Abbott-83] Russell J. Abbott. Program design by informal english descriptions. *Communications of the ACM 26*, 11 (November 1983), 882–894.

[Batini-92] Carlo Batini, Stefano Ceri, and Shamkant B. Navathe. *Conceptual Database Design: An Entity-Relationship Approach*. Redwood City, California: Benjamin Cummings, 1992.

[Coad-95] Peter Coad, David North, and Mark Mayfield. *Object Models: Strategies, Patterns, and Applications*. Englewood Cliffs, New Jersey: Yourdon Press, 1995.

[Coleman-94] D Coleman, P Arnold, S Bodoff, C Dollin, H Gilchrist, F Hayes, and P Jeremaes. *Object-Oriented Development: The Fusion Method*. Englewood Cliffs, New Jersey: Prentice Hall, 1994.

[Jacobson-92] Ivar Jacobson, Magnus Christerson, Patrik Jonsson, and Gunnar Overgaard. *Object-Oriented Software Engineering: A Use Case Driven Approach*. Reading, Massachusetts: Addison-Wesley, 1992.

[Rumbaugh 91] J Rumbaugh, M Blaha, W Premerlani, F Eddy, and W Lorensen. *Object-Oriented Modeling and Design*. Englewood Cliffs, New Jersey: Prentice Hall, 1991.

EXERCISES

8.1a. (3) Figure E8.1 presents requirements for a simple payroll system. Figure E8.2 lists candidate classes. Eliminate the spurious classes and give a reason for each elimination.

Develop software for a simple payroll system. Record the following:

- Employee name and number.
- Date of paycheck.
- Deduction type, amount, and explanation. The payroll system should support both standard and unusual deduction types.
- Income type, amount, and explanation. The payroll system should support both standard and unusual income types.
- Net pay and method of payment (cash, direct deposit, check, . . .).
- Total income, total deductions, and total net pay for each year.

Figure E8.1 Problem statement for a simple payroll system

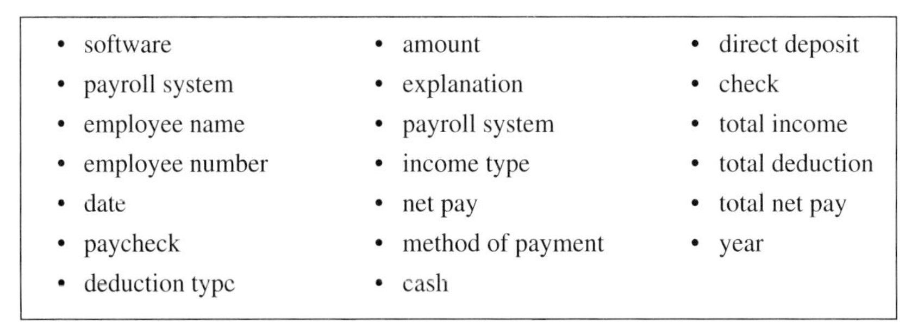

- software
- payroll system
- employee name
- employee number
- date
- paycheck
- deduction typc
- amount
- explanation
- payroll system
- income type
- net pay
- method of payment
- cash
- direct deposit
- check
- total income
- total deduction
- total net pay
- year

Figure E8.2 Candidate classes for a simple payroll system

b. (4) Figure E8.1 does not fully state the relationships for the payroll system. But the problem is so simple that the relationships are implied. Prepare an object model for the payroll system. Be sure to note derived information.

c. (2) Prepare a data dictionary for the classes of the payroll system.

8.2a. (5) Figure E8.3 presents requirements for a sales opportunity manager. Figure E8.4 lists candidate classes. Eliminate the spurious classes and give a reason for each elimination.

b. (5) Figure E8.5 lists candidate associations for the sales opportunity manager. Eliminate the spurious associations and give a reason for each elimination.

c. (7) Construct an object model for the sales opportunity manager.

d. (2) Prepare a data dictionary for the classes of the sales opportunity manager.

e. (5) Use enhanced pseudocode to specify the most recent estimate of revenue from a customer.

f. (5) Use enhanced pseudocode to specify the estimated profit for a primary salesperson.

Develop software for recording and managing sales opportunities. Businesses must carefully track opportunities to make sure they are responsive to potential customers, to anticipate resource demands, and to project revenues and future profits. The software must include the following:

- Record information for opportunities: date, customer name, opportunity name, location, responsible salespersons, primary salesperson, probability of success, estimated revenue, estimated profit, decision makers, competitors, and related opportunities.
- Maintain the "history" of the opportunity—that is, how the status of an opportunity evolves over time.
- Rollup estimated revenue and profit by customer, salesperson, probability of success interval, and date.

Figure E8.3 Problem statement for a sales opportunity manager

- software
- sales opportunity
- business
- opportunity
- potential customer
- resource demand
- revenue
- future profit
- information
- date
- customer name
- opportunity name
- location
- responsible salesperson
- primary salesperson
- probability of success
- estimated revenue
- estimated profit
- decision maker
- competitor
- related opportunity
- history
- status
- time
- profit
- customer
- salesperson
- probability of success interval

Figure E8.4 Candidate classes for a sales opportunity manager

- software for recording and managing sales opportunities
- businesses must carefully track opportunities to make sure they are responsive to potential customers, to anticipate resource demands, and to project revenues and future profits
- record information for opportunities: date, customer name, opportunity name, location, responsible salespersons, primary salesperson, probability of success, estimated revenue, estimated profit, decision makers, competitors, and related opportunities
- maintain the history of the opportunity
- rollup estimated revenue and profit by customer, salesperson, probability of success interval, and date

Figure E8.5 Candidate associations for a sales opportunity manager

8.3a. (4) Figure E8.6 presents requirements for a contact manager. Figure E8.7 lists candidate classes. Eliminate the spurious classes and give a reason for each elimination.

b. (4) Figure E8.8 lists candidate associations for a contact manager. Eliminate the spurious associations and give a reason for each elimination.

c. (6) Construct an object model for the contact manager.

d. (2) Prepare a data dictionary for the classes of the contact manager.

Develop software for recording information about interaction with important business contacts. The software must include the following:

- Descriptive data for contacts, such as name, salutation, job title, employer, manager, subordinates, secretary, addresses (home, office, mailing), phone numbers (home, office, cellular, fax, secretary), email addresses, and personal comments.
- Data for events of interaction, such as phone calls, email, and postal mail. For each interaction event record the date and comments.

Make sure the software is generic with regard to addresses and phone numbers. The user should be able to define a new kind of address or a new kind of phone number without modifying the software.

Figure E8.6 Problem statement for a contact manager

- software
- information
- interaction
- business contact
- descriptive data
- contact
- name
- salutation
- job title
- employer
- manager
- subordinate
- secretary
- address
- phone number
- email address
- personal comment
- data
- event of interaction
- phone call
- email
- postal mail
- date
- comment
- user
- kind of address
- kind of phone number

Figure E8.7 Candidate classes for a contact manager

- software for recording information
- interaction with important business contacts
- data for contacts, such as name, salutation, job title, employer, manager, subordinates, secretary, addresses (home, office, mailing), phone numbers (home, office, cellular, fax, secretary), email addresses, and personal comments
- data for events of interaction, such as phone calls, email, and postal mail. For each interaction event record the date and comments

Figure E8.8 Candidate associations for a contact manager

8.4a. (4) Figure E8.9 presents requirements for course registration and grading software. Figure E8.10 lists candidate classes. Eliminate the spurious classes and give a reason for each elimination.

b. (4) Figure E8.11 lists candidate associations for course registration and grading software. Eliminate the spurious associations and give a reason for each elimination.

c. (6) Construct an object model for the course registration software.

d. (2) Prepare a data dictionary for the classes of the course registration software.

Develop software that tracks course registration and grades at a university. For each course that a student takes, the information system should be able to retrieve the grade, department name, course number, course name, section number, year taken, semester taken, professor name, and credit hours.

The software should be able to compute the total registration for each section of a course. The software should record the starting time and duration of each offered course so it can detect conflicts in student schedules. The software should also record the name and credit hours for each course listed in the department catalog.

Figure E8.9 Problem statement for course registration and grading software

- software
- course registration
- grade
- university
- course
- student
- information system
- department name
- course number
- course name
- section number
- year
- semester
- professor name
- credit hour
- total registration
- section
- starting time
- duration
- offered course
- conflict
- student schedule
- name
- department catalog

Figure E8.10 Candidate classes for course registration and grading software

- software that tracks course registration and grades at a university
- for each course that a student takes the information system should be able to retrieve the grade, department name, course number, course name, section number, year taken, semester taken, professor name, and credit hours
- total registration for each section of a course
- starting time and duration of each offered course
- conflict in student schedules
- name and credit hours for each course listed in the department catalog

Figure E8.11 Candidate associations for course registration and grading software

8.5a. (5) Figure E8.12 presents requirements for managing library loan records. Figure E8.13 lists candidate classes. Eliminate the spurious classes and give a reason for each elimination.

b. (5) Figure E8.14 lists candidate associations for the library loan software. Eliminate the spurious associations and give a reason for each elimination.

c. (8) Construct an object model for the library loan software. (Hint: As you formulate your object model, consider that a person may check out the same library item copy more than once. Similarly, a person may request to be placed on the wait list for a library item more than once.)

d. (2) Prepare a data dictionary for the classes of the library loan software.

e. (6) Write enhanced pseudocode for each of the three library loan searches.

Develop software for tracking library loan records. Library patrons may borrow books, magazines, movies, compact discs, and audio tapes. Each copy of a library item must be managed. For example, a library may have five copies of the book *The Grapes of Wrath*. The software must track borrowers, their address, and phone number. Each borrower is assigned a library card.

Each category of library item has a standard checkout period and number of renewals. For example, children's books may be checked out for a month, while adult books may be checked out for only two weeks. Ordinary books can be renewed once; books with a pending request may not be renewed. The system must record the actual return date and any fine that was paid. The system must enable searches, such as

- Total fine owed by a patron.
- Fine revenue to the library for an interval of time.
- The copies of library items that are grossly overdue. "Grossly" is the number of days specified by the librarian.

The system must also maintain any records of damage and loss and associated payments.

Figure E8.12 Problem statement for library loan software

- software
- library loan record
- library patron
- book
- magazine
- movie
- compact disc
- audio tape
- copy
- library item
- library
- borrower
- address
- phone number
- library card
- category
- standard checkout period
- number of renewals
- pending request
- return date
- fine
- patron
- fine revenue
- librarian
- system
- record of damage and loss
- associated payment

Figure E8.13 Candidate classes for library loan software

- software for tracking library loan records
- library patrons may borrow books, magazines, movies, compact discs, and audio tapes
- copy of a library item
- a library may have five copies of the book
- software must track borrowers, their address, and phone number
- each borrower is assigned a library card
- each category of library item has a standard checkout period and number of renewals
- books may be checked out
- books can be renewed
- books with a pending request
- system must record the actual return date and any fine that was paid
- fine owed by a patron
- fine revenue to the library
- copies of library items that are grossly overdue

Figure E8.14 Candidate associations for library loan software

8.6a. (4) Figure E8.15 presents requirements for a billing system. Figure E8.16 lists candidate classes. Eliminate the spurious classes and give a reason for each elimination.

Develop software for cable TV customer billing. The software must record the following:

- The name, address, phone number, and applicable accounts for each customer. A customer may have multiple accounts; each account pertains to one customer.
- The standard services offered and their current price.
- The recurring services chosen for each customer account and the date of subscription and termination.
- Orders for pay per view events.
- Unusual fees, such as installation and hardware rental.
- Unusual credits, such as promotions and service interruptions.
- Billings for customer accounts—the date of billing, amount, and date due.
- Payments against customer accounts—date, amount, and method of payment.
- Informal comments about the credit experience with accounts.
- The current balance for each account.

Figure E8.15 Problem statement for billing software

b. (4) Figure E8.17 lists candidate associations for the cable TV billing software. Eliminate the spurious associations and give a reason for each elimination.

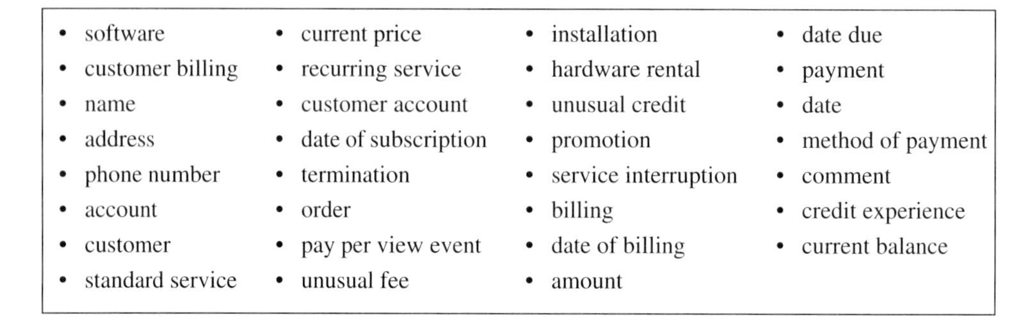

- software
- customer billing
- name
- address
- phone number
- account
- customer
- standard service
- current price
- recurring service
- customer account
- date of subscription
- termination
- order
- pay per view event
- unusual fee
- installation
- hardware rental
- unusual credit
- promotion
- service interruption
- billing
- date of billing
- amount
- date due
- payment
- date
- method of payment
- comment
- credit experience
- current balance

Figure E8.16 Candidate classes for billing software

- software for cable TV customer billing
- name, address, phone number, and applicable accounts for each customer
- a customer may have multiple accounts
- each account pertains to one customer
- the standard services offered and their current price
- recurring services chosen for each customer account
- orders for pay per view events
- billings for customer accounts
- payments against customer accounts
- comments about the credit experience with accounts
- current balance for each account

Figure E8.17 Candidate associations for billing software

c. (6) Construct an object model for the cable TV billing software. You should add class *AdjustmentType* to categorize adjustments to a bill. You will also need to add class *ServiceRequest* since a customer account can request the same standard service for multiple time intervals.

d. (2) Prepare a data dictionary for the classes of the cable TV software.

e. (2) Use enhanced pseudocode to compute the total amount receivable from a customer.

f. (6) Use enhanced pseudocode to compute the proper amount to invoice on the next billing.

8.7 Many business magazines (such as *Business Week*) report stock performance organized by stock sectors (such as banking, retail, mining, computer hardware, computer software, and others). Some mutual funds also use stock sectors to guide investment decisions.

a. (2) Extend the object model for the portfolio manager (Figure 8.11) to include stock sectors.

b. (7) Write enhanced pseudocode that computes the value of portfolio stock holdings organized by stock sector. [Hint: Define operation *Portfolio::computeStockSectorValues (date)* and invoke operation *Portfolio::computeAllAssets (date).*] Figure E8.18 shows an example of some stocks in a portfolio and the computed stock sector values.

Information about a portfolio on some date:

Stock	StockSector	valuePerUnit	quantity
IBM	computer hardware	100	100
Apple Computer	computer hardware	25	200
Microsoft	computer software	150	100
Oracle	computer software	40	500
Hewlett-Packard	computer hardware	90	200

Output from operation:

StockSector	totalValue
computer hardware	33000
computer software	35000

Figure E8.18 Sample input and output for Exercise 8.7b

8.8 (7) Figure E8.19 specifies requirements for a document manager. Figure E8.20 shows our initial object model. Note some flaws in the model.

- There is little difference between subclasses. Is an outline of a paper a "paper" or a "note"? How should we handle a paper that is in both an electronic file and a binder? How should we represent information about slides for talks?
- We would like to handle both standard comments (applicable to many documents and chosen by point and click in a user interface) and custom comments (applicable to one document and specifically typed by the user).
- We should be able to comment on a numbered page without having have sections.

Improve the object model by using generic classes and making it more abstract. Hint: You should have generic classes for location, document properties, and comments. It is adequate to represent document composition with a hierarchy.

Develop software for managing professional records of papers, books, journals, notes, and computer files. The system must be able to record authors of published works in the appropriate order, name of work, date of publication, publisher, publisher city, an abstract, as well as a comment. The software must be able to group published works into various categories that are defined by the user to facilitate searching. The user must be able to assign a quality indicator of the perceived value of each work.

Only some of the papers in each issue of a journal may be of interest. It would also be helpful to be able to attach comments to sections or even individual pages of a work.

Figure E8.19 Problem statement for a document manager

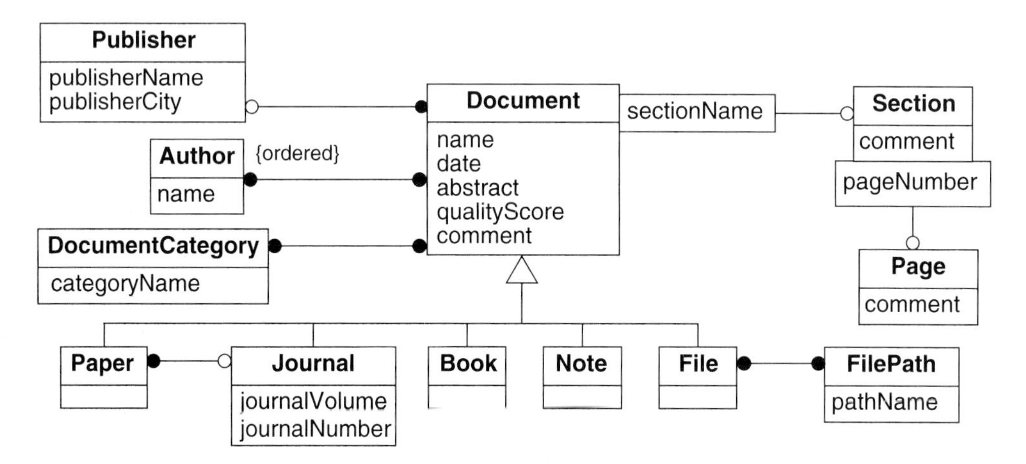

Figure E8.20 Initial class diagram for a document manager

9

System Design

After you have analyzed a problem, you must decide how to approach the design. During analysis, the focus is on *what* needs to be done, independent of *how* it is done. During design, you make decisions about how the problem will be solved, first at a high level (this chapter), then at increasingly detailed levels (the next chapter).

This chapter deals with the broad aspects of software design that we call system design. We begin by describing general design issues and then address specific issues for data-intensive applications. It is at this methodology step that we begin to tailor and augment the general OMT approach.

The focus of this chapter is on the architecture of an individual application running on a single computer. Part 4 contains additional material that complements this chapter; Chapter 18 discusses distributed databases and Chapter 19 discusses the integration of applications.

9.1 OVERVIEW

System design is the first stage of design in which you make high-level decisions for solving the problem. For small and medium-sized problems system design naturally follows analysis; first you should determine the logical requirements and only then fashion a solution approach. However, for large problems (see Section 3.6) this sharp separation between analysis and system design is not practical. For large problems there is a natural interplay between analysis and design: First model the overall system (analysis) and then subdivide into subsystems (system design). Then model the subsystems and continue to subdivide until the ultimate components are manageable and comprehensible. In this book we simplify our explanation by presenting system design as strictly following analysis.

During system design the developer makes strategic decisions with broad consequences for the software as a whole. You must formulate an architecture for the system and choose global strategies and policies to guide the subsequent, more detailed portions of design. You should resolve the following steps during system design. The precise ordering can vary for a particular problem.

- Devise an architecture. [9.2]
- Choose an implementation for external control. [9.3]
- Choose a data management approach. [9.4]
- If using a database, choose a database management paradigm. [9.5]
- Determine opportunities for reuse. [9.6]
- Choose a strategy for data interaction. [9.7]
- Choose an approach to object identity. [9.8]
- Deal with temporal data. [9.9]
- Deal with secondary aspects of data. [9.10]
- Specify default policies for detailed design. [9.11]

9.2 DEVISING AN ARCHITECTURE

The first task for system design is to devise a system architecture. The ***system architecture*** is the high-level plan or strategy for solving the application problem. You will need to formulate architectures for both the overall system and subsystems. Often, for straightforward problems, you can just directly conceive a suitable architecture. For more complex problems the appropriate architecture is not obvious and you should formally collect ideas and evaluate them.

In this section we first discuss architectural design principles that apply to a wide variety of systems and consider the principles for the portfolio manager case study. We then present the following process for devising an architecture. We have found this process to be valuable for assimilating the inputs of multiple persons. Persons can generate ideas in parallel and can evaluate ideas in parallel, letting you consider multiple unbiased opinions.

- Generate candidate architectures. [9.2.2]
- Propose decision criteria and assign weights to them. [9.2.3]
- Quantify the compliance of the architectures against the decision criteria. [9.2.4]
- Compare scores for each candidate architecture. [9.2.5]

We illustrate this process with a simple example from the portfolio manager case study (see Chapter 8). The example illustrates the mechanics, but is too small to convey the complexity of devising a full architecture. Nevertheless, the overall process we present is quite realistic. We have used it to devise architectures for both industrial computer science and engineering problems. We have also observed other software engineers use this kind of process, without any prior influence by us.

9.2.1 Architectural Design Principles

We suggest you consider the following principles in formulating architectures. They are intended as guidelines and may not apply to every application.

- **Distinguish between operational and decision-support applications**. ***Operational applications*** update a database by posting numerous transactions as part of the routine business of an organization. For these applications you will discover most transactions during development and you can embed them in a user interface. In contrast, ***decision-support applications*** tend to have few updates and can involve unpredictable and lengthy queries. Decision-support developers often first load the data into a dedicated database (a data warehouse, see Chapter 19) so that queries do not interfere with the operational systems and degrade their performance.

  The architectural requirements differ greatly for these two kinds of applications. Operational applications must access the current data of an organization. They require fast performance and maximal database availability. Decision-support applications have different trade-offs; they can run more slowly and can tolerate more downtime. However, decision-support applications must let users express new queries readily; it must be easy for users to explore the database to develop ideas and test hypotheses.

  The portfolio manager. Our case study is a decision-support application, but we identified the important queries during analysis. We do not have the data (especially transactions and asset prices) readily available in electronic form, so we will build a user interface for acquiring it. In principle, we could acquire the data by building interfaces to other systems, but we want to limit the size of our case study.

- **Decompose large systems into layers and partitions**. You can often devise a simple and elegant structure by decomposing a system into layers and partitions. A ***layer*** is a subsystem that builds on subsystems at a lower level of abstraction. A ***partition*** is a subsystem that is in parallel to other subsystems. You should end decomposition when you achieve subsystems with clear, crisp themes that are straightforward to understand. Packages for large object models (see Section 3.6) are often organized on the basis of layers and partitions.

  Compilers illustrate the notion of layers. Most compilers consist of successive phases—recognize syntax, recognize semantics, optimize, and emit target code. Each compiler phase depends on the processing of the previous phase. A good way to develop a compiler is to model the input and output of each phase and write a program to convert between pairs of successive models.

  The UNIX operating system uses partitions extensively. Most UNIX commands are decoupled; individual commands interact via a stream of bytes in memory or intermediate files. Since the commands are decoupled, they can be developed separately reducing the complexity of UNIX.

  The portfolio manager. The case study is too small to necessitate the use of layers and partitions. But if we wanted to extend the application, the different assets would provide a natural basis for partitioning.

- **Separate application logic from the user interface**. Normally, you should decouple application logic from the user interface. It is helpful to organize application logic as a library of methods (an application programming interface, or API) and then invoke the methods from a user interface as appropriate. There may be multiple user interfaces that access the application methods; the user interfaces may differ in presentation format and in the information they present and suppress. The separation of application logic and presentation format preserves your options for implementing control (Section 9.3). The discussion of the three-tier client-server architecture in Chapter 18 elaborates this point of separating application logic from the user interface.

  The portfolio manager. Chapter 14 applies this principle in our implementation of a user interface for the portfolio manager.

- **Consider reification**. Sometimes it is helpful to promote information, such as methods, control, rules, and constraints to objects. (See Section 4.1.2.) You can then store information declaratively as data rather than write additional programming code. The declarative representation requires a corresponding metamodel to hold the data. A disadvantage of reification is the intrinsic complexity of metamodels. However, reification lets you evolve and extend a system by merely changing data.

  The portfolio manager. There is no information that would be helpful to reify.

- **Substitute database queries for programming code**. You must strike a proper balance between the role of the DBMS and the role of a programming language. Do not embed all the logic in programming code. Viewing a DBMS as just a low-level store-and-retrieve mechanism analogous to files is short-sighted. Instead, you should offload computation to the DBMS whenever possible. This offloading can greatly improve performance, increase extensibility, reduce development time, and reduce bugs. We have been able to reduce development effort for some applications by more than an order of magnitude by substituting database queries for programming effort.

  The portfolio manager. The portfolio manager provides an excellent illustration of this principle, as we describe in Chapter 14. We present an algorithm that uses database queries, rather than the recursive approach for computing portfolio assets specified by our pseudocode in Section 8.6.3. Sometimes you can facilitate database queries by adding derived information to an object model.

- **Consider major interfaces to persons and other systems**. Your architecture is likely to be affected by the impositions of other systems. Ideally, you should have small and well-defined interfaces between applications (few classes and relationships in common). You should try to minimize the amount of information that a person must enter.

  The portfolio manager. We could acquire data by manual input, automatically from other systems, or by some combination of the two. With manual input we could enter information via forms. With automated input we could dial a stock broker, retrieve asset values and transactions, and then store the data in the portfolio manager database. We provide only manual input for our case study. We don't want to complicate the case study with the details of other systems. Also, it is always handy to provide manual input so that you can thoroughly test a new application. We can always add automated interfaces at a later date.

The remainder of Section 9.2 presents a process you can use to make difficult architecture decisions.

9.2.2 Generating Candidate Architectures

The first step for devising an architecture is a matter of synthesis. It is important to frame your objective carefully as a statement of the problem and the desired results. Our most successful projects have had clearly articulated goals. You must also be creative; actively brainstorm and generate ideas. You will be more successful if you have persons with different backgrounds and experiences generate ideas.

You can generate ideas in many ways—by reflecting on experience, by thinking of ideas on your own, by soliciting the ideas of others, by consulting the literature, and by looking for analogies. Analogies from related problems provide one of the best sources of inspiration. Many inventions are actually old ideas applied to new problems. Carefully weigh experience and lessons learned from similar systems.

Don't be too critical. It is better just to record ideas at first and then separately evaluate them. An idea may seem unattractive at first and become tenable after some reflection. We favor quickly finding an architecture that will work and only then striving to do better. You may need some exploratory work to assess the viability of the various options. Be careful that your level of effort is commensurate with the importance of the system or subsystem.

Keep in mind that automation is not always the best solution. Sometimes it is better to forgo a computer system and rely on manual effort. Furthermore, buying software is often an alternative to custom development.

The portfolio manager. Our statement of the problem is relatively simple, so we can think of only one major architectural issue—whether or not to cache computation. We could compute the portfolio value and return-on-investment for each user request. Alternatively, we could precompute the portfolio values and ROIs and store them in anticipation of need. A hybrid approach would be to perform computations as requested, but record the results for possible reaccess. It is not obvious to us which approach would be best, so we will perform a formal evaluation.

9.2.3 Proposing Decision Criteria and Assigning Weight

The next step is to elicit decision criteria by considering the advantages and disadvantages of each candidate architecture. The requirements statement may mention decision criteria. Examine the needs of different user categories, such as end users, managers, database administrators, and system maintainers. Common criteria are cost, ease of use, development effort, deployment effort, performance, reliability, extensibility, integrity, and security. If you have many criteria, you should group some of them. Keep in mind that it is difficult to gain confidence in the results with more than 10 decision criteria.

Once you have identified the criteria, assign each criterion a weight. You are making architectural decisions as you decide which criteria are important and which are incidental. Take special care to note for each criterion whether it is a "must" or a "want." A "must" is absolutely required or the candidate architecture will be rejected. A "want" is important, but you will have to weigh it against other "wants."

The portfolio manager. We have the following decision criteria, all of which are "wants."

- **Ease of use**. We want to be able to browse the data easily. The ROI is a significant computation and we are not willing to wait for a lengthy evaluation. (High weight)
- **Development effort**. We would like to minimize development effort. Also, we would like to minimize effort in coding a user interface. (High weight)
- **Extensibility**. We want to be able to extend the scope of the portfolio manager easily. (Medium weight)
- **Integrity**. As much as possible, we would like to assure a high quality of data so that we do not infer wrong conclusions. (High weight)

9.2.4 Quantifying Compliance of the Architectures

You can now judge the candidate architectures and assign a score for each criterion. We assign the following scores for the portfolio manager (best score is 10 and worst score is 0).

- **Precomputation**. We assign a score of 3 for ease of use because precomputation can only determine portfolio values and ROIs for fixed dates. The precomputed results are derived data; we would require additional development effort to keep them consistent with the underlying data. For example, we will delete a ROI if an underlying dividend payment is changed (such as a correction of a data entry mistake). We think the additional effort will be modest so we assign a score of 7 for development effort. There appears to be no impediments to extensibility (score of 10). We think the additional code to maintain consistency of derived data would be unlikely to have bugs and compromise data integrity, so we assign a score of 9.
- **Compute on request**. We want to be able to browse the database and study the variation of financial performance with time. It is awkward to study financial data if we must wait for computations. We are reluctant to compromise ease of use; we do not want any inhibition to data browsing (score of 1). This architecture is the easiest to develop because we do not have to maintain derived data (score of 10). There appears to be no impediments to extensibility (score of 10). Since there is no derived data the database is more likely to store consistent data (score of 10).
- **Hybrid approach**. This approach avoids precomputation; we store portfolio values and ROIs as a side effect as they are computed. The software short circuits a request for a portfolio value or ROI if it is found to be cached. The user can choose to compute any sequence of values and then browse.

  We assign a score of 7 for ease of use; the user can inspect portfolio values and ROIs for any dates, and response is quick for previously computed dates. We penalize the score only because response will be slower for uncached dates. Development effort, extensibility, and integrity are similar to precomputation, so we assign the same scores.

Table 9.1 summarizes our ratings. We have converted high and medium weights to numeric values of 10 and 5 respectively. The hybrid approach has the highest total score.

Criteria	Weight	Precompute	Compute on request	Hybrid approach
Ease of use	10	3	1	7
Development effort	10	7	10	7
Extensibility	5	10	10	10
Integrity	10	9	10	9
Total score		240	260	280

Table 9.1 Rating cache computation options against decision criteria

9.2.5 Comparing Scores

At this point you have tentative architectural decisions. Now you need to perform sensitivity analysis to develop confidence in the results. Are you comfortable with the weights and their effect on the total score? Are you comfortable with the relative scoring of the candidate architectures for each criterion? How does a reasonable variation in weights and scores affect the results?

You also need to consider development risk. Your tolerance for risk will depend on your organization and the maturity of the application. Your risk will be mitigated to the extent that you are familiar with similar applications. For example, if you are developing a payroll system for the second time, you will understand many of the architectural issues and their consequences. The odds of success are higher if you purchase an off-the-shelf application from a vendor with experience in your industry. A research application, on the other hand, has high risk and you must be prepared for unforeseen problems.

The portfolio manager. We are satisfied with a hybrid approach to computation; this is not the easiest approach (computation on request requires less development effort), but we greatly value being able to browse the data. There is not much development risk with the portfolio manager; we have performed similar computations for other applications.

9.3 CHOOSING AN IMPLEMENTATION FOR EXTERNAL CONTROL

The next step is to choose a paradigm for ***external control***, that is how an application is controlled at its outermost level and how it interacts with users and other applications. External control contrasts with internal control, which is not visible outside the application. The choice of control paradigm depends on the target platform.

- **Procedure-driven control**. Control resides within a single thread of programming code. Methods issue requests for external input and then wait; when input arrives, control resumes within the method that made the call. The location of the program counter, global variables, and the stack of calls and local variables defines the system state. The major advantage of procedure-driven control is that it is easy to implement with conventional languages; the disadvantage is that you must map the concurrency inherent in objects into a sequential flow of control.

  We prefer procedure-driven control for most applications that lack a substantial user interface.

- **Event-driven control**. Control resides within a dispatcher that the application, language, operating system, or DBMS provides. You must attach application methods to events; the dispatcher calls the methods when the corresponding events occur. All methods return control to the dispatcher, rather than retaining control until input arrives. You cannot preserve system state using the program counter and stack because methods return control to the dispatcher. Methods must use global variables to maintain state. Event-driven control is desirable because it cleanly separates application logic from the user interface. Event-driven control is straightforward to implement when you can use a fourth-generation language (4GL) provided by a DBMS. (See Section 9.7.)

  We often use event-driven control for database applications. Event-driven control is almost mandatory for a polished user interface.

- **Concurrent control**. Control resides concurrently in several autonomous objects that respond to events independently. Parallel hardware or operating system simulation can provide the underlying concurrency. You can use a DBMS to coordinate concurrent objects. Concurrent control has the advantage of directly realizing the concurrency of objects. The disadvantage is conceptual and implementation complexity.

  Do not confuse concurrency *within* an application with concurrency *across* applications. We seldom use concurrency within an application; in contrast, developers routinely use concurrency to coordinate multiple invocations of database applications.

- **Declarative control**. The software derives an implicit flow of control from declarative statements. For example, artificial intelligence systems have implicit evaluation mechanisms that perform forward and backward inference. Declarative control provides a high-level metaphor that can improve the efficiency of system development and maintenance, but at the cost of greater system complexity and lower performance. Another impediment is the effort required to develop an evaluation infrastructure.

  We occasionally use declarative control, especially for rule-based systems.

Table 9.2 summarizes the different paradigms for external control.

The portfolio manager. To make our discussion more complete, we consider three implementations of the portfolio manager: relational DBMS, object-oriented DBMS, and files. We chose event-driven control for the relational DBMS (Chapters 13 and 14) because MS-Access supports this paradigm. We only partially implement the portfolio manager using an object-oriented DBMS (Chapters 15 and 16), so we do not present a control paradigm. Our

External control paradigm	Recommendations
Procedure-driven	Preferred for most applications that lack a substantial user interface.
Event-driven	Preferred for applications that require a polished user interface.
Concurrent	Seldom used within an application. Often used across applications.
Declarative	Occasionally feasible. Consider for rule-based systems.

Table 9.2 External control paradigms

choice of paradigm for this approach would depend on the capabilities of the available class libraries. We also only partially implement the portfolio manager with files (Chapter 12). We normally use procedure-driven control for file-based applications.

9.4 CHOOSING A DATA MANAGEMENT APPROACH

There are several alternatives for data management: in-memory data, files, or a DBMS. Some applications require formal database management services. For other applications memory or files can provide an adequate, simple, and inexpensive solution. Consider the following aspects when choosing an approach:

- **Data persistence**. Both files and DBMSs are strong in this regard. In-memory data requires special kinds of memory or power supplies to prevent loss of data.
- **Purchase cost**. A DBMS can be costly to purchase. However, large corporations often have site licenses, so the incremental use of a preferred DBMS may be inexpensive. Also, application vendors often receive heavy discounts for an embedded DBMS. Persistent in-memory data requires special and more expensive hardware. Files incur no purchase cost because they are intrinsically supported by operating systems.
- **Lifecycle cost**. The cheapest approach depends on the application. In-memory data, files, and DBMSs can all provide the lowest lifecycle cost. Software lifecycle cost reflects purchase, development, deployment, operating, and maintenance costs. A DBMS can reduce development cost, but it may require specially trained support staff, incurring an ongoing operating cost. Files have no purchase cost but may be difficult to maintain, also incurring an operating cost.
- **Large quantities of data**. DBMSs can handle arbitrarily large quantities of data, limited only by secondary storage and incidental software restrictions. Hardware resources limit capacity for in-memory data. Files that are sequentially or randomly accessed may be quite large, constrained only by the operating system. Files that are fully cached are constrained by memory size.

- **Performance**. In-memory structures have the fastest data access. DBMSs have special algorithms and data structures for efficiently handling large quantities of data. Files can provide excellent performance when an application can fully read them into memory, sequentially access them, or randomly access them.
- **Extensibility**. DBMSs promote data independence so that the developer can focus on logical aspects of data and defer most implementation details [Date-86]. With ***logical data independence*** an application is made aware only of relevant portions of the database; this lets you add new applications to the database without disrupting existing applications. ***Physical data independence*** insulates applications from changes in tuning mechanisms; this lets you restructure a database for faster performance without affecting application logic. Files and in-memory data have no counterpart to the robust extensibility of DBMSs.
- **Concurrent access**. Most operating systems can lock files only in their entirety. Record-level locking within files requires a great deal of custom programming. In contrast, in-memory data and DBMSs can provide a fine granularity of concurrency. Most DBMSs can automatically schedule locking and resolve conflicts without any special user actions.
- **Crash recovery**. Backup files can protect against accidental loss from a computer or media failure. DBMSs have sophisticated logic for crash recovery that necessitates duplication of only the updated portion of the database. Shadow memory can protect in-memory data.
- **Integrity**. The database designer can specify rules that data must satisfy. A DBMS can control the quality of its data over and above facilities that application programs may provide. The resulting system is more reliable and easier to debug. The database is more likely to store correct data. Files and in-memory data lack integrity support.
- **Transaction support**. A ***transaction*** is a group of database commands for which concurrency control, crash recovery, and integrity apply. A DBMS ensures that a transaction completely succeeds (an entire transaction is written to the database) or fails (nothing is written). Transactions can fail for a variety of reasons, such as application aborts, computer crashes, disc failures, and communication failures (for distributed DBMSs). Files and in-memory data lack transaction support.
- **Distribution**. Some DBMSs allow data to be available at multiple sites. The DBMS maintains consistency for replicated copies of data. (Chapter 18 discusses distributed databases.) Files and in-memory data lack distribution support.
- **Query language**. Most DBMSs provide a powerful language, so knowledgeable users and programmers can quickly construct queries. Several commercial products provide simple query languages for files. In-memory data lacks query languages.
- **Security**. An application must protect data against unauthorized access. To secure data, you must first determine what kinds of attacks to prevent. DBMSs secure data through means such as passwords, query modification through views, and statistical perturba-

tion. Operating systems provide only simple protection of files. In-memory data typically has little security support.

- **Metadata**. DBMSs explicitly manage the structure of data (metadata) and allow run-time access to it. Compilers typically remove such data structures from in-memory data and files.

Table 9.3 summarizes the features of the data management approaches.

	In-memory data	Files	DBMSs
Data persistence	Requires special hardware	Strong support	Strong support
Purchase cost	Special hardware cost	None	Can be costly
Lifecycle cost	Variable	Variable	Variable
Large quantities of data	Limited by hardware	No limit for sequential and random files; memory limits cached files	No limit
Performance	Very fast	Fast for sequential, random, and cached files	Fast
Extensibility	Limited	Limited	Excellent
Concurrent access	Object locking	File locking	Object/record locking
Crash recovery	Shadow memory	Backup files	Strong support
Integrity	None	None	Designer can specify rules
Transaction support	None	None	Short transactions + sometimes advanced transactions
Distribution	None	None	Sometimes
Query language	None	Partial	Powerful
Security	None	Simple operating system protection	Can be simple or sophisticated
Metadata	None	None	Yes

Table 9.3 Data management approaches

The portfolio manager. Ideally we would like to use a DBMS, primarily to simplify development, improve data integrity, and enhance extensibility. Also, a DBMS would let us export data to spreadsheets, tax software, or other software. We describe the implementations of the portfolio manager—first with files and then with a relational and an object-oriented

DBMS. We did not implement the portfolio manager using in-memory data, although it may be more suitable for other applications.

9.4.1 In-memory data

Computer memory can support persistent data when augmented with special power supplies or hardware. For example, both battery-powered memory and EPROMs retain data. In-memory storage increases performance and simplifies programming. Multiple users and applications may share data. On the down side, access can be inflexible because multiple applications may find it difficult to agree on a format. Furthermore, the rigidity of memory data can inhibit the extension and evolution of large systems. Finally, persistent memory has limited capacity and can be costly.

Data that can best be handled in memory, not with files or a database, includes

- Data for electronic devices (such as cellular telephones, modems, fax machines)
- Data for plug-in cards and cartridges (such as for electronic games)
- Data for applications that cannot tolerate the cost or unreliability of a disc drive.

9.4.2 Files

Applications can directly read and write to sequential or random-access files. Files provide a simple mechanism for achieving data persistence, but their complexity grows rapidly with the quantity and variety of data because all applications must have a common format. Files can be difficult to maintain, extend, and coordinate for concurrent access. Large sequential files are inefficient for accessing small, random portions of data. Random-access files are not readable by a text editor.

Data that belongs in files, not in a database or in memory, includes

- Data that is voluminous and of low information density (such as archival files, debugging dumps, intermediate computations, or detailed historical records)
- "Raw" data that is summarized in the database (such as from data acquisition)
- Modest quantities of data with a simple structure (such as when a DBMS is unavailable or too much of an administrative nuisance)
- Data that is accessed sequentially
- Data that can be fully read into memory (such as desktop publishing files).

9.4.3 DBMSs

A ***database*** is a permanent, self-descriptive repository of data that is stored in one or more files. Self-description is what sets a database apart from ordinary files. A database contains the data structure or ***schema***—description of data—as well as the data. The premise of databases is that the data structure is expected to be relatively static while the actual data may rapidly evolve.

DBMSs provide general-purpose routines and protocols for managing large quantities of data and isolate applications from physical data storage details. DBMSs are well suited for applications in which careful organization of the data and concurrent access are important. Mature products that can handle large quantities of data are readily available. DBMSs are appropriate for both operational and decision-support applications.

You should be aware of several aspects when using a DBMS:

- **Storage space**. Our experience is that most databases require about triple the storage space of the actual data. The extra space is consumed by the data dictionary, indexes, page and object/record headers, logging, and space left empty for future growth.
- **Response time**. Most database applications are I/O bound or communication bound (distributed databases). However, other factors, such as CPU use, locking contention, and delays from frequent screen displays, can also affect the response time of some applications.
- **Performance tuning**. The generic capabilities of a DBMS impose a performance tax that skillful designers can mitigate. DBMSs offer several performance tuning mechanisms, such as indexing, hashing, and physically collocating closely related objects.

  An ***index*** is an inverted tree with a wide fan-out at each node (often a factor of 50 or more). An index can speed the response to certain queries and can enforce uniqueness. Collections with few elements, or collections that are usually sequentially scanned, do not require indexes to improve performance. Indexes have the disadvantage of consuming disc space and slowing most updates.
- **Administrative burden**. Large databases require specially trained support staff to set security policy, manage disc space, prepare backups, monitor performance, adjust tuning, and deal with subtle problems that may arise.
- **Locking modes**. To accommodate concurrent users, many DBMSs support pessimistic locking, optimistic locking, or both. (There are additional approaches for providing concurrency.) A DBMS obtains ***pessimistic locks*** before accessing objects and releases them after object access is complete. You need pessimistic locking when there may be significant contention in accessing the database or when work must not be lost.

  In contrast, a DBMS obtains ***optimistic locks*** at the time of committing a transaction to the database. Reads and writes may freely occur. When activity has been completed and data is ready for committing to the database, the DBMS checks to see if contention has occurred. If there is no conflict (assumed to be the normal situation), writing may occur; otherwise all work in the transaction may be lost.

Data that belongs in a database, not in files or memory, includes

- Data that requires update at fine levels of detail by multiple users
- Data that must be accessed by multiple application programs
- Large quantities of data that must be efficiently handled
- Data that is long lived and highly valuable to an organization
- Data that must be carefully secured against unauthorized and malicious access
- Data that is subject to sophisticated analysis for decision support.

9.4.4 Hybrid Approaches

Occasionally it is helpful to mix data management approaches. For example, a ***product data manager*** combines a DBMS with files. A PDM is an application that organizes data for parts, often mechanical parts. The data includes composition (documented by a bill-of-material), drawings, engineering calculations, marketing literature, and test data. A PDM uses the DBMS to store metadata about parts and to control the execution of applications. The files store detailed information for a part and often serve as input and output to applications. The PDM just holds the files and makes them available. The database contains references to the files (directory and file name) and tries to keep the information in the database consistent with the actual files.

9.5 CHOOSING A DATABASE MANAGEMENT PARADIGM

If you are going to use a DBMS, you must now choose a DBMS paradigm. (If you are using files or persistent memory, you can skip this section.)

The portfolio manager. Chapters 13 and 14 implement the portfolio manager for a relational DBMS. Chapters 15 and 16 implement portions of the portfolio manager for an object-oriented DBMS.

9.5.1 Hierarchical and Network DBMSs

Hierarchical and network DBMSs are now obsolete, so you normally should not use them for new applications. You may still encounter hierarchical and network DBMSs with old legacy systems that must be maintained, integrated, or reverse engineered.

9.5.2 Relational DBMSs

In a ***relational database*** data is logically perceived as tables. A ***relational DBMS (RDBMS)*** manages tables of data and associated structures that increase the functionality and performance of tables. A relational DBMS has three major aspects:

- **Data that is presented as tables**. Tables have a specific number of columns and an arbitrary number of rows.
- **Operators for manipulating tables**. SQL is the standard language.
- **Constraints**. The leading products support referential integrity and simple constraints. Referential integrity ensures that references to values in other tables actually exist.

The basic features of RDBMSs vary little from product to product. RDBMSs have benefitted from a clear definition by an authoritative figure (EF Codd) and the SQL standard [Melton-93]. All RDBMSs support the paradigm of a table and implement a common core of the SQL language. Variations exist for data types, tuning capabilities, programmatic access, and data dictionary schema though the SQL standard is gradually subsuming these areas. Commercial RDBMSs include Oracle, Sybase, Informix, DB2, and MS-Access.

You can readily implement OMT models with an RDBMS. OMT models are largely declarative (you state *what* is desired but not *how* to accomplish it), matching the declarative nature of RDBMSs. Some degradation occurs because you must map all OMT constructs to a single RDBMS construct—the table. RDBMSs provide an adequate basis for implementing associations but lack support for inheritance, so you must use workarounds. The object model maps to data declaration commands (create table, create referential integrity, create index). The functional model maps to data manipulation commands (modify tables, select data from tables). The dynamic model does not directly map to relational constructs, and you must translate it. Chapters 13 and 14 show how to implement an OMT model for an RDBMS.

Although OO-DBMSs are now available, RDBMSs still excel for many applications. RDBMSs and OO-DBMSs have complementary strengths and weaknesses. Without a doubt, RDBMSs are too slow for some demanding applications. However, the performance limits of RDBMSs have been exaggerated. Poor performance is often caused by sloppy design, inadequate attention to tuning, and careless query formulation. The simple paradigm of a table can be deceiving; one table is easy to understand but a group of tables can be difficult to understand. Judicious use of indexes and careful query formulation can compensate for many of the limitations in current RDBMSs.

RDBMSs also have some prominent disadvantages:

- **Slow navigation**. RDBMSs perform slowly for applications with much navigation from object to object. RDBMS joins are inefficient for extensive navigation.
- **Lack of some advanced features**. RDBMSs lack advanced features, such as inheritance, schema evolution, versioning, configurations, long transactions, and change notification.

  Schema evolution is the ability to change the structure of a database that is populated with data; the DBMS automatically performs needed data conversions.

  A ***version*** is an alternative object to some base object. Developers often encounter versions with hypothetical situations, such as different possibilities for an engineering design.

  A ***configuration*** is a set of mutually consistent versions [Cattell-91].

  A ***long transaction*** is a series of DBMS commands that extend over a long period—hours, days, weeks, or even months. A long transaction contrasts with the short transaction supported by conventional DBMSs, which ordinarily resolves within a few seconds. Long transactions arise in cooperative design, in which someone is responsible for a portion of a project and must perform extensive work before sharing the results.

  With ***change notification***, a DBMS broadcasts a notice to registered users when an object changes.
- **Inflexible locking protocols**. RDBMSs automatically lock data for commands. However, the locking behavior is not always what you may want, and RDBMSs do not make low-level primitives available for other kinds of behavior.

- **Few data types**. The user cannot define new data types.
- **Table as the only paradigm**. The data for some applications does not naturally fit within the confines of a table.
- **Unfamiliarity to programmers**. RDBMS technology is unfamiliar to most programmers, who tend to be procedurally oriented. The declarative RDBMS paradigm awkwardly integrates with most programming languages, which tend to be imperative.

RDBMSs have significant strengths, however:

- **Theory and standards**. RDBMSs benefit from a sound theory and the SQL standard.
- **Wide availability**. RDBMSs have been widely deployed. Many corporations have site licenses and trained support staffs for at least one relational product; for these companies there is little infrastructure cost to adding another RDBMS application. Many vendor products incorporate an RDBMS.
- **Much extensibility**. There is latitude to change database schema without affecting existing application programs.
- **Declarative access to data**. Declarative programming is more powerful than imperative programming for skilled practitioners. Declarative statements are more amenable to a compiler's optimization.
- **Data dictionaries**. An RDBMS has a fully integrated data dictionary accessible to the user.
- **Fast associative queries**. If the database is properly indexed (see Chapter 14), RDBMSs can quickly access data for associative queries (computation on a collection of records).
- **Robust security**. The grant and view commands can enforce security.

The following kinds of applications can benefit from RDBMS services:

- **Decision-support applications**. RDBMSs are a good fit for many applications that require a powerful query language and that have ad hoc queries.
- **Ordinary business applications**. RDBMSs also satisfy business applications with modest structural complexity (usually less than several hundred tables) and a large number of instances (thousands, millions, or more records per table). The operations are simple but performed often.
- **Applications that can use a fourth-generation language**. A ***4GL*** is a framework for straightforward database applications that provides screen layout, specification of simple computations, and report preparation. Most 4GLs are built around an RDBMS.
- **Integrated applications**. An RDBMS may be appropriate for an application that must deeply integrate with legacy applications that already use an RDBMS.
- **Conservative implementations**. Many developers prefer to use mature RDBMS technology with proven administration, recovery, and security features.

9.5.3 Object-Oriented DBMSs

You can regard an ***object-oriented database*** as a persistent store of objects created by an object-oriented programming language. With an ordinary programming language, objects cease to exist when the program ends; with an object-oriented database, objects persist beyond the confines of program execution. An ***object-oriented DBMS (OO-DBMS)*** manages the data, programming code, and associated structures that constitute an object-oriented database. In contrast to RDBMSs, OO-DBMSs vary widely in their syntax and capabilities. Commercial OO-DBMSs include ObjectStore, Versant, GemStone, Objectivity, and O2.

You can readily implement OMT models with an OO-DBMS. OO-DBMSs directly represent classes, attributes, methods, and inheritance from the object model; many OO-DBMSs can also reasonably implement simple binary associations. You can implement the dynamic and functional models with OO-DBMS methods. The most significant issue in mapping OMT models to OO-DBMSs is the shift in paradigm—from declarative to imperative. OMT models largely describe properties of data structure and data manipulation. In contrast, most OO-DBMSs adopt an imperative paradigm for manipulating data; typically you must write a sequence of programming commands. Chapters 15 and 16 discuss the implementation of OMT models with an OO-DBMS.

OO-DBMSs have been motivated by the deficiencies in RDBMSs. RDBMSs are too slow for queries that navigate from object to object, their data types are too simple for many applications, and SQL awkwardly combines with programming languages. Unlike RDBMSs, OO-DBMSs support rich data types and cleanly integrate with at least one programming language. OO-DBMSs are much quicker for applications that navigate from object to object. OO-DBMSs support inheritance, object identity, and other advanced features.

Unfortunately, although OO-DBMSs remedy some RDBMS deficiencies, they also lose some of their strengths. A troubling aspect is that OO-DBMSs weaken the declarative paradigm and encourage procedural coding. The more mature OO-DBMSs are mitigating this problem by incorporating declarative query languages.

OO-DBMSs have the following disadvantages:

- **Incomplete theory and standards**. With RDBMSs formal definition preceded implementations; in contrast, OO-DBMSs have been implemented without a formal theory. As a consequence, OO-DBMS products vary widely and it is difficult to develop applications without depending heavily on a specific vendor. The ODMG standard [Cattell-96] is gradually improving this situation.
- **Possible database corruption**. Some OO-DBMSs run in the application process space. Consequently, the database may be subject to security violations or corruption by wild pointers. This potential corruption is a serious issue given the longevity of databases. In contrast, if the DBMS and applications are run in separate processes (as in an RDBMS), some of the performance advantage is lost.
- **Lack of logical extensibility**. Current products lack logical data independence. OO-DBMSs lack an analog to RDBMS views. This is not an intrinsic flaw of OO-DBMS technology, but rather a flaw in current products. If OO-DBMSs incorporated more sup-

port for derived data, each application could be made aware of only the relevant parts of the database. Individual applications could evolve and grow with less effect on related applications.

- **Lack of support for meta applications**. Many C++-based OO-DBMSs lack dynamic binding (binding at run-time) and provide only static binding (binding at compile-time). This restriction arises from language limitations. C++ uses type declarations to generate optimized machine code during compilation and then discards the declarations. In contrast, RDBMSs and most Smalltalk-based OO-DBMSs offer both static and dynamic binding. Lack of dynamic binding complicates framework development. (See Section 4.2.)

OO-DBMSs also have substantial strengths:

- **Fast navigation**. Because of low overhead and smart caching strategies, navigational queries are nearly as fast as accessing data in memory. The performance is not compelling for applications dominated by associative access (computation on a collection of records), however. Current OO-DBMSs often execute associative queries more slowly than RDBMSs. This performance difference is not intrinsic, but is due to the immaturity of current OO-DBMS products. Future OO-DBMSs should have performance comparable to that of RDBMSs for associative queries.
- **Advanced features**. OO-DBMSs have a rich variety of semantic constructs—including inheritance, user-definable data types, and often associations. Some OO-DBMSs also support advanced capabilities, such as schema evolution, versioning, long transactions, and change notification.
- **Flexible locking protocols**. The programmer may access low-level data management functionality. For example, OO-DBMSs provide direct control over the details of concurrent locking. The developer may use clean read locks (transaction consistent data), dirty reads (reading without locks), write locks, long transaction locks, and versions of data.
- **One uniform type system**. An OO-DBMS unifies the DBMS and language type system for at least one language. In contrast, RDBMSs integrate poorly with all conventional languages.

The following kinds of applications can benefit from OO-DBMS services. OO-DBMSs are more appropriate for new applications, rather than for migrating existing applications.

- **Engineering design applications**. OO-DBMSs are a good fit for design applications, such as computer-aided design (CAD), computer-aided manufacturing (CAM), computer-integrated manufacturing (CIM), and computer-aided software engineering (CASE).
- **Multimedia applications**. OO-DBMSs are appropriate for multimedia applications with complex graphics, audio, and video.
- **Knowledge bases**. Expert system rules are difficult to store in a relational database. When a rule is added, the entire rule base must be checked for inconsistencies and redundancies. An OO-DBMS can provide the appropriate low-level access.

- **Applications with demanding distribution and concurrency requirements**. OO-DBMSs allow the necessary access to low-level services.
- **Applications that require advanced features**. OO-DBMSs offer inheritance and user-definable data types. Some OO-DBMSs also have schema evolution, versioning, long transactions, and change notification.
- **Electronic devices with embedded software**. The software content of electronic devices is increasing. Some electronic devices can benefit from data management services, but cannot tolerate the overhead of an RDBMS. An OO-DBMS is a good fit for these kinds of systems because you can remove unneeded overhead.

9.5.4 Other DBMSs

Several other paradigms are at the frontier of DBMS technology.

A ***real-time DBMS*** is a DBMS in which the result must be correct *and* timely. A hard real-time system must meet strict deadlines. In contrast, a soft real-time system will occasionally miss a deadline; the DBMS discards computations that exceed a deadline to free system resources. Applications such as air traffic control, process control, and network management require real-time DBMSs. Real-time DBMSs involve special technology because the general-purpose routines of conventional DBMSs degrade performance. Conventional DBMSs also lack the required semantics—guaranteed response within time limits, interrupt handling, explicit scheduling, and prioritization of commands.

A ***deductive DBMS*** combines conventional data management abilities with the inference capability of an expert system. Hence it must manage data as well as constraints and derivation rules.

An ***active DBMS*** combines the management of data and the management of state. Most active DBMSs adopt an event-driven approach in which the occurrence of an event triggers some database action. An active DBMS provides a powerful paradigm that can directly represent the OMT dynamic model.

Table 9.4 summarizes the appropriate kinds of applications for the different DBMS paradigms.

9.6 DETERMINING OPPORTUNITIES FOR REUSE

You should strive for reuse throughout software development. Reusable software is usually better tested, more reliable, and better documented. You can also apportion software development costs across multiple applications.

Consider reusing generic software, frameworks, complete applications, and models. A DBMS is an example of generic software; for data-intensive applications DBMS commands can substitute for application code, achieving a great deal of reuse. Fourth-generation languages provide a useful framework for many database applications. Reuse of complete applications typically occurs when you purchase an application from a vendor.

DBMS paradigm	Recommended applications
Hierarchical DBMS	• Obsolete for new applications. May be encountered with legacy applications.
Network DBMS	
Relational DBMS	• Decision support applications • Ordinary business applications • Applications that can use a 4GL • Applications that integrate with a legacy RDBMS application • Conservative implementations
OO-DBMS	• Engineering design applications • Multimedia applications • Knowledge bases • Applications with demanding distribution and concurrency • Applications that require advanced features • Electronic devices with embedded software
Real-time DBMS	• Applications for which data must be correct and timely
Deductive DBMS	• Applications that combine data management and inference
Active DBMS	• Applications that combine data management with dynamic behavior

Table 9.4 Recommended applications for DBMS paradigms

Reuse of models is often the most practical form of reuse. The logic in a model can apply to multiple problems; for this reason we have attempted to include many model fragments in this book. Enterprise modeling can also make it easier to reuse models. The enterprise model serves as a carefully managed repository of concepts and model fragments that are important to an organization. You should seed a new application with appropriate fragments from this enterprisewide resource. Furthermore, if you bias application models toward the enterprise model, you will advance the broader goal of application integration. (See Chapter 19.)

Class libraries provide another form of reuse. There are various kinds of class libraries, including data structures (such as collection classes), user interface, window managers, graphics, mathematics, data management (such as that provided by an OO-DBMS), and application libraries. You may not be able to procure application-specific classes but may find them in-house.

[Korson-92] notes several qualities of "good" class libraries:

- **Coherence**. A class library should be organized about a few, well-focused abstract themes.
- **Completeness**. A class library should provide complete behavior for the chosen themes.
- **Consistency**. Polymorphic operations should have consistent names and signatures across classes.

- **Efficiency**. A library should provide alternative implementations of algorithms (such as various sort algorithms) that trade time and space.
- **Extensibility**. The user should be able to define subclasses for library classes.
- **Genericity**. A library should use parameterized class definitions where appropriate.

Unfortunately, problems can arise when integrating class libraries from multiple sources [Berlin-90]. Developers often disperse pragmatic decisions across classes and inheritance hierarchies. Class libraries may adopt policies that are individually sensible, but fundamentally incompatible with those of other class libraries. You cannot fix such pragmatic inconsistencies by specializing a class or adding code. Instead you must break encapsulation and rework the class library's source code. These problems are so severe that they will effectively limit your ability to reuse code from class libraries.

- **Argument validation**. An application may validate arguments as a collection or individually as entered. Collective validation is appropriate for command interfaces; the user enters all arguments and only then are they checked. In contrast, responsive user interfaces validate each argument or an interdependent group of arguments as it is entered. A combination of class libraries, some that validate by collection and others that validate by individual, would yield an awkward user interface.
- **Error handling**. Class libraries use different error-handling techniques. Methods in one library may return error codes to the calling routine, for example, while methods in another library may directly deal with errors.
- **Control paradigms**. Applications may adopt event-driven or procedure-driven control. With event-driven control the user interface invokes application methods. With procedure-driven control the application calls user interface methods. It is difficult to combine both kinds of user interface within an application.
- **Group operations**. Group operations are often inefficient and incomplete. For example, an object-delete primitive may acquire database locks, make the deletion, and then commit the transaction. If you want to delete a group of objects as a transaction, the class library must have a group-delete function or the DBMS must support nested transactions.
- **Garbage collection**. Class libraries use different strategies to manage memory allocation and avoid memory leaks. A library may manage memory for strings by returning a pointer to the actual string, returning a copy of the string, or returning a pointer with read-only access. Garbage collection strategies may also differ: mark and sweep, reference counting, or letting the application handle garbage collection (in C++, for example).
- **Name collisions**. Class names, public attributes, and public methods lie within a global name space, so you must hope they do not collide for different class libraries. Most class libraries add a distinguishing prefix to names to reduce the likelihood of collisions.

The portfolio manager. We achieve reuse for two of our implementations by using generic software (an RDBMS and an OO-DBMS). We will also reuse models across all three implementations (files, RDBMS, and OO DBMS). Application vendors often implement their products across multiple data management platforms.

9.7 CHOOSING A STRATEGY FOR DATA INTERACTION

Over the years, we have identified several strategies for incorporating DBMS capabilities into an application:

- **Batch preprocessor and postprocessor**. Sometimes it is desirable to avoid database traffic within an application program. As Figure 9.1 shows, preprocessors and postprocessors require no changes to existing batch applications. The basic idea is simple: Query the database and create an input file, run the application, and then analyze the output and store the results in the database. However, database interaction via intermediate files can be awkward. You must request all database information before executing the application, and output files with complex formats can be difficult to process.

  The batch strategy is useful for old software or certified software you cannot alter.

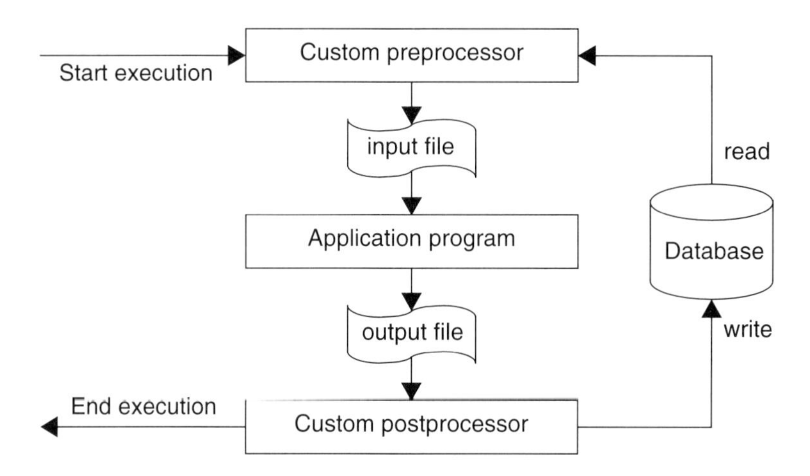

Figure 9.1 Batch preprocessor and postprocessor

- **Script files**. Sometimes an ASCII file of DBMS (relational or OO) commands is all you need to implement a function. For example, typing *@filename* into the interactive SQL (SQL Plus) of the Oracle DBMS causes all commands in *filename* to execute. Control reverts to interactive SQL after the commands in the file are executed. You can use an operating system shell language to execute multiple script files and to control their execution.

  Script files are helpful for simple database interaction, such as creating schema. You can also use script files for prototyping.

- **Embedded OO-DBMS commands**. As Figure 9.2 shows, applications can directly query the database and use the results to control their subsequent execution.

  You can freely embed OO-DBMS commands in OO application code, because an OO-DBMS naturally blends with most OO programming languages. In fact, you may have to adopt this strategy to obtain the best performance. On the down side, if you tightly couple your application to an OO-DBMS, it will be difficult to port to another product. However, the ODMG standard [Cattell-96] is gradually improving portability.

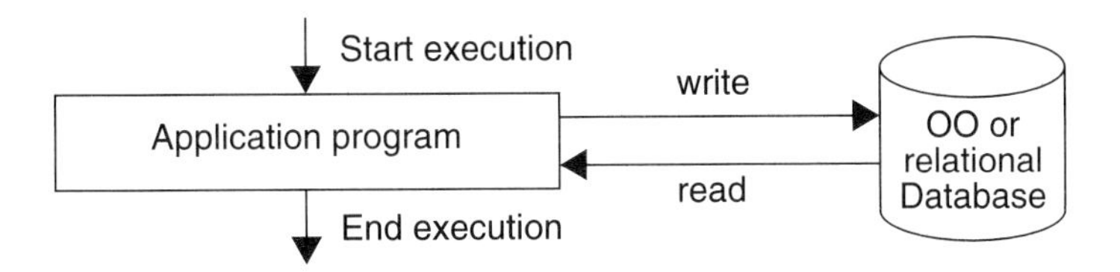

Figure 9.2 Embedded DBMS commands

- **Embedded static SQL.** ***Static SQL*** refers to database code that is known at compile-time (as opposed to ***dynamic SQL***, database code that is not known until run-time). You can intersperse static SQL with application code, a straightforward technique that most commercial products fully support.

  Try to avoid indiscriminate embedding of static SQL code in application programs, because programming code with embedded SQL code is difficult to read. Moreover, the interface between conventional languages and SQL is verbose, unwieldy, and tedious to deal with. The essential problem is that the conceptual basis for a relational DBMS is different than that for most programming languages.

- **Custom application programming interface (API).** Another strategy is to encapsulate database read and write requests within a few application methods. Such methods can help keep derived data consistent with underlying data and enforce transaction scope. The methods collectively provide an interface to the database.

  This strategy has several advantages. The custom API is simple and does not require any special infrastructure, yet you can isolate database access and prevent it from contaminating the application as a whole. This strategy is especially valuable when you need to partition the tasks of data management, application logic, and user interface among development teams.

- **Methods stored in the database.** Stored methods are efficient and provide more opportunities for reuse. Some RDBMSs can bind a named procedure with optional arguments to a block of SQL code stored in the database. For example, Sybase supports the notion of *stored procedures*. Some OO-DBMSs store methods in the database, particularly OO-DBMSs based on Lisp or Smalltalk. Most C++-based OO-DBMSs do not store method code in the database.

  Some RDBMSs require that you use stored procedures for maximum efficiency. You should use stored procedures with caution until they are more fully addressed by the SQL standard.

- **Fourth-generation language (4GL).** Vendors normally provide 4GLs for RDBMSs but could also provide them for OO-DBMSs. 4GLs promote reuse and compensate for the discontinuity between some DBMSs (particularly RDBMSs) and programming languages.

4GLs are widely available and can greatly reduce application development time. They are best for straightforward applications. They are not suitable for applications with complex programming.

- **Generic object-oriented layer**. A generic interface hides the DBMS; you can write application code in terms of the layer and largely ignore the underlying DBMS. A well-conceived interface can serve a variety of applications, and you may be able to substitute a DBMS later on. However, a generic layer can sometimes impede performance.

  Figure 9.3 shows a technique we have used for several projects [Premerlani-90]. Applications can directly access the underlying relational database for global data that must be shared. Or they can use buffered access to in-memory data structures for long transactions. Reads and writes to the underlying RDBMS check buffered data in and out of memory (OO language data structures)—fetch all objects from a section of the database, operate on the data, then write all changed objects back to the database. This technique offers fast buffered updates, simple programming, and the use of mature RDBMS technology. However, initialization is slow, concurrency is coarse, and it is difficult to prevent database corruption by other programs that may circumvent the layer. Several commercial products also provide a generic OO layer for hiding an RDBMS or OO-DBMS.

  An OO layer is most appropriate when you want to use an RDBMS and still have the convenience of OO programming.

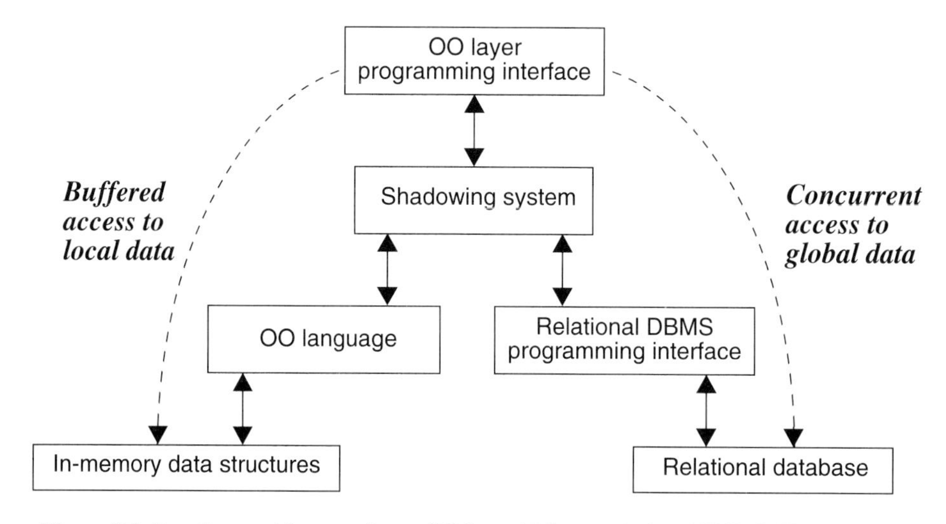

Figure 9.3 Run-time architecture for an OO layer hiding a relational DBMS ([Premerlani-90])

- **Metamodel-driven interaction**. As Figure 9.4 shows, the application indirectly accesses data by first accessing the metadata and then formulating the query to access the data. For example, a DBMS processes commands by first accessing the data dictionary and then accessing the actual data. The metamodel-driven architecture is commonly used with frameworks (see Section 4.2).

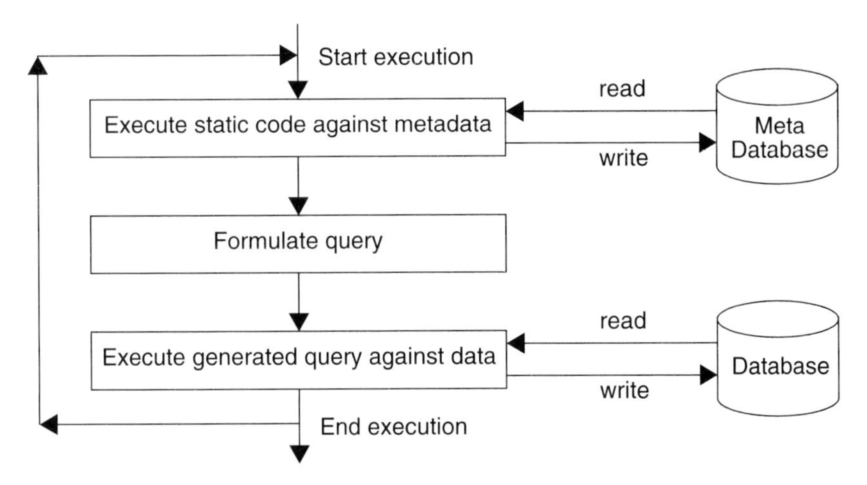

Figure 9.4 Metamodel-driven interaction

Most OO-DBMSs based on C++ do not allow dynamic query construction and consequently cannot support metamodel-driven interaction. This restriction arises from limitations of C++. Both RDBMSs and Smalltalk-based OO-DBMSs can support metamodel-driven interaction.

We have developed several applications with metamodel-driven interaction. [Blaha-90] provides an example of a metamodel-driven application using dynamic SQL with an RDBMS. The resulting applications are complex to develop and understand, but are flexible and powerful. You should consider metamodel-driven interaction if you are developing generic software or a framework.

Sometimes you will find it helpful to combine data interaction approaches. For example, you may use a 4GL and implement some methods with stored procedures. Table 9.5 summarizes the data interaction strategies.

The portfolio manager. We will use a 4GL for the RDBMS implementation of the portfolio manager; MS-Access provides a polished 4GL with event-driven control. We do not need to choose a data interaction strategy for files and OO-DBMSs because we are implementing only portions of the case study.

9.8 CHOOSING AN APPROACH TO OBJECT IDENTITY

The next step in system design is to choose an approach to object identity. OO-DBMSs assume that objects have an ***existence-based identity***, in which a system-generated object identifier (also called an OID, a surrogate, or a pointer) identifies each object. Files and RDBMSs

Data interaction strategy	Recommendations
Batch preprocessor, postprocessor	Preferred for old or certified software.
Script files	Can be helpful for simple database interaction or quick prototyping.
Embedded OO-DBMS commands	May be required for the best performance. Ties application to a specific product.
Embedded static SQL	Discouraged.
Custom API	A good choice, especially when coordinating development teams.
Methods stored in the database	Can yield high efficiency. May be helpful for utility methods to promote reuse. Use stored procedures with caution until they are more fully addressed by the SQL standard.
4GL	A good choice for straightforward applications. Not suitable for applications with complex programming.
Generic OO layer	Appropriate when you want to use an RDBMS and still have the convenience of OO programming.
Metamodel driven	Occasionally useful. Consider if you are developing generic software or a framework.

Table 9.5 Data interaction strategies for DBMSs

give you the choice of existence-based identity or ***value-based identity***, in which some combination of attributes identifies each object. You can mix the two approaches in one application.

Existence-based identity has several advantages. The resulting primary keys are single attribute, small, and uniform in size. An object identifier can also encode information to promote efficient lookup. For many models, especially large and complex models, some classes have no obvious combination of real-world attributes that could provide value-based identity.

Existence-based identity also has disadvantages. Identifiers can be troublesome to generate for files and RDBMSs that lack semantic support. Assigning identifiers may require access to global data that can interfere with concurrent locking. Existence-based identity can complicate inspection of the database or files and debugging of the corresponding application.

Value-based identity has several advantages. Primary keys have intrinsic meaning to the user, facilitating debugging and maintenance of the database. It may be easier to distribute the database and extend the database to support new applications.

On the down side, value-based primary keys can be difficult to change. A change to a primary key may require propagation to many foreign keys (from associations and generalizations). Some RDBMSs do not semantically support this propagation. Furthermore, some objects do not have natural real-world identifiers. Multiattribute primary keys can be lengthy and complicate the generation of foreign key names from role names.

You must use existence-based identity for OO-DBMS applications. OO-DBMSs hide the details of identifier generation and mapping to secondary storage. We recommend that you use existence-based identity for RDBMS and file applications with more than 30 classes. Both existence-based and value-based identity are viable options for small RDBMS and file applications.

Table 9.6 summarizes our recommendations for the approaches to object identity.

The portfolio manager. We will use existence-based identity for our file, RDBMS, and OO-DBMS implementations.

Approach to object identity	Recommendations
Existence-based identity	OO-DBMS applications. RDBMS and file applications with more than 30 classes. Consider for RDBMS and file applications with less than 30 classes.
Value-based identity	Consider for RDBMS and file applications with less than 30 classes
Combination	Consider for RDBMS and file applications with less than 30 classes.

Table 9.6 Approaches to object identity

9.9 DEALING WITH TEMPORAL DATA

Several forms of temporal data arise for the portfolio manager:

- The time when a transaction occurs
- The time when the user records a transaction
- The time of valuation of an asset
- The time a portfolio value is computed
- The time interval (starting time and ending time) for computing ROI.

During analysis we addressed the first two bullets. (See Figure 8.11.) Transactions happen at a single moment in time. Users generally record transactions after they occur.

Time also arises with the value of assets. But in this case, change continually happens. In principle, the value of assets adjusts as trades occur in the marketplace, but the portfolio manager does not record these rapid fluctuations (unless we added an interface to a ticker tape feed).

In Chapter 8 we considered only the current value of an asset. We postponed dealing with the variation in value because we deemed this a solution issue (design) rather than a conceptual issue (analysis). For example, we may implement values differently for an OO-DBMS (use versions) than for an RDBMS (no version support). Also, the temporal change of value is difficult to deal with and we first wanted to grasp the overall problem.

Our solution is to record the value of assets at certain discrete times the user chooses. The burden of manual data entry will necessarily limit the quantity of financial valuation data. The user may compute portfolio value and ROI for the times when all the underlying assets have values.

Figure 9.5 extends the object model in Figure 8.11 to show the variation of value with time (and other changes described in the next section). We can ignore this variation for an OO-DBMS implementation with versioning. We will accommodate structural changes in portfolio composition over time by replaying the transaction log (Section 8.3.12). By replaying the transaction log we can determine the assets in a portfolio at any time.

9.10 DEALING WITH SECONDARY ASPECTS OF DATA

Occasionally you will encounter secondary aspects of attributes and classes. (See Section 3.1.6.) DBMSs ignore secondary data, so you must plan a workaround. Here are some strategies for dealing with secondary data for attribute values:

- **Establish a convention**. Units of measure often arise in scientific applications, for example. You could assume canonical units of measure, that the database or files will always store the specified units for power, length, time, and other measures. You should explicitly note any secondary data conventions in the data dictionary and method documentation.

  You can establish secondary data conventions with files, but avoid such conventions with databases. Assumed units of measure contravene the spirit of database management; the intent of a database is to place documentation with the data (in the data dictionary) without any hidden assumptions.

- **Add attributes**. If you have a small amount of secondary data, you could just add a few attributes or associations to some classes. This approach is attractive because it is simple to implement.

  We prefer this approach for a small amount of attribute secondary data.

- **Extend domain definitions**. This is the most general approach for handling secondary data for attribute values. You can define a structured domain that contains the basic data type as well as additional information and ultimately implement such a domain as a class. (In contrast, you should implement a simple domain as a primitive type, such as integer, real, string, or symbol.) The disadvantage of this strategy is added complexity, increased storage, greater development effort, and reduced performance.

  For example, class *ChemicalSubstance* could have attributes *tradename*, *density*, and *viscosity*. Attribute *tradename* could have the domain of *name* while *density* and *viscosity* could have the domain of *experimentalValue*. The *experimentalValue*, in turn, could consist of *numericValue*, *unitsOfMeasure*, *dataAccuracy*, *dataSource*, and *dateOfLastUpdate*.

We recommend this approach when you have values with several dimensions of secondary data.

Secondary data for classes also requires workarounds:

- **Use multiple inheritance**. You can use multiple inheritance to implement secondary data for classes. The Versant OO-DBMS takes this approach. You declare transient variables that exist only during a program execution in the usual manner. Persistent variables must also inherit from the *PersistentObject* class. Multiple inheritance lets you elegantly mix in orthogonal capabilities. On the down side, multiple inheritance can be complex to implement and class libraries can be awkward to combine. (See Section 9.6.)
  Multiple inheritance is best when you need to reuse complex behavior and the pragmatic and implementation problems of multiple inheritance are not too troublesome.
- **Duplicate classes**. You may introduce logically duplicate classes. For example, you may have one group of classes for transient objects and another group of classes with similar structure for persistent classes.
  In general, you should consider this approach only for a small number of duplicate classes.
- **Add attributes**. You may be able to add an attribute to a class for each kind of class secondary data. For example, objects could have a flag denoting whether or not they are versionable. Similarly, they could have an attribute denoting their section for checkout in a long transaction [Premerlani-90]. These attributes would appear in every class with secondary data. You can define macros that turn these attributes on and off as needed.
  This approach is best when the secondary data applies to many classes.

Table 9.7 summarizes the strategies for dealing with secondary data.

	Strategy	Recommendations
Secondary data for attribute values	Establish a convention	OK for files. Don't use with databases.
	Add attributes	Use for a small amount of secondary data.
	Extend domain definitions	Use for values with several dimensions of secondary data.
Secondary data for classes	Use multiple inheritance	Consider when reusing complex behavior.
	Duplicate classes	Only consider for a small number of duplicates.
	Add attributes	Use when secondary data applies to many classes.

Table 9.7 Strategies for dealing with secondary data

The portfolio manager. We will use the add attributes approach to deal with secondary data for money. We will specify currency (such as dollars, marks, yen) for several classes by adding an association to *Cash*. Then the units of measure for money will be clear throughout the model. We did not notice this oversight in the model until now. Figure 9.5 shows the revised object model for the portfolio manager.

The portfolio manager does not have secondary data for classes.

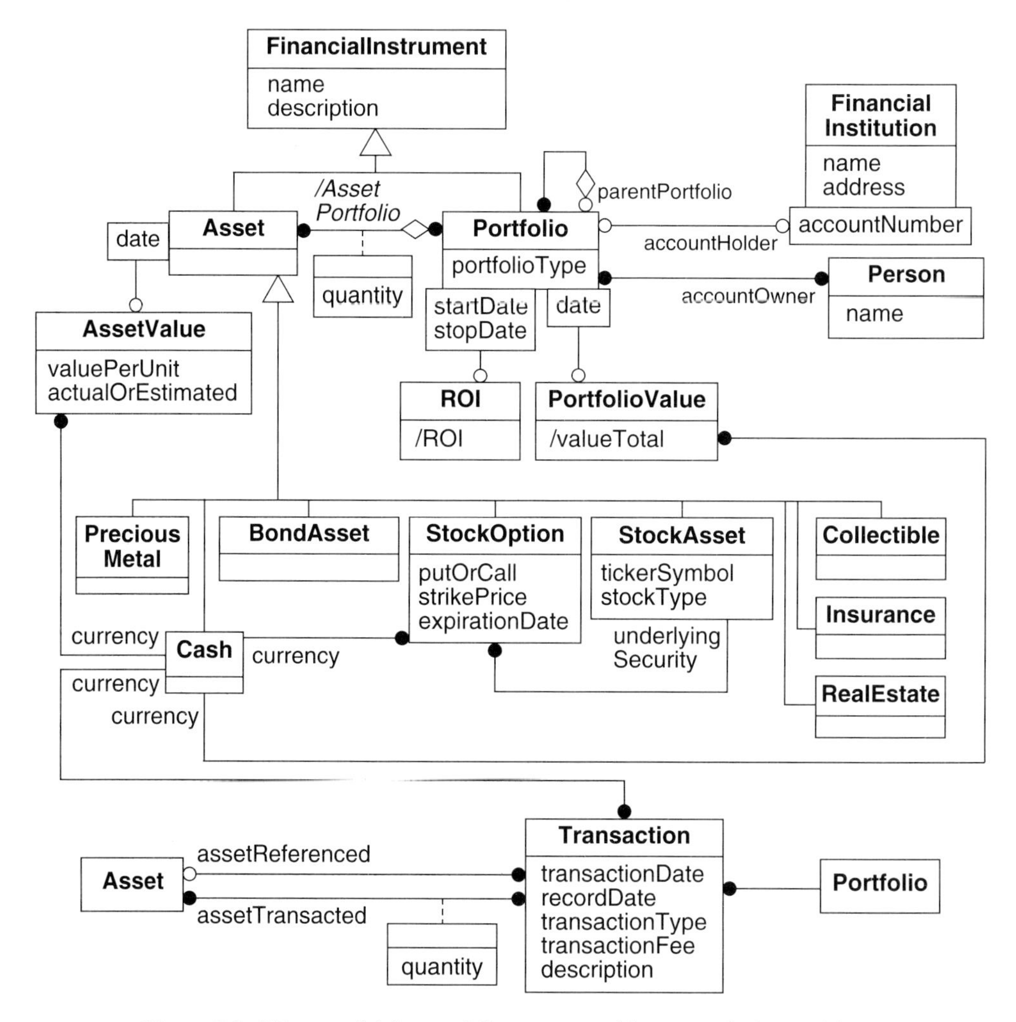

Figure 9.5 Object model for portfolio manager with system design revisions

9.11 SPECIFYING DEFAULT POLICIES FOR DETAILED DESIGN

Many other issues arise in data-intensive applications. During system design you should establish default policies to increase uniformity and simplify the next methodology step, detailed design.

- **Associations**. Choose a basic approach to designing associations and consistently apply this to your application. Deviate only when there is a compelling need, such as specially coding performance bottlenecks. Chapters 12 through 16 present the implementation options. For example, you may promote a one-to-many association to a class or bury it as an attribute in the *many* class.

  Aim for an implementation target that provides semantic support. The ODMG standard [Cattell-96] and some OO-DBMSs support simple binary associations. (ObjectStore has inverse members and Versant has bilinks). OO-DBMSs do not support ternary associations, link attributes, and qualifiers, so you must promote these associations to classes. Relational DBMSs partially support associations with referential integrity on foreign keys. Simple files lack support.

  When the implementation medium does not support associations, you can resort to a workaround. You can always let the application software enforce association semantics. This option is inferior to specification via the database structure, because association semantics are obscured by application semantics. Such code is time-consuming to write, vulnerable to bugs, and a maintenance burden.

  An elegant solution is to implement generic association methods [Rumbaugh-91, page 247]. However, generic code is always more difficult to develop than application code. Generic association methods impose a modest performance overhead (10–20%) compared to custom application code.

  The portfolio manager. Our basic policy depends on the implementation platform. We will discuss this in Chapters 12 through 16.

- **Nulls**. Establish a general policy for null values. ***Null*** is a special value denoting that an attribute value is *unknown* or *not applicable*. You may use the null value supported by the DBMS or designate a special value (a default value) to denote null for each attribute. It may seem easier to use the null logic from the DBMS, but the standard logic for nulls is not always adequate [Date-86].

  The portfolio manager. We will allow null values.

- **Attribute names**. Attribute names may be globally unique or unique within a class or association. Locally unique attribute names require that you specify the context (class or association) for the attribute name. Globally unique attribute names can simplify application software because you need reference only a single name. However, globally unique attribute names can impede concurrent software development by a team.

  The portfolio manager. We will not require that attribute names be globally unique.

- **Role names**. You may interpret role names as a literal name or as a prefix or suffix to be concatenated to the referenced primary key. We have used both strategies, though our usual approach is to treat a role name as a literal name. It is not possible to treat the role as a literal name for composite primary keys.

  The portfolio manager. We will treat role names as literal names.

- **Derived data**. You may compute derived data in advance and cache the results (eager evaluation) or compute derived data as needed (lazy evaluation). A third option is to compute derived data as needed and record the results for possible reaccess. Few OO-

DBMSs support derived data. RDBMS views let you read derived data, but often cannot support updates. When using files or an inadequate DBMS, you must compensate with application software.

The portfolio manager. In Section 9.2 we chose a hybrid solution to data caching, in which portfolio values and ROIs are stored as a side effect as they are computed. Any change to the underlying data causes the stored values to be deleted.

- **Security/access control**. Often it is convenient to define privileges for prototypical users. For example, you may wish to define privileges for an engineer, a manager, a secretary, and a database administrator. As you authorize new users, you can assign them the privileges of their prototype. The Sybase RDBMS adopts this approach to access control. You can also use RDBMS views and stored procedures, as well as OO-DBMS derived data and methods, to enhance security.

  The portfolio manager. We will ignore security and access control issues. We have conceived the portfolio manager as a single user application.

9.12 CHAPTER SUMMARY

After you have analyzed a problem, you must decide how to approach the design. During analysis, the focus is on *what* needs to be done, independent of *how* it is done. During design, you make decisions about how the problem will be solved, first at a high level (system design), then at increasingly detailed levels (detailed design).

You should begin system design by formulating a system architecture, the high-level plan or strategy for solving the application problem. We presented several principles to guide your formulation. For difficult problems, you should first generate candidate architectures and evaluate them in a separate step. You will promote creative thinking and more effectively assimilate ideas from multiple persons by temporarily suspending critical review. We have often had ideas at first we were inclined to reject, but found more promising after reflection.

You should choose a paradigm for controlling your application. You must choose from four strategies. With procedure-driven control, control resides within a single thread of programming code. In contrast, with event-driven control, a dispatcher retains control and calls application code as needed. Concurrent control directly realizes the concurrency of objects. With declarative control, the software derives an implicit flow of control from declarative statements.

The next step is to choose a data management approach: in-memory data, files, or a DBMS. Developers seldom use persistent in-memory data, because it requires special power supplies or hardware. Files are a viable option for archival data, data that is accessed sequentially, or data that can be fully read into memory. DBMSs are an appropriate choice for large, complex, and demanding applications.

If you are using a DBMS, you must choose the specific DBMS paradigm. Hierarchical and network DBMSs are now obsolete, and normally you should no longer use them for new applications. Relational DBMSs and object-oriented DBMSs are viable choices. RDBMSs have the advantages of maturity and widespread use; they are well suited for straightforward business applications. OO-DBMSs resolve some limitations of RDBMSs, but have different shortcomings. You should consider OO-DBMSs for applications with a complex data structure or demanding distribution and concurrent user requirements.

You should actively seek reuse during software development. Consider reusing models, generic software, frameworks, complete applications, and class libraries. A DBMS is an example of generic software; for data-intensive applications DBMS commands can substitute for application code, achieving a great deal of reuse.

There are many strategies for interacting between a DBMS (or files) and an application: batch interface, script files, embedded DBMS commands, an application programming interface, methods stored in the database, fourth-generation languages, generic layers, and metamodel-driven interaction. The appropriate strategy depends on the application.

The next step in system design is to choose an approach to object identity. Existence-based identity is the implementation of identity with system-generated object identifiers (also called OIDs, surrogates, or pointers). In contrast, with value-based identity some combination of attribute values identifies each object. OO-DBMS applications, as well as large relational DBMS and file applications, should use existence-based identity. Either approach is viable for small applications (fewer than 30 classes) that use relational DBMSs or files.

You are likely to encounter temporal data. For the portfolio manager the appropriate solution differs for OO-DBMSs (use versions if available) and RDBMSs (no version support).

You will occasionally encounter secondary aspects of attributes and classes. Since DBMSs ignore secondary data, you must plan a workaround. Three strategies for dealing with secondary data for attribute values are to establish a convention, add secondary attributes to the model, and extend domain definitions. To cope with secondary data for classes you may use multiple inheritance, duplicate classes, or add attributes that cut across classes.

Finally, in data-intensive applications many miscellaneous issues are likely to arise. To make detailed design easier, you must choose a basic approach to designing associations and decide whether or not to use nulls. Attribute names may be globally unique or unique to a class or association. You may interpret role names as a literal name or as a prefix or suffix to be concatenated to the referenced primary key. You must decide whether to compute derived data in advance or on demand. We find it convenient to control access by defining privileges for prototypical users.

Figure 9.6 lists the key concepts for this chapter.

application programming interface	fourth-generation language	relational DBMS
architecture	in-memory data	reuse
data interaction	layer	schema
database	logical data independence	secondary data
existence-based identity	OO-DBMS	system design
external control	partition	temporal data
files	physical data independence	value-based identity

Figure 9.6 Key concepts for Chapter 9

BIBLIOGRAPHIC NOTES

[Rumbaugh-91] describes several canonical application architectures. Most database applications or the subsystem that provides database services can be characterized by the transaction manager architecture. The other canonical architectures seldom apply to database applications.

[Blanchard-90] presents a broad treatment of systems engineering of which the material in Section 9.2 is only a small subset. Blanchard's treatment is oriented toward mechanical and aerospace systems, but many of the same principles apply to software.

REFERENCES

[Berlin-90] Lucy Berlin. When objects collide: experiences with reusing multiple class hierarchies. *ECOOP/OOPSLA 1990 Proceedings*, October 21–25, 1990, Ottawa, Ontario, Canada, 181–193.

[Blaha-90] MR Blaha, WJ Premerlani, AR Bender, RM Salemme, MM Kornfein, and CK Harkins. Bill-of-material configuration generation. *Sixth International Conference on Data Engineering*, February 5–9, 1990, Los Angeles, California, 237–244.

[Blanchard-90] Benjamin S. Blanchard and Wolter J. Fabrycky. *Systems Engineering and Analysis*. Englewood Cliffs, New Jersey: Prentice Hall, 1990.

[Cattell-91] RGG Cattell. *Object Data Management: Object-Oriented and Extended Relational Database Systems*. Reading, Massachusetts: Addison-Wesley, 1991.

[Cattell-96] RGG Cattell, editor. *The Object Database Standard: ODMG-93, Release 1.2*. San Francisco, California: Morgan Kaufmann, 1996.

[Date-86] CJ Date. *Relational Database: Selected Writings*. Reading, Massachusetts: Addison-Wesley, 1986.

[Korson-92] Tim Korson and John D. McGregor. Technical criteria for the specification and evaluation of object-oriented libraries. *Software Engineering Journal* (March 1992), 85–94.

[Melton-93] Jim Melton and Alan R. Simon. *Understanding the New SQL: A Complete Guide*. San Francisco, California: Morgan Kaufmann, 1993.

[Premerlani-90] WJ Premerlani, MR Blaha, and JE Rumbaugh. An object-oriented relational database. *Communications ACM 33*, 11 (November 1990), 99–109.

[Rumbaugh-91] J Rumbaugh, M Blaha, W Premerlani, F Eddy, and W Lorensen. *Object-Oriented Modeling and Design*. Englewood Cliffs, New Jersey: Prentice Hall, 1991.

EXERCISES

9.1 (4) For the following applications discuss the relative merits of files, relational DBMSs, and object-oriented DBMSs.

- **a.** A personal computer chess game that maintains a history of past games and the sequences of actual moves.
- **b.** A payroll processing program.
- **c.** A three-dimensional piping layout program.

9.2 (4) For Exercise 9.1 discuss the relevance of the DBMS features listed in Section 9.4.

9.3 Figure E9.1 presents requirements for a prescription drug information system. Figure E9.2 shows the corresponding object model. Figure E9.3 explains the object model.

Develop a drug description system to assist a pharmacist in filling prescriptions.

A drug is a medical substance used in the treatment of a disease, having a name, chemical formula, and medical action. A drug may come in many forms, including pills, capsules, gels, ointments, tinctures, suspensions, and powders, for example. A compound is a particular form and strength of a drug or combination of drugs produced by a particular manufacturer. Compounds may be purchased through distributors. There may be various compounds of a drug. A compound may have secondary ingredients besides the drugs, which are the primary ingredients.

Every drug has a controlled substance classification relevant to its addictive properties and a functional classification, which is a broad categorization of the purpose of the drug, such as antibiotics, antifungals, and sedatives.

Finally, there are possibly harmful drug interactions and allergies to consider. Drug interactions usually involve a pair of drugs, but might involve three or more. Allergies are summarized with allergy identifiers, which specify allergic reaction mechanisms.

The system should assist with the following tasks:

- Detect allergies and drug interactions.
- Manage information about drugs and compounds.
- Determine distributors and manufacturer of a compound.

Figure E9.1 Problem statement for a prescription drug information system

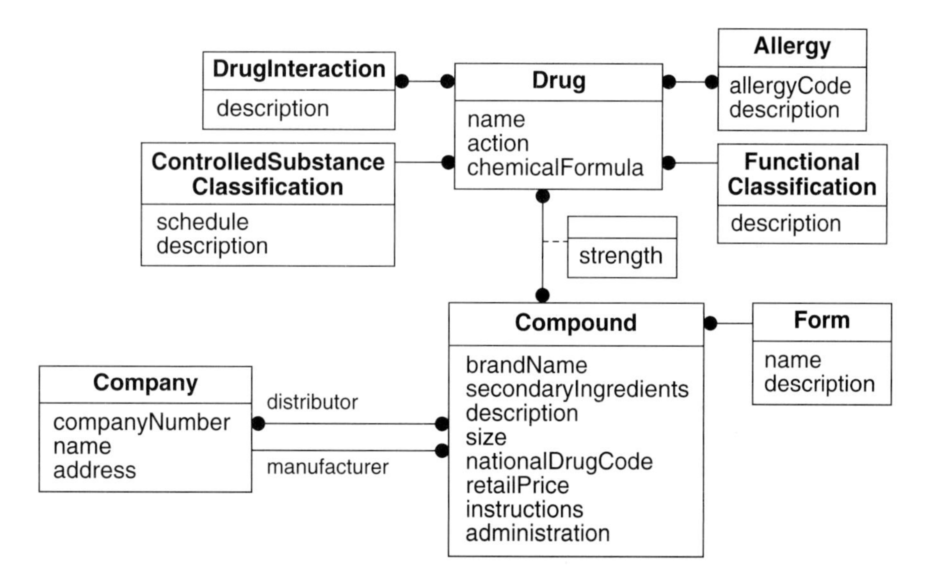

Figure E9.2 Object model for a prescription drug information system

A compound is a particular form, size, and strength of a drug or combination of drugs, produced by a manufacturer and sold by distributors. A compound often has several secondary, inert ingredients besides the drugs, which are the primary ingredients. A compound may have a brand name assigned by the manufacturer, such as Tylenol, MetroGel 0.75%, or Retin-A Gel 0.025%. The strength is the concentration or amount of a drug in a given compound. Thus, the compound MetroGel 0.75% has a 0.75% concentration of metronidazole. In the case of pills, strength is the amount of the drug in the compound, expressed in grams.

Compounds come in many forms, such as capsule, gel, liquid, pill, ointment, tincture, suspension, suppository, or powder. The size of a compound is the amount of the compound. The meaning of size depends on the form. For example, Retin-A Gel comes in both a 45-gram tube and a 15-gram tube. Other forms, such as pills, are sold individually.

A drug has a unique name, such as acetaminophen, metronidazole, or tretinoin. Every drug has a controlled substance classification as well as a functional classification. The controlled substance classification is referred to as the schedule of the drug. The term "schedule" arises from the fact that there are published "schedules," or lists of drugs, and that there are regulatory controls associated with a schedule. There are five possible values for schedule in the United States: I, II, III, IV, and V. Schedule I drugs have high potential for abuse and no accepted medical use. Schedule II, III, IV, and V drugs have an accepted medical use. Schedule II drugs have a high potential for abuse. Schedule V drugs have a low potential for abuse. Schedules III and IV have intermediate potential for abuse. The functional classification of a drug broadly refers to its purpose. Classifications include antibiotics, antifungals, antihistamines, sedatives, beta-blockers, and tranquilizers, for example.

Multiple drugs taken at the same time may produce an interaction. An interaction usually involves a pair of drugs, but may in some situations consist of three or more drugs. A drug may be assigned any number of allergy codes to indicate an allergic mechanism for vulnerable patients. Most drugs do not have any allergy codes.

In the United States, each prescription compound is assigned a national drug code (NDC), which is a structured code consisting of three parts. The first part is the manufacturer code. The second part is a drug code assigned by the manufacturer. The last part is a code for the size of the compound. For example, the NDC for a 30-gram tube of MetroGel 0.75% gel manufactured by Galderma is 00299-3835-28. (Yes, the size code for a 30-gram tube is 28.)

A compound has an intended route of administration. Some examples are oral, subcutaneous (injection under the skin), intramuscular (injection into a muscle), intravenous (into a vein), sublingual (under the tongue), or transdermal. Some routes are normally used in a hospital, but the information system could be used in a hospital pharmacy.

Figure E9.3 Explanation of object model for a prescription drug information system

a. (3) Assume that the prescription drug information system will be used by a chain of pharmacies. There are at least three architecture choices for storing data—locally at each pharmacy, at several regional sites, or at a single corporate facility. With local storage each pharmacy could use a personal computer for all data storage and processing. The regional and corporate facilities could use file servers to store and process data; personal computer clients at each store would provide the user interface and access data over telephone lines. Assign weights and explain your reasoning for each of the following decision criteria: performance and ease of use, integrity and accuracy, development effort, extensibility, ease of maintenance, legal compliance.

b. (4) Rate the choices using the decision criteria and explain your ratings. (Your documentation should be similar to that in Section 9.2.4.) Which architectural choice would you favor?

c. (3) Choose an implementation for external control, consistent with the architecture that you favored in part b. Explain your decision.

d. (3) Choose a data management approach and explain your decision.

e. (3) If you were to use a DBMS, which would be the most appropriate database management paradigm?

f. (4) Suggest an appropriate strategy for data interaction. Give both your first and second choices.

g. (3) Choose an approach to identity and explain your choice.

h. (3) Note that the units of measurement for strength and size are implicit. For example, strength is stated as percentage for a gel and weight for a pill. This is an example of secondary data for attributes. Adjust Figure E9.2 to make this secondary data explicit.

9.4 Figure E9.4 presents requirements for prescription refill software. Figure E9.5 shows the corresponding object model. Figure E9.6 explains the object model.

> Extend the drug information system to assist a pharmacist in filling and refilling prescriptions by storing and retrieving prescription and customer information.
>
> A prescription is a dated, written order for a drug compound issued for a customer by a physician. Information on a prescription must include a single specification of the compound, quantity, and instructions. Instructions are required, even if they are simply "take as directed." Some of the information may be in Latin. A prescription may optionally include a number of refills, diagnostic identifier, route of administration, and an indication to fill generically or as written.
>
> The system should assist with the following tasks:
>
> - Assign a prescription number to the prescription for identification purposes.
> - Translate Latin phrases.
> - Fill or refill a prescription, including printing labels and instructions as well as information that is attached to and filed with the original written order.
> - Track the number of refills left and expiration of the prescription.
> - Verify that the drug is appropriate for the condition being treated.
> - Track customer information.
> - Locate prescription information by customer name and/or prescription number.
> - Screen for drug interactions and allergies.

Figure E9.4 Problem statement for prescription refill software

a. (4) The prescription drug system in Exercise 9.3 is a source of standard information about drugs that is often read but seldom written. In contrast, the prescription refill system in this exercise must process numerous updates as a pharmacy dispenses prescriptions.

Assume that pharmacies will be using both the prescription drug information system and the prescription refill software. Reconsider your choice of architectures from Exercise 9.3b for the combined system.

b. (3) Reconsider your approach to data management and explain your decision.

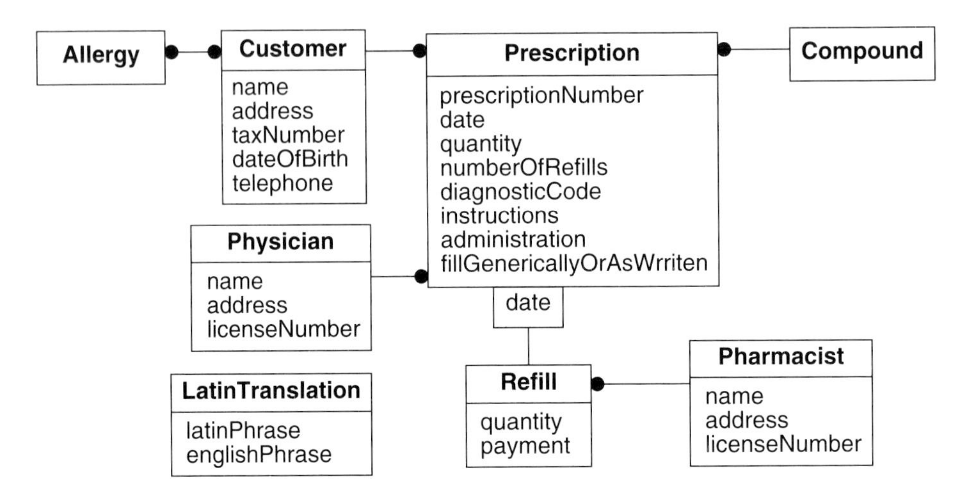

Figure E9.5 Object model for prescription refill software

A customer brings a prescription to a pharmacy to have it filled. The pharmacy assigns an number to each prescription and files it. Prescriptions expire one year after they are written but are held at a pharmacy for five years.

A customer may have many active prescriptions. A prescription is written by a physician for a compound. The prescription may specify a route of administration for the compound. The prescription must specify instructions the customer should follow in taking the compound, even if the instructions are simply "take as directed."

The class *LatinTranslation* is a dictionary of Latin phrases and abbreviations to help the pharmacist understand the Latin used by a doctor in a prescription. The doctor's text may contain a mix of English (or another appropriate native language), Latin, and abbreviations. For example *p.o.* is the abbreviation for the Latin phrase *per os*, which means by mouth. Pharmacists are familiar with Latin phrases and abbreviations, but the dictionary is helpful as a double check on rarely used phrases and as an aid in filling prescriptions quickly and accurately.

If a prescription does not have instructions, the pharmacist may call the doctor and will typically add notes on the original prescription. The physician may optionally provide a diagnostic code for the purpose of the prescription. The pharmacist can use the diagnostic code to verify that the compound is being taken for an appropriate condition. For example, if a medication for high blood pressure was prescribed with a diagnostic code for low blood pressure, the pharmacist might check with the doctor. Each compound has default instructions and a route of administration, which the pharmacist might check against the specifications of the doctor. The prescription has an authorized number of refills, each of which must be tracked. This model treats the first filling of the prescription as a refill for simplicity. Each refill is handled by a pharmacist.

Figure E9.6 Explanation of object model for prescription refill software

Also, a physician may check one of two boxes on the prescription, either "fill generically" or "fill as written." Fill generically can save money by allowing a generic substitution for a specified compound. Fill as written indicates that the physician believes that a particular compound is needed and has an advantage over the generic compound.

An allergy is a code that identifies an allergic reaction to a drug or family of drugs. Customer allergies are identified by a pharmacy as a list of allergy codes. A customer is allergic to any compound with the same code. Thus, when new compound becomes available, the manufacturer can identify vulnerable individuals by determining their allergy codes. Allergies for inert ingredients are generally not tracked.

Figure 9.6 (continued) Explanation of object model for prescription refill software

c. (3) If you were to use a DBMS, which would be the most appropriate database management paradigm?

d. (4) Note that class *LatinTranslation* in Figure E9.5 is "detached" and has no associations or generalizations with the other classes. Comment on this. When is it appropriate to have such a "detached" class?

9.5 We further extend the pharmacy software to handle insurance claims. Figure E9.7 presents requirements for the insurance claim software. Figure E9.8 shows the corresponding object model. Figure E9.9 explains the object model.

a. (3) Characterize each of the three systems we have presented in Exercises 9.3 through 9.5 as operational or decision support.

b. (3) How might you organize the software described in Exercises 9.3 through 9.5 into layers and partitions?

Extend the software to assist a pharmacist in obtaining electronic insurance authorization for prescriptions.

A customer may have a primary insurance carrier and any number of secondary carriers that may pay for a prescription. A customer may have dependents who are also customers and covered by the insurance carrier. In some cases, an insurance carrier may approve direct payment to the pharmacy on behalf of a customer. There are reimbursement rules governing the insurance coverage, including specification of deductible amounts and copayments. Payment approval for a refill is done electronically and may require several transactions with the insurance carrier. Each insurance transaction includes a transaction number, which identifies and authenticates the transaction, and an approval code, which indicates the response of the insurance carrier.

The system should assist with the following tasks:

- Entering and editing customer insurance information.
- Determining if a given prescription is covered by the customer's insurance.
- Electronic negotiation of payment for a refill with a customer's insurance carrier.
- Determining how much of the price of the refill is covered by insurance.

Figure E9.7 Problem statement for insurance claim software

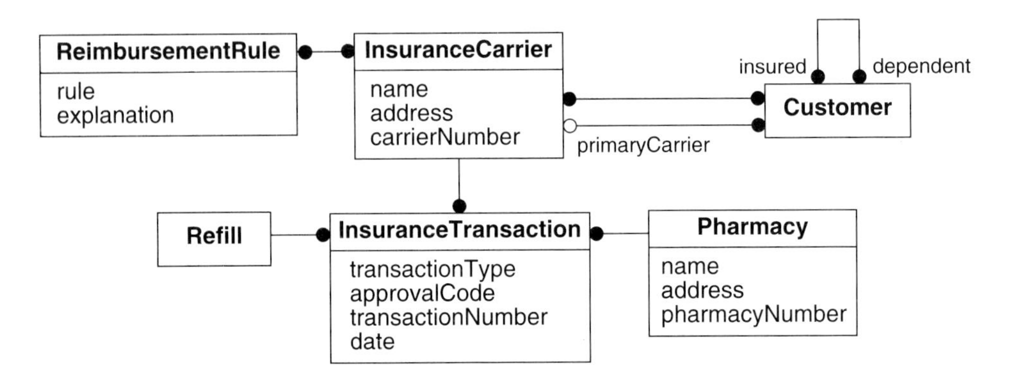

Figure E9.8 Object model for insurance claim software

The amount the customer pays for a refill is not the same as the retail price of the compound and will depend on the reimbursement rules of the insurance carrier. The customer payment is derived from the retail price of the compound and the reimbursement rules. Some insurance plans pay a certain percentage of the retail price. Some plans have a fixed co-pay for each prescription. Many plans have annual deductible amounts per person and per family.

A customer may have a primary insurance carrier and any number of secondary carriers. A customer may have any number of dependents who are also customers. The object model allows multiple levels of dependents, but in practice pharmacists will enter dependent information for only one level of dependents.

There are reimbursement rules governing the insurance coverage. Each health insurance carrier has many rules, and the rules may be shared by more than one carrier. Approval of payment for a prescription refill may involve several insurance transactions. Each transaction concerns a particular refill from a particular pharmacy. (Several transactions may be processed in a single session between a particular pharmacy and an insurance carrier.) We have shown *InsuranceTransaction* as a class, not a ternary, because it is possible to have more than one transaction for a given combination of refill, pharmacy, and insurance carrier. For example, if an initial request for approval is rejected, the pharmacist may call the insurance carrier to receive an override, which itself is a transaction. The transaction type indicates the kind of transaction being requested. Approval code indicates the response of the insurance carrier and transaction number is the identifier generated by the insurance carrier to authenticate the transaction.

Figure E9.9 Explanation of object model for insurance claim software

9.6 (6) Prepare an object model to manage inventory for pharmacies. This model extends the software described in Exercises 9.3 through 9.5. A pharmacy must track orders that are sent to distributors. Your model should record the following: periodic counts for each compound in a store, date of inventory, the person taking an inventory, the discrepancy between measured and computed inventory, the compounds that comprise an order, the quantity of each compound ordered, the date an order is sent, and the date an order is received.

9.7 (4) Add wholesale price information to the pharmacy models. The unit price for a compound varies by distributor, lot size, and effective date.

9.8 (6) Consider Figure 4.6. In the model an *Attribute* has an optional *Domain*. This model is a compromise. What we really want is that the domain of an attribute should be optional during analysis, but required at the completion of design. How could you accomplish this with a computer-aided software engineering (CASE) tool that implements the model?

This illustrates a common problem: Often certain attributes and associations are optional while data are being entered, but are mandatory upon completion of data entry.

9.9 (3) Characterize the applications in Exercises 8.1 through 8.6 as operational or decision support.

9.10 (3) Some commercial software programs combine the notions of sales opportunity management and contact management as described in Exercises 8.2 and 8.3. How would you combine these separate systems using layers and partitions?

9.11 (4) Suppose a large corporation is developing a sales opportunity manager. (See Exercise 8.2.) They have adopted an architecture in which a central server manages records. Each salesperson has client software on a laptop computer that manages the user interface and exchanges data with the central server.

There is an architectural issue with regard to access of records. In principle, a salesperson could have useful information or contribute to any sale across the company. One approach is for each salesperson to have a copy of every opportunity on his or her laptop. This is easy for the salesperson to use but may compromise security and could lead to excessive network traffic with propagating updates. Another approach is that a salesperson could just have a copy of the data within his or her area of geographic or product responsibility. Such assignment of data involves more deployment effort since assignment of opportunities to users must also be managed. Use the techniques of Section 9.2 to evaluate these two approaches. The decision criteria are security, performance, deployment effort, and ease of use.

9.12 (3) There is another architectural issue for the sales opportunity manager. The client software for a salesperson could lock records before making an update (pessimistic concurrency). Or the client software could wait to check for contention until the time of update (optimistic concurrency). Discuss the advantages and disadvantages of each approach. (Section 9.4.3 discusses optimistic and pessimistic concurrency.) Which would you prefer for the situation of a large corporation developing a sales opportunity manager?

10

Detailed Design

During system design you made broad strategic and architectural decisions for your application. Now it is time to begin addressing the fine issues and make decisions for individual classes and methods. In addition, you may want to adjust your analysis model to simplify implementation and improve execution.

During detailed design you make decisions that apply regardless of the data management approach. Issues that are specific to files, relational databases, and object-oriented databases are postponed until implementation. Our distinction between detailed design and implementation is consistent with the overall OMT philosophy of deferring decisions until they are needed.

10.1 OVERVIEW

During detailed design you reconsider the models from analysis with an eye toward ease of implementation, ease of maintenance, and good performance. You will complete the following tasks during detailed design:

- Use transformations to simplify and optimize the object model from analysis. [10.2]
- Elaborate the object model. [10.3]
- Elaborate the functional model. [10.4]
- Evaluate the quality of the detailed design model. [10.5]

10.2 OBJECT MODEL TRANSFORMATIONS

We begin this section by defining transformation concepts. Section 10.2.2 explains how transformations enrich and deepen the OMT software development process. Section 10.2.3

presents examples of primitive transformations. Section 10.2.4 shows a simple demonstration of transformations for the portfolio manager case study. We conclude with a discussion of mathematical properties. Some of the material in this section is from [Blaha-96].

10.2.1 Transformation Concepts

A ***transformation*** is a mapping from the domain of object models to the range of object models. You can think of a transformation as accepting a source object model pattern and yielding a target object model pattern. (Section 4.3 discusses object model patterns.) You apply a transformation to an object model by aligning the source pattern with the model and then substituting the instantiated target pattern.

By applying a series of transformations, you can simplify and optimize a model. In principle, transformations provide a systematic way for evolving from an analysis model to a design model. (Most software development tools lack support for transformations. Nevertheless, it is useful to think in terms of transformations. Eventually the tools will become more capable.) At any point in a model's evolution, several transformations may apply. The choice of transformations to apply is not obvious; the decisions about the choice and sequence of transformations to apply are largely up to the designer.

Transformations are also important for their impact on maintenance. You can evolve an application to meet new requirements without inadvertently losing satisfaction of existing requirements.

Figure 10.1 shows two transformations—depending on the direction in which you read the diagram. The angle brackets denote a placeholder for a class that you must instantiate when you use the transformation.

- Read from left to right, Figure 10.1 shows a transformation for partitioning a class. A designer may split a class (*X*) into two smaller classes (*X1*, *X2*) by apportioning and replicating information.
- Read from right to left, Figure 10.1 shows a merge transformation. Under some circumstances a designer can combine *X1* and *X2* to form *X*.

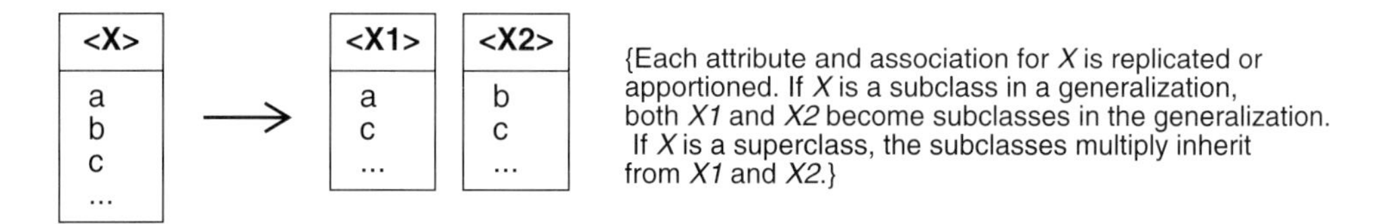

Figure 10.1 Transformations: Partition a class or merge classes

The arrow indicates whether a transformation gains or loses information. A transformation in the direction of an arrow yields a model with equal or less semantic content. A designer need not add information to allow a transformation to occur in the direction of an arrow. Conversely, transformation in a direction without an arrow yields a model with increased semantic content. The designer must assert some additional information to allow such a transformation to proceed. A bidirectional arrow denotes an equivalence transformation.

Figure 10.2 shows an example for the transformations. A database may store both personal and business information for a person; you can represent this information as one class or split the information into two classes. Both models are correct; the choice between models would depend on the purpose and scope of an application. If you have much personal and business information, it's a good idea to separate them. If there is only a modest amount of information, it may be easier to combine them.

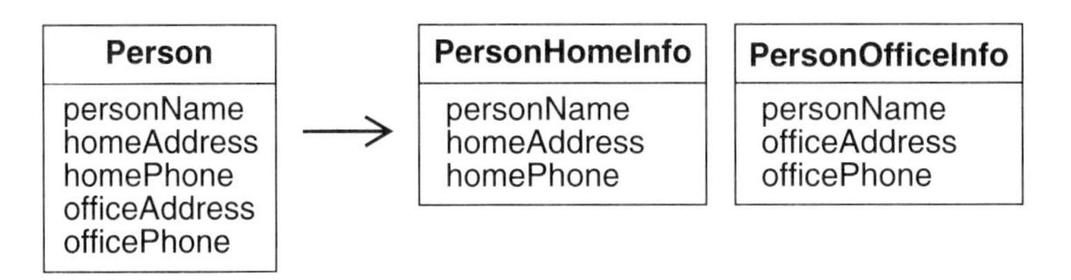

Figure 10.2 Example of partitioning a class or merging classes

Transformations fall into three categories depending on the gain or loss of semantic information:

- **Equivalence**. The source and target object models describe sets of instances in which each source instance corresponds to exactly one target instance. Similarly, each target instance corresponds to exactly one source instance. An equivalence transformation may lose or gain some incidental information from the object model, such as association names or role names, but there is no loss or gain in the described instances. A special case of an equivalence transformation is a renaming transformation. In this case, the designer substitutes a name for a modeling construct.
- **Information losing**. The source model is more constrained than the target model. The target model can describe all instances of the source model; the source model cannot describe some instances of the target model. The transformation in Figure 10.1 loses information when applied from left to right (partition transformation). The *X* class has an implicit relationship between the attributes that the separate classes may not enforce.
- **Information gaining**. The source model is less constrained than the target model. The source model can describe all instances of the target model; the target model cannot describe some instances of the source model. The transformation in Figure 10.1 gains information when applied from right to left (merge transformation). The software developer must assert some correspondence between *X1* and *X2* for the merge transformation to occur.

All transformations have a ***local effect*** on a model—that is, a transformation affects only the constructs that align with the source pattern and the associations and generalizations that connect to aligned classes.

10.2.2 The Use of Transformations in the OMT Methodology

Transformations provide a powerful metaphor for thinking about object models. With transformations you can rigorously adjust a model and be aware of the full consequences in terms of the model's expressability and the data it will store and reject. The object model from analysis should meet the requirements of an application. With transformations you can adjust a model to satisfy implementation concerns without inadvertently losing the satisfaction of requirements.

We recommend that you forgo transformations during analysis (except perhaps near the end). The rigor of transformations is unnecessary when a model is incomplete, and transformations may even be distracting. It is better to be more flexible and conceive a model without inhibition. Once you understand the requirements, you can evolve a model in a more disciplined manner during design. You do not want accidentally to lose information about requirements as you address implementation needs.

Transformations sharpen the distinction between methodology phases and strengthen the OMT software development process. The purpose of detailed design is to shift focus from the real world toward the computer implementation. Transformations let you consider implementation needs independent of the language and database. Transformations directly manipulate the object model and simplify the mappings for implementation.

In principle, transformations could apply to all three OMT models, but here we consider only the object model. The object model is clearly the dominant model for database applications; in addition, we understand object model transformations better than transformations for the dynamic and functional models.

In this chapter we discuss the use of transformations in forward engineering—simplifying and optimizing an analysis model as a prelude to implementation. Transformations also apply to other aspects of the OMT methodology. We briefly mention these other uses of transformations here and discuss the first two bullets further in Chapters 19 and 20.

- **Schema integration**. Transformations are useful for mapping an enterprise model to application models. Such mappings are a prerequisite for allowing common data to flow between applications.
- **Reverse engineering**. The reverse engineer takes implementation artifacts and deduces the underlying logical intent. (You go backward from an existing implementation toward analysis.) Transformations let you note assumptions clearly.
- **Schema evolution**. Transformations also make it easier to convert data. You can associate each transformation with commands that migrate data from the source to the target. Schema evolution often arises when deploying a new application or evolving an existing application.

10.2.3 Examples of Primitive Transformations

A ***primitive transformation*** is a transformation that cannot be decomposed into lesser transformations. Table 10.1 lists the primitive transformations we have identified. We do not claim that our list is complete, and if you discover a new transformation, please email one of us ({blaha, premerlani}@ acm.org). Figure 10.1 is part of *T7* in our transformation list. (*T7* also includes transformations for partitioning an association or an attribute and merging associations or attributes.)

Category	Number	Description
Transformations on a single construct type	T1	Remove or add a construct.
	T2	Assert that a construct is derived or not derived.
	T3	Degrade or restrict multiplicity.
	T4	Transform a multivalued attribute.
	T5	Reorder attributes.
	T6	Transform an enumeration attribute.
	T7	Partition a construct or merge constructs.
	T8	Compose associations.
Transformations on multiple construct types	T9	Merge associated classes or partition a class.
	T10	Move an attribute across an association.
	T11	Move an attribute across generalization.
	T12	Move an association across generalization.
Modifying inheritance	T13	Remove or add a subclass.
	T14	Push subclass information up or specialize.
	T15	Fragment multiple inheritance.
	T16	Factor multiple inheritance.
Conversions	T17	Convert generalization to/from exclusive-or associations.
	T18	Convert a qualifier to/from a link attribute.
	T19	Convert a link attribute to/from an object attribute.
	T20	Convert an association to/from a class.

Table 10.1 Primitive transformations

The remainder of Section 10.2.3 presents examples of some of the primitive transformations in Table 10.1. [Blaha-96] explains all the transformations. Section 10.2.4 applies some transformations to our case study.

- **Remove or add a construct** (*T1*). The transformations in Figure 10.3 modify a model by removing or adding a construct. You may remove or add a class, association, object attribute, or link attribute. The notation *m1..n1* specifies the minimum multiplicity (*m1*) and maximum multiplicity (*n1*) of a role. The transformations from left to right are information losing, since they remove information from a model. In contrast, the transformations from right to left are information gaining, since the designer must assert the new information before the transformations can proceed.

Such transformations are useful because they can help you rigorously track the incidental additions and removals that are part of evolving an object model.

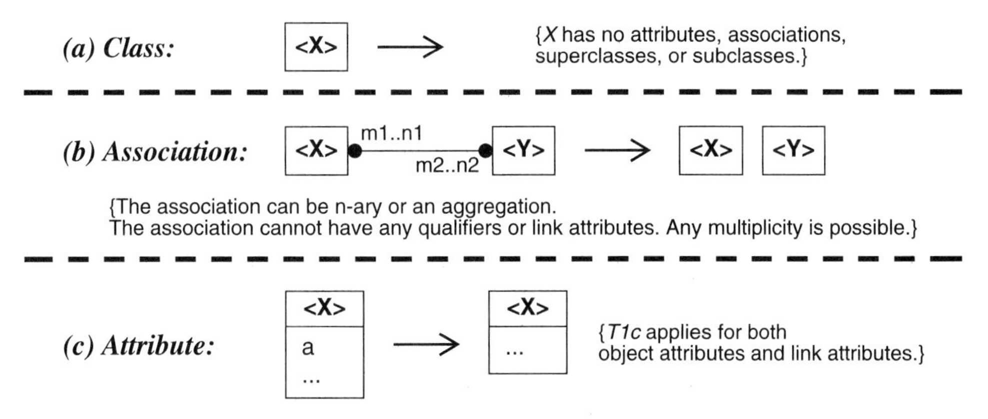

Figure 10.3 *T1* transformations: Remove or add a construct

- **Assert that an association is derived or not derived** (*T2b*). (The *T2* transformation refers to three constructs—class, association, and attribute. The *T2b* transformation specifically concerns derivation of an association. We use a letter suffix to refer to a portion of a transformation from [Blaha-96].) The transformations in Figure 10.4 let you mark an association as derived or not derived. The transformation from left to right loses information because the derived association is constrained to store information consistent with the derivation rule. In contrast, the ordinary association on the right can store any information.

  Like the transformations in Figure 10.3, the transformations in Figure 10.4 are helpful for fine-tuning an object model.

{The association can be n-ary or an aggregation and have qualifiers and link attributes.}

Figure 10.4 *T2b* transformations: Assert that an association is derived or not derived

- **Degrade or restrict multiplicity** (*T3a*). As Figure 10.5 shows, you may degrade or restrict the multiplicity of an association. The transformation from left to right loses information because more instances can conform to the broader multiplicity ranges in the right model. It is common to adjust multiplicity during design.
- **Merge associated classes or partition a class** (*T9a*). As Figure 10.6 shows, you can sometimes combine two classes and an intervening association. The multiplicity *m2..1*

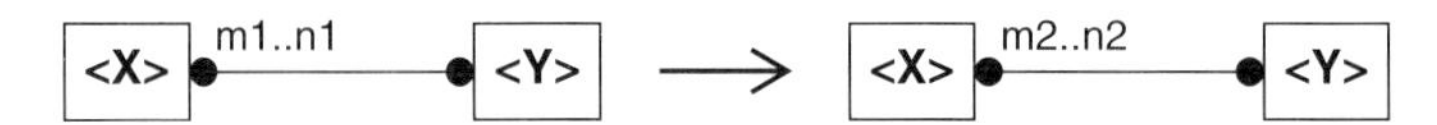

{m1>=m2, n1<=n2.
The associations can be n-ary or an aggregation and have qualifiers and link attributes.}

Figure 10.5 *T3a* transformations: Degrade or restrict multiplicity

in the left model means that each *Y* object corresponds to at most one *X* object (*m2* is 0 or 1); thus we can fold *X* information into the *Y* class.

We use the "*" in two senses in Figure 10.6. The "*" next to the *Y* multiplicity ball is the UML notation denoting that multiplicity is unbounded. (See Section 2.2.2.) In contrast, the "*" in the *Y* class denotes multiplication.

You can use this transformation to simplify a model and reduce the number of classes for implementation. In the example, it would be appropriate to eliminate *City* if it had no additional relationships and was conceptually unimportant to the model.

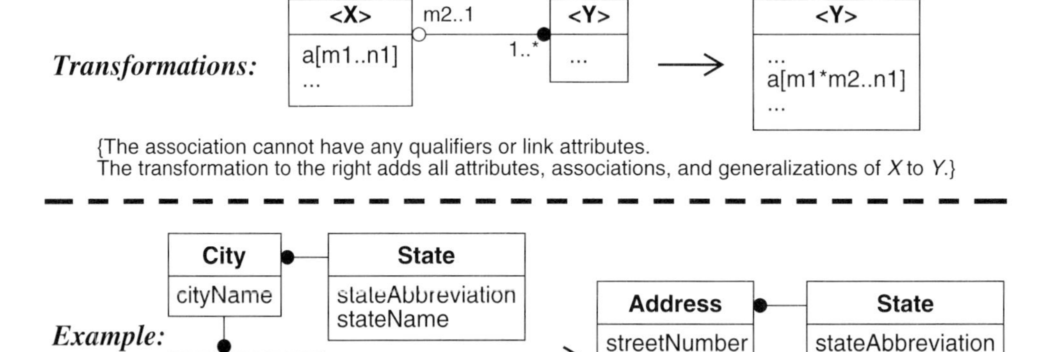

Figure 10.6 *T9a* transformations: Merge classes and intervening association or partition a class

- **Move an attribute across generalization** (*T11a*). The transformations in Figure 10.7 move an attribute down or up a generalization level. The transformation from left to right loses information, because the subclass attributes do not necessarily correspond. You can use this transformation in fine-tuning the division of attributes between a superclass and its subclasses.
- **Move an association across generalization** (*T12a*). The transformations in Figure 10.8 move an association down or up a generalization level. The transformation from left to right loses information, because the multiplicities of the subclass associations may differ. This transformation can be helpful in moving an association from the superclass to the subclasses. With subsequent transformations, you can then refine the multiplicities of the subclass associations.

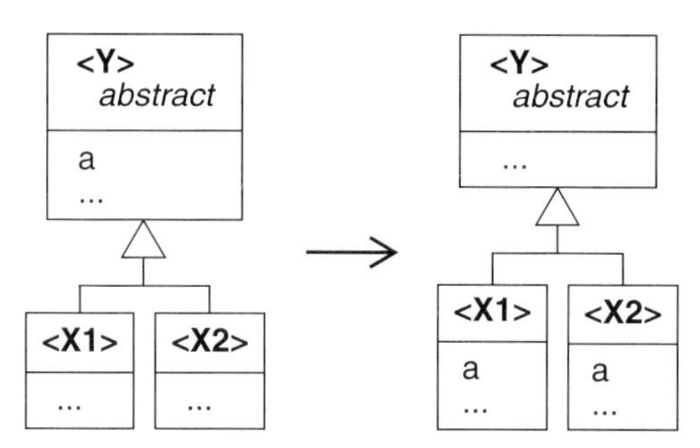

{There may be an arbitrary number of subclasses.}

Figure 10.7 *T11a* transformations: Move an attribute across generalization

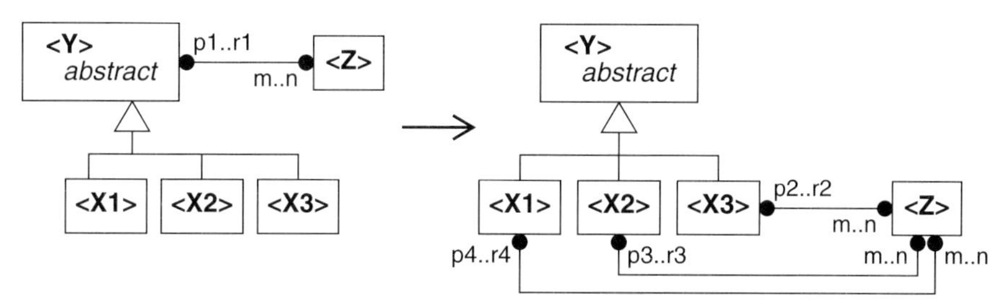

{There may be an arbitrary number of subclasses. Class *Z* may be one of the subclasses.
p1>=p2+p3+p4. r1<=r2+r3+r4.
All associations can be n-ary or aggregations and may have qualifiers and link attributes.
All associations must have the same qualifiers and link attributes for a transformation to the left.}

Figure 10.8 *T12a* transformations: Move an association across generalization

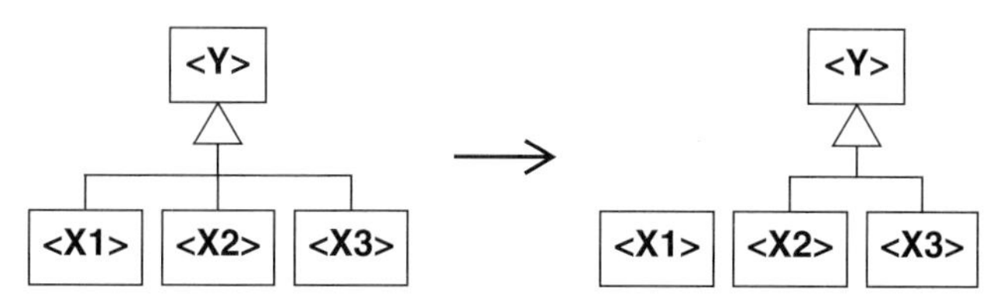

{There may be an arbitrary number of subclasses. For a transformation to the right, all the attributes, associations, and superclasses of *Y* must be added to *X1*. For a transformation to the left, *X1* must have attributes, associations, and superclasses that match those of *Y*.}

Figure 10.9 *T13* transformations: Remove or add a subclass

- **Remove or add a subclass** (*T13*). Figure 10.9 removes or adds a subclass. By successively applying the left-to-right transformation, you can push all the information of a superclass down to its subclasses. Thus if *A* is the superclass of *C* and *C* is the superclass of *D* and *E*, two successive applications of *T13* can collapse the inheritance hierarchy and make *A* a direct superclass of *D* and *E*.

- **Convert qualifier to/from link attribute** (*T18*). In Figure 10.10 the transformation from left to right loses information. The qualifier constrains the maximum multiplicity of *Y* to be *n2*, which is smaller and more restrictive than the maximum multiplicity of *n3* without a qualifier.

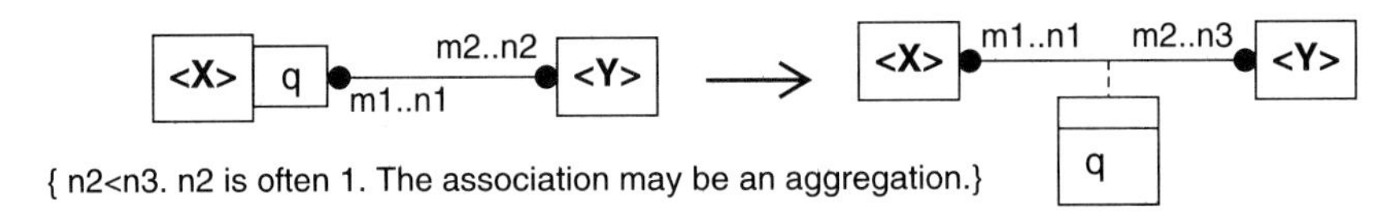

Figure 10.10 *T18* transformations: Convert qualifier to/from link attribute

- **Convert link attribute to/from object attribute** (*T19*). In Figure 10.11 the transformation from left to right loses information because the link attribute is more precise than the object attribute. You can use *T19* and *T18* together to refine a model—converting an object attribute first to a link attribute and then to a qualifier.

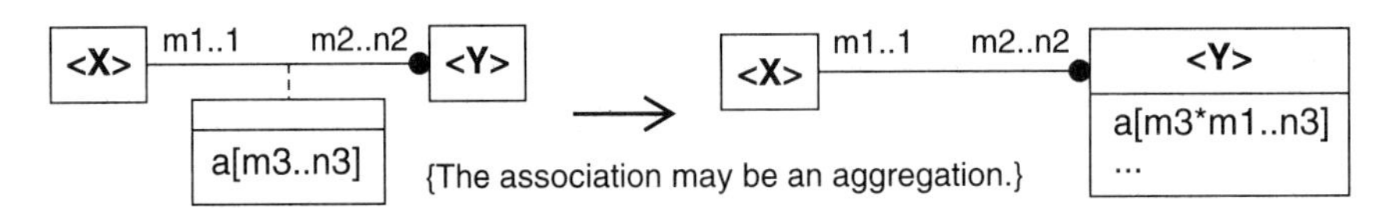

Figure 10.11 *T19* transformations: Convert link attribute to/from object attribute

- **Convert association to/from class** (*T20*). In Figure 10.12 the transformation from left to right promotes an association to a class and loses the derivation of identity for the association class. We will use this transformation in Chapters 15 and 16 when we promote an advanced association to a class so that we can implement it for an object-oriented database.

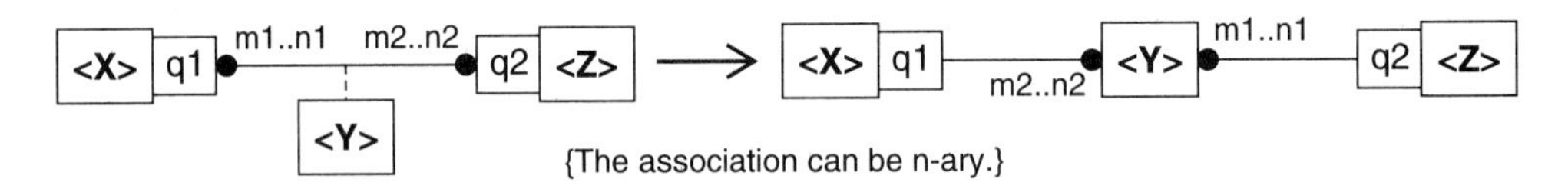

Figure 10.12 *T20* transformations: Convert association to/from class

10.2.4 Use of Transformations for the Portfolio Manager

In Chapter 8, we refined the object model for the portfolio manager (evolved from Figure 8.9 to Figure 8.10) by introducing the notion of a transaction. At that time, we just made the changes. With transformations we can rigorously fine-tune the object model. For example, Figure 10.13 uses transformations to adjust the *FinancialInstrument—Portfolio* aggregation.

The first transformation (*T12a*) moves the aggregation between *FinancialInstrument* and *Portfolio* down the generalization. The *L-to-R* denotes that the transformation is applied left to right. *T12a* loses information when applied from left to right, because the aggregations between *Portfolio—Asset* and *Portfolio—Portfolio* are no longer constrained to have the same meaning. The next transformation, *T1c*, removes the *quantity* link attribute for *Portfolio—Portfolio*, because it does not apply. *T1c* from left to right also loses information. The third transformation, *T3a*, from right to left gains information; we assert that each portfolio has at most one parent portfolio. And the final transformation, *T2b*, from right to left also gains information. We have realized that we can derive the assets comprising a portfolio.

Figure 10.13 is just an example. We could apply different transformations or apply them in a different order than we show. For example, we could apply *T12a L-to-R*, *T3a R-to-L*, *T2b R-to-L*, and finally *T1c L-to-R*.

We chose the transformations as follows. We had the initial object model in Figure 8.9. Based on our understanding of transactions, we had the target model in Figure 8.10. We then scanned the list of transformations in Table 10.1 and chose a sequence of transformations that would evolve from the source model to the target model. Given the modest number of primitive transformations and their organization into categories, it was not difficult to devise a suitable sequence.

Transformations are cumbersome to apply manually, as we have done in Figure 10.13. Application would be much easier with support from modeling tools. With tool support you can better realize the benefits of transformations—rigorous traceability from analysis through design and implementation, better documentation, and migration of data from the schema implementing the source model to the schema implementing the target model. At least one tool already provides some transformations [Hainaut-94], and we expect that more tools will in the future.

Earlier we stated that you should confine your use of transformations to design and not use them during analysis when a model is tentative. The example in Figure 10.13 is from analysis, but the model was not tentative because we were nearly finished with analysis and had a solid understanding of the application's requirements. Our point is that there is no hard line between analysis and design, so you may use transformations whenever you feel the model is logically sound. Just remember that the goal of transformations is to iterate and refine the model.

10.2.5 Mathematical Properties of Transformations

In this section we present a partial theory of transformations. We do not have a full theory (nor does anyone else we know of), although the idea of transformations is intuitive and clearly relevant for design. (Other methodologists have noted the importance of transformations, as we describe in the Bibliographic Notes at the end of this chapter.)

First, transformations are difficult to specify. We specify transformations in terms of source and target object model patterns. We informally state the preconditions that the source pattern must satisfy and the postconditions that the transformation guarantees to hold for the

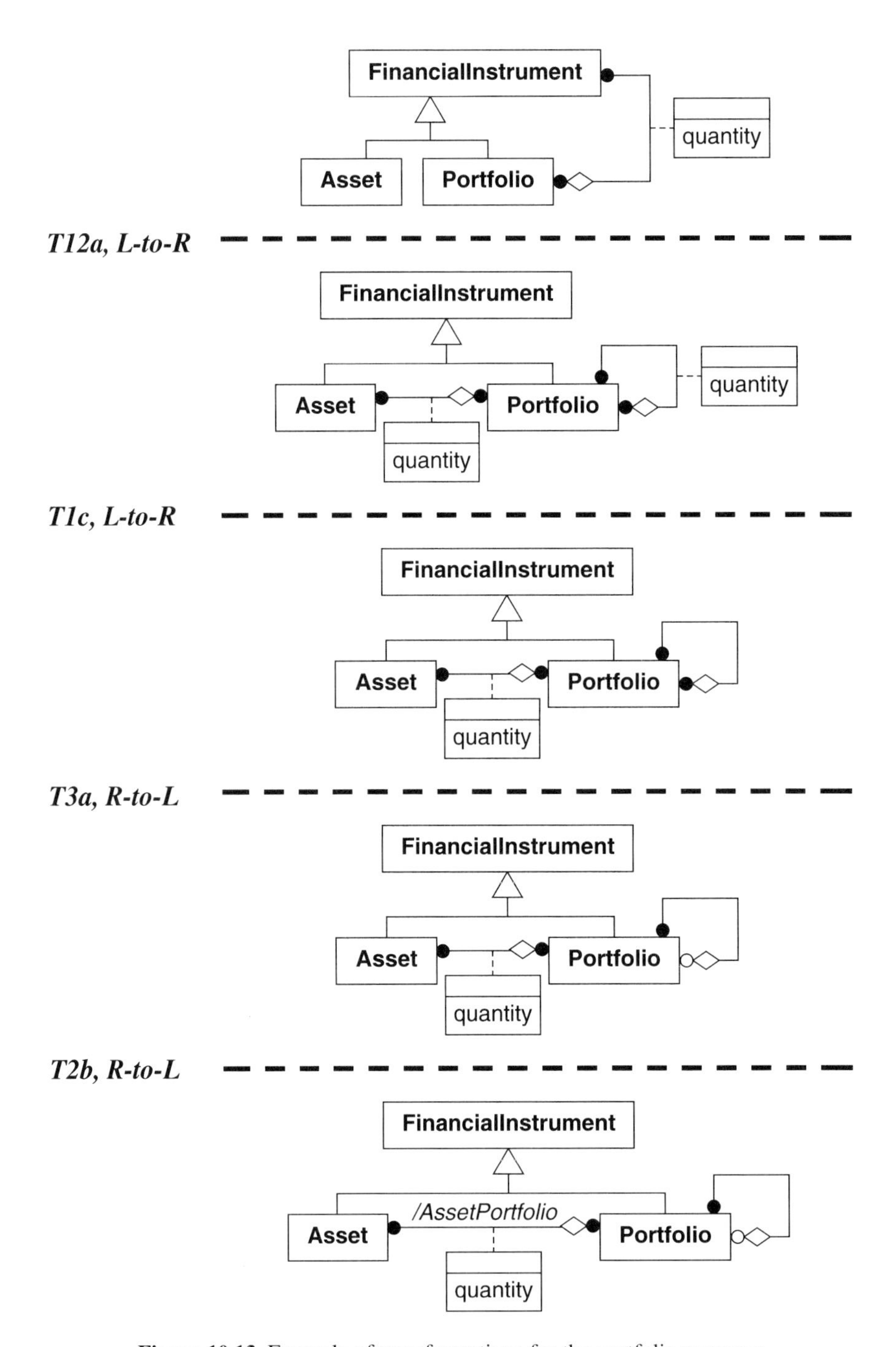

Figure 10.13 Example of transformations for the portfolio manager

target pattern. It is always difficult with preconditions and postconditions to specify fully and avoid oversights.

In general, transformations are not commutative and not associative. The target model depends on the order in which you apply transformations. Figure 10.14 shows some non-commutative transformations.

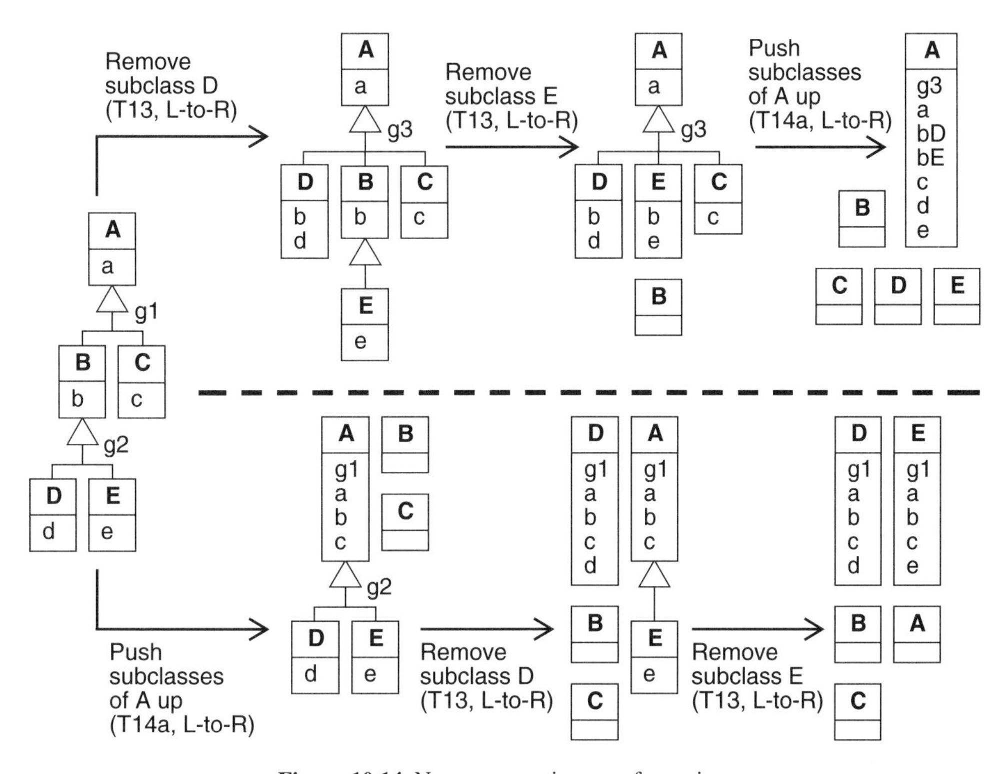

Figure 10.14 Noncommutative transformations

A set of primitive transformations is ***complete*** if it meets two conditions:

- The designer must be able to build any target model from any source model by applying a sequence of primitive transformations from the set.
- The designer must be able to choose an optimal sequence of transformations. The designer must be able to add a minimal amount of information via assertions to evolve a source model to a target model.

The first condition is relatively easy to meet because you can always evolve a schema by removing and adding constructs. The second condition is more onerous because a model's information content is difficult to quantify. Also, we seem to discover new transformations as we encounter new problems.

For a ***minimal*** sct of transformations all thc transformations must bc orthogonal and not overlap. In essence, proving the minimality of a set of transformations is tantamount to proving that each transformation is primitive. By definition, if a transformation is primitive, it cannot be decomposed into lesser transformations and there is no overlap. However, a minimal set of transformations is not necessarily complete. We attempted to present a minimal set of transformations in Table 10.1; we did not attempt to present a complete set of transformations.

You must also be careful when applying transformations that merge classes (such as Figure 10.6) when a class diagram has a cycle, because there may be a cycle in the instance diagram. Figure 10.15 shows two examples of models with cycles.

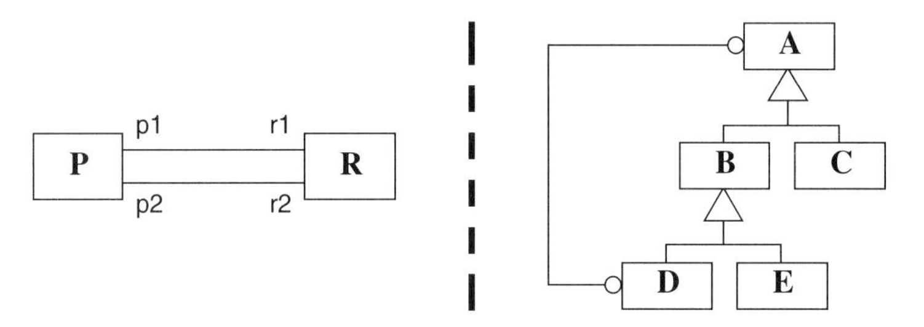

Figure 10.15 Relationship cycles

By definition, transformations preserve closure of the object model; the input to a transformation is an object model and the output is an object model.

10.3 ELABORATING THE OBJECT MODEL

After you have finished transforming the object model, you should elaborate it by adding missing detail. Tasks include adding candidate keys to classes and associations, assigning a domain to each attribute; specifying attribute multiplicity, especially nulls; and roughly estimating the physical storage demands for each class and many-to-many association.

- **Add candidate keys**. You should review your object model and add candidate keys. Candidate keys for classes often arise from names that are unique within the scope of an application. Candidate keys for associations are normally captured by the structure of an object model, but you will need to specify candidate keys for ternary associations.
- **Assign domains**. As we explained in Section 3.1.5, domains convey more information than a mere data type. More important, domains promote the consistency of data types and make mass changes easier. You can also use domains to validate queries partially; a domain provides some indication if a comparison between attributes is meaningful. For example, it usually does not make sense to compare a length to a weight even though both may be stored as numbers.
- **Specify nulls**. You may want to avoid using null values because they can complicate queries. For example, the SQL language of relational databases requires the syntax *attribute = value* to test a non-null value and *attribute is not null* to test a null value. Sometimes you can avoid using null values by substituting a default value.

As you specify nulls, consider that DBMSs enforce constraints at the end of each transaction. A non-null constraint may apply at some initial and final state, but not at intermediate commit points. For example, you may use one program to mass load some data, a second program to scrub the data, and a third program for normal operation. Non-null constraints may be violated after the mass load, but honored after scrubbing and during normal operation. Sometimes there is no choice but to relax non-null constraints. This is unfortunate because you then weaken the DBMS schema and shift the burden of non-null enforcement to application programs.

- **Estimate physical storage**. Estimate the following information for each class and many-to-many association: the number of occurrences, growth rate, and retention period. You do not need to estimate for one-to-one and one-to-many associations because their storage consumption will correspond to that of a related class. During implementation, you use this information to compute the required physical storage space.

The portfolio manager. We add several candidate keys for classes to the object model: *FinancialInstrument.name*, *FinancialInstitution.name*, and *Person.name*. We could show domains and null information on the object model, but we decided against that. Instead we will specify this information for the portfolio manager with separate tables (Table 10.2 and Table 10.3) to reduce clutter in the object diagram.

Class	Attribute	Domain	Nulls permitted?
FinancialInstrument	name	name	n
	description	longString	y
FinancialInstitution	name	name	n
	address	address	y
Portfolio	portfolioType	enumPortfolioType	n
Person	name	name	n
AssetValue	valuePerUnit	money	n
	actualOrEstimated	enumActualOrEstimated	n
ROI	ROI	shortNumber	n
PortfolioValue	valueTotal	money	n
StockOption	putOrCall	enumPutOrCall	n
	strikePrice	money	n
	expirationDate	date	n
StockAsset	tickerSymbol	shortName	y
	stockType	enumStockType	n

Table 10.2 Attribute specifications for portfolio manager classes

Class	Attribute	Domain	Nulls permitted?
Transaction	transactionDate	date	n
	recordDate	date	n
	transactionType	enumTransactionType	n
	transactionFee	money	n
	description	longString	y

Table 10.2 (continued) Attribute specifications for portfolio manager classes

Link attribute or qualifier	Attribute	Domain	Nulls permitted?
qualifier	accountNumber	mediumString	n
link attribute	quantity	longNumber	n
qualifier	date	date	n
qualifier	startDate	date	n
qualifier	stopDate	date	n

Table 10.3 Attribute specification for portfolio manager associations

Table 10.4 estimates physical storage parameters for the portfolio manager. The number of occurrences and growth rate are very rough estimates. We intend the portfolio manager for a single user, so storage demands can vary greatly from person to person. We will retain basic data until it is no longer referenced. After 10 years we will delete old computations and archive old values and transactions if there is a need to recover disc space. We plan to record asset values monthly; thus we expect to have 12 values per asset and per year. Similarly, we specify *ROI*, *PortfolioValue*, and *Transaction* per portfolio and year. The growth rate of each *FinancialInstrument* is the same as the corresponding *Asset* or *Portfolio*. Similarly, the growth rate of the *Asset* subclasses is the same as *Asset*.

10.4 ELABORATING THE FUNCTIONAL MODEL

During detailed design, you should elaborate the functional model so that you will be fully prepared to implement the code. Tasks include the following:

- **Design algorithms for the various methods**. You need not address every method in the implementation, but you should consider all the difficult methods. For these, you must be sure to have an algorithm with acceptable computational complexity as a fall-back option. If you find a better algorithm during implementation, that's a bonus, but you don't want to encounter a problem suddenly during implementation and have to

Class / Association	Number of occurrences	Growth rate	Retention period
FinancialInstrument	number of assets + number of portfolios	—	until no longer referenced
FinancialInstitution	5	0	until no longer referenced
Asset	100	10%/year	until no longer referenced
Portfolio	25	5%/year	until no longer referenced
Person	5	0	until no longer referenced
AssetValue	12 / asset-year	0	archive after 10 years
ROI	4 / portfolio-year	0	delete after 10 years
PortfolioValue	4 / portfolio-year	0	delete after 10 years
PreciousMetal	5	—	until no longer referenced
Cash	5	—	until no longer referenced
BondAsset	5	—	until no longer referenced
StockOption	5	—	until no longer referenced
StockAsset	65	—	until no longer referenced
Collectible	5	—	until no longer referenced
Insurance	5	—	until no longer referenced
RealEstate	5	—	until no longer referenced
Transaction	20 / portfolio-year	5%/year	archive after 10 years
Portfolio@ accountOwner	40	5%/year	until constituent objects are no longer retained
Transaction@ assetTransacted	50 / portfolio-year	5%/year	

Table 10.4 Physical storage for the portfolio manager

reach back into your models to correct it. As you design algorithms, you may need to transform the object model. [10.4.1]

- **Decide which classes will own the methods**. Earlier (in Chapters 5 and 8) we did this by intuition. Here you can use more rigorous techniques to determine ownership in less obvious situations. [10.4.2]
- **Reconcile the conflict of encapsulation versus query optimization**. Encapsulation limits a method's scope of knowledge. Query optimization, on the other hand, encourages developers to state queries broadly. [10.4.3]
- **Begin to think about user interaction**. Sketch out user interaction screens and any needed reports. The goal is not to seek perfect forms but to look for reasonable, serviceable forms that a human factors expert could improve. [10.4.4]

- **Address other design principles**. Consider principles that transcend an implementation medium including portability, minimizing software restrictions, and use of assertions. [10.4.5]

10.4.1 Designing Methods

In designing methods, you should consider the following issues:

- **Transaction modeling**. Some applications, such as complex aerospace systems, require that you perform ***transaction modeling***, in which you determine the kinds of queries likely to be executed and obtain data for each query. This data includes how often each query occurs, the number of read and write I/Os, the required response time, the number of concurrent users, and the database bottlenecks. Transaction modeling is intended to increase performance, but the resulting data is often difficult to estimate, and the estimates are often wrong. Moreover, commercial information systems seldom need this level of detail. Our experience is that you can often obtain very good performance by merely tuning a database for fast navigation of the object model. The object model implies that the traversal of associations is reasonably efficient, so you must design your schema accordingly.

  We prefer to deliver a database *with* tuning. Some persons recommend that you first get the database working and then add tuning information. We disagree with this strategy for two reasons. First, an untuned database may be painfully slow, confusing users and causing them to lose confidence in the database design. Second, is it is relatively easy to deliver an approximate tuning. In the implementation chapters we discuss the details of tuning for the different data management paradigms.

- **Transaction locks**. Avoid holding read and write locks while waiting for user input. Otherwise, the performance of the affected methods could be awful because user response time is unpredictable. Normally, you should obtain transaction locks before or after user input and not hold them while waiting. If you have a transaction that must wait for user input, you may be able to use optimistic concurrency or dirty reads. (A dirty read does not lock the database, but merely reads the database regardless of any update activities that may be underway.) Another option is to time out a lock for excessive user delay.

- **Algorithmic code versus lookup tables**. DBMSs provide an alternative mechanism for implementing methods. You may implement a function (both discrete and continuous) or a constraint with algorithmic code or via lookup in a data structure. For example, engineers often use precomputed tables of values and interpolate if the desired value is not listed. As another example, you may specify income tax by formula or by a table that maps income brackets to the corresponding tax. The advantage of using an algorithm is that it can cover many instances. An explicit listing may be simpler for a small number of instances, however.

The portfolio manager. The major calculations are to determine all assets in a portfolio, portfolio value, and ROI. The ROI calculation involves straightforward iteration: Assume a rate of return and compare the discounted value of the portfolio cash flow with the current portfolio value. Then adjust the rate of return until the sum of the discounted cash flow matches the current value. The iteration for ROI is a nuisance to program, but has satisfactory computational complexity. From experience, we know computation will converge within a fraction of a percent of the correct answer after several iterations.

The determination of all assets in a portfolio is more difficult—the assets contained in a portfolio on a given date are the sum of the direct assets and the assets of lesser portfolios. This computation intrinsically involves recursion because the total assets are defined in terms of lesser portfolios. We wanted to devise an algorithm that avoids recursion for two reasons. First, recursion is awkward for some implementations, notably SQL. Second, we want to provide a list in the user interface (a pick list) for assigning a parent portfolio to a portfolio; we want the pick list to display only the parent portfolios that will not cause a cycle. The pick list computations must be quick so that the user interface is not encumbered.

Thus we decided to precompute the transitive closure of the portfolio tree.* We add a derived *Portfolio@Portfolio* association with roles *ancestorPortfolio* and *descendantPortfolio* to our object model. For any changes to the base *parentPortfolio@childPortfolio* association we adjust the derived association. Chapter 14 elaborates the calculation of the transitive closure for the portfolio tree.

Given the set of assets in a portfolio, the computation for portfolio value is straightforward—scan the set of assets and sum the quantity per asset multiplied by asset value.

10.4.2 Ownership of Methods

The most appropriate owner class is not always obvious. We recommend the following guidelines for determining the appropriate owner class:

- **Query for an origin object**. Many methods are queries that retrieve information given some origin object. Assign these methods to the class for the origin object. Nearly all our methods in Chapters 5 and 8 were queries for an origin object.
- **Query for an origin collection**. For example, some methods are queries on an extent for a class. (Section 2.1.2 explains the notion of an extent.) If all objects in the collection belong to the same class, assign ownership to the class. Otherwise, it is difficult to assign ownership for the method, and you should try to adjust the method or your object model so that the query involves only a single class.
- **Update to an object**. Assign ownership of the method to the class for the updated object.
- **Update to a collection of objects**. You may occasionally have a method that updates multiple objects, possibly from different classes. For example, a banking application

* A ***tree*** is a set of nodes, such that each node has at most one *parent* node and zero or more *child* nodes. Furthermore, a tree may not have any cycles. It must not be possible to start at a node, traverse a sequence of parent-child links, and arrive at the original node. For the portfolio manager, a portfolio corresponds to a node. The ***transitive closure*** is the set of nodes that can be reached, directly or indirectly, from some starting node.

could have a method that credits a checking account and debits a savings account. You may want to adjust your object model so that the method updates a single class. Thus for a banking application, you could define a transaction that relates to multiple accounts.

- **Update to a link**. Assign ownership to one of the related classes.
- **Update to a collection of links**. Try to assign ownership to one of the related classes. You may have to adjust the object model.

The portfolio manager. Ownership of the methods is straightforward, as we will discuss in Chapter 14.

10.4.3 Encapsulation versus Query Optimization

There is a fundamental, irreconcilable conflict between the goals of encapsulation and the goals of query optimization. This conflict is more prominent for RDBMSs than OO-DBMSs, because RDBMSs emphasize nonprocedurality with the SQL language.

The law of Demeter [Lieberherr-88] eloquently states the principle of encapsulation: An object should only access objects that are directly related (directly connected by an association or generalization). Indirectly related objects should be accessed indirectly via the methods of intervening objects. Encapsulation increases the resilience of an application; local changes to an application model cause local changes to the application code.

On the other hand, DBMS optimizers take a logical request and generate an efficient execution plan. If queries are broadly stated, the optimizer has greater freedom for devising an efficient plan. For example, RDBMS performance will usually be best if you join multiple tables together in a single SQL statement, rather than disperse logic across multiple SQL statements.

Thus encapsulation boosts resilience but limits optimization potential. On the other hand, broadly stating queries makes optimization easier, but a small change to an application can affect many queries.

For an application that uses an OO-DBMS, you should usually tilt toward encapsulation—writing methods that access only directly related objects. This is because OO-DBMSs provide direct semantic support for the associations and generalizations in an object model. Also, most OO-DBMSs currently have weak query optimizers, so there is less incentive for broadly stating queries.

For RDBMS applications, there is no simple resolution of this conflict. There are three different situations:

- **Complex programming**. You should encapsulate your code if the programming is intricate and performance degradation is not too severe.
- **Easy programming and good query performance**. You should broadly state queries if doing so improves RDBMS performance and the programming code and queries are relatively easy to write—and rewrite if the object model changes.
- **Easy programming and poor query performance**. Somewhat paradoxically, you can sometimes improve performance by fragmenting queries. Query optimizers are imper-

fect, and occasionally you will need to guide the optimizer manually. Section 14.8.2 elaborates this point for RDBMSs.

The portfolio manager. For the RDBMS implementation we chose to state most of our queries broadly, because it is easier for us to write the code. In contrast, most of our code for the file and OO-DBMS implementations honors encapsulation.

10.4.4 User Interaction

The user interface of an application depends heavily on the available resources. RDBMSs tend to have closely bundled languages for quickly developing forms and reports (fourth-generation languages, also called 4GLs). 4GLs are especially convenient for routine applications that are dominated by I/O and have only simple calculations. In principle, OO-DBMSs could also have 4GLs, but OO-DBMSs are less mature than RDBMSs and thus tend to have less robust ancillary software. Powerful user interface libraries are available for OO-DBMSs, but the combination of a user interface library and an OO-DBMS is often less cohesive than an RDBMS and a 4GL. This loss of cohesion is a drawback for simple applications, but the flexibility can facilitate advanced applications.

The portfolio manager. Chapter 14 discusses the user interface of the portfolio manager for the RDBMS implementation.

10.4.5 Other Considerations

Other design principles for functionality include the following:

- **Design for portability**. It is good practice to limit system dependencies to a few basic methods. You can then more easily port software to other hardware platforms and operating systems.

  For example, despite the SQL standard, there are fine differences between RDBMS products. This is one reason we dislike dispersing SQL code with application code (Section 9.7). We prefer to encapsulate SQL code with an application programming interface. If we then port to another RDBMS, we are less likely to rewrite code beyond the API.

- **Minimize arbitrary limits**. Try to minimize arbitrary restrictions, assumptions, and limits in your software. Introduce these only as a last resort. It is often easier to write more complex software than to explain limits.

  For example, we want the portfolio manager to accept a hierarchy of any depth. This flexibility may complicate our development task, but simplifies the software for the user. Our attitude is to regard any added complexity as a requirement and just deal with it.

- **Use assertions**. Liberally place assertions in your code. Performance is not a consideration, because you can always bracket assertions with optional compile-time switches to disable the additional code. Assertions let you design for testability and make debugging easier.

10.5 EVALUATING THE QUALITY OF A DESIGN MODEL

In Chapters 2 through 5 we presented some practical tips for constructing object and functional models. These tips focused on the quality of the analysis model. In this section we provide some additional tips for evaluating the quality of a design, specifically a database design. These criteria apply to files, relational databases, and OO databases.

- **Identity**. Is there a systematic approach to realizing identity? Users and applications must be able to find objects uniquely and simply. You may identify objects with existence-based identity, value-based identity, or possibly some combination (see Section 9.8), but regardless you should have a consistent strategy throughout your database.
- **Integrity**. Does the design specify candidate keys and incorporate important constraints? The structure of a model must enforce constraints when reasonably possible. DBMSs have features for maintaining integrity beyond the facilities in application programs.
- **Redundancy**. Does the database structure exhibit any "accidental" (careless, thoughtless) redundancy such as that proscribed by normal forms? If there is careless redundancy, your model is likely flawed.
- **Performance**. Is the response time and computation burden of anticipated queries acceptable? You should pay careful attention to the order of complexity of algorithms.
- **Extensibility**. Can the database design be extended for future demands with local modifications? Maintainers can then modify the database to remedy design flaws and extend it for new requirements.
- **Security**. Does the model permit resolution of security concerns? If you group related data, it is easier to handle security and oversights are less likely to occur.
- **Data distribution**. Does the database design permit data distribution and parallel processing? You may partition a database across various sites, organizations, and hardware platforms. As Chapter 1 discusses, a rational approach to database design (entity-based design) enables deeper insights into the underlying data structure than an empirical approach (attribute-based design).

10.6 CHAPTER SUMMARY

The inputs to detailed design are the OMT models from analysis and the architectural decisions from system design. During detailed design you adjust and elaborate the OMT models, postponing database-specific issues until implementation.

The first step in detailed design is to use transformations to simplify and optimize the object model from analysis. You can also use transformations to facilitate subsequent maintenance of the software. A transformation is a mapping from the domain of object models to the range of object models. You can think of a transformation as accepting a source object model pattern and yielding a target object model pattern. You apply a transformation to an object model by aligning the source pattern with the model and then substituting the instantiated target pattern.

Transformations fall into several categories depending on whether they gain or lose information, which is indicated by the direction of an arrow. A transformation in the direction of an arrow yields a model with equal or less semantic content. Conversely, the designer must assert some additional information to allow a transformation to proceed in a direction without an arrow. A bidirectional arrow denotes an equivalence transformation, which neither gains nor loses information.

A primitive transformation is a transformation that cannot be decomposed into lesser transformations. Our list of primitive transformations is extensive but not complete; you will occasionally encounter situations requiring a transformation that we have not covered.

After you have finished transforming the object model, you should add missing detail. You should add candidate keys to classes and associations. You should choose a domain for each attribute and indicate whether each attribute can be null. You should roughly estimate the physical storage demands for each class and association.

During detailed design you should also elaborate the functional model. You should devise algorithms for the various methods and determine their computational complexity. You need not address every method in the ultimate implementation; however, you should consider all the difficult methods. Our experience is that you can design most algorithms without performing rigorous transaction modeling. Often you can obtain very good performance by merely tuning a database for fast navigation of the object model. The object model implies that traversal of associations is reasonably efficient, so you must design your schema accordingly.

Finally, you should check the quality of the detailed design. Figure 10.16 lists the key concepts for this chapter.

designing methods	object model transformation
detailed design	ownership of methods
domain	physical storage
elaborating the functional model	primitive transformation
elaborating the object model	quality of a database design
encapsulation	query optimization
nulls	

Figure 10.16 Key concepts for Chapter 10

BIBLIOGRAPHIC NOTES

Much of Section 10.2 is taken from [Blaha-96]. In this chapter we have made several corrections to the transformations that were originally published in [Blaha-96].

Hainaut has written a number of excellent papers about transformations and database application development. [Hainaut-92] describes the TRAMIS database software development tool. TRAMIS is notable for its semantic support of simple transformations.

[Hainaut-94] discusses the philosophy of the DB-MAIN tool, which is the successor to TRAMIS. DB-MAIN supports forward engineering (progression through analysis, design, and implementation), reverse engineering, and the evolution of populated databases. DB-MAIN incorporates many transformations and can replay transformations against a model; replay of transformations is helpful when a model changes or the developer wants to reconsider the evolution of a model.

[Hainaut-92] and [Rosenthal-87] describe equivalence transformations, transformations with no loss or gain of information. Hainaut and Rosenthal use only equivalence transformations so that the designer cannot lose information from analysis. In contrast, [Batini-92] and ourselves adopt a more permissive viewpoint; designers may use transformations that add or lose information, but should be aware of the consequences.

[Batini-92] is a very good book and has greatly influenced our thinking. [Batini-92] presents one of the more comprehensive list of transformations we have seen in the literature. [Blaha-96] presents additional transformations. [Batini-92] favors the use of transformations throughout the entire modeling process. In contrast, we favor a hybrid approach. We use transformations for refinement and optimization, but do not use them during analysis when we are discovering basic information.

Chapter 9 of [Halpin-95] discusses schema transformations for the ORM approach to modeling. ORM is an extension of the NIAM approach to modeling [Verheijen-82].

REFERENCES

[Batini-92] Carlo Batini, Stefano Ceri, and Shamkant B. Navathe. *Conceptual Database Design: An Entity-Relationship Approach*. Redwood City, California: Benjamin Cummings, 1992.

[Blaha-96] Michael Blaha and William Premerlani. A catalog of object model transformations. *Third Working Conference on Reverse Engineering*, November 1996, Monterey, California, 87–96.

[Hainaut-92] JL Hainaut, M Cadelli, B Decuyper, and O Marchand. TRAMIS: a transformation-base database CASE tool. *Proceedings 5th International Conference on Software Engineering and Applications*, Toulouse, France, December 7–11, 1992, EC2 Publishing, 1992, 421–431.

[Hainaut-94] JL Hainaut, V Englebert, J Henrard, JM Hick, and D Roland. Database evolution: the DB-MAIN approach. *Proceedings of the Entity-Relationship Conference*, Manchester, UK, 1994, 112–131.

[Halpin-95] Terry Halpin. *Conceptual Schema and Relational Database Design*, 2nd edition. Sydney, New South Wales: Prentice Hall Australia, 1995.

[Lieberherr-88] K Lieberherr, I Holland, and A Riel. Object-oriented programming: an objective sense of style. *OOPSLA'88 as ACM SIGPLAN 23*, 1 (November 1988), 323–334.

[Rosenthal-87] Arnon Rosenthal and David Reiner. Theoretically Sound Transformations for Practical Database Design. *Proceedings of the Entity-Relationship Conference*, 1987, 97–113.

[Verheijen-82] G Verheijen and J Van Bekkum. NIAM: an information analysis method. Unpublished paper. Information Systems Department, Control Data Corporation, The Netherlands, 1982.

EXERCISES

10.1 Figure E10.1 shows alternative models for a directed graph. These are the same models that we presented in the exercises for Chapter 2. Apply a sequence of transformations to make the following conversions. You will need to obtain a copy of [Blaha-96] to do this exercise. (See our Web site, www.omtassociates.com.)

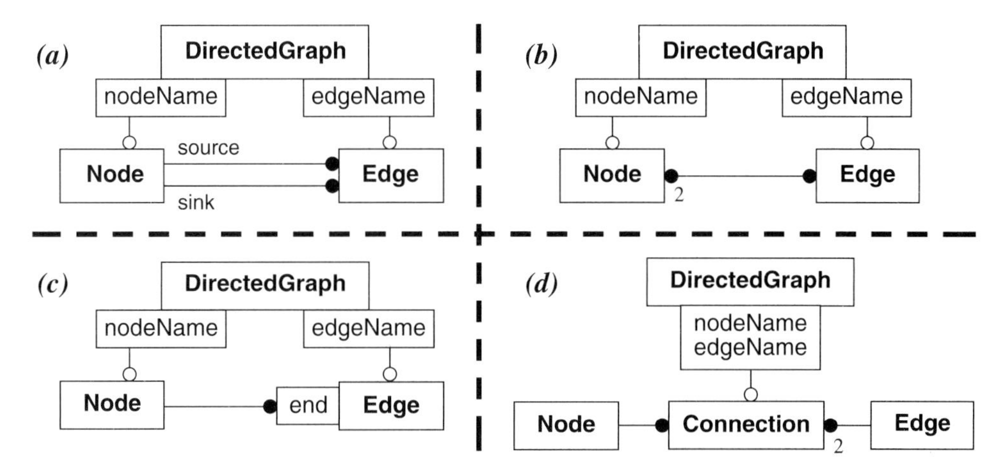

Figure E10.1 Alternative object models for a directed graph

a. (5) Convert Figure E10.1b to Figure E10.1a.
b. (5) Convert Figure E10.1c to Figure E10.1a.
c. (7) Convert Figure E10.1b to Figure E10.1d.

10.2 (8) Apply a sequence of transformations to convert Figure E10.2a to Figure E10.2b. (Hint: Section 10.2.3 contains all the transformations you will need.)

10.3 (6) Earlier in Section 5.4.2 we updated the model for airline flight reservations to fix an error. Figure 5.10 shows an excerpt of the original model and the corresponding revised model. Apply a sequence of transformations to fix this error. (Hint: You will need to obtain a copy of [Blaha-96] to do this exercise.)

10.4 Consider the prescription drug information system described in Exercise 9.3.

a. (2) Assign domains to all attributes. You can either add domains to the object model or use supplementary tables, as we did in this chapter.
b. (2) Specify attribute multiplicity.
c. (2) Estimate physical storage requirements.
d. (6) We will add three candidate keys to the object model: *ControlledSubstanceClassification.schedule*, *Drug.name*, and *FunctionalClassification.description*. Write enhanced pseudocode for a method that inserts a new drug. The arguments to the method are *name*, *action*, *chemicalFormula*, *schedule*, and *functionalClassification*. Your method should enforce the candidate key for *Drug* and use the other two candidate keys. You can assume that the appropriate *FunctionalClassification* and *ControlledSubstanceClassification* objects already exist.

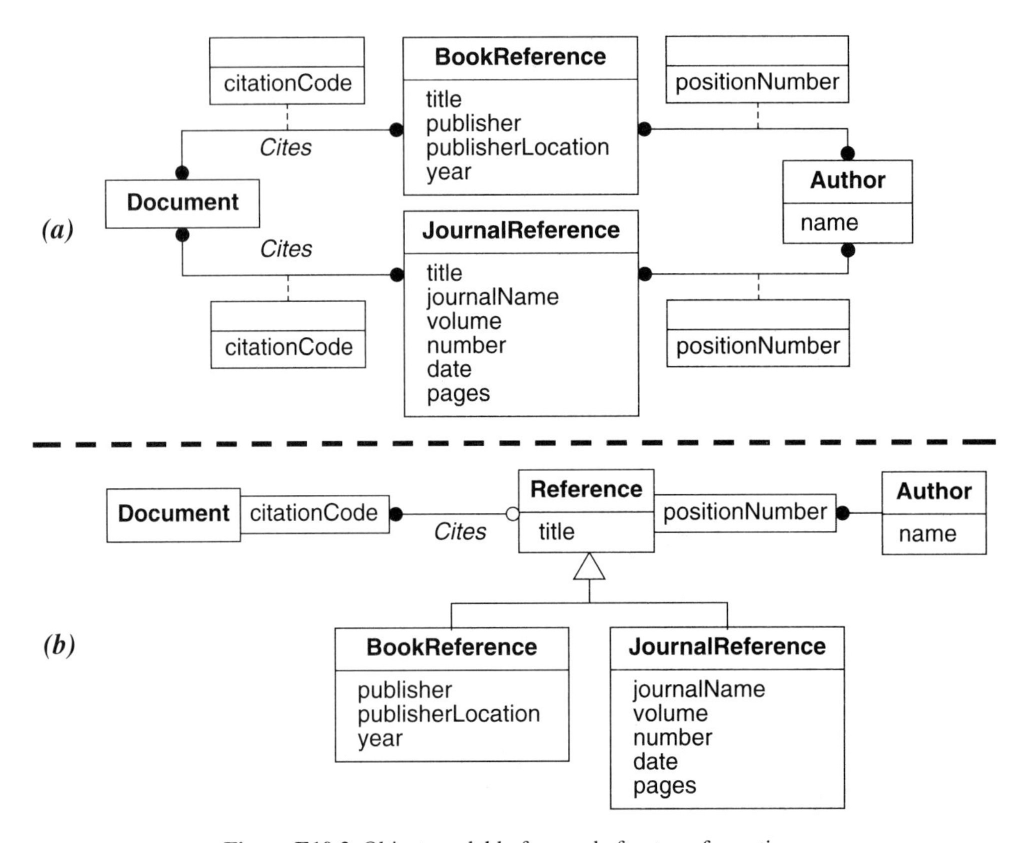

Figure E10.2 Object model before and after transformations

- **e.** (3) Specify any other candidate keys that should be added to the object model.
- **f.** (2) Write enhanced pseudocode for a method that adds a new distributor for a compound.
- **g.** (5) Write enhanced pseudocode for a method that finds all compounds that contain a given pair of drugs.
- **h.** (4) Write enhanced pseudocode for a method that finds all compounds from a given manufacturer, in any form, that contain a given drug.

10.5 Consider the prescription refill software described in Exercise 9.4.

- **a.** (3) Add candidate keys to the object model.
- **b.** (2) Assign domains to all attributes. You can either add domains to the object model or use supplementary tables, as we did in this chapter.
- **c.** (2) Specify attribute multiplicity.
- **d.** (2) Estimate physical storage requirements.
- **e.** (4) Write enhanced pseudocode for a method that inserts a new customer into the database. Be sure to check to check if the customer is already in the database. Assume that *taxNumber* is a candidate key of *Customer*.
- **f.** (2) Write enhanced pseudocode for a method that adds an allergy for a customer.
- **g.** (2) Write enhanced pseudocode for a method that deletes an allergy for a customer.

h. (8) Write enhanced pseudocode for a method that fills a new prescription. Check for allergies and drug interactions. Check that the customer does not already have an existing prescription for the same compound. You must also create a new *Refill* object, since we are treating the first filling of a prescription as a refill.

i. (6) Write enhanced pseudocode for a method that refills an existing prescription. Check to see that there are refills left and that the prescription is not older than one year.

10.6 Consider the insurance claim software described in Exercise 9.5.

a. (3) Add candidate keys to the object model.

b. (2) Assign domains to all attributes. You can either add domains to the object model or use supplementary tables, as we did in this chapter.

c. (2) Specify attribute multiplicity.

d. (2) Estimate physical storage requirements.

10.7 Consider the sales opportunity manager from Exercise 8.2.

a. (3) Add candidate keys to the object model.

b. (2) Assign domains to all attributes. You can either add domains to the object model or use supplementary tables, as we did in this chapter.

c. (2) Specify attribute multiplicity.

d. (2) Estimate physical storage requirements.

10.8 Consider the contact manager from Exercise 8.3.

a. (3) Add candidate keys to the object model.

b. (2) Assign domains to all attributes. You can either add domains to the object model or use supplementary tables, as we did in this chapter.

c. (2) Specify attribute multiplicity.

d. (2) Estimate physical storage requirements.

10.9 Consider the course registration and grading software from Exercise 8.4.

a. (3) Add candidate keys to the object model.

b. (2) Assign domains to all attributes. You can either add domains to the object model or use supplementary tables, as we did in this chapter.

c. (2) Specify attribute multiplicity.

d. (2) Estimate physical storage requirements.

10.10 Consider the library loan software from Exercise 8.5.

a. (3) Add candidate keys to the object model.

b. (2) Assign domains to all attributes. You can either add domains to the object model or use supplementary tables, as we did in this chapter.

c. (2) Specify attribute multiplicity.

d. (2) Estimate physical storage requirements.

10.11 Consider the billing software from Exercise 8.6.

a. (3) Add candidate keys to the object model.

b. (2) Assign domains to all attributes. You can either add domains to the object model or use supplementary tables, as we did in this chapter.

c. (2) Specify attribute multiplicity.

11

Process Review

We have presented the front portion of the OMT process—as customized for database applications. In this chapter, we briefly recapitulate the treatment of analysis and design from Part 2 and then preview the coverage of implementation in Part 3.

11.1 CONCEPTUALIZATION

Conceptualization is the first phase of the OMT process for software development and deals with the genesis of an application. Initially some person thinks of an idea for an application and prepares a statement of intent.

11.2 ANALYSIS

The purpose of analysis is to understand the problem requirements thoroughly and devise a model of the real world. The goal of analysis is to specify *what* needs to be done; during design we determine *how* it is done. Only once you understand a problem are you prepared to attempt a solution. The same OMT analysis process applies uniformly to both programming and database applications, although the relative importance of the three models varies. Analysis has the following steps:

1. Collect raw materials to serve as modeling grist. These raw materials include the problem statement from conceptualization, existing artifacts from prior systems, use cases, and advice from problem experts.
2. Build an object model.
   - Add classes.
   - Add associations between classes.

- Add attributes for objects and links.
- Use generalization to note similarities and differences.
- Test access paths.
- Iterate and refine the model. Consider shifting the level of abstraction.
- Organize the object model using packages.

3. Construct a data dictionary. The data dictionary should define all modeling entities and explain the rationale for key modeling decisions.
4. Build a dynamic model. This step is seldom important for data management applications.
5. Build a functional model.

- Specify use cases. Use cases provide a concise way of specifying themes for interacting with a system.
- Choose one or more functional model paradigms.
- Specify operations from the object model.
- Specify operations from the dynamic model. This step is unimportant for most data management applications.

6. Verify the entire analysis model.

11.3 SYSTEM DESIGN

The purpose of system design is to devise a high-level strategy—the system architecture—for solving the application problem. You must also establish policies to guide the subsequent detailed design. The design and implementation of databases differs greatly from the design and implementation of programming code. Thus at this phase the general OMT process begins to mold to the specific requirements of data management applications. You should resolve the following steps:

1. Devise an architecture.
2. Choose an implementation for external control.
3. Choose a data management approach.
4. If using a database, choose a database management paradigm.
5. Determine opportunities for reuse.
6. Choose a strategy for data interaction.
7. Choose an approach to object identity.
8. Deal with temporal data.
9. Deal with secondary aspects of data.
10. Specify default policies for detailed design.

11.4 DETAILED DESIGN

During detailed design you begin to address the fine issues and make decisions for individual classes and methods. You carry forward your model from analysis and adjust it to simplify implementation and improve execution. Detailed design deals with decisions that are platform independent. You should perform the following steps:

1. Use transformations to simplify and optimize the object model from analysis.
2. Elaborate the object model by adding missing detail.
3. Elaborate the functional model, primarily by devising algorithms for the most important methods.
4. Evaluate the quality of the detailed design model.

11.5 OVERVIEW OF IMPLEMENTATION

In the next several chapters we implement the model from detailed design and write the actual programming and database code. We discuss files, relational databases, and object-oriented databases, because all three are being widely used for new systems development. The implementation is simplified because you have already addressed most difficult issues. Sophisticated software development tools can automate part of the implementation.

PART 3: IMPLEMENTATION

12

Files

This chapter shows how to implement an OMT design with files. We emphasize the object model, because files intrinsically involve data structure issues, and the role of the object model is to describe data structures and capture structural constraints. We also cover one aspect of functionality, the saving and loading of files. To express additional behavior, you would need to use a programming language with files, which is beyond our scope here. We assume that readers are familiar with the notion of a grammar and with BNF notation.

A file-based implementation should not attempt to deliver complete data management. Implementors should not attempt to incorporate concurrency, transaction support, and data distribution, in particular, because such things can be purchased at a fraction of their development cost.

12.1 INTRODUCTION TO FILES

Files provide a reasonable approach to data persistence for simple problems. They incur no purchase cost and can manage modest quantities of data for stand-alone applications.

A ***sentence*** is a finite sequence of symbols from a specific alphabet, such as ASCII characters. A ***language*** is any set of sentences formed from an alphabet; examples are C and Smalltalk. A ***file*** is an instance of a particular language, that is a sentence for a particular language. A large body of theory applies to programming languages as well as file structure [Aho-86]. Here we are interested in the theory of file grammars.

A ***grammar*** specifies the allowable sentences for a language. You can express some grammars as production rules. A ***production rule*** defines one language element in terms of other elements in the same language. Figure 12.1 shows three production rules. The first specifies that a *FinancialInstrument* is an *Asset* or a *Portfolio*. The second and third define an *Asset* and a *Portfolio*. (We explain production rule syntax later in this chapter.)

A ***context-free grammar*** can be defined in terms of production rules whose right side can be substituted for the left side in any context. In Figure 12.1, for example, we can sub-

```
FinancialInstrument  :  Asset | Portfolio ;
Asset                :  '<' 'Asset'
                        financialInstrumentID
                        name
                        description '>' ;
Portfolio            :  '<' 'Portfolio'
                        financialInstrumentID
                        name
                        description
                        portfolioType '>' ;
```

Figure 12.1 An example of production rules

stitute the definition to the right of the colon in any production rule where *Portfolio* appears. A context-free grammar, which normally describes ASCII files, has several advantages:

- **Coherence**. It is precise, easy to understand, and extensible.
- **Tool availability**. You can generate an efficient parser with standard tools, such as *Lex* [Lesk-75], *Yacc* [Johnson-75], and *Yacc++* [Compiler-92] [Avotins-95]. *Lex* parses a file into a sequence of tokens specified by a grammar. *Yacc* accepts the output of Lex and executes actions associated with the grammar. *Yacc++* is a newer, object-oriented variation of Yacc.
- **Correspondence to object models**. The structure implied by a context-free grammar is homomorphic to the structure of objects. As a consequence, you can readily map an object model to a context-free grammar, as we show later.

The Backus-Naur Form (BNF) is one of the most popular notations for expressing a context-free grammar. All the production rules in this chapter use the BNF dialect of *Lex*, *Yacc*, and *Yacc++*.

Because a context-free grammar recursively defines a language through production rules, you can apply the rules repeatedly to a file to produce a parse tree. A grammar that produces more than one parse tree is ambiguous and generally not desirable. The use of a context-free grammar does not *guarantee* a precise language, but it does make it easier to define one.

A context-free grammar cannot describe all files. Some constructs in binary files, such as indirection, counts, and sizes, do not have a context-free description. There are few (if any) parser generation tools for non-context-free grammars, and the scientific community knows less about them than context-free grammars. Nevertheless, non-context-free grammars can be practical for simple languages.

12.2 IMPLEMENTING THE OBJECT MODEL

The first step in implementing an application is to map the object model to file constructs, using the following procedure. (The numbers in brackets refer to sections that explain these tasks in detail.)

- **Organize data into files**. A ***file type*** organizes the classes and associations of an application. The distinction between file and file type is analogous to that between object and class—a file is an instance of a file type. Decide which classes and associations must be persistent and group them into file types. Then for each file type decide how to apportion application data among one or more files. [12.3]
- **Select file approach(es)**. For each file type, choose a file approach. Use of sequential-access ASCII files is the simplest and most common approach. [12.4]
- **Implement identity**. Implement the approach to identity chosen during system design. [12.5]
- **Select mapping strategies**. Implement the domains, classes, associations, and generalizations for the file types. In general, the mappings for associations and generalizations are not commutative; the net result depends on their application order. We discuss mapping strategies only for the sequential-access ASCII file approach, because this is the approach you are likely to use most often. [12.6–12.11]

We use the excerpt of the portfolio manager model in Figure 12.2 for most examples in this chapter.

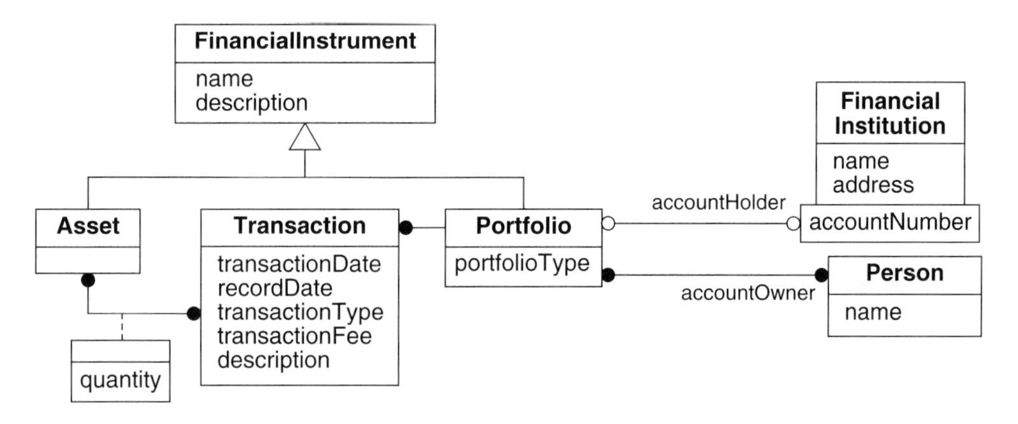

Figure 12.2 Excerpt of the portfolio manager model used for examples

12.3 ORGANIZING DATA INTO FILES

12.3.1 Organizing Data by File Type

Some applications (such as a text file editor) require only a single file type; others can involve several file types (such as a desktop publishing system that has separate file types for documents and spelling information). There are several bases for organizing classes and associations into file types:

- **By package**. A package is a group of classes, associations, generalizations, and lesser packages with a common theme (Chapter 3). Sometimes it is helpful to define file types according to packages.
- **By file approach**. Section 12.4 discusses several file approaches: sequential-access ASCII file, sequential-access binary file, random-access binary file, and memory-image binary file. Each file type must have only one approach. If you want to use multiple file approaches, you must define multiple file types. For example, a CAD application might store general information in an ASCII file, but actual drawings in binary files.
- **By application**. If you have several coupled applications, you can define a separate file type for data that is private to each application. This is simpler than mixing private data with shared data.
- **By access characteristics**. You can partition file types according to access characteristics. For example, you might separate data that is normally read from data that is frequently updated. You might also partition a model when data quantities per class differ widely.

If you have multiple file types, you must relate them. Normally, associations and generalizations appear in a single file type. A class, however, can appear in multiple file types, and bind them.

You can relate files by either replicating or referring to objects. Sometimes you can defer the decision to the user. For example, a document in many desktop publishing systems can include a graphics file either by referring to another file or by replicating it—the user decides. Each method of relating has its own strengths. Data replication avoids file dependencies; references consume less disc space. References yield a "live" connection to data that can still evolve; replication yields a snapshot of the data that is not affected by changes.

The portfolio manager. In this chapter we only partially implemented the portfolio manager for files. That is, we did not use the full model in Figure 9.5. However, for the full model we would define four file types: transactions, asset values, ROIs and portfolio values, and all other data. We would place transactions and asset values in separate file types because the volume of this data can be large. We would place ROIs and portfolio values in another file type because they are derived and could be reconstructed from other data, if necessary. "All other data" would comprise miscellaneous data that is infrequently updated, such as portfolio structure, asset description, and portfolio ownership. This data is in contrast to transactions and asset values, which change relatively often.

We would relate the four file types via object references, because we are using existence-based identity and need not propagate updates to references. Furthermore, we would implement functionality so that we do not have to propagate deletions. For example, we would not allow an asset (in *other* file type) to be deleted if it had corresponding transactions.

12.3.2 Assigning Data to Files

For each file type you must decide whether to store all data in a single file or apportion data across multiple files. The desktop publishing system with one file type for documents and another for spelling dictionaries may store each document in a separate document file, but combine spelling entries for all documents in a single spelling file.

You should also determine which files will require backup copies. You should always backup fundamental data, but you need backup derived data only when it is tedious to recompute. The Framemaker desktop publishing software, for example, preserves the prior copy of a document as a backup file before writing a new one. However, it does not store the ASCII interchange file, because it can easily reconstruct such a file from the binary format.

The portfolio manager. We would use one file per file type. There is no obvious basis for partitioning the data for a file type, and we expect only a modest amount of data, since the portfolio manager is for a single user.

12.4 SELECTING A FILE APPROACH

You can use any of several kinds of files to manage data, depending on the application:

- **Sequential-access ASCII file**. By "sequential access" we mean that you must access each data item in turn. You cannot read or update an item in the middle of the file without accessing all prior items. By "ASCII," we mean that the file uses only printable characters and is human readable. You can describe most sequential-access ASCII files with a context-free grammar. Typically, the approach is to read the entire file into memory upon a *load* command or application startup, operate on the data, and then write the entire file upon a *save* command or application termination.

  For example, PCs use ".bat" files to store commands for later execution. The UNIX operating system stores the recent commands for each user in a ".history" file. X Windows uses a file with the ".Xdefaults" extension to store default settings for each user. Framemaker can save documents in ASCII format as ".mif" files for file interchange.

  Sequential-access ASCII files have several advantages. One of the most important is that they are usually based on a context-free grammar, which means you can use tools to generate code for file loading and readily implement an object-oriented model. Also, a knowledgeable programmer can read and debug the file. You can readily transmit ASCII files through a network or across hardware platforms.

  There are also disadvantages. Even a slight modification requires bringing all data into memory. The startup and save delay can be substantial for large files. Also, to provide crash resistance, the computer must copy the entire file. Thus sequential ASCII files are not suitable for fine concurrent access. Memory size can limit the amount of persistent data. Users can tamper with an ASCII file via a text editor, instead of executing update methods.

- **Sequential-access binary file**. This file encodes its data, normally with the same binary format used for data in memory. You can describe a sequential-access binary file with the process for writing its data. As with a sequential ASCII file, the typical approach is to read the entire file into memory upon a *load* command or application startup, operate on the data, and then write the entire file upon a *save* command or application termination.

Sequential-access binary is the second most common file approach. Framemaker normally stores documents in this format, for example. PCs and UNIX machines use sequential binary files to store digital music, sounds, and bit-mapped pictures. CAD software also normally stores geometric data in sequential binary files.

Sequential-access binary files are more compact than ASCII files, which means that loading and saving are much faster. Also, users are not tempted to tamper with the file's contents. Finally, unlike a random-access file, you can readily organize sequential files into physical records with different lengths.

Sequential-access binary files have many of the disadvantages of ASCII files. Even a slight modification requires that you bring all data into memory. The computer must copy the entire file before writing to provide crash resistance. Sequential binary files are also not conducive to fine concurrent access. Memory size limits the amount of persistent data.

Sequential-access binary files have additional disadvantages, however. Normally, you cannot define a context-free grammar because such files use pointers, arrays, counts, and sizes. So sequential-access binary files lose the benefits of a context-free grammar, including a deep theoretical base and automated parser generation tools. Moreover, a binary format requires special debugging software. The presence of binary data complicates file transmission through a network or across hardware platforms.

- **Random-access binary file**. This approach is similar to the sequential-access binary file, except the file is divided into fixed-sized records accessible by physical record number. Random-access binary files are normally used for special applications, for which it is not possible or desirable to read the contents of an entire file into memory.

  DBMSs use random-access binary files. Sequential files are not acceptable because a database can be huge and a DBMS must be able to access only a small portion of the underlying files. One benefit of a DBMS is that it hides the complexity of finding records and accessing them by physical record number.

  Random-access binary files are very efficient. Memory size does not limit the amount of persistent data, since records are accessed individually. Thus, random-access binary files can provide incremental crash resistance by processing a few records at a time. Users can concurrently access different records within a file. The obscure format dissuades users from tampering with files.

  The biggest disadvantage of random-access binary files is complexity. Again, you cannot define a context-free grammar because such files use pointers, arrays, counts, and sizes. Furthermore, physical record numbers are meaningless to persons, so your application must translate between the user's view and the low-level physical mechanism. Random-access binary files often require a search mechanism (such as an index) to find physical record numbers. Your application must also map objects with variable-sized attributes to fixed-sized records. The presence of binary data complicates transmission of files through a network or across hardware platforms.

- **Memory-image binary file**. This approach is similar to the sequential-access binary file, except you need not encode and decode when moving data between memory and file storage. A *save* command writes the literal contents of memory to the disc, even absolute pointer addresses. A *load* command overwrites memory with the contents of a binary file.

Smalltalk and Lisp illustrate this approach. Each can save the current state of computation by writing a memory image to disc. The memory image approach is less common than the three other file approaches, however.

The most significant benefit of this approach is probably its efficiency and understandability. Memory-image binary files also provide a straightforward technique for adding persistence to an object-oriented language and are tamper resistant.

A memory image has major drawbacks. For example, there is no easy way to preserve portions of memory and still load the binary file. The memory image file requires special decoding software for debugging. Furthermore, memory size limits the amount of persistent data, and you must copy the entire file to provide crash resistance. It is difficult to transmit memory-image binary files through a network or across hardware platforms. We normally discourage this approach.

Table 12.1 summarizes the advantages and disadvantages of the different file approaches. In the remainder of this chapter we assume the use of a sequential-access ASCII file.

	Sequential-access ASCII file	**Sequential-access binary file**	**Random-access binary file**	**Memory-image binary file**
Automation	Tools can generate code	Developer must write custom code	Developer must write custom code	Operating system may generate image
Load/save performance	Slow	Medium	Fast	Medium
Maximum file size	Limited by memory size	Limited by memory size	Not limited by memory size	Limited by memory size
Concurrency	Must lock entire file	Must lock entire file	Can lock record within file	Must lock entire file
Crash resistance	Must copy entire file	Must copy entire file	Can copy records	Must copy entire file
Extensibility	Straightforward	Difficult	Difficult	Straightforward
Portability	Straightforward	Difficult	Difficult	Difficult
Debugging	Human readable	Unreadable	Unreadable	Unreadable
Tamper resistance	User can tamper with file	Tamper resistant	Tamper resistant	Tamper resistant
Recommendation	Our preferred approach	Our second choice	Sometimes chosen for speed and concurrency	Discouraged

Table 12.1 Comparison of file-based approaches to persistent data

12.5 IMPLEMENTING IDENTITY

To implement identity, you begin by defining a primary key for each class. A ***primary key*** is an arbitrarily chosen candidate key that is used to reference instances preferentially; each class should normally have exactly one primary key. Primary keys let you clearly refer to objects in a file. Some implementations of associations (Sections 12.9 and 12.10) and generalizations (Section 12.11) even mandate that objects have a primary key. Primary keys also give you more flexibility to change the implementation of associations and generalizations and make it easier to evolve schema (Section 12.15).

As we described in system design (Chapter 9), identity can be existence based or value based. For existence-based identity you must generate an artificial unique number as the primary key for each object. With value-based identity some combination of attribute values identifies each object. We recommended existence-based identity for applications with more than 30 classes; either approach is viable for files with small applications.

Generating identifiers is straightforward for sequential-access ASCII files that stand alone and are read entirely into memory. However, you must be careful when associations can span files; an association in one file could point to an object in another. In this case we recommend that you ensure uniqueness by using an identifier prefix that varies by file.

We assume existence-based identity for the remainder of this chapter.

12.6 IMPLEMENTING DOMAINS

You should already have selected a domain for each attribute during detailed design (Chapter 10). As we discussed in Chapter 3, assigning a domain is preferable to directly assigning a data type because domains help ensure that data types are uniformly assigned to attributes. They have an additional purpose for files, specifying both the in-memory encoding and the in-file encoding for attributes.

Encoding data in memory and encoding data in files are often different. For example, many compilers represent character strings in memory as null-terminated, variable-length character arrays. In a file it may be more appropriate to delimit a string with quotes. Strings can have embedded quotes, and you must take care to distinguish such quotes from the end of a string. One practice is to precede an embedded quote with a special character such as "\". Another is to represent an embedded quote with successive quotes.

Number encoding is another difference between data encoding for memory and files. Most compilers store numbers with a binary format for compactness and computational efficiency. In contrast, a sequential ASCII file stores numbers as strings so that they are readable.

You should adopt a uniform encoding policy across data types, rather than remember special cases. For example, it is not a good idea to represent embedded quotes for some strings as \" and for other strings as "".

Figure 12.3 shows several production rules for domains. In the BNF dialect of *Lex*, *Yacc*, and *Yacc++*, a semicolon terminates each production rule. The token is to the left of the co-

```
logicalInteger           :  INTEGER ;
logicalString            :  INTEGER | STRING ;
logicalMoney             :  INTEGER | STRING ;

idDomain                 :  logicalInteger ;
nameDomain               :  logicalString ;
longStringDomain         :  logicalString ;
enumPortfolioTypeDomain  :  logicalString ;
moneyDomain              :  logicalMoney ;
```

Figure 12.3 Domain implementation

lon and the definition to the right. The vertical bar denotes alternation. Primitive elements of the grammar are in uppercase; we use white space and indentation to improve readability.

The first three production rules map logical data types to physical data types. For example, a *logicalString* is an *integer* or a *string*. The processing of physical data types into logical data types bridges the difference in representation between data encoding for memory and files. We do not show the code here, but we could define *integer* as a sequence of digits, optionally preceded by a plus or minus sign. We could define *string* as a sequence of characters that contains at least one letter or is enclosed with double quotes.

The next five production rules in Figure 12.3 define domains in terms of logical data types. For example, *nameDomain* is a *logicalString*, and so can be *aPensionAccount*, *"antique car,"* or *123456*.

There are several special kinds of domains:

- **Identifiers**. Files do not intrinsically support identifiers, so you must write extra application code. We normally represent identifiers as integers.
- **Enumerations**. You can implement an enumeration either with strings or an encoding. For example, you could represent color with the strings *red*, *green*, and *yellow* or the encoding *0*, *1*, and *2*. With ASCII files, we normally recommend enumeration strings so that you can read enumeration values for debugging.
- **Structured domains**. You can implement structured domains with a hierarchy of production rules. For example, in Figure 12.4 *address* consists of *street*, *city*, *state*, and *zipCode*, each of which would have its own production rules.

```
address      : street city state zipCode ;
```

Figure 12.4 Structured domain implementation

- **Multivalued domains**. You can readily implement multivalued domains with a list. Figure 12.7 shows the use of a list of objects for implementing a class.

12.7 IMPLEMENTING CLASSES

For each class you should define a production rule that includes attributes, buried associations, and physically included classes. (Section 12.9 discusses the second and third items.) For example, Figure 12.5 shows production rules for *FinancialInstrument*. You should also include a literal string to identify the class, as we do in Figure 12.5 ('FinancialInstrument').

```
FinancialInstrument   :  'FinancialInstrument'
                         financialInstrumentID
                         name
                         description ;

financialInstrumentID:   idDomain;
name                  :  nameDomain ;
description           :  longStringDomain ;
```

Figure 12.5 Class implementation: An approach

Note that we have added the attribute *financialInstrumentID* in Figure 12.5; you should define a primary key for each class. (See Section 12.5.) We prefer that tokens in the grammar have the same name as constructs in the model. Thus in Figure 12.5 the token *FinancialInstrument* is analogous to the class *FinancialInstrument* in the object model.

You can more easily implement a grammar if you enclose the contents of a class. For example, Figure 12.6 restates the production rule for *FinancialInstrument* to use '<' and '>' as enclosing delimiters. The enclosing delimiters give the parser cues for avoiding ambiguities. If a file has an error, the parser can detect a problem more quickly and give a more meaningful error message. The delimiter also makes the file more human readable. We use enclosing delimiters for the remainder of this chapter.

```
FinancialInstrument   :  '<'
                         'FinancialInstrument'
                         financialInstrumentID
                         name
                         description
                         '>' ;
```

Figure 12.6 Class implementation: With enclosing delimiters

The production rule in Figure 12.6 labels each instance of the class *FinancialInstrument* with the string *FinancialInstrument*. Alternatively, you can store all instances of a class together, as Figure 12.7 shows. This approach saves space, but at the expense of a less readable file, especially if a class has many instances. The asterisk is additional notation indicating that you can repeat the preceding token, but the token must occur zero or more times. Thus

```
FinancialInstruments :  '<' 'FinancialInstruments'
                        FinancialInstrument * '>' ;
FinancialInstrument  :  '<'
                        financialInstrumentID
                        name
                        description
                        '>' ;
```

Figure 12.7 Class implementation: Storing instances of a class together

the *FinancialInstruments* production rule can invoke multiple occurrences of the *FinancialInstrument* production rule. We recommend you use production rules like those in Figure 12.7 for implementing a class.*

Figure 12.8 shows an excerpt of a file corresponding to the production rules in Figure 12.7.

```
< FinancialInstruments
  < 1 "antique car" "1955 Thunderbird">
  < 2 "retirement fund" "Carl's pension">
  < 3 "college fund" "Carl's daughter's college savings"> >
```

Figure 12.8 An excerpt of a file for the production rules in Figure 12.7

12.8 IMPLEMENTING ATTRIBUTES

The production rules in Figure 12.5–Figure 12.7 list attributes in sequential order. These rules differentiate attributes by their relative location, so the file must store the attributes in the correct order. Alternatively, you can write production rules that are insensitive to order by including the attribute name. Figure 12.9 shows unordered production rules. Figure 12.10 shows an excerpt of a file corresponding to the production rules in Figure 12.9.

Note that the production rules in Figure 12.9 permit missing and duplicate attributes, unlike those in Figure 12.5–Figure 12.7. For example, Figure 12.9 would allow *FinancialInstrument* to have several *names* and no *description*. You might want to allow missing attributes as an implementation of nulls. However, duplicate *names* can create a problem unless the software for writing files avoids duplicates.

Another issue concerns the uniqueness of attribute names. In object models attribute names have local scope for uniqueness (local to a class or association). In contrast, for a con-

* *Yacc++* supports the asterisk syntax for repeating a token, but *Yacc* does not. Instead, you must use recursion. *FinancialInstrument* * would be replaced with a reference to *FinancialInstrument_group*, which is defined as *FinancialInstrument_group* : *FinancialInstrument* | *FinancialInstrument_group FinancialInstrument*.

```
FinancialInstruments :  '<' 'FinancialInstruments'
                        FinancialInstrument * '>' ;
FinancialInstrument  :  '<'
                        FinancialInstrumentAttribute *
                        '>' ;

FinancialInstrumentAttribute:financialInstrumentID
                     |  name
                     |  description ;

financialInstrumentID:  '<' 'financialInstrumentID'
                        idDomain '>' ;
name                 :  '<' 'name' nameDomain '>' ;
description          :  '<' 'description'
                        longStringDomain '>' ;
```

Figure 12.9 Attribute implementation: Unordered storage

```
< FinancialInstruments
  < < financialInstrumentID 1 >
    < name "antique car" >
    < description "1955 Thunderbird"> >
  < < name "retirement fund" >
    < financialInstrumentID 2 >
    < description "Carl's pension" > >
  < < financialInstrumentID 3 >
    < name "college fund" >
    < description "Carl's daughter's college savings"> > >
```

Figure 12.10 An excerpt of a file for the production rules in Figure 12.9

text-free grammar the names of tokens have a global scope. Consequently, you must take care to avoid potential name collisions when mapping a model to a grammar. For example, we used the attribute *name* in several places in the portfolio manager object model. If the same production rule applies to *name*, there is no problem. Otherwise, you should modify the attribute names in the object model during design to make them globally unique. For example, you might add a prefix of the class name to attribute names.

12.9 IMPLEMENTING SIMPLE ASSOCIATIONS

You can use three techniques to implement associations.

- **Explicit links**. You can explicitly represent links, as long as each associated class has a primary key. The production rule in Figure 12.11 specifies explicit links for the association between *Portfolio* and *Person* from Figure 12.2. You may use a role name to refer to an object or you may use the class name if there is no ambiguity. *Portfolio* is a subclass of *FinancialInstrument*, so this example also reflects an implementation of inheritance. (Section 12.11 discusses generalization.) In Figure 12.11 we generated a name for the association, *Portfolio__accountOwner*.

```
Portfolio__accountOwner: '<' 'Portfolio__accountOwner'
                         financialInstrumentID
                         personID '>' ;
```

Figure 12.11 Association implementation: Explicit links

Figure 12.12 shows some alternative production rules that store explicit links together. Figure 12.12 is analogous to Figure 12.7, which stores all instances of a class together. Figure 12.13 shows an excerpt of a file corresponding to the production rules in Figure 12.12. The example has two portfolios and two persons. The person with ID *4* owns the portfolios with IDs *2* and *3*. The person with ID *5* is joint owner of the portfolio with ID *3*.

```
Portfolio__accountOwners:  '<' 'Portfolio__accountOwners'
                           Portfolio__accountOwner * '>' ;
Portfolio__accountOwner :  '<' financialInstrumentID
                           personID '>' ;
```

Figure 12.12 Association implementation: Storing explicit links together

```
< Portfolio__accountOwners
  < 2 4 >   < 3 4 >   < 3 5 > >
```

Figure 12.13 An excerpt of a file for the production rules in Figure 12.12

For sequential-access ASCII files, we recommend that you normally implement associations with explicit links. Explicit links provide a uniform technique applicable to any kind of association. They also make it easier to evolve schema, because you can change the multiplicity of an association without changing the grammar. In addition, explicit links let you simplify the logic for saving and loading files. (See Section 12.14.)

```
Portfolio       :  '<' 'Portfolio'
                   financialInstrumentID
                   name
                   description
                   portfolioType
                   financialInstitutionID
                   accountNumber '>' ;
```

Figure 12.14 Association implementation: Buried reference

```
< Portfolio
  2 "retirement fund" "Carl's pension" pension 6 2173 >
```

Figure 12.15 An excerpt of a file for the production rule in Figure 12.14

- **Buried reference**. The second way to implement an association is to bury a pointer or set of pointers in an associated class. The referenced class must have a primary key. In Figure 12.14 *financialInstitutionID* in *Portfolio* refers to *FinancialInstitution*. (Figure 12.14 also has examples of implementing generalization and of using qualifiers, both of which we describe later.) Figure 12.15 shows an excerpt of a file that corresponds to Figure 12.14. The portfolio is held by the financial institution with ID *6* and has account number *2173*.

  In the object model, a *Portfolio* has an optional *FinancialInstitution* holding the account. Thus we would need to include a special value to represent a null pointer in the production rule for *financialInstitutionID*. Alternatively, we could use unordered labeled attributes (Figure 12.9) and simply omit *financialInstitutionID*.

  You should not bury an association in both classes. Dual references are redundant and unnecessary, since ASCII files are only loaded and saved, not directly queried. Thus, for example, you could implement the *Portfolio__accountOwner* association by burying a list of *Person* references in *Portfolio*, or burying a list of *Portfolio* references in *Person*. But you should not bury lists in both *Portfolio* and *Person*.

  Furthermore, we recommend that you completely avoid buried references for files. For files it is easier to implement associations with explicit links. Buried references yield more complex code for saving and loading files, as we show in Section 12.14.

- **Physical inclusion**. You can also implement an association by physically including one class in another. This technique applies when the included class cannot exist without the parent class and no associations or generalizations reference the included class. For example, you might use physical inclusion to implement physical aggregation. (Chapter 3 discusses physical aggregation.)

  We normally discourage inclusion, because it restricts cycles in the instance diagram in subtle ways, and such an implementation is difficult to extend. The full portfolio manager object model (Figure 9.5) does not have any associations for which we could use physical inclusion.

For all practical purposes, it is difficult to enforce multiplicity constraints with the structure of files. Consequently, you should enforce multiplicity constraints with your code for loading and saving files. You should put multiplicity logic in *both* the save and load code to reduce the chance of oversight.

12.10 IMPLEMENTING ADVANCED ASSOCIATIONS

We also recommend that you implement advanced associations with explicit links.

- **Link attributes**. You can add link attributes to the production rule for the explicit link.
- **Association classes**. When implementing these with explicit links, you must propagate identity from the related classes. The portfolio manager does not have any association classes, so we show another example in Figure 12.16.

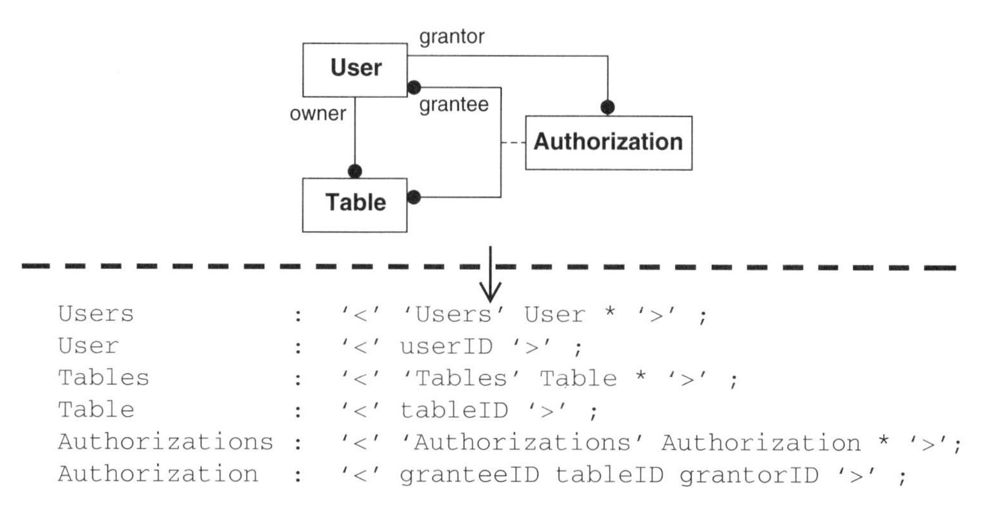

Figure 12.16 Association implementation: Association class

- **Qualified associations**. Figure 12.17 shows production rules for the qualified association between *FinancialInstitution* and *Portfolio*. Effectively, you should treat the qualifier as a link attribute. Your save and load code must enforce any uniqueness constraints implied by qualification. For example, the combination of a *FinancialInstitution* and an *accountNumber* identifies a unique *Portfolio*.

  Figure 12.18 shows an example for Figure 12.17. The financial institution with the ID of *6* holds two portfolios. The first portfolio has an ID of *2* and an account number of *2173*. The second portfolio has an ID of *3* and an account number of *5644*.

```
Portfolio__accountHolders: '<' 'Portfolio__accountHolders'
                           Portfolio__accountHolder * '>' ;
Portfolio__accountHolder:  '<' financialInstrumentID
                           financialInstitutionID
                           accountNumber '>' ;
```

Figure 12.17 Association implementation: Qualified association

```
< AccountHolder__portfolios
  < 2 6 2173 >   < 3 6 5644 > >
```

Figure 12.18 Example of links for the qualified association in Figure 12.17

- **Cascaded qualified associations**. You can implement these as you would qualified associations, except that for value-based identity you must propagate primary keys.
- **Ternary associations**. Implementation of a ternary association is the same as a binary association, except that you must include three primary keys instead of two.
- **Ordered associations**. You can either use explicit links and add a sequence number attribute or bury a list that orders association references.

  Figure 12.19 illustrates the explicit link technique. A *TripReservation* consists of a sequence of *FlightReservations*. The *sequenceNumber* is tantamount to a qualifier, where a *TripReservation* and a sequence number yield a *FlightReservation*.

  Figure 12.20 illustrates the use of a list to order association references for the flight reservations model. The grammar does not enforce the order of the attributes, so you must be careful with save and load routines.

```
TripReservations   :  '<' 'TripReservations' TripReservation
                      * '>' ;
TripReservation    :  '<' tripReservationID
                      flightReservationID * '>' ;

FlightReservations:   '<' 'FlightReservations'
                      FlightReservation * '>' ;
FlightReservation :   '<' flightReservationID '>' ;
```

Figure 12.20 Association implementation: Using a list for an ordered association

- **Symmetric association**. A ***symmetric association*** is an association between objects of the same class that have interchangeable roles. You can implement these as you would simple associations.
- **Aggregations**. Aggregation is a kind of association and observes the same mapping rules.

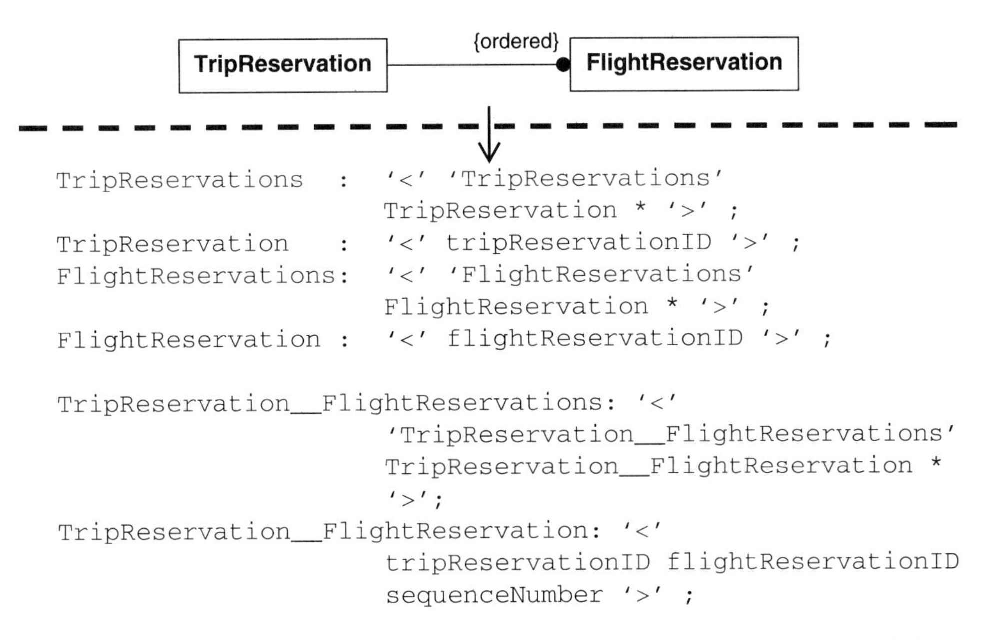


```
TripReservations   :  '<' 'TripReservations'
                      TripReservation * '>' ;
TripReservation    :  '<' tripReservationID '>' ;
FlightReservations:   '<' 'FlightReservations'
                      FlightReservation * '>' ;
FlightReservation :   '<' flightReservationID '>' ;

TripReservation__FlightReservations: '<'
                      'TripReservation__FlightReservations'
                      TripReservation__FlightReservation *
                      '>';
TripReservation__FlightReservation: '<'
                      tripReservationID flightReservationID
                      sequenceNumber '>' ;
```

Figure 12.19 Association implementation: Using a sequence number for an ordered association

12.11 IMPLEMENTING GENERALIZATIONS

You can implement generalizations by replicating superclass attributes for the subclasses, as Figure 12.21 shows.

```
FinancialInstrument  :  Asset
                     |  Portfolio ;
Portfolio            :  '<' 'Portfolio'
                        financialInstrumentID
                        name
                        description
                        portfolioType '>' ;
Asset                :  '<' 'Asset'
                        financialInstrumentID
                        name
                        description '>' ;
```

Figure 12.21 Generalization implementation: Replicating superclass attributes

Figure 12.22 shows a better way to implement generalizations. You can define a production rule for the superclass and reference it by the production rules for the subclasses. For the portfolio manager, *Asset* has no attributes beyond those it inherits. Figure 12.23 shows an example of a portfolio corresponding to the production rules in Figure 12.22.

```
FinancialInstrument   :  Asset | Portfolio ;
financialInstrumentAttributes:financialInstrumentID
                         name
                         description ;
Portfolio             :  '<' 'Portfolio'
                         financialInstrumentAttributes
                         portfolioType '>' ;
Asset                 :  '<' 'Asset'
                         financialInstrumentAttributes '>';
```

Figure 12.22 Generalization implementation: Referencing superclass attributes

```
< Portfolio
  2 "retirement fund" "Carl's pension" "pension" >
```

Figure 12.23 Example of a portfolio for the production rules in Figure 12.22

We recommend that you implement generalization by referencing superclass attributes (Figure 12.22). With replication you must incorporate attributes in the leaf subclasses for every superclass in the generalization lattice. In contrast, the reference technique is more extensible, because a subclass depends only on immediate superclasses. You can readily implement multiple inheritance by including the attributes for each superclass.

12.12 SUMMARY OF OBJECT MODEL MAPPING RULES

Table 12.2 summarizes the mappings for implementing the object model with files. Figure 12.24 shows a complete grammar for Figure 12.2 using ordered attributes. (Our Web site, www.omtassociates.com, has a complete grammar for Figure 9.5.) We suggest you define your grammar in the order of logical data types, domains, attributes, classes, and associations for readability.

12.13 IMPLEMENTING THE DYNAMIC MODEL

There are no dynamic modeling issues for files. All dynamic modeling issues concern programming, which is outside the scope of this book.

Concept	Object model construct	Production rule
Domain	Simple domain	Map each domain to a logical data type. Then map logical data types to physical data types (Figure 12.3).
	Enumeration	Enumeration string or encoding. We recommend strings.
	Structured	Hierarchical production rules (Figure 12.4).
	Multivalued	Production rule with asterisk after token for an attribute.
Class	Class	One production rule to define objects and another production rule to store all instances of a class together (Figure 12.7). Optionally, you might choose unordered storage of attributes (Figure 12.9).
Association	All associations	Explicit links, buried references, or sometimes physical inclusion. We recommend explicit links. You should store all links of an association together (Figure 12.12).
	Aggregation	Same as association.
Generalization	Single inheritance	Replicate or reference superclass attributes for the subclasses. We recommend referencing superclass attributes (Figure 12.22).
	Multiple inheritance	Same as single inheritance.

Table 12.2 Summary of object model mapping rules for ASCII files

12.14 IMPLEMENTING THE FUNCTIONAL MODEL

File grammars do not involve much functional behavior either. Essentially, applications load file(s), process the data, and save file(s). Here we discuss saving and loading only, because the other processing is purely a programming matter and thus outside the scope of this book. We do not show an implementation of the Object Navigation Notation because this is a matter of programming and not file structure. As before, we recommend that you use a context-free grammar so that you can use tools to generate parsers for loading files. This section describes the mechanics of saving and loading sequential-access ASCII files using *Lex* and *Yacc*.

```
logicalInteger              :INTEGER ;
logicalString               :INTEGER | STRING ;
logicalDate                 :STRING ;
logicalMoney                :INTEGER | STRING ;
logicalReal                 :STRING ;

idDomain                    :logicalInteger ;
nameDomain                  :logicalString ;
longStringDomain            :logicalString ;
enumPortfolioTypeDomain     :logicalString ;
dateDomain                  :logicalDate ;
enumTransactionTypeDomain   :logicalString ;
moneyDomain                 :logicalMoney ;
addressDomain               :logicalString ;
longNumberDomain            :logicalReal ;
mediumStringDomain          :logicalString ;

financialInstrumentID       :idDomain ;
financialInstrumentName     :nameDomain ;
description                 :longStringDomain ;
portfolioType               :enumPortfolioTypeDomain ;
transactionID               :idDomain ;
transactionDate             :dateDomain ;
recordDate                  :dateDomain ;
transactionType             :enumTransactionTypeDomain ;
transactionFee              :moneyDomain ;
financialInstitutionID      :idDomain ;
financialInstitutionName    :nameDomain ;
address                     :addressDomain ;
personID                    :idDomain ;
personName                  :nameDomain ;
quantity                    :longNumberDomain ;
accountNumber               :mediumStringDomain ;

financialInstrumentAttributes:financialInstrumentID
                        financialInstrumentName
                        description ;
Portfolio             :  '<' financialInstrumentAttributes
                        portfolioType '>' ;
```

Figure 12.24 Full BNF grammar for Figure 12.2

```
Portfolios           :  '<' 'Portfolios'
                           Portfolio * '>' ;
Asset                :  '<' financialInstrumentAttributes '>';
Assets               :  '<' 'Assets' Asset * '>' ;
Transaction          :  '<' transactionID
                        transactionDate
                        recordDate
                        transactionType
                        transactionFee
                        description '>' ;
Transactions         :  '<' 'Transactions' Transaction * '>' ;
FinancialInstitution:'<' financialInstitutionID
                        financialInstitutionName
                        address '>' ;
FinancialInstitutions:'<' 'FinancialInstitutions'
                        FinancialInstitution * '>' ;
Person               :  '<' personID personName '>' ;
Persons              :  '<' 'Persons' Person * '>' ;

Transaction__Asset:     '<' transactionID
                        financialInstrumentID
                        quantity '>' ;
Transaction__Assets:    '<' 'Transaction__Assets'
                        Transaction__Asset * '>' ;
Transaction__Portfolio:'<' transactionID
                        financialInstrumentID '>' ;
Transaction__Portfolios:'<' 'Transaction__Portfolios'
                        Transaction__Portfolio * '>' ;
Portfolio__accountOwner:'<' financialInstrumentID
                        personID '>' ;
Portfolio__accountOwners:'<' 'Portfolio__accountOwners'
                        Portfolio__accountOwner * '>' ;
Portfolio__accountHolder:'<'
                        financialInstrumentID
                        financialInstitutionID
                        accountNumber '>' ;
Portfolio__accountHolders:'<'
                        'Portfolio__FinancialInstitutions'
                        Portfolio__FinancialInstitution *
                        '>' ;
```

Figure 12.24 (continued) Full BNF grammar for Figure 12.2

12.14.1 Saving Files

You can readily save files by applying methods:

- **Objects**. Define a *save* method for each class that is not a superclass. The *save* method takes a file object as an argument and writes an object to a file. You will also need to define a *saveAttributes* method to write the attributes for each superclass. We are assuming all superclasses are abstract. (See Chapter 3.)

  Scan all objects and apply the *save* method to every persistent object. The *save* method for an object first writes an opening delimiter and the name of the class, then invokes the *saveAttributes* method for any superclasses, and finally writes the attribute values for the object, followed by a closing delimiter. Make sure each object has a primary key. If you save all instances of a class together, write the name of the class once for each class rather than for each instance.
- **Links**. The details depend on your mapping strategy. If you implement associations with explicit links, use the following logic: Define a *save* method for each association, with a file object as an argument. Then scan all links and apply the *save* method to every persistent link. The *save* method for a link first writes an opening delimiter and the name of the association and then writes the primary key values for the related objects, any link attribute values, and finally a closing delimiter. If you save all links of an association together, write the name of the association once for each association, rather than for each link.

You will need to adjust these methods if you have implemented some associations with buried references. Then you will need to write the references as well as the attribute values when saving objects. You will also need to adjust the algorithm if you have used physical inclusion.

12.14.2 Loading Files

You can use tools such as *Lex*, *Yacc*, and *Yacc++* to generate code for loading a file or you can write the code yourself. We recommend that you generate code for loading a file, because it is difficult to write a correct, robust, extensible, and efficient parser.

You can simplify the parser by keeping the representation in memory homomorphic to the representation in a file. The *Lex* grammar reflects the representation of a file, and the *Yacc* or *Yacc++* actions create the representation in memory. For example, if a file implements associations with explicit links, you should also represent associations in memory with explicit links. If a file implements associations with buried references, you should bury associations in memory.

You can always load a file with a two-pass approach. The first pass creates objects and links and builds an index that maps each primary key to an object identifier. The second pass uses the index to convert foreign key references to in-memory pointers.

First Pass

During the first pass you read all the data from a file and store it in memory. You must capture domain values, create objects, and set attribute values in each object. As you create each object, you can also place an entry in the index mapping the primary key value to the object identifier. *Yacc* lets you attach actions to production rules and executes the actions as it recognizes the elements.

In Figure 12.25 we have extended a portion of the grammar in Figure 12.3 with actions for capturing domain values. The braces are *Lex* and *Yacc* notation, delimiting the actions that the generated parser will execute when it encounters the token to the left of the colon. The first production rule shows how to construct logical integers from physical integers. The variable *yytext* is a string buffer provided by *Lex* and *Yacc*. You must provide the function *convert_string_to_integer* to scan the characters of a string and compute an integer. Upon successful execution, this production rule stores the logical integer in the global variable *logicalInteger.*

```
logicalInteger :  INTEGER { logicalInteger =
                  convert_string_to_integer(yytext);} ;
logicalString  :  INTEGER { logicalString =
                  convert_string_to_string(yytext) ;}
               |  STRING { logicalString =
                  convert_string_to_string(yytext) ;} ;

idDomain       :  logicalInteger
                  { idDomain = logicalInteger ;} ;
nameDomain     :  logicalString
                  { nameDomain = logicalString ;} ;
```

Figure 12.25 First-pass actions for file loading: Capturing domain values

You do not have to use global variables with *Lex* and *Yacc*. These tools have pseudovariables (similar to *yytext*) that refer to values passed back by the actions of a production rule. Although pseudovariables are preferred for compilers, global variables are easier to use in passing values between actions.

The second production rule shows how to construct logical strings from physical integers or physical strings. Upon successful execution, this production rule stores the logical string in the global variable *logicalString*. You must provide the function *convert_string_to_string* to convert a string from the file format to the memory format. The production rule executes as follows. If the parser is expecting a logical string and *Lex* recognizes the current token in the file (stored in *yytext*) as an integer or a string, then the function converts the contents of *yytext* to a logical string. Otherwise the *logicalString* production rule fails and the current token must be handled by some other part of the grammar.

The production rule for *idDomain* copies the value in the global variable *logicalInteger* to the global variable *idDomain*. Similarly, the production rule for *nameDomain* copies the value in the global variable *logicalString* to the global variable *nameDomain*.

For a full implementation of the portfolio manager, we would need a production rule for each logical data type and domain. Figure 12.24 has five logical data types and 10 domains for the excerpt of the portfolio manager model in Figure 12.2. The production rules in Figure 12.25 concern simple domains; all we must do is store the logical data type. For other kinds of domains (enumerated, structured, multivalued) you must define additional functions that convert logical data types to domains.

Lex and *Yacc* next execute the production rules in Figure 12.26 to copy the global domain values to the global attribute variables. For example, the first production rule copies the value in global variable *idDomain* to global variable *financialInstrumentID*. You can handle associations that are implemented as buried references as you would attributes and just store foreign key values during the first pass; during the second pass you can then make the foreign key values pointers to objects.

```
financialInstrumentID:'<' 'financialInstrumentID'
                  idDomain
                  { financialInstrumentID = idDomain ;}
                  '>' ;
name            : '<' 'name' nameDomain
                  { name = nameDomain ;} '>' ;
description     : '<' 'description' longStringDomain
                  { description = longStringDomain ;} '>' ;
```

Figure 12.26 First-pass actions for file loading: Capturing attribute values

```
FinancialInstruments :  '<' 'FinancialInstruments'
                        FinancialInstrument * '>' ;

FinancialInstrument  :  '<'
                        {  financialInstrument =
                           FinancialInstrument_NEW () ;}
                           FinancialInstrumentAttribute *
                        '>' ;
```

Figure 12.27 First-pass actions for file loading: Creating an object

You must also have actions that create objects. For example, the production rules in Figure 12.27 elaborate Figure 12.9 and create an object when a new financial instrument is first encountered in a file. We must provide the implementation of the *financialInstrument_NEW* function. We must also record the identifiers of important objects that serve as entry points for navigation. This is true of any application. You can include an action that adds entry points to master list(s) as you recognize objects.

Some languages may require a casting operator for generalization. Otherwise, you do not have to take any special actions. You can simply create an object for the leaf class and populate attributes as the parser encounters them.

For explicit associations, you must build objects in memory that hold the foreign key values. During the second pass you can convert the foreign key values to object references.

The final step of the first pass is to copy the global attribute values to object attributes, as Figure 12.28 shows. You must then implement the functions to set the attributes.

```
FinancialInstrumentAttribute:
                financialInstrumentID
                {  financialInstrument_setID (
                   financialInstrument ,
                   financialInstrumentID ) ;}
              | name
                {  financialInstrument_setName (
                   financialInstrument , name ) ;}
              | description
                {  financialInstrument_setDescription (
                   financialInstrument , description ) ;} ;
```

Figure 12.28 First-pass actions for file loading: Setting attribute values in an object

Second Pass

The first-pass actions for file loading were to load objects and build an index that maps primary key values to object identifiers. In the second pass you scan the foreign keys and use the index to resolve foreign key references to in-memory references. This conversion of foreign keys is straightforward programming, so we do not show it here.

The second pass is not always needed. You can use a single pass if you do not have forward references; that is, no foreign key value references an object that has not yet been read into memory. You can avoid forward references if you do not use buried references, have no association classes, and store all links after all objects.

12.15 OTHER IMPLEMENTATION ISSUES

You should also consider several miscellaneous issues related to files:

- **Locking**. Ordinarily, you should lock only at the file level. You can, of course, write software to support record-level locks, but it will be complex and difficult to develop. If you require record-level locks, you should really use a DBMS.

  In principle, you could permit both read and write locks for entire files. However, most software supports only write locks. You can lock files with operating system commands or by writing a sentinel file to denote the current holder of a lock token.

- **Crash resistance**. Backup files can provide some crash resistance. You can also use multiple backup files to provide primitive versioning. The VAX VMS operating system takes this approach. Once again, we advise that you do not try to program crash resistance into your applications; instead, you can save time, effort, and cost by using a DBMS.
- **Constraints**. The object model implies constraints that the file structure cannot readily enforce. For example, you can define grammars that support mandatory single-valued attributes (attribute multiplicity of *[1]*) and grammars that support optional multivalued attributes (attribute multiplicity of *[0..*]*), but files do not gracefully enforce other kinds of attribute multiplicity. They also have no intrinsic ability to enforce candidate keys and referential integrity.

  Our advice is to hope for the best, but plan for the worst. Although you may intend for an application to enforce constraints, oversights can always occur. Implement applications to enforce constraints on a write and to recover gracefully if there is a constraint violation during a read.
- **Schema evolution**. The format of a file may change with versions of an application. Consequently, you should include the version number and the date in the file grammar.

  New applications versions should be able to read old file versions, but the converse need not be true. To handle old versions you can either use stand-alone "filters" to convert between pairs of versions or include the capability to read any previous version. Filters are easier to implement and can also import data from other applications. However, a built-in capability for reading previous versions is easier to operate. We prefer to include the capability to read any previous version.

12.16 CHAPTER SUMMARY

Files are useful for simple applications. If your application does not need the full services of a database management system, you should consider using files for data persistence.

We recommend that you try to confine your use of files to sequential-access ASCII files, because you can then normally use a context-free grammar. Context-free grammars are supported by a sound theoretical framework and code-generation tools for file loading. In addition, context-free grammars promote extensibility for migrating to new versions of an application.

You can map an object model to a grammar with the rules we have described. You must implement logical data types, domains, classes, attributes, associations, and generalizations. You can readily save objects to a file by defining a *save* method for each class and association. You can load objects from a file by binding actions to the production rules for the grammar.

Figure 12.29 lists the key concepts for this chapter.

context-free grammar	object model, implementing domains
file	object model, implementing generalizations
file type	parser
grammar	parser generation tool
language	production rule
loading files	random-access binary file
memory-image binary file	saving files
object model, implementing associations	schema evolution
object model, implementing attributes	sequential-access ASCII file
object model, implementing classes	sequential-access binary file

Figure 12.29 Key concepts for Chapter 12

BIBLIOGRAPHIC NOTES

[Schreiner-85] provides a good introductory treatment for applying *Lex* and *Yacc*. He emphasizes practical advice without dwelling on the theory. You may find the original articles on *Lex* [Lesk-75] and *Yacc* [Johnson-75] to be of historical interest. [Aho-86] presents a more theoretical treatment and discusses deep implementation issues. [Graver-92] presents a good explanation of concepts for mapping objects to files. Graver's approach is similar to ours.

An object-oriented tool called *Yacc++* is commercially available from Compiler Resources, Inc., Hopkinton, Massachusetts. [Avotins-95] describes some recent work on object-oriented compiler generators.

REFERENCES

[Aho-86] AC Aho, R Seth, and JD Ullman. *Compilers: Principles, Techniques, and Tools*. Reading, Massachusetts: Addison-Wesley, 1986.

[Avotins-95] J Avotins, C Mingins, and H Schmidt. Yes! An Object-Oriented Compiler Compiler (YOOCC), *Technology of Object-Oriented Languages and Systems Tools 17* (Proceedings of the 17th International Conference TOOLS SANTA BARBARA, 1995) Englewood Cliffs, New Jersey: Prentice Hall, 1995.

[Compiler-92] YACC++ and the Language Objects Library Reference Guide, Compiler Resources, Inc., 85 Main St. Suite 310, Hopkinton, Massachusetts 01748, 1992.

[Graver-92] JO Graver. T-gen: a string-to-object translator generator, *Journal of Object-Oriented Programming 5*, 5 (September 1992), 35–42.

[Lesk-75] ME Lesk. Lex—A Lexical Analyzer Generator, CSTR 39, Bell Laboratories, Murray Hill, New Jersey, 1975.

[Johnson-75] SC Johnson. Yacc—Yet Another Compiler Compiler, CSTR 32, Bell Laboratories, Murray Hill, New Jersey, 1975.

[Schreiner-85] AT Schreiner and HG Friedman. *Introduction to Compiler Construction with Unix*. Englewood Cliffs, New Jersey: Prentice Hall, 1985.

EXERCISES

12.1 Figure E12.1 specifies requirements for a two-dimensional graphics editor. Figure E12.2 shows the corresponding object model. Figure E12.3 specifies a domain and multiplicity for each attribute. Assume that all data will be stored in sequential-access ASCII files. Note that it is quite reasonable for a graphics editor to use files rather than a database.

> Develop software for a simple two-dimensional graphics editor. Assume you are given a library of standard shapes identified by name. All standard shapes are represented by a bitmap. Provide the following functionality:
>
> - The software instantiates a standard shape each time the user places it in a diagram. Record the *x* and *y* scaling, centerpoint, and orientation. Also record the fill color, fill pattern, outline color, outline thickness, and outline pattern.
> - For textual contents, record the font, size, color, centerpoint, and orientation.
> - Order the diagram components to resolve display of overlap.

Figure E12.1 Problem statement for a simple two-dimensional graphics editor

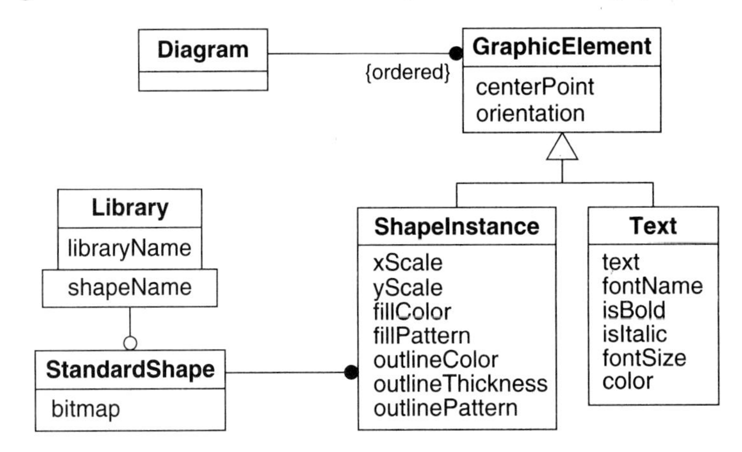

Figure E12.2 Object diagram for a simple two-dimensional graphics editor

a. (3) Organize the data into two file types. Explain the rationale for your answer.
b. (2) Assign the data to files and explain your answer.
c. (3) Choose an approach to implementing identity and explain the reasons for your decision. How would you handle associations for which the classes are stored in different files?
d. (3) Would any kind of locking be needed? How could this be implemented simply?
e. (4) Define production rules for implementing domains.
f. (7) Define production rules for implementing classes, generalization, and ordered attributes.
g. (8) Revise your production rules from (f) to use unordered attributes. Are there any benefits of using unordered attributes for this problem?
h. (7) Define production rules for implementing associations. Use explicit links.

Class	Attribute	Domain	Attribute multiplicity
GraphicElement	centerPoint	coordinate	1
	orientation	number	1
Library	libraryName	name	1
StandardShape	bitmap	binaryString	1
ShapeInstance	xScale	number	1
	yScale	number	1
	fillColor	color	1
	fillPattern	pattern	1
	outlineColor	color	1
	outlineThickness	number	1
	outlinePattern	pattern	1
Text	text	longString	1
	fontName	name	1
	isBold	boolean	1
	isItalic	boolean	1
	fontSize	number	1
	color	color	1
...qualifier...	shapeName	name	1

Figure E12.3 Attribute specifications for the two-dimensional graphics editor

i. (6) Show the contents of the files for the grammar with ordered attributes (defined by parts *e*, *f*, *h*) for the following simple example.

There are two libraries, one for geometric shapes and one for aircraft icons. The geometric shapes in the library include a circle, a square, and a line. The aircraft icons include a plane, a jet, and a helicopter.

There is one diagram, containing a green ellipse (circle with scaling) and a red square. Both the ellipse and the square have no fill and no pattern, with a black outline. The outline has a thickness of 1 and a solid pattern. The ellipse is centered at (−2,0) with x and y scale factors of 1 and 2. The square is centered at (2,0) with x and y scale factors of 2 and 2. There are two black text elements, one centered in the ellipse and one centered in the square, both using size 12 bold helvetica font. The text in the ellipse is "ellipse" and the text in the square is "square." The ellipse, the square, and the text are not rotated.

Show values of all attributes except bitmap, which you may ignore. Assign arbitrary values for IDs that are consistent with the object diagram.

j. (7) Repeat part (i) for the grammar with unordered attributes (defined by parts *e*, *g*, *h*).

12.2 (6) Given the following data, construct a file for the grammar in Figure 12.24.

Main bank on 102 State Street holds two portfolios with account numbers 9687 and 7465. Joe and Mary jointly own account 9687, which is a normal passbook savings account. Mary also owns account 7465, which is a retirement account.

The passbook account is relatively new and only has a few transactions. On February 12, 1996 the account was opened with an initial deposit of $100. On June 29, 1996, $2.00 of interest was credited. And on August 5, 1996, $75 was withdrawn.

The retirement account is also relatively new. On April 15, 1996 it was opened with an initial deposit of $700. On June 29, 1996 Mary purchased 10 shares of Johnson Controls stock at a cost of $60 per share and incurred a $30 commission. On December 31, 1996 the account was assessed a $15 retirement account fee.

12.3 (5) Define production rules to implement Figure 2.17. Use ordered attributes and existence-based identity.

12.4 (7) Define production rules to implement Figure 2.19. Use ordered attributes and existence-based identity.

12.5 (6) Define production rules to implement Figure 3.10. Use ordered attributes and existence-based identity.

12.6 (6) Define production rules to implement the left model in Figure 2.21. Use ordered attributes and existence-based identity.

12.7 Consider the prescription drug information system described in Exercise 9.3. We would prefer to store prescription drug data in a database (see the exercises in Chapters 13–16), but for illustration the next several exercises will concern implementation with files. Assume that all data will be stored in sequential-access ASCII files.

a. (3) Organize the data into file types. Explain the rationale for your answer.
b. (2) Assign the data to files and explain your answer.
c. (3) Choose an approach to implementing identity and explain the reasons for your decision.
d. (4) Define production rules for implementing domains.
e. (7) Define production rules for implementing classes, generalization, and ordered attributes.
f. (8) Revise your production rules from (e) to use unordered attributes.
g. (7) Define production rules for implementing associations. Use explicit links.

12.8 Repeat Exercise 12.7 for the prescription drug information system described in Exercise 9.4. Assume that all data will be stored in sequential-access ASCII files.

12.9 Repeat Exercise 12.7 for the prescription drug information system described in Exercise 9.5. Assume that all data will be stored in sequential-access ASCII files.

12.10 Repeat Exercise 12.7 for the sales opportunity manager described in Exercise 8.2. Assume that all data will be stored in sequential-access ASCII files. Several commercial products implement an opportunity manager for a single user. For such an application, it could be quite reasonable to use files.

12.11 Repeat Exercise 12.7 for the contact manager described in Exercise 8.3. Assume that all data will be stored in sequential-access ASCII files. Several commercial products implement a contact manager for a single user. For such an application, it could be quite reasonable to use files.

13

Relational Databases: Basics

The object-oriented paradigm provides a useful approach to designing relational database applications, because object-oriented models are expressive, concise, and easy to develop. This chapter shows basic implementation of OMT constructs with a relational database. Chapter 14 describes advanced issues. The principles in both chapters apply to various relational database managers, including Oracle, Sybase, Informix, DB2, and MS-Access. The core relational features are standard, so we can perform much implementation in a manner that transcends individual products.

We have fully implemented our case study with MS-Access 2.0. Our implementation is only a prototype, but it does illustrate many of the software engineering principles contained in this book. We chose to use MS-Access because it is so widely available. Also, MS-Access has a powerful 4GL (fourth-generation language) that makes implementation easier. Chapter 14 presents some code. Our Web site, www.omtassociates.com, has the complete code.

We are placing all of our implementation code for the portfolio manager in the public domain. You may study the code and use it without restriction. You may freely distribute the code, and you may use all or any portion of the case study code for your own work, even for a commercial product. Obviously, we are not responsible for any errors because our implementation is intended only as a prototype.

We invite you, the readers, to submit your implementations of the portfolio manager case study for other platforms. You can contact either of us via email ({blaha, premerlani}@ acm.org). If we receive enough submissions, we will publish them and acknowledge the authors. We would also be interested in metrics such as development time, number of bugs, and execution performance.

The relational database literature uses the term "relation." We substitute the word "table" with an intended equivalent meaning to avoid confusion with the word "relationship" from Chapter 2. We assume you are familiar with relational database technology.

13.1 INTRODUCTION TO RELATIONAL DATABASES

A ***relational database*** is a database in which the data is logically perceived as tables. A ***relational database management system*** (RDBMS) manages tables of data and associated structures that increase the functionality and performance of tables. An RDBMS has three major aspects:

- **Data that is presented as two-dimensional tables**. Tables have a specific number of columns and an arbitrary number of rows with a value stored at each row-column intersection. The columns represent attributes; the rows, data occurrences. RDBMSs use various devices to speed access because literal tables are much too slow for practical needs. These tuning devices are usually transparent and not visible in the commands for reading and writing to tables.
- **Operators for manipulating tables**. SQL [Melton-93] is the standard language for accessing data. SQL commands are set oriented and operate on entire tables, as opposed to being record oriented and operating on individual rows. An important property of the relational model is closure; all data manipulation commands accept tables as input and yield a table as output.
- **Integrity rules on tables**. The leading commercial RDBMSs support ***referential integrity*** and simple constraints. Referential integrity ensures that the values referenced in other tables really exist.

Figure 13.1 shows a simple object model and several database tables you could use to implement it, populated with some data. We use bold type to indicate the primary key of each table, the combination of attributes that uniquely identifies each row in a table.

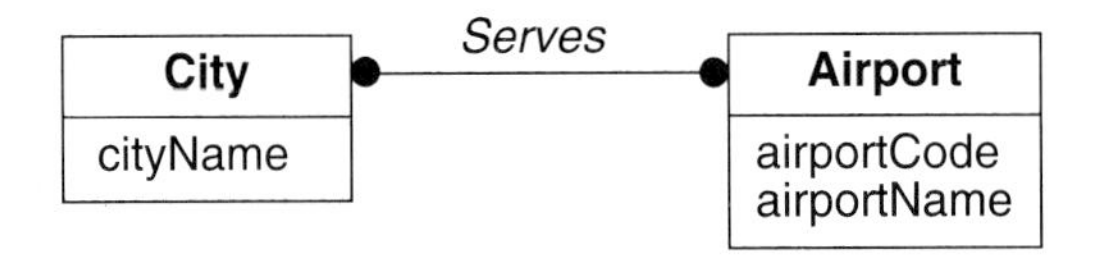

City table

cityName
Houston
Albany
Schenectady
St. Louis

Airport table

airportCode	airportName
IAH	Intercontinental
HOU	Hobby
ALB	Albany county
STL	Lambert

Serves table

cityName	**airportCode**
Houston	IAH
Houston	HOU
Albany	ALB
Schenectady	ALB
St. Louis	STL

Figure 13.1 Sample object model and sample relational database tables

An important feature of RDBMSs is their support for data independence. With ***logical data independence*** an application is made aware only of relevant portions of the database; this lets you add new applications to the database without disrupting existing applications. ***Physical data independence*** insulates applications from changes in tuning mechanisms; you can then restructure a database for faster performance without affecting application logic. The developer describes what is needed; the RDBMS decides how to access the data physically.

You must carefully model a relational database to achieve acceptable performance. If you do not construct models, you are not fully considering the competing demands on your database. As a consequence, performance will be unpredictable and your database will be vulnerable to errors and difficult to program against.

In addition, you must tune your database; tuning consists mostly of defining the appropriate indexes and properly phrasing SQL code. The simple notion of a table can be misleading; RDBMSs appear to be easy to use and are advertised as such, but in reality careful modeling and tuning is required to obtain excellent performance.

13.1.1 Metamodel

Figure 13.2 presents a simplified object metamodel for RDBMSs. This metamodel is similar to the metamodel RDBMSs incorporate, called the ***data dictionary***. Users and applications may freely read but not directly write to the data dictionary. An RDBMS updates metadata in its data dictionary as a side effect of commands. A relational database is self-descriptive, since the data dictionary tables contain data that describe themselves.

In Figure 13.2 a database has many authorized users. Each user is uniquely identified by name, has a password and certain privileges, and may own multiple tables. The class *User* is a surrogate for a person. Any person may log in as a database user as long as he or she knows a user name and password. Domain, table, and index names must also be unique within the scope of a database.

A user may access tables owned by other users. This granting of permissions forms a tree. For example, user *A* may grant permissions to users *B* and *C* for a table; user *B* may, in turn, grant permissions to user *D* for the table. If *A* revokes permissions for *B*, the RDBMS must also revoke the permissions *B* granted to *D*. The *Authorization* association class supports cascading permissions.

A table contains many attributes ordered from left to right as they would appear in a printout. Each attribute has a domain and a name that is unique within the scope of a table. Base attributes (attributes that are physically stored) also have a default value and a flag that indicates whether null values are permitted. Base tables are associated with various types of keys: candidate keys, primary keys, and foreign keys.

A ***candidate key*** consists of one or more attributes that uniquely identifies the instances within a table. The collection of attributes in a candidate key must be minimal; if you discard an attribute from the candidate key, you destroy its uniqueness. No attribute in a candidate key can be null. A ***primary key*** is an arbitrarily chosen candidate key used to reference instances preferentially; normally each table should have exactly one primary key. A primary key can be an artificially generated identifier or a combination of attributes that arise from

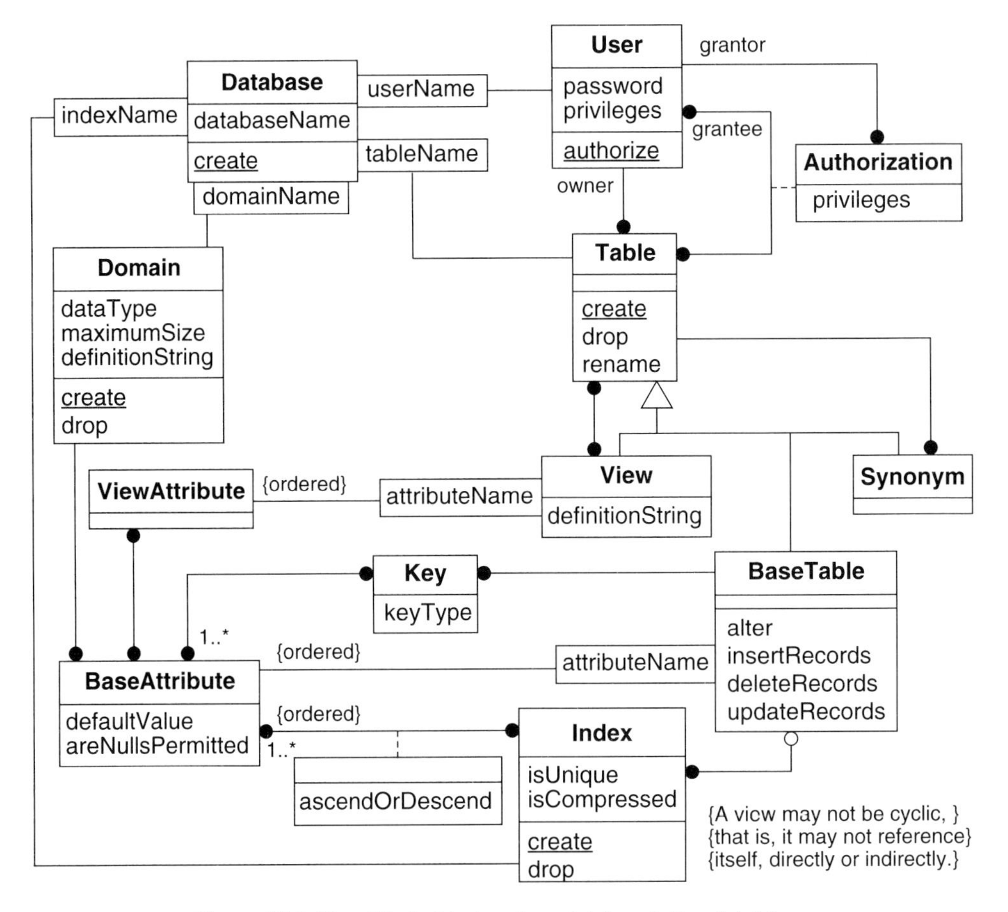

Figure 13.2 Simplified object metamodel for relational databases

application knowledge. A ***foreign key*** is a reference to a candidate key. However, we strongly advise that all foreign keys refer only to primary keys, not to other candidate keys. (See Section 13.3.) A foreign key must include a value for each constituent attribute or it must be wholly null. The metamodel does not show all this detail, but simply shows the class *Key*.

A ***view*** is a table that the RDBMS dynamically computes from a query stored in the database. A synonym provides an alias for a table. We abstract *View*, *BaseTable*, and *Synonym* into *Table*. The association between *View* and *Table* indicates that a view is defined in terms of base tables and other views. Similarly, each view attribute is ultimately defined in terms of base attributes.

You may define indexes for each base table. The model does not specify the purpose of these indexes, though typically developers use them to improve the performance of certain queries. You can also use indexes to enforce the uniqueness of keys. An index is built from multiple attributes that are ordered in terms of their significance in building the index hierarchy. Each index attribute may be in ascending or descending sort order.

Figure 13.2 is incomplete; we could note more attributes and operations. We could also construct additional metamodels that describe other aspects of RDBMSs. We could build an object model of physical structures, for example, that includes clusters, rollback segments, and files.

13.1.2 SQL

SQL [Melton-93] is the standard RDBMS language that unifies data definition, data manipulation, and data control. You can use it to access data both interactively and within a program. SQL data definition commands create, drop, and rename database constructs such as tables, views, synonyms, and indexes. Data manipulation commands can insert, update, delete, and retrieve rows. Data control commands authorize new users and grant and revoke privileges for accessing tables and views.

A major objective of SQL is to provide data independence. The goal is to decouple application logic from access paths and let the RDBMS automatically optimize queries; database administrators can then tune performance without disrupting application code. With SQL this goal is largely, but not fully, realized. SQL optimizers do not always recognize the equivalence of queries phrased in different ways. Developers must consider the available physical access paths and likely optimization results when phrasing a query. Nevertheless, SQL still offers substantial data independence and freedom to add indexes or restructure a populated database.

From the perspective of language theory [Date-86], however, SQL is an imperfect language. For example, it is not computationally complete. You can express conditionality with predicates in SQL queries, but this capability is limited. SQL can provide iteration by processing sets of data, but you cannot combine this iteration with arbitrary user-defined functions. There is no natural way to save state between the execution of complementary queries, aside from populating an intermediate table.

To be fair, other modern languages also have weaknesses. For example, C++ has a difficult grammar and Smalltalk's weak typing is not conducive to programming-in-the-large.

In general, SQL is widely accepted, and it is gradually increasing in power.

13.1.3 Normal Forms

Normal forms are guidelines for relational database design that increase the consistency of data. As tables satisfy higher normal forms, they are less likely to store redundant data. A table is in ***first normal form*** if each row-column combination stores a single value (rather than a collection of values). A table is in ***second normal form*** if it is in first normal form and all attributes depend on the entire primary key. A table is in ***third normal form*** if it is in second normal form and no attribute transitively depends on the primary key. (See Chapter 17 of [Rumbaugh-91] for a more detailed explanation of normal forms.)

We see normal forms as an anachronism, largely irrelevant to OMT modeling. Normal forms are an artifact of attribute-based design. As we explained in Chapter 1, OMT is an entity-based approach. If you directly describe objects and do not intermix different things, you should not need normal forms. We mention them only so that you can see how they fit into the context of OMT design.

An object-oriented database does not require normal forms because it can directly represent an object-oriented model. In contrast, a relational database provides more primitive, more permissive, and consequentially more dangerous structures that can store both meaningful and nonsensical data. Relational databases achieve data independence by introducing possibly dangerous structural redundancies. Normal forms restrict the use of relational database tables to avoid unfortunate side effects. Our mappings in Sections 13.5 through 13.9 also limit usage to meaningful RDBMS constructs.

Some of our optimizations improve performance by violating normal forms. We note these violations as they occur.

13.1.4 Overview of Microsoft Access 2.0

MS-Access 2.0 is an impressive RDBMS suitable for implementing many small to medium-sized data management applications.* Furthermore, you can quickly prototype large applications before implementing them with a more capable RDBMS. For the portfolio manager, we generally avoided the MS-Access programming language, instead using SQL and macros. Performance was good because the interpretation overhead for macros is not significant. (Performance bottlenecks stem primarily from interaction with the database.)

MS-Access has many strengths. Its updatable views, polished user interface, and event-driven forms are especially commendable.

- **Support for most of SQL DML**. MS-Access supports much of the SQL data manipulation language (DML). DML commands insert, update, delete, and retrieve rows. The MS-Access query optimizer is effective and query execution is fast.
- **Support for views**. You can treat MS-Access queries as views. You can directly execute queries, reference them with other queries, and use them as the basis for forms and reports.
- **Update through views**. You can update data through single table views, propagating the changes to the underlying table. You can also update many views that combine tables with the *inner join* clause. As we describe in Chapter 14, update through views lets you preserve the coherence of objects represented by multiple records across a generalization hierarchy.
- **Support for referential integrity**. MS-Access partially supports referential integrity through a graphical interface.
- **Polished user interface (4GL)**. MS-Access has powerful and friendly graphical languages for constructing queries, forms, and reports.

---

* We roughly characterize "small to medium-sized" applications as having fewer than 100 tables, less than 100 MB of data, and up to nine concurrent users.

- **Event-driven forms**. The forms include various events to which you can attach behavior. Occurrence of an event triggers execution of the attached code. You may write code in MS-Access or you can invoke macros. We were able to implement most of our case study with macros and SQL, enabling fast development and few errors.
- **Support for object identity**. MS-Access partially supports object identity through the counter data type.
- **Client-server support**. MS-Access supports a two-tier architecture. You can define tables in one database (the *server*), and then *attach* them with a second database (the *client*) that contains application and user interface logic. The server holds tables, indexes, and referential integrity actions. The client holds queries, programming code, forms, and reports. The attachments are transparent, and you can write logic in the client database as if the tables were physically stored there.
- **Import and export**. MS-Access can read and write many database and file formats.

MS-Access also has some significant weaknesses:

- **Incomplete support for SQL DDL**. MS-Access does not support much of the SQL data definition language (DDL). DDL commands create, drop, and rename database constructs such as tables and indexes. You must edit forms to set default values, *not null* constraints, and other *check* constraints. MS-Access graphically supports the *on delete cascade*, *on delete no action*, *on update cascade*, and *on update no action* referential integrity actions, but does not support *on delete set null* and *on delete set default*. Also you cannot directly drop a table that is involved in referential integrity; first you must remove the referential integrity and then you can drop the table.
- **No support for SQL DCL**. MS-Access lacks support for the SQL data control language (DCL). MS-Access provides its own approach to security instead of the SQL *grant* and *revoke* commands.
- **Lack of a three-tier architecture**. The two-tier client-server support is helpful, but MS-Access cannot further separate application logic from the user interface. Chapter 18 describes the three-tier architecture in detail.
- **No domains**. MS-Access does not support domains, even though domains are an important aspect of RDBMS theory. Unfortunately, many other RDBMSs share this flaw.
- **Incomplete support for transactions**. MS-Access SQL lacks *set transaction*, *commit*, and *rollback* commands. You can define transactions only by using the programming language.
- **Limited concurrency**. Normally, MS-Access relies on optimistic concurrency control, checking for contention when it writes a record. You may also lock a table and lock a page within a table (pessimistic locking), but you cannot lock an individual record.
- **Incomplete crash recovery**. MS-Access lacks a transaction log for recovering from a storage media error.

The most serious flaws, omissions and deviations in the implementation of SQL, make it harder to write portable code (the reason for having an SQL standard in the first place). We compensated for this by using features specific to MS-Access where necessary.

13.2 IMPLEMENTING THE OBJECT MODEL

The first step in implementing an application is to map the object model to relational database constructs with the following tasks. (The numbers in brackets refer to subsections that explain these tasks in detail.)

- **Implement identity**. In Chapter 9 (system design) you chose a strategy for identity—existence-based or value-based. We recommended existence-based identity for applications with more than 30 classes; either approach is viable for a small data management application. In this task, follow through and implement whichever strategy for identity you chose during system design. [13.3]
- **Implement domains**. Implement the domains added during detailed design. For simple domains, merely substitute the corresponding data type and size. Complex domains (identifier, enumeration, structured, and multivalued) require more effort. For each attribute that uses a domain, add an SQL check clause for each domain constraint. [13.4]
- **Define tables**. Map the object model (classes, associations, generalizations) to tables. Because you have already used transformations to refine and optimize the object model during detailed design, you should need only a few mappings. We use table skeletons (listings of tables and their columns) to illustrate the mappings, because they are easier to read than SQL code. [13.5–13.10]

  Implementing the object model also involves defining constraints on the tables, creating indexes to improve database navigation, and creating a schema. We discuss these tasks in Chapter 14 because they involve more advanced aspects of using an RDBMS.

13.3 IMPLEMENTING IDENTITY

In theory primary keys are the sole record-addressing mechanism for a relational database. For this reason you should normally define a primary key for each table, though there are exceptions. For example, we sometimes use temporary tables to store intermediate results during program execution; such ephemeral data may not require a primary key.

We implement associations and generalizations with foreign keys. (See Sections 13.6 through 13.9.) Each foreign key should refer only to a table's primary key (not some other candidate key). There is no disadvantage and there are several advantages to doing this. Multiple referents complicate the schema, and it is good style to have consistency in your implementation strategies. Searching and comparisons are also more difficult if you must consider multiple access routes to a record.

13.3.1 Existence-Based Identity

Implementation is relatively easy for existence-based identity. You must add an object identifier attribute to each class table and make it the primary key. The primary key for each association table consists of the identifiers from one or more related classes. Ideally, you should use the same identifier value for an object throughout an inheritance hierarchy. You need not show the identifiers on the object model, but you must include them in the table schema. Some RDBMSs provide semantic support for identifiers (see Section 13.4.1).

13.3.2 Value-Based Identity

If you chose value-based identity, you should now add flow of identity notation to your object model—to specify the primary key for each class. The flow of identity notation applies only to applications with value-based identity. You must indicate for each class whether it has intrinsic identity or identity derived from some other classes. Figure 13.3 shows flow of identity for an excerpt of the portfolio manager. (We show flow of identity for illustration, even though we chose existence-based identity.)

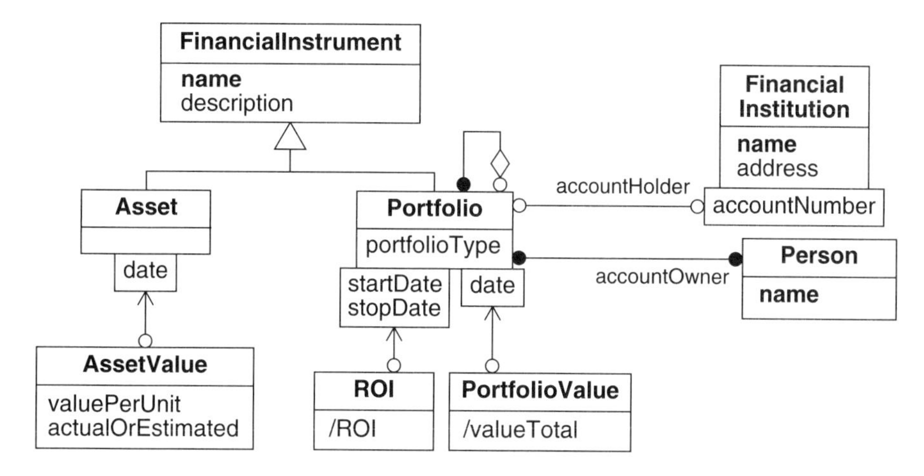

Figure 13.3 Flow of identity for an excerpt of the portfolio manager

FinancialInstrument, *FinancialInstitution*, and *Person* all have a primary key of *name*. *Asset* and *Portfolio* inherit the primary key of *FinancialInstrument*. The primary key of *AssetValue* is the primary key of *Asset* combined with the qualifier *date*. The primary key of *ROI* is the primary key of *Portfolio* and the qualifiers *startDate* and *stopDate*. The primary key of *PortfolioValue* is the primary key of *Portfolio* combined with *date*.

We use the following notation to indicate flow of identity:

- **Independent classes**. The objects of an ***independent class*** are distinguished with one or more attributes intrinsic to the class. Attributes that comprise the primary key of each independent class are in boldface. In Figure 13.3 *FinancialInstrument*, *FinancialInstitution*, and *Person* are independent classes.
- **Dependent classes via association**. The objects of a ***dependent class*** derive identity from other objects. An arrowhead indicates the associations that derive identity. It is placed at the identity source, where the multiplicity must always be "one." The arrow tail may have "many" multiplicity if a class derives identity from two or more sources.

In Figure 13.3 *AssetValue*, *ROI*, and *PortfolioValue* are dependent classes that derive identity from a source class via a qualified association.

- **Dependent classes via generalization**. Normally, the primary key of a superclass propagates to the subclass(es). Otherwise, the attribute(s) of the subclass that serve as the primary key are in boldface.

13.4 IMPLEMENTING DOMAINS

In Chapter 10 (detailed design) we added domains to the portfolio manager's object model. Even though domains are lacking in many RDBMSs, you should still use them in database design. Domains are an important aspect of RDBMS theory and the OMT methodology, as we described in Chapter 3. Domains promote a more consistent design than direct use of data types and make it easier to port an application.

You can define constraints on domain values, for example, to enforce enumerations. You can also use constraints to enforce attribute multiplicity. For each attribute that uses a domain, you must add an SQL *check* clause for each domain constraint.

Simple domains, such as *name*, *longString*, and *money*, are straightforward to implement; you merely substitute the corresponding data type and size. More complex domains (identifier, enumeration, structured, multivalued) require more specialized implementations. Table 13.1 specifies domains for the portfolio manager.

13.4.1 Identifier Domain

Most RDBMSs facilitate existence-based identity with identifier domains. For example, you can define named sequences for the Oracle RDBMS (syntax: *CREATE SEQUENCE sequenceName;*). When you insert a record, you list the sequence name as a data value and Oracle supplies the next integer. The underlying Oracle implementation is straightforward; a data dictionary record contains the highest number allocated for each sequence. As Oracle allocates a new sequence number, it increments this number. (For distributed databases, Oracle also incorporates a site prefix as part of the ID.) Because Oracle stores sequences as very large integers, the available numbers are practically inexhaustible.

MS-Access has a similar approach. As you define tables, you assign each identifier a data type of *counter* and the corresponding foreign keys a data type of *long*. (If you use existence-based identity and all foreign keys reference a primary key, all foreign keys will reference a *counter* and have a data type of *long*.) As you insert records into tables, MS-Access assigns the next number in the sequence and stores it as the identifier.

13.4.2 Enumeration Domains

An enumeration domain restricts an attribute to a set of listed values (Chapter 3). We can think of four ways to implement enumeration domains.

Domain	Data type	Maximum size	Constraints
ID (primary key)	counter		
ID (foreign key)	long		
shortName	text	5	
name	text	30	
address	text	80	
mediumString	text	30	
longString	text	255	
money	currency		
shortNumber	single		
longNumber	double		
date	dateTime		
enumPortfolioType	text	6	in (“normal”, “401K”, “IRA”, “other”)
enumActualOrEstimated	text	9	in (“actual”, “estimated”)
enumPutOrCall	text	4	in (“put”, “call”)
enumStockType	text	10	in (“common”, “preferred”, “closed-end”, “other”)
enumTransactionType	text	19	in (“purchase”, “sale”, “deposit”, “withdrawal”, “barter”, “journal”, “dividend”, “interest”, “return of principal”)
financialInstrumentType	text	9	in (“Asset”, “Portfolio”)
assetType	text	13	in (“PreciousMetal”, “BondAsset”, “StockOption”, “RealEstate”, “Cash”, “StockAsset”, “Insurance”, “Collectible”)

Table 13.1 Domain specifications for the portfolio manager using MS-Access data types

- **Enumeration string**. In this approach, you simply store an enumeration attribute as a string. If the RDBMS allows, you can define a constraint to restrict the enumeration to the allowed values; otherwise, enforce enumeration values with application code.

  Enumeration strings are simple, and the controlled vocabulary enables meaningful searches. However, the performance impact of checking strings can be excessive for a large enumeration. Also, an RDBMS may not allow the lengthy constraint required for an enumeration of 100 or 1000 values.

Despite these drawbacks, we normally use enumeration strings. We used this implementation approach for the portfolio manager.

- **One flag per enumeration value**. You can define a boolean attribute for each enumeration value. Thus you could implement the enumeration *color* with values *red*, *yellow*, and *blue* as three boolean attributes.

  Boolean attributes circumvent a nasty naming problem. Often, the values are meaningful to application experts, but it is difficult to name the enumeration itself. For example, the meaning of *portfolioType* and *stockType* was not clear until we listed the values in the data dictionary (Chapter 8). Enumeration flags can incorporate the value in the name (such as *redOrNotRed*), allowing more meaningful names. Enumeration flags also handle multiple enumeration attributes. For example, a car may be both red and blue.

  Verbosity is the obvious drawback of enumeration flags. Having one attribute per value can quickly clutter your model and the resulting database schema. Also, enumeration flags do not enforce exclusivity of values, as is normally required.

  We use enumeration flags when enumeration values are not mutually exclusive and multiple values may simultaneously apply.

- **Enumeration table**. Another option is to store enumeration definitions in a table. You may have one table per enumeration or one table for all enumerations, essentially a metatable. Application software inspects the enumeration table to find the values to enforce.

  Enumeration tables can efficiently handle large enumerations. Also, the enumeration values are visible in the database. This is consistent with the strategy of prominently storing business rules in the database rather than burying them in application code. Enumeration tables allow maintainers to define new enumeration values without changing application code.

  Enumeration tables have the drawback of being cumbersome for incidental use. It is easier to use an enumeration string and an SQL constraint. You must write special generic software to read the enumeration table and enforce values.

  The enumeration table is appropriate when you have many enumeration values or unknown enumeration values during development.

- **Enumeration encoding**. A final alternative is to encode enumeration values as ordinal numbers. Application software automatically encodes and decodes enumeration values upon writing and reading. You can use an enumeration table to store both the value and the encoding.

  Enumeration encoding conserves disc space; an integer takes less space than a string. The user interface can still display enumeration values; it is straightforward to encode and decode enumerations transparently within application software, and the performance impact is modest. Enumeration encoding is helpful when you have software that must deal with multiple languages.

  However, encoding has a subtle and severe disadvantage—it complicates maintenance and debugging. You cannot easily inspect the database, because you must join to the enumeration metatable to decode each enumeration. For example, the number *3* has no intrinsic meaning; you must translate it into the underlying value. We discovered the disadvantage of encoding the hard way—by using it in an application and painfully experiencing the debugging and maintenance drawbacks.

For this reason, we do not recommend enumeration encodings unless you are dealing with multiple languages.

13.4.3 Structured Domains

You can expand a structured domain (Chapter 3) into lesser domains that are either also structured or simple. Figure 13.4 shows three ways to implement structured domains. All three are useful; the choice depends on the application. With concatenation you discard the substructure and store the domain as a string. With multiple columns you replace the structured domain by its constituent elements. Alternatively, you may define a table to hold domain substructure. The *references* clause documents the referent of the foreign key (the primary key of the referenced table).

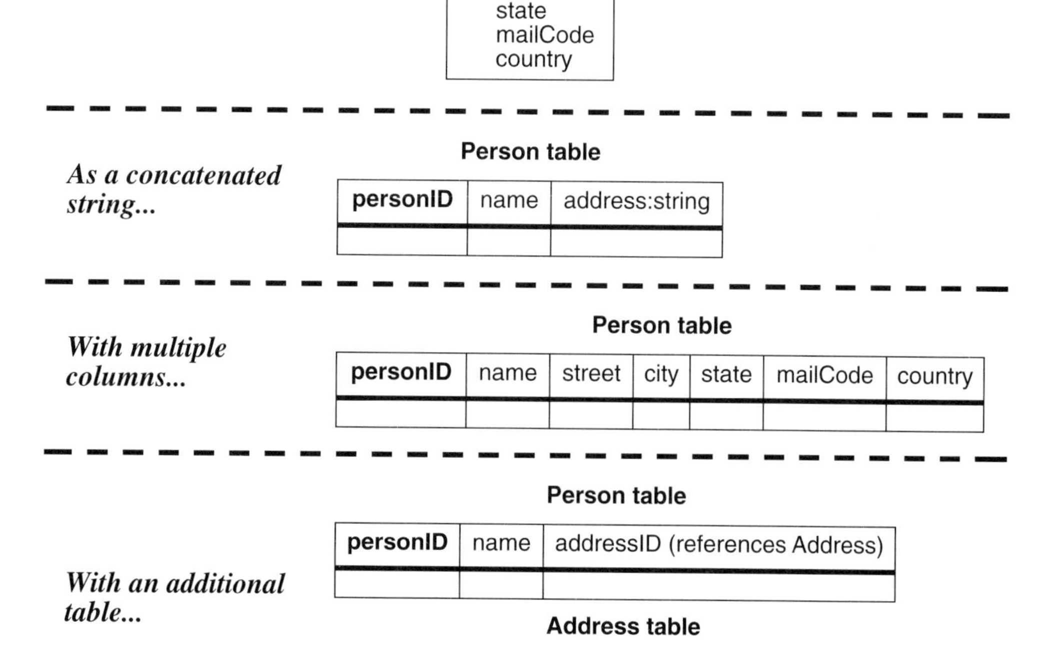

Figure 13.4 Approaches to implementing a structured domain

13.4.4 Multivalued Domains

In Chapter 3 we discussed attribute multiplicity, the number of values for an attribute. You may specify a mandatory single value *[1]*, an optional single value *[0..1]*, an unbounded collection with a lower limit *[lowerLimit..*]*, or a collection with fixed limits *[lowerLimit..upperLimit]*. You can readily implement an optional attribute with null values. However, RDBMSs cannot directly express a multivalued domain. You must use the techniques for structured domains to implement multivalued domains (concatenation, multiple columns, or an additional table).

13.5 IMPLEMENTING CLASSES

Normally we map each class to a separate table, in which each attribute is a column. You may require additional columns for a generated identifier (existence-based identity), buried associations (Section 13.6), and generalization discriminators (Section 13.8).

13.6 IMPLEMENTING SIMPLE ASSOCIATIONS

You can use several approaches to implement associations. Because some are better than others, we organized these into recommended, alternative, and discouraged mappings. Recommended mappings are sufficient for implementing an object model and you should always consider them first. Sometimes, however, for performance, extensibility, or style reasons, you may decide to use an alternative mapping. You should always avoid the discouraged mappings.

13.6.1 Recommended Mappings

- **Distinct table for many-to-many associations**. You should map each many-to-many association to a distinct table, as Figure 13.5 shows. The primary key of the association is the combination of the primary keys from each table. Note that we used the role name *accountOwner* in the association table. When you implement associations, role names become foreign key attribute names, prefixes, or suffixes.

  The order of the classes does not matter for the primary key. Thus you can define the primary key of the association as either *portfolioID* + *accountOwner* or *accountOwner* + *portfolioID*. You should also define a secondary index for the association on the lesser class in the primary key to speed searching. Thus if the primary key for the association is *portfolioID* + *accountOwner*, you should define a secondary index on *accountOwner*.

  An association table cannot enforce a minimum multiplicity of one, since there is no entry in the table if either role lacks a value. In that case, you must write additional application code. This is not a problem in Figure 13.5 because both roles have a minimum multiplicity of zero.

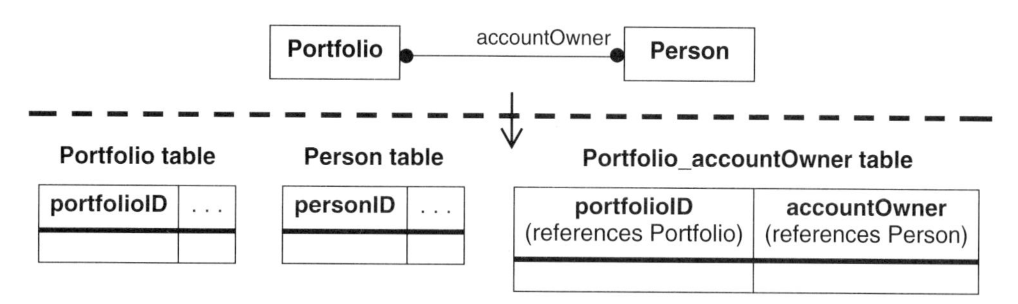

Figure 13.5 Recommended implementation: Many-to-many association

- **Buried one-to-many associations**. You can implement a one-to-many association with a buried foreign key, as Figure 13.6 shows. To do so, bury the foreign key in the class with the "many" role. In Figure 13.6 each transaction must have a portfolio; thus the *portfolioID* foreign key in the *Transaction* table cannot be null (although we have not specified this in the skeleton table). If the multiplicity of *Portfolio* had been "zero-or-one" in the object model, we would have allowed the *portfolioID* foreign key to be null.

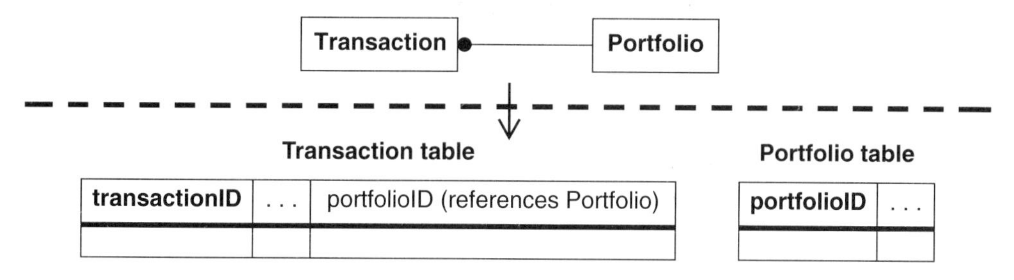

Figure 13.6 Recommended implementation: Buried one-to-many association

- **Buried zero-or-one-to-exactly-one association**. Bury the foreign key in the class with the "zero-or-one" role. Define the buried foreign key as not null.
- **Buried other one-to-one associations**. Bury the foreign key in either class. Whether or not the foreign key is null depends on the minimum multiplicity of the target role.

13.6.2 Alternative Mapping

- **Distinct table**. We recommend that you bury one-to-many and one-to-one associations, but you can also implement them with a distinct table, as Figure 13.7 shows. For a one-to-many association, choose the foreign key for the "many" class as the primary key for the association table. For a one-to-one association, you can choose either.

  Distinct association tables improve the uniformity of design and yield a more extensible implementation. If you change the multiplicity of the association, you need change only the constraints, not the table structure.

On the downside, association tables fragment the database and increase navigation. You must join additional tables to navigate an object model, which complicates SQL code and reduces performance. Also, a distinct association table cannot enforce a minimum multiplicity of one, since a nonexistent link is not entered in the table. For example, the tables in Figure 13.7 cannot enforce that every transaction must have a portfolio. We would have to write additional application code to check the constraint.

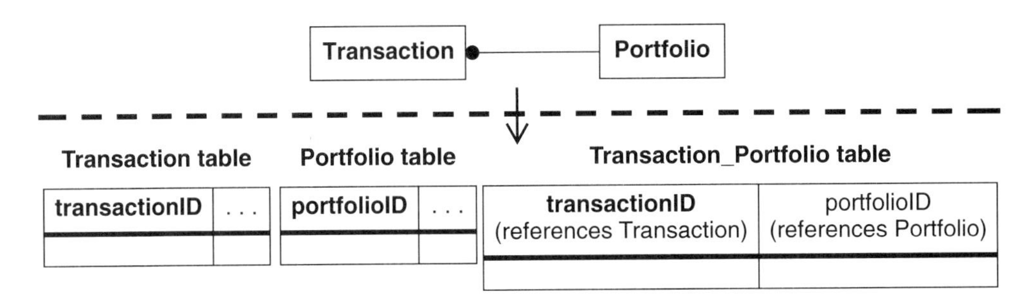

Figure 13.7 Alternative implementation: Distinct tables for one-to-many association

13.6.3 Discouraged Mappings

- **Combination**. Figure 13.8 combines multiple classes and the intervening association into a single table. This combination yields a single table, which contrasts with our recommended mapping (Figure 13.6) that yields two tables.

  Combining classes has substantial disadvantages. The combination table can violate third normal form, depending on the meaning of a model. Furthermore, it is philosophically troublesome to combine multiple classes into a single table. Classes are atomic units of object-oriented thinking and encapsulation; an implementation is more understandable and extensible if it honors these semantics.

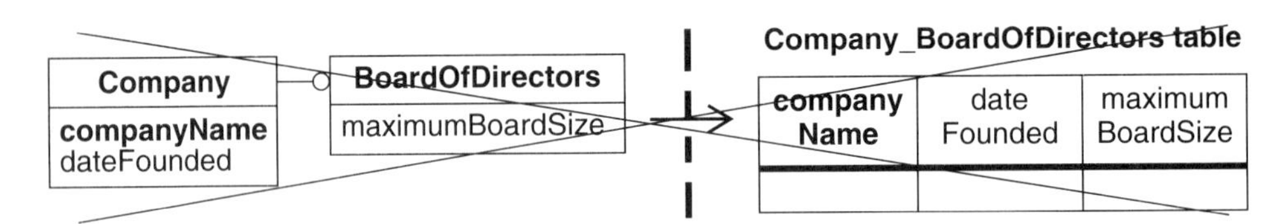

Figure 13.8 Discouraged implementation: Combining classes with an intervening association

- **Double buried one-to-one associations**. You can redundantly implement an association with two foreign keys—one buried in each related class. For example, if we implemented the object model in Figure 13.8 with existence-based identity, we could bury *boardOfDirectorsID* in *Company* and *companyID* in *BoardOfDirectors*.

Double-buried associations have few advantages, because one-to-one associations seldom occur in models. For most queries that involve both tables, you will need to construct a join anyway. Moreover, RDBMSs will not maintain the consistency of redundant associations. Consequently, you must write additional application code, which is prone to error.

13.7 IMPLEMENTING ADVANCED ASSOCIATIONS

13.7.1 Recommended Mappings

- **Link attributes**. As a rule, use distinct tables to implement associations with link attributes.
- **Association classes**. As a rule, implement association classes with distinct tables. Instances of association classes receive identity propagated from the related classes. In Figure 13.9 the primary key of *Authorization* is the combination of the primary keys of the *Table* and *User* classes.

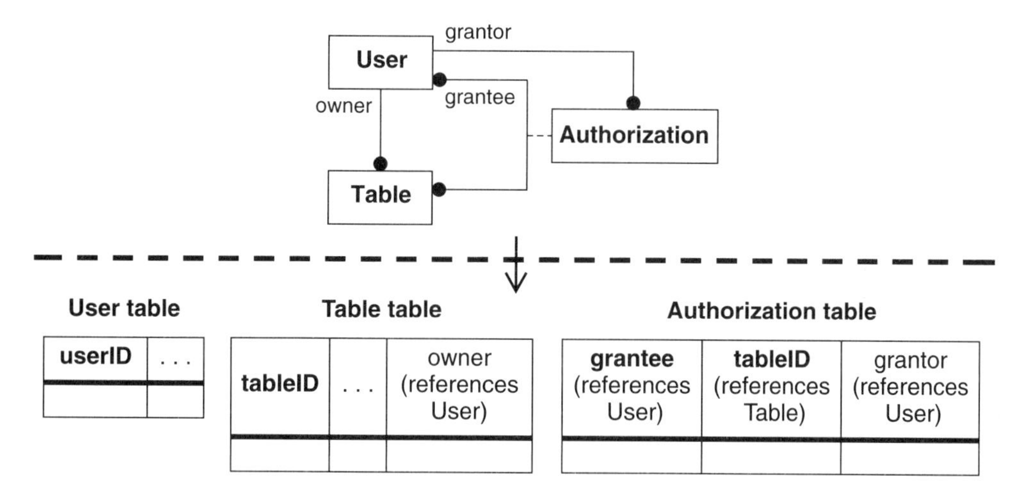

Figure 13.9 Recommended implementation: Association class

- **Ternary associations**. As a rule, implement each ternary association with a distinct table.
- **Qualified associations**. Qualified associations follow the same mappings as the corresponding simple association without the qualifier. We recommend that you implement the qualified associations in Figure 13.10a and Figure 13.10b by burying them in the *B* class. Examples (c) and (d) require distinct association tables, since the underlying association is many-to-many without the qualifier. The primary key of the (d) association is the combination of the *A* primary key, *B* primary key, and *q*.

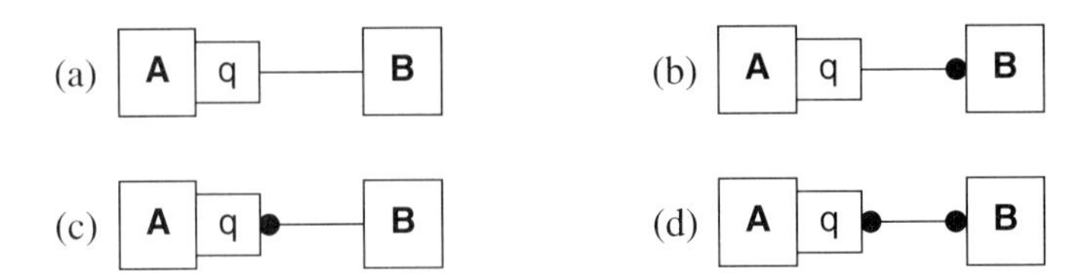

Figure 13.10 Possible multiplicities for qualified associations

- **Ordered associations**. You can implement ordered associations by introducing a sequence number attribute. In Figure 13.11 attributes are ordered according to their significance for building an index. For example, an index for a phone book would treat *last name* as more significant than *first name*. The *IndexAttribute* table has two candidate keys: *indexName* + *attribName* and *indexName* + *sequenceNumber*. We arbitrarily chose *indexName* + *attribName* as the primary key. The *CK1* annotation documents the candidate key of the *IndexAttribute* table that is not the primary key.

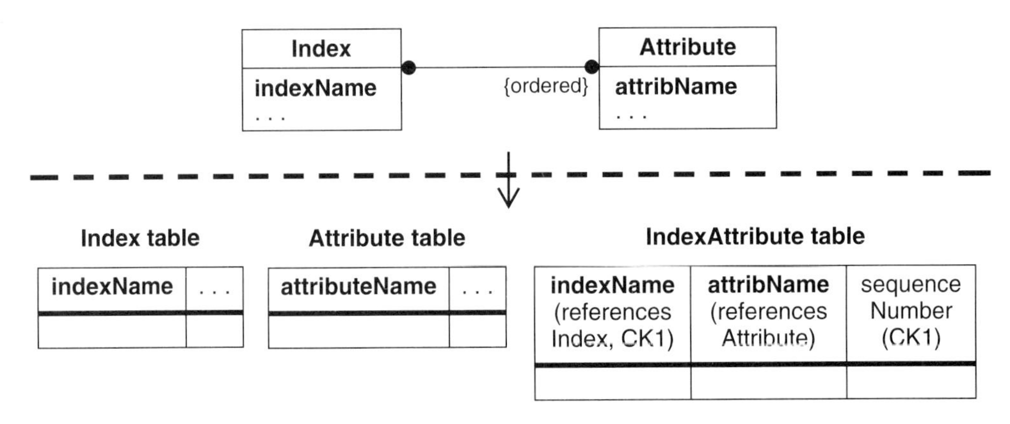

Figure 13.11 Recommended implementation: Ordered association

- **Symmetric associations**. A symmetric association is an association between objects of the same class that have interchangeable roles. In Figure 13.12 a contract may be associated with other contracts. Contracts can be related in many ways—such as in terms of legal consequences, complementary business, or underlying technology.

  Fortunately, symmetric associations seldom occur. When they do, they pose nasty implementation problems. For the *AssociatedContract* table, if you don't break the symmetry, you must store the data twice or search the table twice. If contract *A* is associated with contract *B*, you can store two records (*A*, *B*) and (*B*, *A*) and have redundancy. Or to find all contracts related to *A*, you can search for records with *firstContractID* equal to *A* and then search for records with *secondContractID* equal to *A*. If possible, try to change your model to avoid the symmetry.

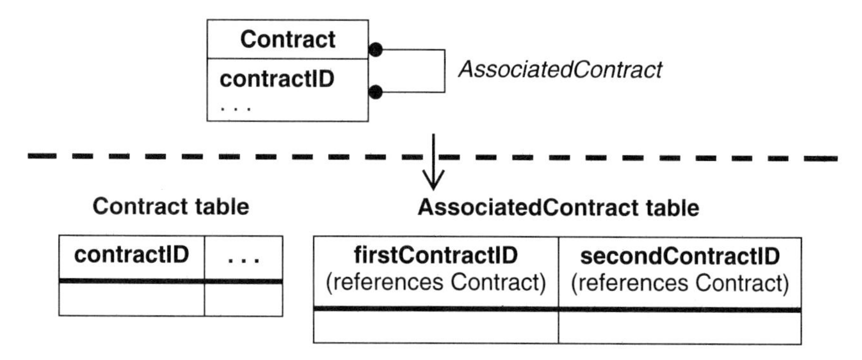

Figure 13.12 Recommended implementation: Symmetric association

- **Aggregations**. Aggregation is a kind of association, so you can implement aggregations as you would associations.

13.7.2 Alternative Mappings

- **Link attributes**. You can implement one-to-one and one-to-many associations with link attributes using a buried foreign key. You must then bury the link attributes in addition to the foreign key.

  The buried foreign keys decrease the tables you must navigate and let you enforce a minimum multiplicity of one with the database schema. However, buried link attributes violate second normal form for a one-to-many association and third normal form for a one-to-one association. For this reason, we recommend that you normally implement associations with link attributes as distinct tables.

- **Qualified associations**. You can use distinct tables to implement qualified associations that are one-to-many with the qualifier removed. This can be helpful for enforcing uniqueness. For example, a distinct *Portfolio@FinancialInstitution* table could enforce the uniqueness of *FinancialInstitution + accountNumber*.

13.7.3 Discouraged Mapping

- **Qualified associations**. Do not implement the "many" role of an association by burying parallel attributes. Parallel attributes most often arise for qualified associations with an enumerated qualifier. In Figure 13.13, for example, we prefer separate *FiscalWeek* and *DailyReceipt* tables to a table with an attribute for each business day of the week.

  About the only advantage of parallel attributes is that you have fewer tables to search. However, parallel attributes can be difficult to search. For example, the parallel attributes table would require complex programming to determine the day with the highest receipts. The parallel attributes approach is also less extensible; if you add a value to the enumeration, you must add a column to the table. In contrast, the *DailyReceipt* table requires no changes to the schema when new enumeration values are introduced.

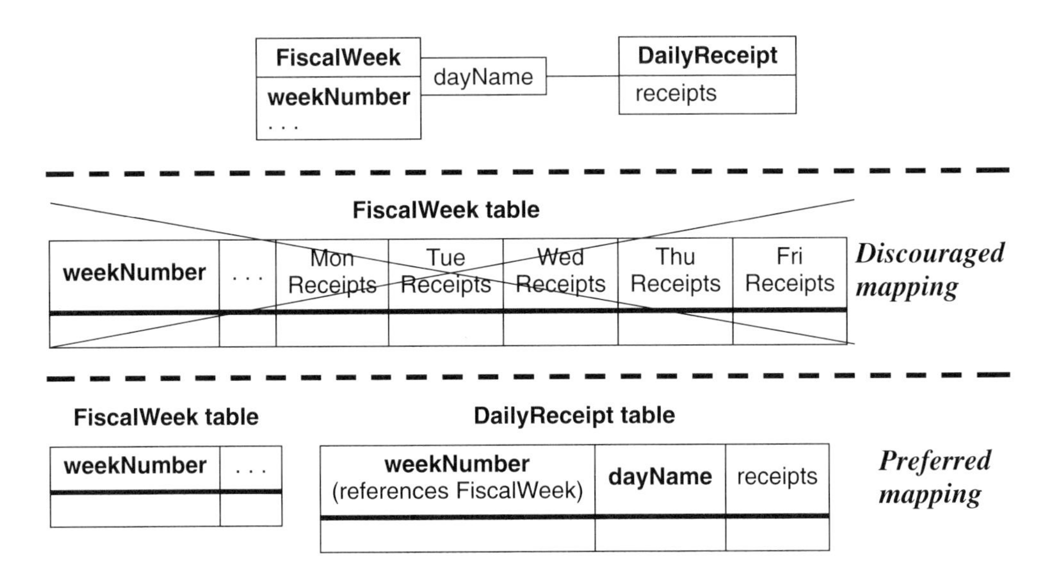

Figure 13.13 Discouraged implementation: Parallel attributes for an association

13.8 IMPLEMENTING SINGLE INHERITANCE

13.8.1 Recommended Mapping

We usually implement generalization with separate superclass and subclass tables, as Figure 13.14 shows. Ideally, an object should have the same primary key value throughout an inheritance hierarchy. The discriminator (*financialInstrumentType*, which is implied though not shown in the object model) indicates the appropriate subclass table for each superclass record. For a multiple-level generalization, you should apply the mappings one level at a time. (Section 13.10 shows a multiple-level generalization for the portfolio manager, *FinancialInstrument* specializing to *Asset* specializing to various subclasses.)

Referential integrity can ensure that each subclass record corresponds to a superclass record. However, an RDBMS cannot ensure that each superclass record corresponds to some subclass record. It also cannot enforce the generalization partition, that each superclass record must be further described by the table the discriminator indicates. You must enforce the missing generalization semantics with application code.

You may wish to define a view for each subclass to consolidate inherited data and make object access easier. Figure 13.15 defines a view for *Portfolio* based on Figure 13.14.

13.8.2 Alternative Mappings

- **Elimination**. You can optimize away classes that have no attributes other than a primary key. In Figure 13.16 the *assetType* discriminator indicates whether each *Asset* object

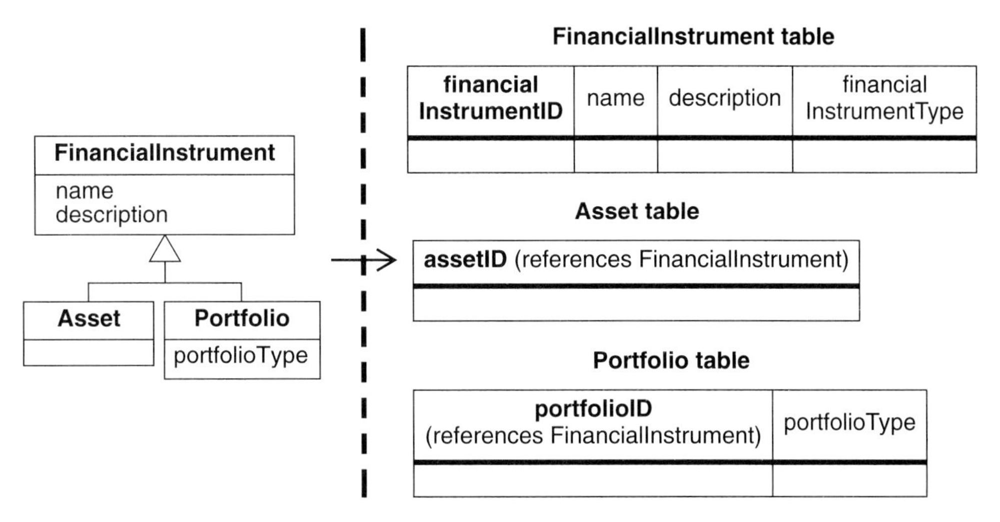

Figure 13.14 Recommended implementation: Separate superclass and subclass tables

```
CREATE VIEW view_Portfolio AS
   SELECT portfolioID, name, description, portfolioType
   FROM FinancialInstrument AS FI JOIN Portfolio AS P
      ON FI.financialInstrumentID = P.portfolioID;
```

Figure 13.15 A view can unify objects fragmented by generalization tables (based on Figure 13.14)

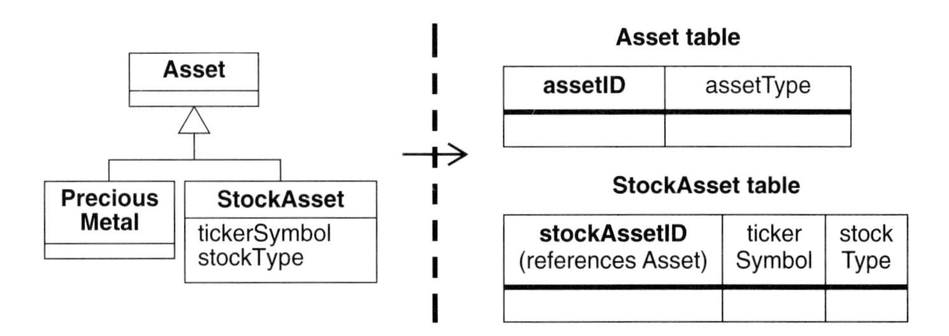

Figure 13.16 Alternative implementation: Eliminate incidental subclass tables

is in the class *PreciousMetal* or *StockAsset*. *PreciousMetal* has no attributes, buried associations, or discriminator (see the full portfolio manager model in Figure 9.5), so you can eliminate its table.

The advantage of elimination is that you have one less table in the database. The disadvantage is that your implementation becomes less regular. Navigation code depends on whether or not a subclass has a corresponding table. Also, such an implementation is less extensible; you will need to add a table if you elaborate an eliminated subclass.

- **Push superclass attributes down**. You can also implement generalization by eliminating the superclass table and replicating superclass attributes for each subclass (Figure 13.17). You can then describe an object in a single table, rather than spreading the description across tables for each level of the hierarchy.

  A benefit is that each object is described in a single table, so you need not join tables to reconstitute an object. However, although this approach satisfies normal forms, it introduces schema redundancy. Also, you may need to search multiple subclass tables to find an object. If you push attributes down across multiple generalization levels, your implementation loses the structuring of description that is intrinsic in generalization.

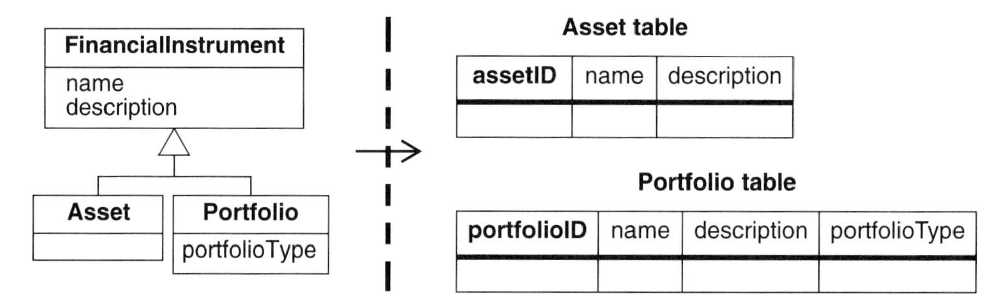

Figure 13.17 Alternative implementation: Push down superclass attributes

- **Push subclass attributes up**. Another option is to push attributes up to the superclass and eliminate the subclass tables (Figure 13.18). The subclass attributes that do not apply to a given object are set to null.

  This approach also lets you describe each object in a single table. However, pushing attributes up violates second normal form because some attributes do not fully depend on the primary key. If you push attributes up across multiple generalization levels, your implementation again loses the structuring of description.

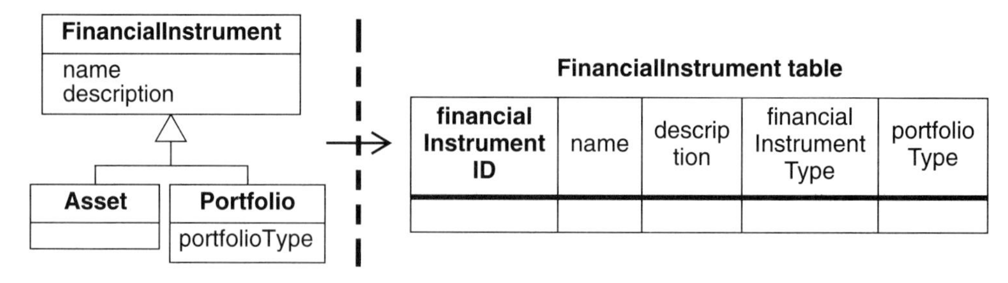

Figure 13.18 Alternative implementation: Push up subclass attributes

- **Hybrid approach**. A hybrid approach is occasionally helpful. You may push superclass attributes down the hierarchy and keep a superclass table for navigating the subclasses (Figure 13.19).

  You can again describe an object in a single table, and the superclass table indicates the table that holds each object. However, some of the same disadvantages apply: schema redundancy and loss of description structuring.

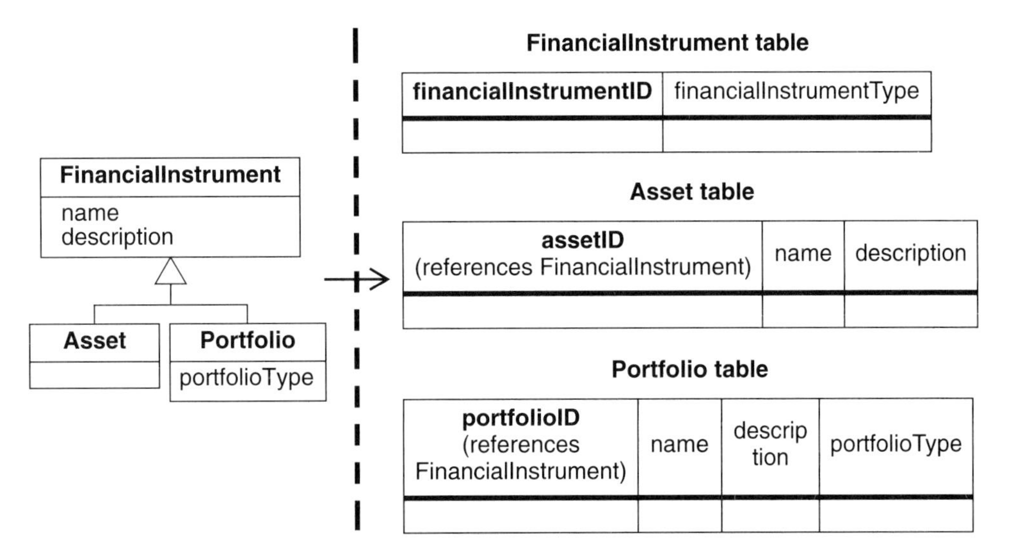

Figure 13.19 Alternative implementation: Hybrid generalization approach

- **Generalization table**. A final alternative is to use one superclass table, multiple subclass tables, and a generalization table. The generalization table binds the superclass primary key to the subclass primary key.

  This approach is similar to our recommended mapping, but the generalization table makes the schema and database more verbose. However, this technique can be helpful when merging databases. (See Chapter 19.)

13.9 IMPLEMENTING MULTIPLE INHERITANCE

We recommend that you use separate superclass and subclass tables to implement multiple inheritance (Figure 13.14). Chapter 3 discusses different forms of multiple inheritance.

13.10 SUMMARY OF OBJECT MODEL MAPPING RULES

Table 13.2 summarizes our recommendations for implementing the object model with RDBMS constructs. Figure 13.20 applies the recommendations for the portfolio manager.

The *CKn* notation documents the candidate keys that we specified in Section 10.3. We also define tables for ROIs and portfolio values, which we decided to cache in Section 9.2.

Concept	Object model construct	Recommended RDBMS mapping
Domain	Simple domain	Map to a data type and size.
	Identifier	Use available RDBMS-specific features.
	Enumeration	Typically store enumeration attribute as a string.
	Structured	Use concatenated string, multiple columns, or additional table.
	Multivalued	
Class	Class	Map each class to a table.
Association	Many-to-many association	Distinct table
	One-to-many association	Buried foreign key
	One-to-one association	
	Link attribute	Distinct table
	Association class	
	Ternary association	
	Qualified association	Follow recommendation for association without qualifier.
	Ordered association	Introduce a sequence number attribute and then follow recommendation for qualified association.
	Symmetric association	Try to break the symmetry.
	Aggregation	Follow recommendation for association.
Generalization	Single inheritance	Use separate superclass and subclass tables.
	Multiple inheritance	

Table 13.2 Summary of recommended object model mapping rules for RDBMSs

13.11 IMPLEMENTING THE DYNAMIC MODEL

The dynamic model is seldom important for database applications, aside from the user interface, which we discuss in Chapter 14.

FinancialInstrument table

financialInstrumentID	name (CK1)	description	financialInstrumentType

FinancialInstitution table

financialInstitutionID	name (CK1)	address

Person table

personID	name (CK1)

Asset table

assetID (references FinancialInstrument)	assetType

Portfolio table

portfolioID (references FinancialInstrument)	portfolioType	parentPortfolio (references Portfolio)	accountHolder (references FinancialInstitution)	account Number

Portfolio_accountOwner table

portfolioID (references Portfolio)	**accountOwner** (references Person)

AssetValue table

asset ValueID	valuePer Unit	actualOr Estimated	assetID (CK1, references Asset)	date (CK1)	currency (references Cash, CK1)

ROI table

ROIID	ROI	portfolioID (CK1, references Portfolio)	startDate (CK1)	stopDate (CK1)

Figure 13.20 RDBMS tables for portfolio manager case study

PortfolioValue table

portfolioValueID	valueTotal	portfolioID (CK1, references Portfolio)	date (CK1)	currency (references Cash, CK1)

BondAsset table

bondAssetID (references Asset)

Cash table

cashID (references Asset)

StockOption table

stockOptionID (references Asset)	putOrCall	strikePrice	expiration Date	underlyingSecurity (references StockAsset)	currency (references Cash)

StockAsset table

stockAssetID (references Asset)	tickerSymbol	stockType

RealEstate table

realEstateID (references Asset)

PreciousMetal table

preciousMetalID (references Asset)

Collectible table

collectibleID (references Asset)

Insurance table

insuranceID (references Asset)

Transaction table

tran sac tionID	transac tion Date	re cord Date	transac tion Type	transac tion Fee	de scrip tion	asset Referenced (references Asset)	portfolioID (references Portfolio)	currency (references Cash)

Transaction_assetTransacted table

assetTransacted (references Asset)	**transactionID** (references Transaction)	quantity

Figure 13.20 (continued) RDBMS tables for portfolio manager case study

13.12 IMPLEMENTING THE FUNCTIONAL MODEL

You can implement the functional model by writing code for the methods determined during analysis and design. You can also use tables to implement parts of the functional model, as we describe in Chapter 14. Here we describe the implementation of the Object Navigation Notation with SQL code.

You can map all ONN expressions to SQL code, but you cannot represent all SQL code with ONN expressions. For example, the ONN does not encompass SQL set functions such as *count*, *max*, and *sum* as well as existential subqueries. But it can express the most common queries—traversing an object model and retrieving values. And you can use the SQL code that results from an ONN expression as the nucleus for more complex code. The mapping of ONN expressions to SQL code depends on both the ONN construct and the mapping to tables chosen for the object model. In our implementation, we consider only recommended and alternative mappings.

In our applications, we emphasize the use of SQL queries rather than programming. Often you can substitute a query (a single well-considered declarative statement) for a method (possibly lengthy custom code). You can write a query much faster than a method, and a query is less prone to error and more extensible. Furthermore, the performance of a skillfully written query can be difficult to surpass with custom code.

13.12.1 Traversal of Simple Binary Association

The *objectOrSet.targetRole* construct traverses an association to a target role. The *objectOrSet.~sourceRole* construct traverses an association from a source role. You can start traversal with a single object or a set of objects. Both expressions may yield a single object or a set of objects. The resulting SQL code depends on how the association is implemented.

- **Buried association.** In Figure 13.21 we gave each class a separate table and buried the association in the table for the "many" role. Given a transaction, we would like to retrieve the type for its portfolio. The corresponding SQL code involves a single join. This join and all the joins in this section will be efficient if you define foreign keys and indexes. (See Chapter 14.)

  Note that our SQL code uses the condition join [Melton-93]. We could have used the natural join. But the natural join works only when the object model does not have role names and when foreign key attribute names match the corresponding primary key attribute names. In contrast, the condition join always applies.

  You should also forgo use of the obsolete construct that places the join logic in the where clause *select. . . from table A, B where join condition*. An SQL optimizer is more likely to generate efficient code when you explicitly state a join in the from clause, rather than implicitly state a join in the where clause. Furthermore, for complex queries the meaning of a join predicate in the where clause can be ambiguous [Celko-95].

- **Distinct association table.** Figure 13.22 implements the classes and the association with distinct tables. Our SQL code involves two joins: join *Portfolio* to the association table and then join *Transaction* to the prior result.

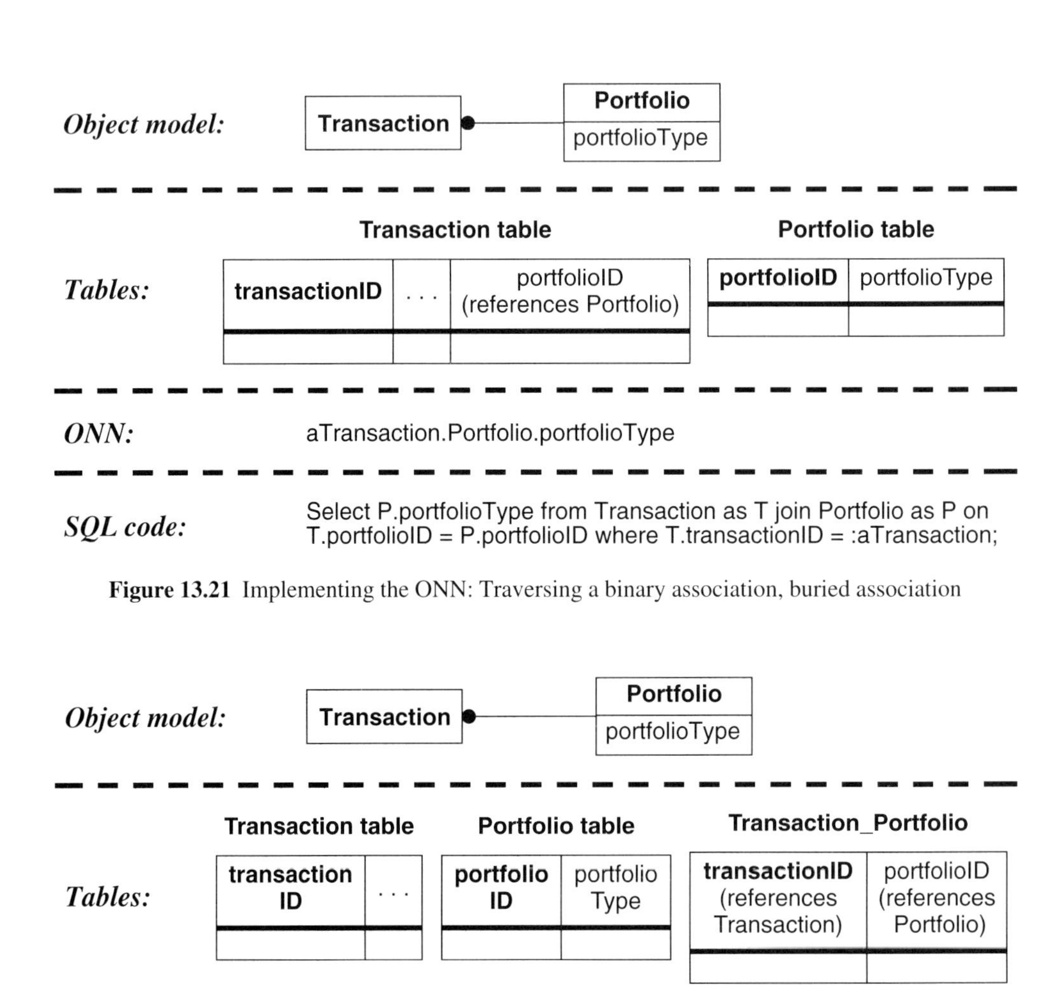

Figure 13.21 Implementing the ONN: Traversing a binary association, buried association

ONN: aTransaction.Portfolio.portfolioType

SQL code:

```
Select P.portfolioType from (Transaction_Portfolio as TP
join Portfolio as P on P.portfolioID= TP.portfolioID)
join Transaction as T on T.transactionID = TP.transactionID
where T.transactionID = :aTransaction;
```

Figure 13.22 Implementing the ONN: Traversing a binary association, distinct tables

If you started with a primary key of one class, you could just perform a single join. Thus, in Figure 13.22 we could join *Portfolio* to *Transaction_Portfolio* and in the where clause set *TP.transactionID = :aTransaction*. However, it is often wise to avoid such optimization, because it degrades the implementation's regularity; for some queries on association tables you can optimize and use a single join; others may require a double join.

Furthermore, an ONN expression can involve additional traversals, which necessitate additional join clauses.

13.12.2 Traversal of Qualified Association

You can always traverse a qualified association without specifying the qualifier. You should implement this the same as simple binary associations described in Section 13.12.1.

Or you can use a qualifier in an ONN expression (*objectOrSet.role[qualifier = value]*). The corresponding SQL query must specify the qualifier value in the where clause. In Figure 13.23 given a financial institution and an account number, we find the portfolio type. The qualifier restricts the collection of objects that satisfies a query. In the example, we have buried the qualified association in the table for the *Portfolio* class. We cannot define a candidate key on *accountHolder* + *accountNumber* because a portfolio need not be held by a financial institution.

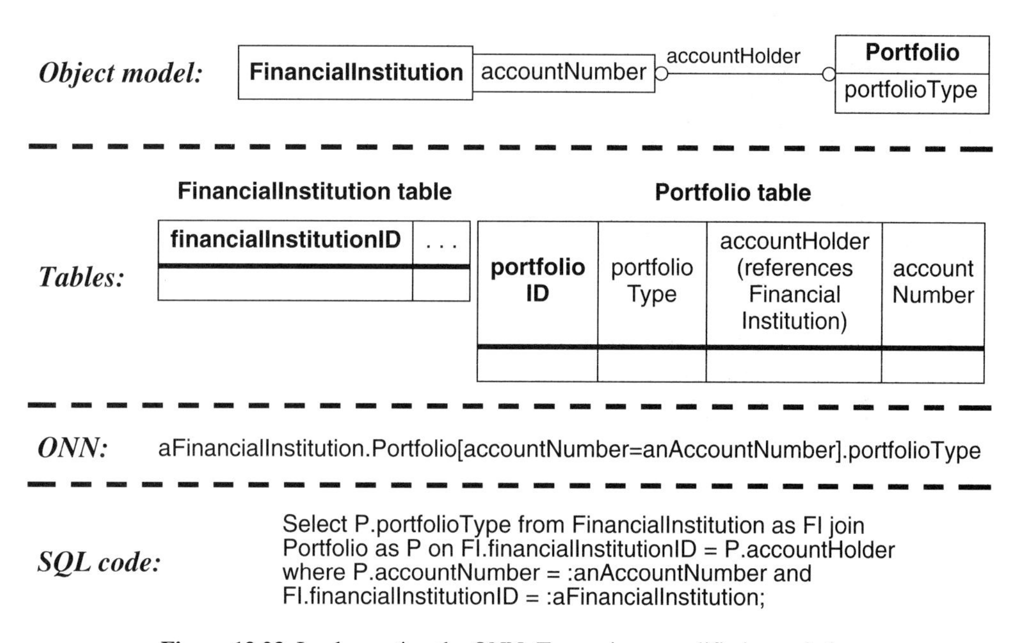

Figure 13.23 Implementing the ONN: Traversing a qualified association

13.12.3 Traversal of Generalization

The SQL code used to traverse a generalization (*objectOrSet:subclass* and *objectOrSet:superclass*) depends on how the object model is implemented.

- **Separate superclass and subclass tables**. Your query should join the superclass table to a subclass table. In Figure 13.24 we find the names of the portfolios owned by a person. Once again, we systematically traverse from foreign keys to primary keys and do not try to eliminate joins that may appear superfluous. The resulting code is easier to understand if we use uniform mappings, and we are more likely to benefit from RDBMS optimization.

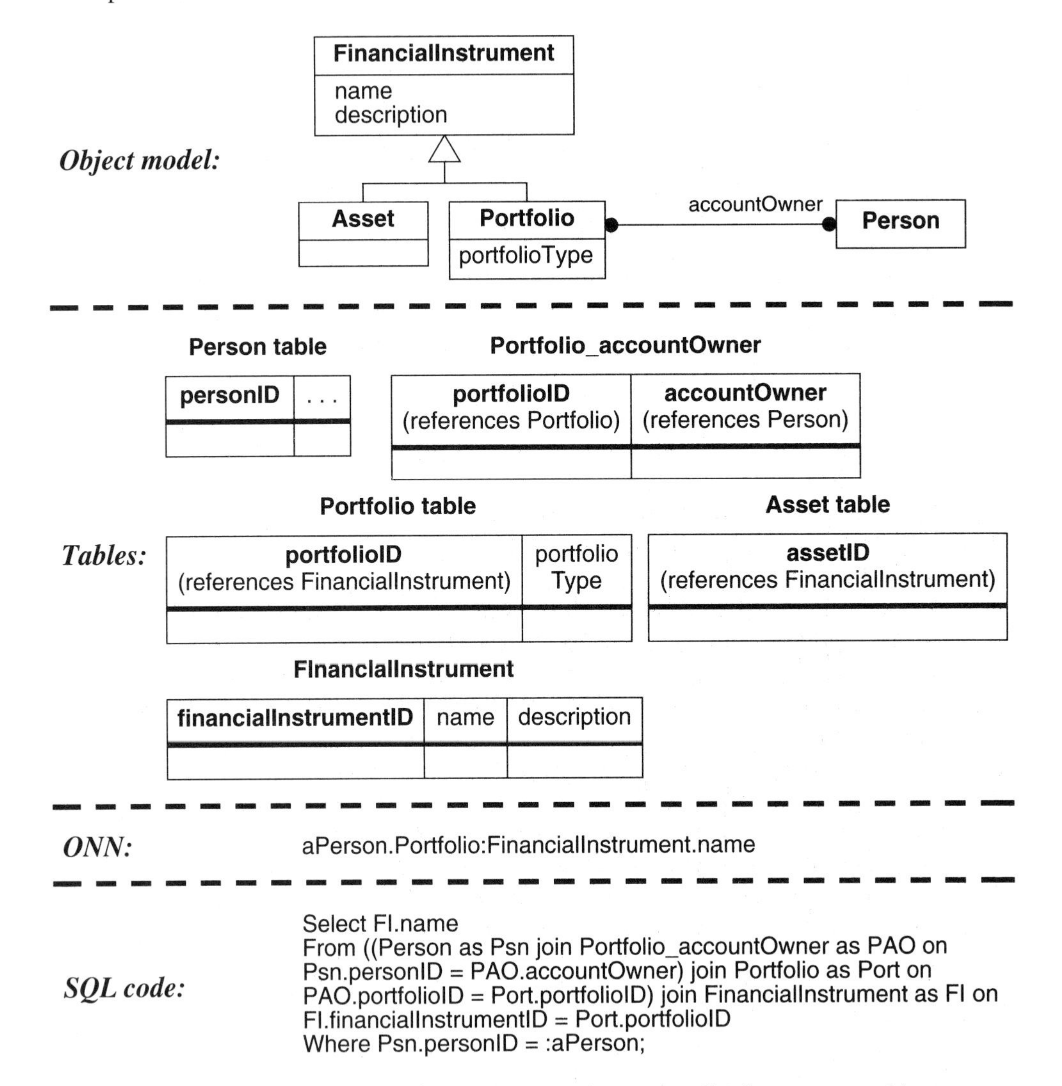

Figure 13.24 Implementing the ONN: Traversing a generalization, separate tables

- **Push down superclass attributes**. You should query the appropriate subclass table. In Figure 13.25 we find the names of the portfolios owned by a person. As we explained in Section 13.8.2, this mapping is an alternative, not a recommendation.

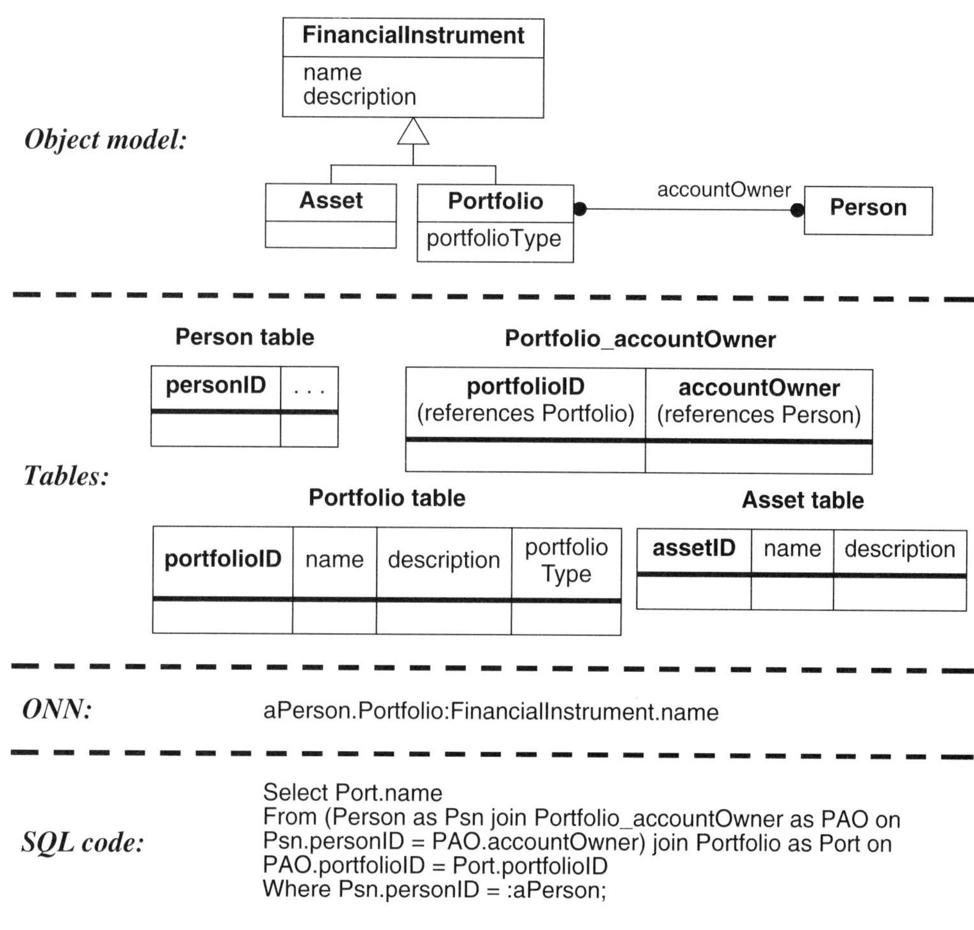

Figure 13.25 Implementing the ONN: Traversing a generalization, push down superclass attributes

- **Push up subclass attributes**. You should query the superclass table. In Figure 13.26 we again find the names of the portfolios owned by a person. This mapping is also an alternative.

13.12.4 Traversal from Link to Object

The ONN expression *linkOrSet.role* is straightforward to implement for an RDBMS. How this is done depends on the implementation of the object model. If you have buried the association in the table for the *role* class, you can merely select from the *role* table. If you have placed the association in a separate table from the *role* class, you need perform only a single join, similar to Figure 13.21.

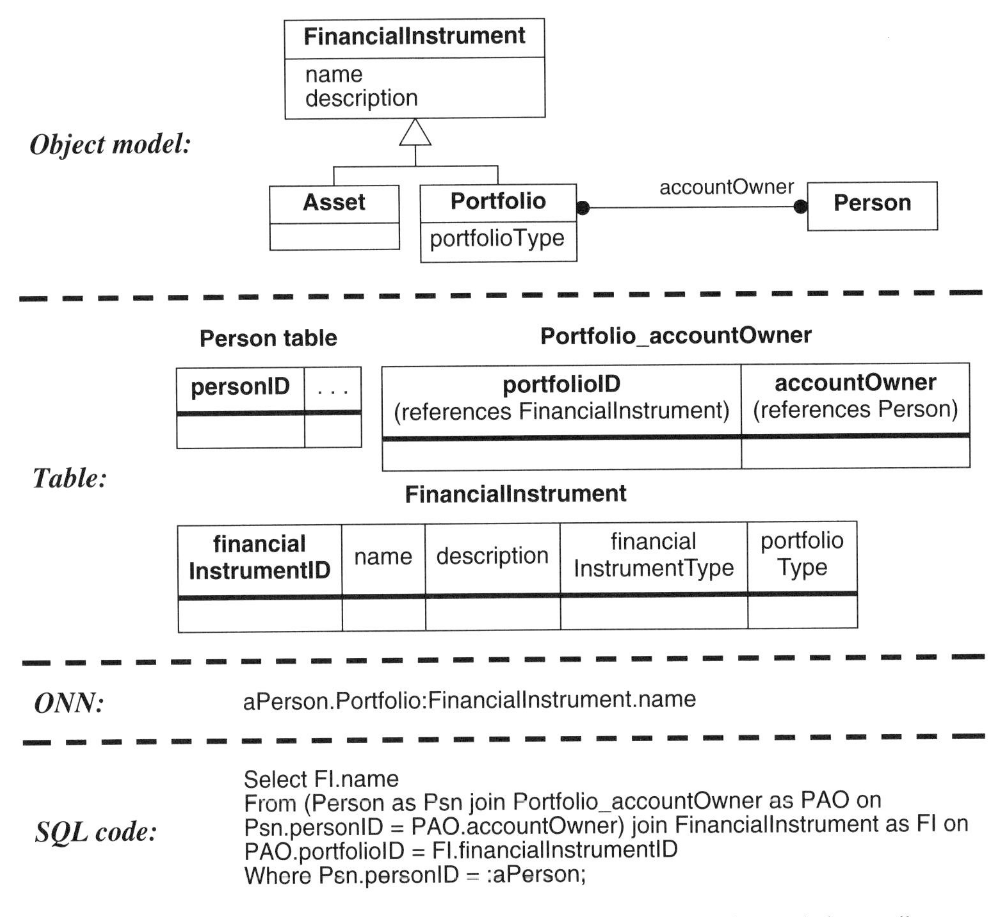

Figure 13.26 Implementing the ONN: Traversing a generalization, push up subclass attributes

13.12.5 Traversal from Object to Link

The ONN provides two ways to retrieve links:

- **Using multiple roles**. The construct *{ role1 = ObjectOrSet1, role2 = ObjectOrSet2, ..., associationName }* finds links between objects using multiple roles. You must include the parameter of *associationName* when the role names are not sufficient to identify a link. Ordinarily, you would use this construct only to access links for a many-to-many association or ternary association (that is, an association with a primary key composed of references to two or more classes). You would normally implement such an association with a distinct table and just query the table to find the appropriate links.
- **Using only one role**. The construct *objectOrSet@role* lets you retrieve links by specifying only one role. How this expression is implemented depends on the implementation

of the object model. Essentially, you directly query a single table or query the join of a class table to an association table.

13.12.6 Filtering

The construct *objectOrSet[filter]* lets you specify a general expression to winnow a set of objects. You can implement this by simply including the filter in the where clause of the SQL code that computes *objectOrSet*. Similarly, *linkOrSet[filter]* lets you winnow a set of links. You can include the filter in the where clause of the SQL code that computes *linkOrSet*.

13.12.7 Traversal from Object to Value

The construct *objectOrSet.attribute* lets you access an attribute value of an object or a set of attribute values for a set of objects. This expression is easy to implement for an RDBMS. You just query the table for *objectOrSet* and specify the desired attribute in the SQL *select* list. Thus *aFinancialInstrument.name* yields the SQL code *select name from FinancialInstrument where financialInstrumentID = :aFinancialInstrument*. We used attribute access in our earlier examples for implementing the ONN.

13.12.8 Traversal from Link to Value

The construct *linkOrSet.attribute* is similar to *objectOrSet.attribute*, except the values are scanned from links. If the object model is implemented by burying the association, you can query the table for the class and specify the link attribute in the SQL *select* list. Otherwise, you can query a distinct association table and specify the link attribute in the SQL *select* list.

13.12.9 Summary of ONN Mapping Rules

Table 13.3 summarizes the ONN mapping rules. In the table *A* and *B* denote tables for classes *A* and *B*. *AB* denotes a distinct table that implements an association between classes *A* and *B*.

13.13 CHAPTER SUMMARY

A relational database is a database in which the data is logically perceived as tables. A relational database management system (RDBMS) manages tables of data and associated structures that increase the functionality and performance of tables. This chapter shows how to implement an OMT design with an RDBMS.

There are two approaches to realizing identity: existence-based and value-based. Implementation is straightforward for existence-based identity—just add an object identifier attribute to each class table and make it the primary key. Some RDBMSs can sequentially allocate numbers to facilitate existence-based identity. With value based identity you must explicitly choose one or more values to be the primary key for each class.

<table>
<tr><th>Concept</th><th>Object model implementation</th><th>ONN construct</th><th>ONN implementation</th></tr>
<tr><td rowspan="2">Traverse binary association</td><td>Buried association</td><td rowspan="2">objectOrSet.role</td><td>select . . . from table A, B on join condition</td></tr>
<tr><td>Distinct association table</td><td>select . . . from table A, B, AB on 2 join conditions</td></tr>
<tr><td>Traverse qualified association</td><td></td><td>objectOrSet.role [qualifier=value]</td><td>similar to binary association except also specify qualifier value in where clause</td></tr>
<tr><td rowspan="3">Traverse generali-zation</td><td>Separate superclass and subclass tables</td><td rowspan="3">objectOrSet: subclass and objectOrSet: superclass</td><td>join subclass table to superclass table</td></tr>
<tr><td>Push down superclass attributes</td><td>query appropriate subclass table</td></tr>
<tr><td>Push up subclass attributes</td><td>query superclass table</td></tr>
<tr><td rowspan="2">Traverse from link to object</td><td>Buried association</td><td rowspan="2">linkOrSet.role</td><td>select . . . from table</td></tr>
<tr><td>Distinct association table</td><td>select . . . from table B, AB on join condition</td></tr>
<tr><td rowspan="3">Traverse from object to link</td><td>Distinct association table</td><td>{role1=objectOrSet1, role2= objectOrSet2, . . . , associationName}</td><td>select . . . from table</td></tr>
<tr><td>Buried association</td><td rowspan="2">objectOrSet@ role</td><td>select . . . from table</td></tr>
<tr><td>Distinct association table</td><td>select . . . from table B, AB on join condition</td></tr>
<tr><td>Filter objects</td><td></td><td>objectOrSet[filter]</td><td rowspan="2">specify filter in where clause</td></tr>
<tr><td>Filter links</td><td></td><td>linkOrSet[filter]</td></tr>
<tr><td>Traverse object to value</td><td></td><td>objectOrSet. attribute</td><td rowspan="2">specify desired attribute in SQL select list</td></tr>
<tr><td>Traverse link to value</td><td></td><td>linkOrSet.attribute</td></tr>
</table>

Table 13.3 Summary of ONN mapping rules for RDBMSs

You can readily map OMT object models to RDBMS constructs. Table 13.2 summarizes our recommended mappings. You can generally trade off complexity, performance, integrity, and extensibility.

Also, we showed how to map the ONN to SQL, as Table 13.3 summarizes. The mapping depends on both the navigation construct and the implementation of the object model.

Figure 13.27 lists the key concepts for this chapter.

candidate key	object model, mapping domains
data independence	object model, mapping generalizations
foreign key	primary key
functional model, mapping the ONN	relational database
implementing identity	relational DBMS
object model, mapping associations	SQL
object model, mapping classes	

Figure 13.27 Key concepts for Chapter 13

BIBLIOGRAPHIC NOTES

[Date-82] and [Elmasri-94] are good introductory texts that explain RDBMSs. Most other books that discuss database methodology (such as [Shlaer-88], [Teorey-90], [Rumbaugh-91], and [Bruce-92]) deal only with structural aspects. They do not show techniques for dealing with functional behavior. Consequently, the following discussion is limited to database structure.

In principle, the object model in [Shlaer-88] has similar expressive power to the OMT object model. However, their exposition has a needless bias toward relational databases. In contrast, OMT models are well suited for programming, files, relational databases, object-oriented databases, and other purposes.

The mapping of entity-relationship models to RDBMS tables in [Teorey-90] is comparable to our mapping of the OMT object model to tables. Teorey advocates constructing logical ER models and systematically mapping them to RDBMS tables. The difference is that ER models are less expressive than the OMT object model. Also, Teorey advocates a separate step for dealing with normal forms, which we believe to be unnecessary. Teorey does not discuss some of the more recent aspects of the SQL standards, especially referential integrity. (This is not surprising given the copyright date of the Teorey book.)

The IDEF1X notation [Bruce-92] provides an alternative notation to the table skeletons we have presented. IDEF1X has the advantage of being a standard, familiar notation. It is useful for relational database *design* because it concisely documents design decisions. However, it is an ineffective approach for *analysis* because it reflects a premature commitment to a relational database precluding other implementation platforms. The attention to fine design details with IDEF1X also inhibits the abstract conceptualization so important to analysis.

[Melton-93] contains a thorough explanation of the SQL-92 standard.

REFERENCES

[Bruce-92] Thomas A. Bruce. *Designing Quality Databases with IDEF1X Information Models*. New York: Dorset House, 1992.

[Celko-95] Joe Celko. *SQL for Smarties: Advanced SQL Programming*. San Francisco, California: Morgan Kaufmann, 1995.

[Date-82] CJ Date. *An Introduction to Database Systems*. Reading, Massachusetts: Addison-Wesley, 1982.

[Date-86] CJ Date. *Relational Database: Selected Writings*. Reading, Massachusetts: Addison-Wesley, 1986.

[Elmasri-94] Ramez Elmasri and Shamkant Navathe. *Fundamentals of Database Systems*. 2nd edition. Redwood City, California: Benjamin Cummings, 1994.

[Melton-93] Jim Melton and Alan R. Simon. *Understanding the New SQL: A Complete Guide*. San Francisco, California: Morgan Kaufmann, 1993.

[Rumbaugh-91] J Rumbaugh, M Blaha, W Premerlani, F Eddy, and W Lorensen. *Object-Oriented Modeling and Design*. Englewood Cliffs, New Jersey: Prentice Hall, 1991.

[Shlaer-88] S Shlaer and SJ Mellor. *Object Oriented System Analysis: Modeling the World in Data*. Englewood Cliffs, New Jersey: Yourdon Press, 1988.

[Teorey-90] TJ Teorey. *Database Modeling and Design: The Entity-Relationship Approach*. San Francisco, California: Morgan Kaufmann, 1990.

EXERCISES

13.1 Consider the object model in Figure 2.23.

a. (7) Prepare table skeletons for the object model. Use the recommended mappings and existence-based identity.

b. (6) Write SQL code to implement the following ONN expressions from Section 5.4.1:

- theStLouisAirport.~origin.destination
- theStLouisAirport.~origin.destination.~origin.destination
- aFrequentFlyerAccount@Airline.accountNumber
- aFlight.FlightDescription.AircraftDescription.modelNumber
- aTripReservation.FlightReservation.Flight.FlightDescription.Airline

c. (7) Write SQL code to implement the following methods from Section 5.4.2:

- Airport::findZeroOneStops
- TripReservation::hasOnlyAisleSeats
- Airline::calcFractionLate (month, year)
- TravelAgency::calcMonthlySales (month, year)
- TripReservation::setFrequentlFlyerAccount (aFrequentFlyerAccount)

13.2 Consider the object models in Figure E2.7.

a. (6) Prepare table skeletons for the object model in Figure E2.7a. Write SQL code (combined with pseudocode as necessary) to implement the following methods in Exercise 5.7:

- Find all edges leaving a node.
- Find all edges with the same node as source and sink—that is, the direct cycles in the graph.
- Find all nodes that are reachable from a source node—that is, the transitive closure.

b. (8) Repeat part (a) for the object model in Figure E2.7b and the methods in Exercise 5.8.

c. (7) Repeat part (a) for the object model in Figure E2.7c and the methods in Exercise 5.9.

d. (8) Repeat part (a) for the object model in Figure E2.7d and the methods in Exercise 5.10.

13.3 (5) Prepare table skeletons for your answer to Exercise 2.19. Use the recommended mappings and existence-based identity. Show populated tables for the sample data given in Exercise 2.19.

13.4 Consider the object model in Figure E13.1. We have arbitrarily modified Figure 3.17 for this exercise.

a. (5) Prepare table skeletons for the object model. Use the recommended mappings and existence-based identity.

b. (5) Prepare table skeletons for the object model. Use the recommended mappings and value-based identity.

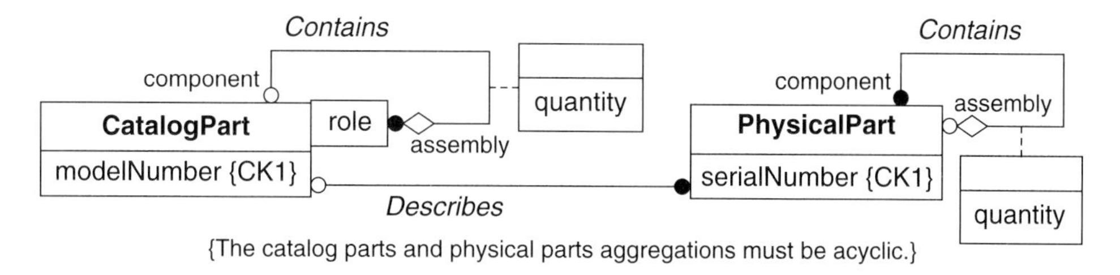

Figure E13.1 Object model for catalog versus physical aggregation

13.5 (7) Prepare table skeletons for your answer to Exercise 3.9. Use the recommended mappings and existence-based identity. Show populated tables for the sample data given in Exercise 3.9.

13.6 Consider the object model in Figure 4.6.

a. (8) Prepare table skeletons for the object model. Use the recommended mappings and existence-based identity.

b. (5) Show populated tables for the sample model given in Exercise 4.4.

c. (6) Show populated tables for the sample model given in Exercise 4.5.

d. (7) Show populated tables for the sample model given in Exercise 4.6.

e. Write SQL code (combined with pseudocode as necessary) to implement the following methods from Exercise 5.11:

- (2) Find all the direct superclasses for a subclass.
- (2) Find all the direct subclasses for a superclass.
- (5) Find all the components for an assembly.
- (5) Find all the assemblies that include a component.
- (5) Find all the attributes that describe the objects of a class. (Compute the transitive closure of inherited attributes for a class.) You need traverse only up the inheritance lattice.

13.7 (5) Prepare table skeletons for your answer to Exercise 4.9. Use the recommended mappings and existence-based identity. Show populated tables with the lasagna recipe.

13.8 (7) Prepare table skeletons for your answer to Exercise 4.12. Use the recommended mappings and existence-based identity. Show populated tables with the census data.

13.9 Consider your answer to Exercise 8.2.

a. (6) Prepare table skeletons for the object model. Use the recommended mappings and existence-based identity.

b. (6) Write SQL code to implement your answer to Exercise 8.2c.

c. (6) Write SQL code to implement your answer to Exercise 8.2f.

13.10 (6) Prepare table skeletons for your answer to Exercise 8.3c. Use the recommended mappings and existence-based identity.

13.11 (6) Prepare table skeletons for your answer to Exercise 8.4c. Use the recommended mappings and existence-based identity.

13.12 Consider your answer to Exercise 8.5.

a. (6) Prepare table skeletons for the object model. Use the recommended mappings and existence-based identity.

b. (7) Write SQL code to implement your answer to Exercise 8.5e.

13.13 Consider your answer to Exercise 8.6.

a. (6) Prepare table skeletons for the object model. Use the recommended mappings and existence-based identity.

b. (5) Write SQL code to implement your answer to Exercise 8.6e.

c. (7) Write SQL code to implement your answer to Exercise 8.6f.

13.14 (6) Prepare table skeletons for the object model in Exercise 9.3. Use the recommended mappings and existence-based identity.

13.15 Consider Exercise 9.4.

a. (6) Prepare table skeletons for the object model. Use the recommended mappings and existence-based identity.

b. (6) Write enhanced pseudocode for computing the number of refills remaining for a prescription. Also write SQL code for this.

c. (7) Write enhanced pseudocode and SQL code for finding if a customer has a problem with allergies for a prescription. You will need to compare allergies recorded for a customer with those recorded for a drug.

d. (7) Write enhanced pseudocode and SQL code for finding if a new customer prescription will cause drug interaction problems with past prescriptions.

13.16 (5) Prepare table skeletons for the object model in Exercise 9.5. Use the recommended mappings and existence-based identity.

13.17 (8) Prepare table skeletons for the object model in Figure 13.2. Use the recommended mappings and existence-based identity.

14

Relational Databases: Advanced

An RDBMS can do more than just store and retrieve data. It can also partially assure data quality and perform much computation, substituting for programming code. This chapter builds on Chapter 13 by describing advanced aspects of implementing the object, dynamic, and functional models. Throughout, we present details of our portfolio manager implementation.

14.1 IMPLEMENTING THE OBJECT MODEL

Implementing the object model consists of the tasks outlined in Section 13.2 as well as the following:

- **Define constraints on tables**. You can use RDBMS constraints to enforce the structure of the object model. Chapter 13 addressed primary and candidate keys. You should also define referential integrity for foreign keys. [14.2]
- **Define indexes**. Tune the database so that navigation is fast, as the object model implies. Indexes are the primary means for tuning a relational database. [14.3]
- **Allocate storage**. Most RDBMSs let you allocate space for each table and index. You can compute the necessary storage space from the physical storage requirements (Chapter 10) and the implementation of domains (Chapter 13). [14.4]
- **Create a schema**. Prepare and run the SQL code to create the actual application database. [14.5]

14.2 DEFINING CONSTRAINTS ON TABLES

Both developers and maintainers benefit if constraints (often business rules) are explicitly noted in the database, not buried in programming code. RDBMS constraints also reduce the amount of programming code that must be written. Constraints expressed in the schema are enforced across all applications. Normally, you should let the RDBMS enforce primary and candidate keys.

14.2.1 SQL Options for Referential Integrity

Similarly, you should let the RDBMS referential integrity mechanism enforce foreign keys. Many RDBMSs can propagate the effects of deletion and updates for foreign keys. The SQL standard [Melton-93] provides the following referential integrity options for deletions. As we describe later, you can include these options in the schema that implements the object model.

- **Cascade**. With a mandatory (not null) foreign key, the deletion of a referent record may imply the deletion of all referencing records. For example, deletion of an *Asset* record implies deletion of all corresponding *AssetValue* records.
- **No action**. You may forbid the deletion of a referent record if there are referencing records. For example, we do not want the user to be able to delete an asset that is referenced by transactions.
- **Set null**. With an optional (null allowed) foreign key, the deletion of a referent record implies that the references are set to null. Thus deletion of a financial institution implies that there is no account holder for all corresponding portfolios (account holder is set to null).
- **Set default**. You may set a foreign key to a default value instead of to null. Deletion of the referent record then implies that the references are set to a default value. We did not use this option for the portfolio manager.

SQL also provides referential integrity for updates. For value-based identity you should either let updates cascade or disallow updates that would otherwise cause a cascade. For example, if we had used value-based identity for the portfolio manager, renaming an asset would imply changing the asset names buried in the *AssetValue* table. With existence-based identity, you need not consider update for foreign keys; IDs are artificial numbers that need not change.

14.2.2 Implementing Referential Integrity

Foreign keys arise during the implementation of associations and generalizations; the referential integrity constraints ensure that there are no dangling foreign keys. The semantics of updating are straightforward, so we do not discuss them further. We now specify guidelines for the SQL deletion options that apply for both existence-based and value-based identity.

- **Generalization**. Always cascade deletions for foreign keys that arise from the implementation of generalization. For example, *StockAsset* is a subclass of *Asset* and *stockAssetID* references the primary key of *Asset*. Upon deletion of an *Asset* record, the RDBMS must also delete the corresponding *StockAsset* record. Since we implemented generalizations with separate superclass and subclass tables for the portfolio manager, we specified *on delete cascade* for each subclass reference to a superclass.

The RDBMS can propagate deletion downward from the superclass to the subclasses. However, an RDBMS cannot propagate deletion upward from the subclass toward the superclass. For example, we could have the deletion of a *Stock* cause the deletion of any associated *StockOption* records. But deleting a *StockOption* record will not ripple up the inheritance hierarchy. Consequently, we must first delete the *FinancialInstrument* records that are stock options for a stock and then delete the *FinancialInstrument* record that is the stock.

- **Buried association, minimum multiplicity of zero**. Normally, set the foreign key to null, but sometimes you may forbid deletion with the *no action* clause. For example, upon deletion of a *FinancialInstitution* we set the *accountHolder* to null. However, we forbid the deletion of an *Asset* that is an *assetReferenced* for a *Transaction*.
- **Buried association, minimum multiplicity of one**. You can cascade the effect of a deletion. For example, each *AssetValue* record must refer to an *Asset*; the deletion of an *Asset* record should cascade to cause the deletion of referencing *AssetValue* records. Otherwise, forbid the deletion. For example, we forbid the user to delete a *Stock* with corresponding *StockOption* records.
- **Association table**. Normally, we cascade deletions to the records in an association table. For example, deletion of a *Person* record or a *Portfolio* record causes the corresponding *Portfolio_accountOwner* records to be deleted. However, sometimes we forbid a deletion. For example, we forbid an *Asset* to be deleted if it is involved in some *Transaction*.

With our guidelines, we cannot fully infer referential integrity actions for associations. Instead we list the specific actions for each foreign key in Table 14.1.

Some RDBMSs do not support referential integrity. Also, some vendor applications forgo referential integrity and instead support the "lowest common denominator" so that they can run with various RDBMSs. Furthermore, many legacy applications were never designed to use referential integrity. In these cases you can use an audit approach to check for violations of referential integrity after the fact. You can write a battery of SQL commands that finds referential inconsistencies in the database. You can then periodically run the commands and repair the violations they discover.

14.3 DEFINING INDEXES

Performance tuning is also important for a relational database. With an RDBMS you have wide latitude to adjust database structures without changing application code and database semantics. You must properly tune a relational database or users will be disappointed with the slow response and your application will fall short of its potential.

Indexes are the primary means for tuning a relational database. Normally, you define a unique index for each primary and candidate key. (Most RDBMSs create unique indexes as a side effect of SQL primary key and candidate key constraints.)

Association reference	On delete action	Explanation
Portfolio.parentPortfolio	no action	The user cannot delete a portfolio with child portfolios.
Portfolio.accountHolder	set null	If a *financialInstitution* is deleted, nullify *accountHolder* for the associated portfolios.
Portfolio_accountOwner. portfolioID	cascade	If a portfolio is deleted, delete the corresponding links with *Person*.
Portfolio_accountOwner. accountOwner	cascade	If a person is deleted, delete the corresponding links with *Portfolio*.
AssetValue.assetID	cascade	If an asset is deleted, delete its values.
AssetValue.currency	no action	The user cannot delete a *Cash* object that is referenced.
PortfolioValue.currency		
StockOption.currency		
Transaction.currency		
ROI.portfolioID	cascade	If a portfolio is deleted, delete its ROIs.
PortfolioValue.portfolioID	cascade	If a portfolio is deleted, delete its values.
StockOption. underlyingSecurity	no action	The user cannot delete a *stockAsset* that has *stockOptions*.
Transaction.portfolioID	no action	The user cannot delete a referenced portfolio.
Transaction. assetReferenced	no action	The user cannot delete an asset that is involved in a transaction.
Transaction_assetTrans acted.assetTransacted	no action	
Transaction_assetTrans acted.transactionID	cascade	If a transaction is deleted, delete the *assetTransacted* links.

Table 14.1 Referential integrity specifications for portfolio manager associations

You should also create an index for each foreign key that is not subsumed by a primary key or candidate key constraint; foreign keys implement associations and generalizations. These indexes for foreign keys are important. Users and developers of object-oriented applications expect fast navigation, so you must provide the appropriate underlying data structures. With proper indexing, traversal of an association instance (a link) involves an SQL select (or is part of an SQL join) and is of order *log n* (where *n* is the number of records in a table) and close to order *1*. Without indexes, traversal of a link involves a sequential scan and is of order *n*. Proper indexing can dramatically improve performance by reducing the order of computation. The update cost of foreign key indexes is inconsequential compared to their benefit for traversing object models.

Once your software is running, the database administrator can determine if it needs additional indexes. Some persons advocate forecasting the mix of queries that will execute against a database, ***transaction modeling***, and using these statistics to compute the most effective combination of indexes. (See Chapter 10.) For example, you could predict the frequency of the various kinds of queries that read and write from each table and their I/O activity for different combinations of indexes. In practice, we have seldom needed transaction modeling.

Some RDBMSs support additional data structures for improving performance. Oracle, for example, has the *cluster* feature for physically collocating data. With an Oracle cluster you can group records from different tables on the same disc page according to a common value. For example, for a physical aggregation you may wish to group each assembly object with its component objects. The rest of this chapter considers only indexes for performance tuning.

14.4 ALLOCATING STORAGE

Many RDBMSs let you allocate storage space to each table and index. Oracle, for example, has a storage clause that lets you specify the initial and subsequent space allocations for tables and indexes. The storage clause lets you resolve competing demands. Large blocks of disc space permit contiguous placement of data, which can improve performance. However, small blocks enable more efficient use of disc space.

You can compute storage space from storage requirements and the implementation of domains. In Table 14.2 we calculate that each *Portfolio* table record will require 60 bytes of storage space. We are assuming that an ID will occupy about 8 bytes of space. In Section 10.3 we estimated that we will need 25 portfolios, which will then consume $60 \times 25 = 1500$ bytes. Most RDBMSs require that you allocate storage in 1K or 2K blocks, so we could allocate 2K initial storage and 1K subsequent storage for the portfolio table.

Column	Domain	Data type	Bytes of storage
portfolioID	ID	counter	8
portfolioType	enumPortfolioType	text(6)	6
parentPortfolio	ID	long	8
accountHolder	ID	long	8
accountNumber	mediumString	text(30)	30
Total			60

Table 14.2 Computation of storage space for a portfolio record

We will not consider further storage for the portfolio manager because MS-Access does not let the user allocate storage.

14.5 CREATING A SCHEMA

The final object-modeling task is to prepare SQL code from the table skeletons, domain specifications, referential integrity actions, and storage requirements. We created MS-Access schema for the portfolio manager with the following steps:

- **Write SQL code for each table**. We substituted MS-Access data types for domains, but otherwise wrote generic SQL code without regard for the limitations of MS-Access. Figure 14.1 shows some of the *Transaction* table code. Our Web site, www.omtassociates.com, has the complete code and the resulting MS-Access database. We separated foreign key definitions from table definitions to avoid circularity.

  We specified nullability for attributes. The *counter* data type of MS-Access has an implicit *not null* constraint. We required that all discriminators and attributes of candidate keys have a value. Chapter 10 specifies nullability for the other attributes.

  We were careful to avoid reserved words and other words with confounding side effects. *Currency*, *date*, and *value* are reserved words. *Name* is not a reserved word, but has a special meaning for reports. We renamed all these attributes in the schema. We used *PrimaryKey* as the name of our primary keys. Some MS-Access commands become confused if another name is used.

- **Write SQL code for each index**. We defined an index for each foreign key that is not subsumed by a primary key or candidate key constraint. (MS-Access creates an index as a side effect of primary key and candidate key constraints.) For example, because the *Portfolio_accountOwner* table has a primary key of *portfolioID* + *accountOwner*, we defined an additional index on *accountOwner*.

- **Create MS-Access queries**. We pasted the SQL *create table* and *create index* statements into MS-Access, one SQL statement per MS-Access query. We deleted constraints and default value clauses because MS-Access could not execute this SQL code. We converted tabs into spaces for readability.

- **Write a schema creation macro**. We wrote a macro that sequentially executes the queries with the *create table* and *create index* commands. We executed the macro.

- **Enter constraints and default values**. We used the MS-Access table properties form to enter manually constraints and default values that were functionally equivalent to the generic SQL code that MS-Access could not execute.

- **Enter foreign keys**. MS-Access also does not fully support SQL commands for foreign keys. So we manually entered foreign keys with the *relationships* graphical interface. MS-Access does not support the *on delete set null* clause, so we had to write additional application code.

After these steps we had an empty MS-Access database that provided a sound basis for further implementation.

```
CREATE TABLE Transaction
(transactionID      counter,
 transactionDate    dateTime
        CONSTRAINT nn_transaction1 NOT NULL,
 recordDate         dateTime,
        CONSTRAINT nn_transaction2 NOT NULL,
 transactionType    text(19)
        CONSTRAINT nn_transaction3 NOT NULL,
 transactionFee     currency DEFAULT 0
        CONSTRAINT nn_transaction4 NOT NULL,
 description        text(255),
 assetReferenced    long,
 portfolioID        long,
        CONSTRAINT nn_transaction5 NOT NULL,
 transactionCurrency long
        CONSTRAINT nn_transaction6 NOT NULL,
CONSTRAINT PrimaryKey PRIMARY KEY (transactionID));

ALTER TABLE Transaction ADD CONSTRAINT fk_transaction1
FOREIGN KEY (assetReferenced) REFERENCES Asset
ON DELETE NO ACTION;

ALTER TABLE Transaction ADD CONSTRAINT fk_transaction2
FOREIGN KEY (portfolioID) REFERENCES Portfolio
ON DELETE NO ACTION;

ALTER TABLE Transaction ADD CONSTRAINT fk_transaction3
FOREIGN KEY (transactionCurrency) REFERENCES Cash
ON DELETE NO ACTION;

ALTER TABLE Transaction ADD CONSTRAINT enum_transaction1
CHECK (transactionType IN ('purchase', 'sale', 'deposit',
'withdrawal', 'barter', 'journal', 'dividend', 'interest',
'return of principal'));

CREATE INDEX index_transaction1 on Transaction
(assetReferenced);

CREATE INDEX index_transaction2 on Transaction (portfolioID);

CREATE INDEX index_transaction3 on
Transaction(transactionCurrency);
```

Figure 14.1 Portion of SQL code to create schema for portfolio manager

14.6 IMPLEMENTING THE DYNAMIC MODEL

The dynamic model shows control as well as constraints on the life history of objects. A database is essentially a giant repository of state that records the cumulative effects of actions on a system. In principle, you could use triggers to implement simple state diagrams, but few RDBMSs support triggers. (A ***trigger*** performs an action upon the occurrence of the specified event.) Triggers are adequate for implementing simple state diagrams, but lack the structure needed for handling complex state diagrams. Fortunately, the dynamic model is seldom important for database applications, aside from the user interface.

14.6.1 Implementing the User Interface

When implementing data management applications, you should separate application logic from user interface logic. (Chapter 18 amplifies this point with a discussion of the three-tier client-server architecture.) Although the dynamic model contributes little to the understanding of data structure and behavior, it is important for the user interface. Form-based interfaces are common for data management applications, due largely to the widespread availability of powerful 4GL languages. (See Section 9.5.2.) We made several decisions of philosophy and style for the portfolio manager's user interface.

- **Use a form hierarchy**. We organized the forms into a simple hierarchy (Figure 14.2) according to structure and functionality. Command buttons let the user switch between forms. The user may update and inspect data for various asset types. The user may also specify portfolio data and analyze the financial performance of portfolios. The other two portions of the interface manage transactions and miscellaneous data.

- **Make command buttons conspicuous**. We underlined all command buttons and only command buttons. We also gave command buttons a distinct color. Command buttons switch to a child form, exit a form, delete an object, and invoke reports.

- **Organize forms about objects**. Forms are *object oriented*, not *class oriented*. This distinction is important for generalization hierarchies. Our mapping of the object model to tables fractures an object into multiple records, one for each level of a generalization hierarchy. As we described in Section 13.8.1, we defined a view for each kind of object that reconstitutes the fragments for presentation on forms. Furthermore, we were able to define writable views with MS-Access; as the user interacts with a form, MS-Access automatically translates view updates into table updates.

  Figure 14.3 shows the *StockAsset* form, which combines records from the *FinancialInstrument*, *Asset*, and *StockAsset* tables. The *StockAsset* form also includes a subform with stock asset values (see subsequent bullet *collection of values*).

- **Suppress IDs and discriminators**. Object identifiers are irrelevant to the user other than for debugging and maintenance. Consequently, our user interface does not show any IDs. This complicated some of our pick lists, because we had to include additional logic to convert names to IDs. Similarly, we had to set the values of discriminators automatically without making them apparent to the user.

```
Main menu
   Asset menu
      Bond asset
         Bond details
      Stock asset
         Stock details
      Stock option
      Real estate
      Precious metal
      Cash
      Collectible
      Insurance
   Portfolio information
      Portfolio owners
   Portfolio analysis
      Portfolio direct assets
      Portfolio direct + indirect assets
   Transaction filter
      Transaction information
   Miscellaneous menu
      Person information
      Financial institution information
```

Figure 14.2 Hierarchy of forms for the portfolio manager's MS-Access user interface

Stock asset information

Stock asset name: Mexico Fund Inc. **Ticker symbol:** MXF

Description: closed-end fund for Mexican stocks

Stock type: closed-end

Stock asset value versus time

date	value	currency	source
30-Jun-1995	$16.50	US dollars	actual

Record: 1 of 1

Stock asset details **Exit stock asset information** **Delete stock asset**

Figure 14.3 *StockAsset* form for portfolio manager

- **Order records**. We defined views to impose a meaningful order on the records for all forms; none of our forms directly access tables. We ordered data using attributes that are meaningful for the application—qualifiers or candidate keys.
- **Ensure correctness by construction**. We prefer to keep data correct by construction rather than by checking for user errors. We explain how we enforced this preference in the next two bullets and in our discussion of portfolio cycles in Section 14.7.3.
- **Make forms dynamic**. With MS-Access, you can make the visibility of a form's elements depend on an expression. We used visibility to customize the *Transaction information* form according to transaction type. The *assetReferenced* association is appropriate only for certain transaction types (dividend, interest, and return of principal). This is consistent with our philosophy of ensuring correctness by construction.
- **Control deletion**. We disable the delete button when delete is inappropriate. For example, we do not want the user to delete an asset that a transaction references. We could have let referential integrity intercept such an attempt (we still did specify the referential integrity), but prevention of an improper deletion, rather than an error response, is more consistent with our philosophy of correctness by construction.
- **Enforce clean delete behavior.** MS-Access makes it difficult to do this for both committed and uncommitted records. The SQL *delete* command works only for committed records (that is, records written to the database). For uncommitted records, MS-Access provides the *undo record* command. But *undo record* complains if there is not actually something to undo, so we had to recognize an empty record separately. A further complication is that MS-Access occasionally seems to confuse empty strings with null values.
- **Put constraints in the database, not the form**. When possible, we put constraints in the database rather than in forms. Then the user does not subvert constraints by accessing the database through a different interface. Much of our programming effort was directed toward enforcing constraints and ensuring data quality.

  For the association between *Portfolio* and *FinancialInstitution*, the semantics of the qualifier require that both *accountHolder* and *accountNumber* be null or not null. We enforced this with a table constraint. We built several constraints involving *transactionType* into the user interface, some of which are difficult to express in SQL. MS-Access permits only one table constraint per table, so we had to enforce the *transactionType* semantics with forms.

  Our implementation led to one additional constraint that is difficult to express with the object model. The value of a *FinancialInstrument* is unique given a date *and* a currency. We had to express this with a form.

We found it helpful to categorize the fields on forms.

- **Value**. A value is a piece of data without identity. (See Section 2.1.3.) Values instantiate attributes of objects and links. For example, *stock asset name, ticker symbol, description*, and *stock type* in Figure 14.3 describe values for the portfolio manager. The user may enter a value by filling a slot on a form or by choosing a constant from a pick list.

- **Collection of values**. Collections of values arise for some one-to-many associations; the user can access objects in the "many" role only within the context of a parent object. For example, in Figure 14.3 a user can access a stock asset value only by first specifying a stock asset. In a sense such objects lack intrinsic identity and are more like multiple values. We implemented collections of values with subforms. The subform displays the collection of child records associated with the parent record. The user may freely type into the subform.
- **Reference**. In contrast to a value, a reference concerns an object. A single reference may arise for a one-to-one or a one-to-many association. In Figure 14.4 *Transaction.portfolioID* is an example of a reference. The transaction is the focus of the form, and the user must choose the appropriate portfolio. The portfolios are objects in their own right and are defined in another form. We provide a pick list for each reference.

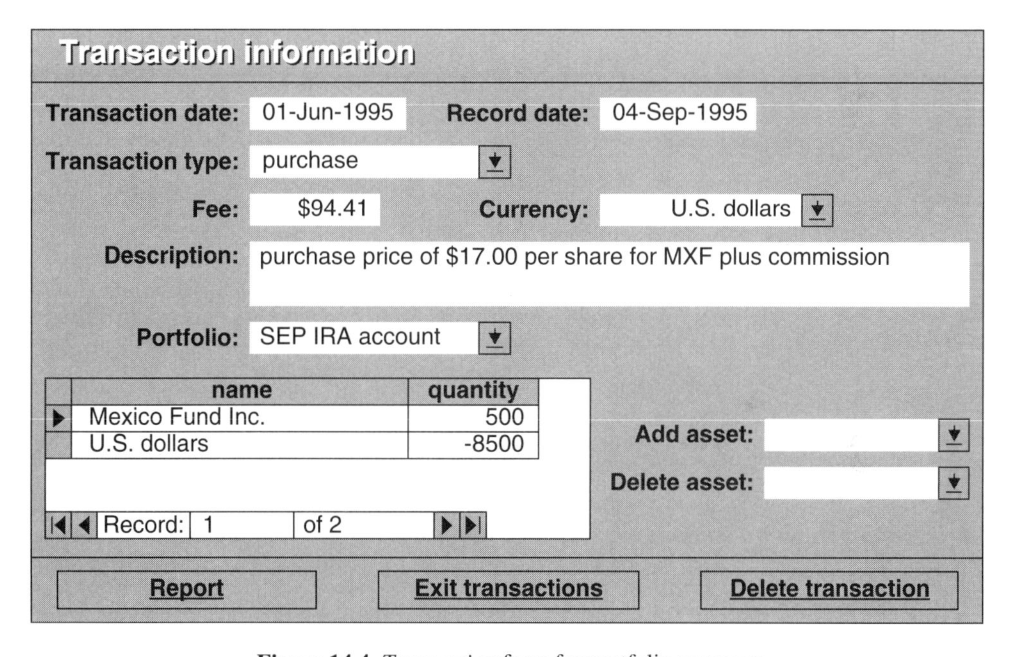

Figure 14.4 *Transaction* form for portfolio manager

- **Collection of references**. Collections of references arise for some one-to-many associations and for all many-to-many associations. For example, in Figure 14.4 we implemented *Transaction.assetTransacted* as a collection of references. As for a collection of values, we use a subform to display the child records. For a collection of references, however, we must control the input. The *add asset* pick list can add an asset not already in the subform. The *delete asset* pick list can delete an asset in the subform.
- **Qualifier**. Qualifiers add constraints to the categories just described. For an association that is one-to-one after qualification, the combination of the source object plus the qual-

ifier is a candidate key. Thus *assetID* + *date* is a candidate key of *AssetValue*. We cannot define a candidate key for an association that is one-to-optional after qualification, such as the *accountHolder* association. However, for such an association another constraint applies: Both *FinancialInstitution* and *accountNumber* are either null or not null.

With MS-Access, you can easily define a pick list for enumerated values. You can hard-code values in the definition of the pick list or soft-code values in one or more auxiliary database tables. For simplicity, we hard-coded enumeration pick lists.

You can also easily define a pick list for mandatory references. We defined a view to query the underlying tables and collect the valid choices for presentation in the pick list. For optional references we finessed MS-Access by using the SQL *union* command to add a null record to the list of valid references. The user either chooses an entry from the list or enters a null reference.

14.7 IMPLEMENTING THE FUNCTIONAL MODEL

Most PC application developers seem to do a lot of low-level programming. Much of this is unnecessary with MS-Access, because you can quickly deliver functionality with SQL and macros. The problem with many business applications is not performance, but how to develop quickly an application that is correct.

In this section we describe many of the portfolio manager's implementation details. You may find it helpful to study the code posted on our Web site and experiment with the portfolio manager as you read this section.

14.7.1 Operations for the Portfolio Manager Case Study

In addition to the improvements we made in implementing the object and dynamic models for the portfolio manager, we made several functional simplifications:

- **No currency conversion**. We allow the value of assets to vary, except for *Cash* assets. Consequently, we are not supporting currency conversion.
- **A single portfolio currency**. We restrict the calculation of portfolio value to a single currency. The portfolio manager checks for mixed currencies in a portfolio and issues an error message if this occurs.
- **Interest on cash**. Many brokers roll cash into a money fund that pays interest. We don't want to have to enter transactions to convert between cash and the money fund, so we allow interest to be paid on cash.

Figure 14.5 shows the object model for the portfolio manager with the major operations we implemented. We include it to show how the operations involve the various objects. The MS-Access 4GL implicitly provides many more operations than we implemented. For example, creation of most records is implicit with a 4GL. All operations for the *FinancialInstrument* class are abstract and have different implementations for *Portfolio*, *PreciousMetal*, *Cash*, and the other concrete subclasses. We now explain the operations.

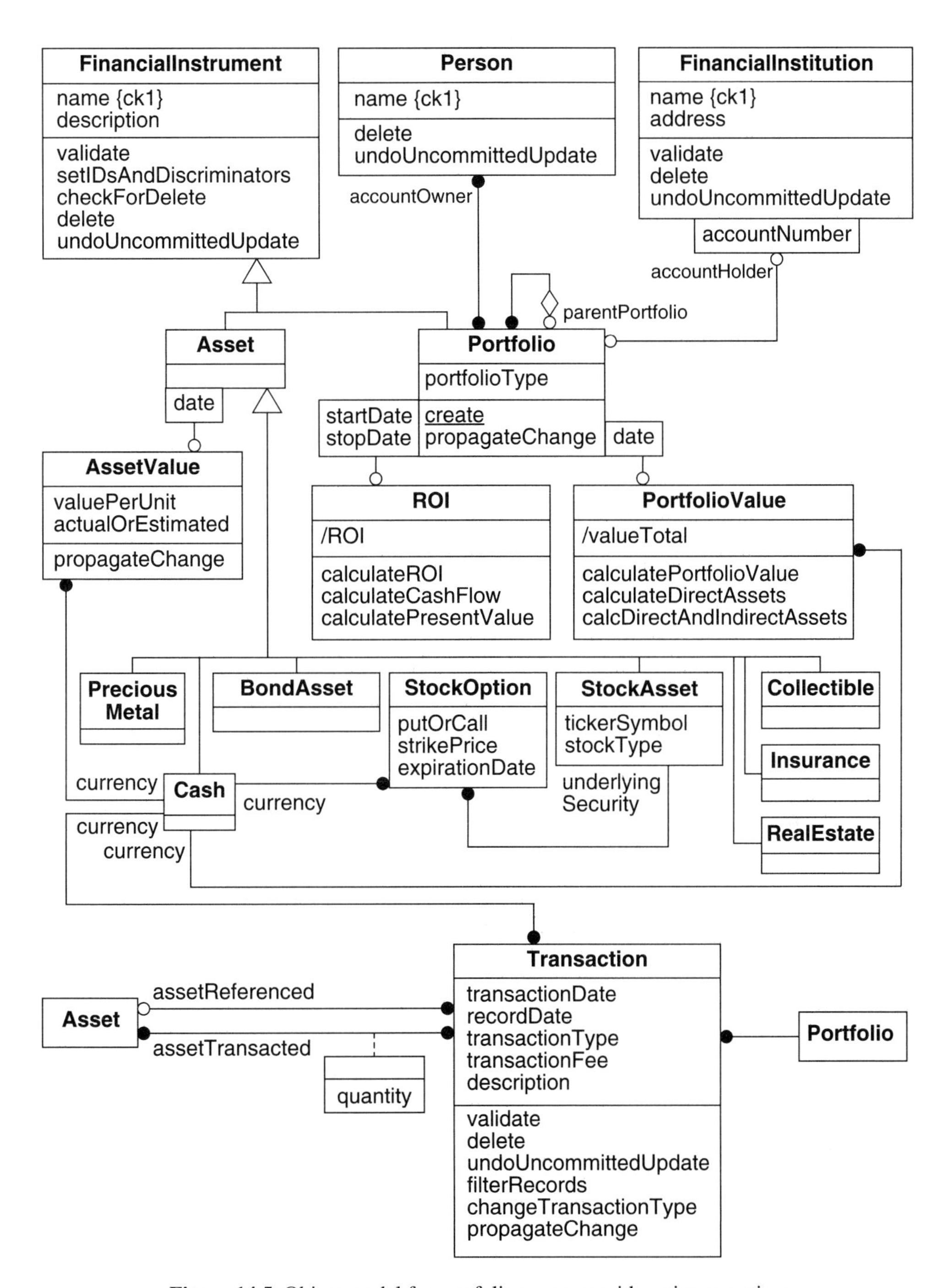

Figure 14.5 Object model for portfolio manager with major operations

- **validate**. Complain if some *not-null* attribute lacks a value on record commit. (The constraints in the schema ensure that all mandatory attributes have a value. However, the validate operation provides a more informative message to the user and moves the cursor to the offending field.)
- **setIDsAndDiscriminators**. Set discriminators and foreign key IDs for an object dispersed across tables that implement a generalization hierarchy.

  For example, for a *BondAsset* object, MS-Access automatically sets the *financialInstrumentID* (*counter* data type). Upon commit of the *BondAsset* object presented by the *BondAsset* form, this operation sets the *assetID* and *bondAssetID* to the same value as the *financialInstrumentID*. Similarly, this operation sets the *financialInstrumentType* to "*Asset*" and the *assetType* to "*BondAsset*."
- **checkForDelete**. Determine whether the *delete* command button should be enabled. We do not allow the user to delete a *FinancialInstrument* that is referenced by a *Transaction*.
- **delete**. Delete an object in the database. Most often the implementation of this operation is a simple SQL *delete* statement. Occasionally it is more complex.

  For example, the *delete* method for *StockAsset* must first delete any *StockOption* objects. We cannot rely on referential integrity to delete the *StockOption* objects. We could have deletion of a *StockAsset* cascade and delete the dependent *StockOption* records, but then we would have dangling *Asset* and *FinancialInstrument* records for the deleted *StockOptions*. As we described in Section 14.2.2, referential integrity cascades down an inheritance hierarchy but does not cascade upward.
- **undoUncommittedUpdate**. Delete an object that has not been committed to the database.

  MS-Acccss provides different delete behavior for committed and uncommitted records. The SQL *delete* command deletes a committed record. The *undo record* command deletes a record that exists only in memory. The *delete* button on the forms has the appropriate logic to delete a record, regardless of whether it has been committed.
- **create**. Create a *Portfolio* object. (This operation is underlined because it is a class operation. See Section 3.1.2.) Creation and deletion of *Portfolio* objects is special, because we must keep the derived *PortfolioTransitiveClosure* table consistent with the underlying parent-child links. Each portfolio may have a parent portfolio, allowing the portfolio manager to support portfolio trees. The transitive closure table precomputes information for traversing the tree. Section 14.7.3 provides more details.
- **propagateChange**. Delete portfolio values and ROIs affected by a change in the underlying data. Portfolio value and ROI are derived and depend on portfolio composition, asset values, and transactions. (Actually, our implementation of this operation may occasionally delete a few extra portfolio value and ROI records. But this is not a problem because they are derived.)

  The MS-Access programming language supports transactions, but MS-Access SQL does not. We compensate by first deleting the portfolio values and ROIs and then changing the fundamental data. Thus the portfolio manager guarantees that all portfolio values and ROIs are correct and consistent with the underlying data (aside from our possible implementation errors).

- **calculateROI**. Calculate the ROI for a portfolio for an interval of time. This is one of our basic requirements for the portfolio manager from analysis.
- **calculateCashFlow**. Calculate the value of assets flowing into (*deposit*) and out from (*withdrawal*) a portfolio for an interval of time. We treat the portfolio value on the starting date as a cash flow in and the portfolio value on the stopping date as a cash flow out. Cash flow is an intermediate calculation for ROI.
- **calculatePresentValue**. Calculate the present value of a series of cash flows, given an assumed ROI. When we have assumed a correct ROI, the present value is zero. This operation is also an intermediate calculation for ROI.
- **calculatePortfolioValue**. Calculate the value of a portfolio on a given date. This is one of our basic requirements for the portfolio manager from analysis.
- **calculateDirectAssets**. Calculate the collection of assets that belong to a portfolio on a specified date. This operation is an intermediate calculation for portfolio value and is also significant in its own right.
- **calcDirectAndIndirectAssets**. Calculate the collection of assets that directly or indirectly belong to a portfolio. Essentially we add the assets in a portfolio to the assets in all descendants of a portfolio. This operation is an intermediate calculation for portfolio value and is also significant in its own right.
- **filterRecords**. Filter the portfolio manager's transaction records on the basis of affected portfolio, date interval, or both. The filter helps the user search a large number of transactions.
- **changeTransactionType**. Change the transaction type and make any necessary adjustments to the screen and associated data. For example, a *dividend* transaction must have a *StockAsset* as an *assetReferenced*. A *withdrawal* does not have an *assetReferenced*. The data dictionary in Section 8.4 specifies the allowable transaction types.

14.7.2 Implementing the ONN with MS-Access

You can readily implement ONN expressions with MS-Access using SQL and the graphical design interface. First press the *query*, *new*, *new query*, *close*, and *design view* buttons in succession to open a graphical query window. Then pick the tables being traversed by repeatedly executing the *add table* command. MS-Access automatically creates a join when both a foreign key and its primary key referent are in the displayed tables. These joins implement the traversal of associations and generalizations. You can delete an automatic join when the ONN expression lacks the corresponding traversal. You can then indicate desired values and miscellaneous predicates graphically or by switching back to the *SQL view*.

14.7.3 Major Algorithms

We devised several important algorithms for the portfolio manager.

Subtracting Tables

The expression *table A EXCEPT table B* returns all records of *A* that are not in *B*. (*Except* is the SQL-92 keyword that denotes table subtraction.) We used table subtraction in several methods. For example, the *add account owners* pick list on the *portfolio owners* form presents the list of persons, except those who already own the portfolio that is the subject of the form. Many RDBMSs, including MS-Access, do not support table subtraction so you must use a workaround.

One possibility is to use a subquery—return the list of persons that do not exist in the *Portfolio_accountOwner* table for the given portfolio. We used the following code in our implementation. (Recall that we must also decode IDs to names for presentation to the user. Also we renamed *Person.name* to *Person.personName* to avoid an MS-Access reserved word.) The brackets delimit a value that is provided from the user interface forms.

```
SELECT personName, personID
FROM Person
WHERE NOT EXISTS
  (SELECT * FROM Portfolio_accountOwner AS PAO
   WHERE Person.personID = PAO.accountOwner AND
      PAO.portfolioID = [theSubjectPortfolio])
ORDER BY personName;
```

Alternatively, you can express subtraction by using a left join. MS-Access uses this technique in its *find unmatched query* wizard. It is more cumbersome for the *add account owners* pick list because we must define an intermediate view. We define query *PAO* with the following code:

```
SELECT portfolioID, accountOwner
FROM Portfolio_accountOwner
WHERE portfolioID = [theSubjectPortfolio];
```

The left join implementation for table subtraction is then (recall that you can treat an MS-Access query as a view)

```
SELECT personName, personID
FROM Person LEFT JOIN PAO ON personID = accountOwner
WHERE accountOwner is NULL;
```

Avoiding Portfolio Cycles

A pick list on the *portfolio information* form allows the user to assign a parent portfolio to a portfolio. As Figure 14.6 shows, the user can form a portfolio tree* by performing assignments for several portfolio records. The database stores a collection of portfolio trees.

The portfolio manager structurally enforces that a portfolio can have at most one parent; the *Portfolio* table has a single *parentPortfolio* attribute that may be null. However, nothing in the schema will prevent the user from forming a cycle. For example, the user could assign

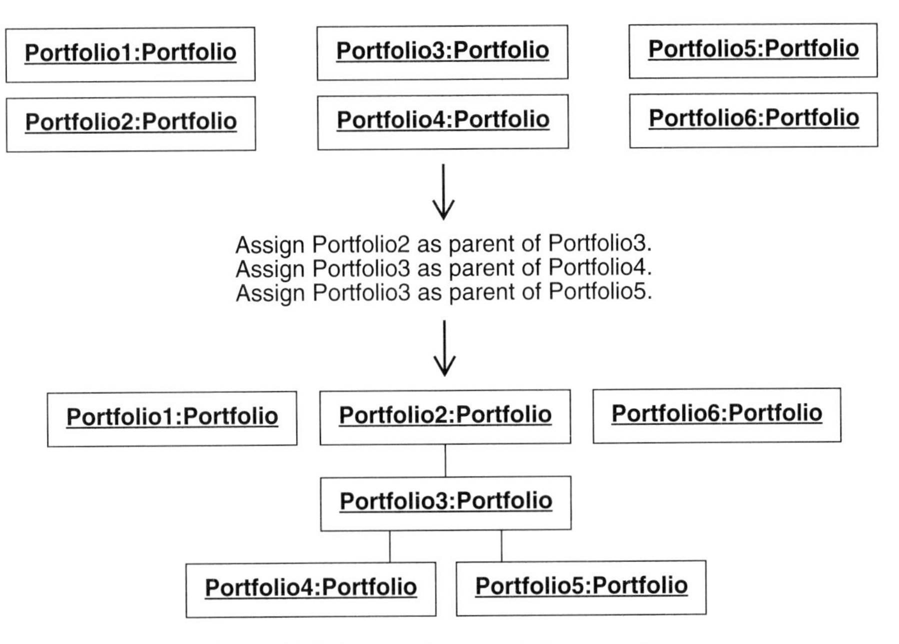

Figure 14.6 An example of portfolio composition

Portfolio5 as the parent of *Portfolio2*. Cycles clearly violate the definition of a tree and are inconsistent with the computation of portfolio value and ROI.

An obvious solution would be to let the user add a new parent-child link and then check for cycles. We could traverse from the children to the grandchildren to the great-grandchildren to other descendants and eventually determine if we arrived back at the original portfolio. We would disallow the new link if it caused a cycle. Otherwise, we would write the new link to the database. This approach has two problems.

The first is philosophical. This approach is not preventive; the user can proceed to take an illegal action. We would rather restrict data entry than let the user make such a choice and force the portfolio manager to deal with the error. This is consistent with our "correctness by construction" philosophy.

The other problem is computational; SQL cannot traverse an arbitrary number of links. (SQL cannot compute the transitive closure.[†]) Traversal of a link corresponds to an RDBMS *join* clause. SQL supports only a fixed number of join clauses, not an indeterminate number

[*] The ***tree***, a term from graph theory, is a set of nodes, such that each node has at most one *parent* node and zero or more *child* nodes. A tree may not have any cycles; you cannot start at a node, traverse a sequence of parent-child links, and arrive back at the original node. In the portfolio manager, a portfolio corresponds to a node.

[†] Like "tree," ***transitive closure*** is a term from graph theory. It is the set of nodes that can be reached, directly or indirectly, from some starting node.

of join clauses. Thus we cannot directly express cycle detection with SQL. For the first solution we would have to perform additional programming to compensate for SQL limits.

A better solution is to compute the pick list for the parent portfolio so that the user cannot form cycles. To make this computation easier, we maintain a derived table with the transitive closure of portfolios. The *PortfolioTransitiveClosure* table has two attributes, *ancestorPortfolio* and *descendantPortfolio*, and stores all the possible combinations. For efficiency, we added the self cycle (portfolioID, portfolioID) to the table. We can incrementally update the derived table as the user updates parent portfolios.

We implemented the following methods for the *portfolio information* form. We had to invoke the SQL commands for the first four methods from the programming language so that we could execute each method as a transaction.

- **Create portfolio**. Create the portfolio and then insert the following record into the transitive closure table.

```
INSERT INTO PortfolioTransitiveClosure
   (ancestorPortfolio, descendantPortfolio)
VALUES ([theSubjectPortfolio], [theSubjectPortfolio]);
```

- **Delete portfolio**. Delete from the transitive closure table and then delete the portfolio. Two SQL commands delete from the transitive closure table.

```
DELETE FROM PortfolioTransitiveClosure
WHERE ancestorPortfolio = [theSubjectPortfolio];
DELETE FROM PortfolioTransitiveClosure
WHERE descendantPortfolio = [theSubjectPortfolio];
```

- **Add parent portfolio**. The following code adds the records that combine the ancestors of the parent with the child portfolio (the record that is the current subject of the *portfolio information* form). Then it adds the records that combine the parent with the child's descendants. When the user changes an existing parent portfolio to some other portfolio, two commands occur for the *portfolio information* form: First delete the existing parent portfolio (next bullet) and then add the new parent portfolio.

```
INSERT INTO PortfolioTransitiveClosure
   (ancestorPortfolio, descendantPortfolio)
SELECT ancestorPortfolio, [theSubjectPortfolio]
FROM PortfolioTransitiveClosure
WHERE descendantPortfolio = [theParentPortfolio] AND
   ancestorPortfolio <> descendantPortfolio;
INSERT INTO PortfolioTransitiveClosure
   (ancestorPortfolio, descendantPortfolio)
SELECT [theParentPortfolio], descendantPortfolio
FROM PortfolioTransitiveClosure
WHERE ancestorPortfolio = [theSubjectPortfolio];
```

- **Delete parent portfolio**. Delete the records that combine the parent's ancestors with the child portfolio (the record that is the current subject of the *portfolio information* form). Then delete the records that combine the parent with the child's descendants.

```
DELETE * FROM PortfolioTransitiveClosure
WHERE descendantPortfolio = [theSubjectPortfolio] AND
   ancestorPortfolio <> descendantPortfolio;
DELETE * FROM PortfolioTransitiveClosure
WHERE ancestorPortfolio = [theParentPortfolio] AND
   descendantPortfolio IN
  (SELECT descendantPortfolio
   FROM PortfolioTransitiveClosure
   WHERE ancestorPortfolio = [theSubjectPortfolio]);
```

- **Parent portfolio pick list**. The following code computes the list of portfolios that can be added as a parent without causing a cycle. The potential parent portfolios cannot be descendants of the child portfolio. (We renamed *FinancialInstrument.name* to *FinancialInstrument.financialInstrumentName* to avoid an MS-Access reserved word.)

```
SELECT financialInstrumentName AS portfolioName,
   financialInstrumentID AS portfolioID
FROM FinancialInstrument
WHERE financialInstrumentType = "Portfolio" AND
  (financialInstrumentID = [theParentPortfolio] OR
   financialInstrumentID NOT IN
     (SELECT descendantPortfolio
      FROM PortfolioTransitiveClosure
      WHERE ancestorPortfolio = [theSubjectPortfolio]))
ORDER BY financialInstrumentName
UNION
SELECT NULL, NULL FROM FinancialInstrument;
```

The cycle-avoidance logic is suitable only for certain kinds of problems. We expect the logic can handle up to several thousand portfolios, which is much higher than we would expect to encounter in practice. For more portfolios, the transitive closure table could become quite large and performance may degrade. Also, our code is not intended for deep trees. For wide trees (five or more children per parent), the transitive closure table has $n \log(n)$ records. However, for a deep degenerate tree in which every parent has one child, the transitive closure table would be large and have $n \times n$ records.

Calculating Portfolio Value

Our computation for portfolio value avoids recursion by using the transitive closure table. Our algorithm is much different from that implied by the analysis specification, but analysis specifies only *what* is desired, not *how* to accomplish it. The latter is decided during design and implementation when algorithms crystallize.

We compute portfolio value for a specified date. First we collect the *assetTransacted* and *transactionFee* records that occur before the specified date for the subject portfolio and its descendants. The sum of all these records is the net composition of the portfolio on the specified date, which we store in a temporary table. We then insert the net quantity multiplied by asset value for noncurrency assets into a second temporary table. To this second temporary table we add the net quantity for currency assets. Finally, we sum the data from the second temporary table and insert a record into the portfolio value table.

We also check for missing computation date, missing computation currency, and an asset that is missing a valuation on the computation date. Our use of the temporary tables is motivated by MS-Access restrictions and a desire to simplify calculations.

Calculating Return on Investment

In this algorithm, we first build a cash flow table that lists the value of assets added or removed from a portfolio between the specified dates. The table also lists the dates of addition or removal as well as the value of the portfolio on the starting and ending dates. We then assume an ROI of 0% and compute the present value of the cash flow. We next assume an ROI of 20% and again compute the present value of the cash flow. We iteratively compute more accurate ROIs with the approximation that present value is proportional to ROI. Ultimately we terminate computation when the present value of the cash flow is within 0.1% of the ending portfolio value. For an exact ROI the present value of the cash flow would be zero. The portfolio manager returns an error message if the computation does not converge within 10 iterations.

Ensuring That Derived Data Is Correct

The portfolio manager stores some derived data—portfolio value and ROI. Earlier during system design, we decided to store portfolio value and ROI to make browsing easier. Whenever the *Transaction* table, *AssetValue* table, or portfolio composition are updated, the portfolio manager deletes any portfolio values and ROIs affected. In this way, it ensures that all derived data stored in the database is correct.

For portfolio value and ROI, the portfolio manager stores only one number as the result of each calculation. In contrast, the calculation of portfolio composition on a given date yields a collection of assets. Because it is multivalued, portfolio composition is not stored; instead the portfolio manager computes it upon request by replaying the transaction log.

14.8 OTHER FUNCTIONAL MODELING ISSUES

14.8.1 Functional Model Tables

Most tables arise from the object model, but you can also use tables to implement portions of the functional model. For example, you can readily implement decision tables with tables.

You can also implement a function by precomputing values. Mathematical and engineering handbooks often tabulate selected values for complicated functions. Furthermore, you can define the inputs to a function to be a candidate key, ensuring that each combination of inputs yields a single output. Often you can extend a table's usefulness by applying interpolation and extrapolation to determine values not explicitly listed.

14.8.2 Performance Tuning

You can use several techniques to improve the performance of RDBMS applications. Performance has many dimensions, including elapsed time, CPU time, I/O time, disc space consumption, memory consumption, and communications delay. Aim to implement an application simply at first and consider performance tuning only when bottlenecks become apparent.

- **Create indexes**. Indexes are the primary data structure RDBMSs provide for tuning performance. Section 14.3 presented guidelines for defining indexes.
- **Finesse queries**. RDBMSs can choose different algorithms for different phrasings of the same conceptual query. In principle, RDBMSs are intended to be nonprocedural; you state what is desired and the RDBMS determines which algorithm to use in accessing the data. However, in practice, RDBMS optimizers are imperfect and you can improve an application by finessing the optimizer to choose the best algorithm for a query. The next three bullets elaborate this point.
- **Use explicit joins rather than nesting**. Most RDBMSs optimize explicit joins better than equivalent nested queries. For example, as Figure 14.7 shows, we could use joins to find the portfolios that are held by a financial institution. Alternatively, we could find the portfolios with a nested query, but a nested query often performs more slowly than an equivalent query using only joins. Use nesting only to express queries you cannot reasonably state with joins.

Recommended query with explicit joins:

```
SELECT financialInstrumentName AS portfolioName
FROM (FinancialInstrument JOIN Portfolio ON
   financialInstrumentID = portfolioID) JOIN
   FinancialInstitution ON accountHolder =
   financialInstitutionID
WHERE financialInstitutionName = :anInstitutionName
```

Discouraged query with nested queries:

```
SELECT financialInstrumentName AS portfolioName
FROM FinancialInstrument JOIN Portfolio ON
   financialInstrumentID = portfolioID
WHERE accountHolder IN
  (SELECT financialInstitutionID
   FROM FinancialInstitution
   WHERE financialInstitutionName = :anInstitutionName);
```

Figure 14.7 Use explicit joins instead of nested queries where possible

- **Define intermediate tables**. You can influence the choice of algorithm by defining intermediate tables. For example, *sometimes* it is helpful to decompose an SQL query manually with a join of three or more tables into multiple queries with intermediate tables. When an RDBMS processes joins, it spawns intermediate tables without indexes. By subdividing a query, you can index the intermediate tables and possibly speed execution. However, since temporary tables clutter your code and database schema and complicate multiuser access, use them only when the performance improvement is compelling.
- **Use precomputation**. You may be able to increase efficiency by precomputing a common portion of several queries. For example, you may decide to maintain derived data, as we have done with the portfolio value and ROI calculations for the portfolio manager.
- **Check the RDBMS query optimizer**. Many RDBMSs will let you check their decisions for evaluating SQL expressions.
- **Avoid insert lock contention**. Your may experience lock contention if you frequently insert records and the RDBMS writes to the same database page. For example, Oracle's default behavior is to write records to the last page allocated for a table. This contention can be especially troublesome with page locking. You can resolve this problem by clustering the table on one or more attributes that spread inserted records across pages.
- **Use functions carefully**. You should carefully use functions on attributes because they can preempt the use of indexes during query optimization. Most RDBMSs will forgo the use of indexes for attributes that are subject to a function in a *where* clause.
- **Do physical tuning**. The database administrator can manipulate various RDBMS parameters that affect an application's efficiency. For example, the DBA can adjust the size of the buffers for caching disc I/O. The DBA can also define additional indexes to support frequent value-based queries. The tasks for database administration vary greatly and are outside the scope of this book.

14.8.3 Static versus Dynamic SQL

Most RDBMSs support both static and dynamic SQL. Normally, you should use static SQL. It has several advantages:

- **Simplicity**. Static SQL is simpler to understand and program than dynamic SQL. Most RDBMSs have a straightforward protocol for interfacing static SQL variables and statements with application programming code. In contrast, dynamic SQL is cumbersome to use because you must manually bind variables and invoke a sequence of procedures to execute an SQL statement.
- **Locking**. The throughput of many RDBMS applications is limited by locking contention. Often the bottleneck is not the actual data, but rather the data description in the data dictionary. Static SQL checks much of the data dictionary at compile-time. Dynamic SQL defers checking to run-time and is consequently more likely to suffer delay.
- **Performance**. Static SQL is faster than dynamic SQL. An RDBMS preprocesses static SQL commands at compile-time. The preprocessing varies but usually consists of pars-

ing the commands, validating permissions, optimizing the commands, and choosing the execution plan. At run-time the RDBMS need execute only the chosen algorithms. In contrast, an RDBMS defers all processing of dynamic SQL commands until run-time.

Dynamic SQL has compelling advantages for some complex applications:

- **Flexibility**. Dynamic SQL is more flexible than static SQL. Because the application need not provide the query until run-time, the formulation of an SQL statement can reflect the current program context.
- **Generic code**. Dynamic SQL enables generic code. You can write generic methods that inspect metatables and use the metadata to construct queries for accessing the actual data. A few powerful generic methods can substitute for the many specific methods necessary with static SQL.

  For example, [Premerlani-90] describes a simple OO-DBMS that we built on top of an RDBMS. The OO-DBMS supports a variety of operations, including *createNewObject*, *putAttributeValueInObject*, and *getAttributeValueFromObject*. We implemented these operations with static SQL code by writing a compiler that generated a method for each class and operation. Eight operations were defined for classes (and more operations for associations and miscellaneous purposes). Thus for an application with 50 classes, we generated 400 methods, each with its own static SQL code.

  We considered an alternative approach that required only one generic method per operation. We would have passed the class name as an argument to the generic methods and dynamically constructed the SQL string for execution. For an application with 50 classes, we would have needed only a total of eight methods for the classes. We chose the static SQL approach because the intended application was an interactive diagram editor that required fast execution.

You will need dynamic SQL if you are developing an application framework. (See Chapter 4.) You will also need it if you have operations that pertain to many classes and you choose to write a few generic methods rather than many specific static SQL methods.

14.8.4 SQL Style Guidelines

Your SQL code should observe the following style guidelines:

- **Explicitly list target attributes**. Do not use *select* * in an application program. Explicitly listing the target attributes for an SQL query makes it easier to see how your code depends on the RDBMS schema. Explicit target attributes also isolate the program from minor schema changes, such as reordering or adding attributes.

  Also, explicitly list the target attributes for an *insert* command.
- **Use SQL-92 join syntax**. Use the modern syntax that places join logic in the *from* clause, rather than the obsolete construct that places the join logic in the *where* clause.
- **Establish a naming convention**. Distinguish programming variables from database variables. In our Oracle applications, for example, all Oracle variables, and only Oracle variables, begin with the prefix *O_*.

- **Set multiple attributes with an update.** Try to avoid issuing *update* commands that set one attribute at a time for an object. Instead try to set all relevant attributes for an object at once.
- **Filter within SQL.** As much as possible, apply predicates within SQL, not within an application program. For example, a reasonable program could contain the statement *select portfolioType from Portfolio where portfolioID = :aPortfolio*. In contrast, only a poorly conceived program would run the statement *select portfolioType from Portfolio* and then successively test each record to find the *aPortfolio* record.

  Filtering within SQL reduces communication between the RDBMS and the application program, improving performance. It also simplifies your program.
- **Use views.** Sometimes you can use one or more views to divide a complicated SQL query into several more understandable statements. Views are not evaluated until run-time, so you are still giving the RDBMS broad latitude for optimization.

  You can also use views to finesse some of the limitations of SQL. SQL-92 allows a table expression in the *from* clause of a query; a select expression itself can be the target of a *from* clause. Many current products do not support this syntax yet, but you can always define a view and use the view as the target of a *from* clause.
- **Use synonyms.** Use synonyms (indirection) in referring to tables rather than directly burying ownership in DBMS queries.

14.9 PHYSICAL IMPLEMENTATION ISSUES

The material in this book will help you significantly in developing RDBMS applications. Nevertheless, we do not address all the issues you must consider. Specifically, physical implementation issues are outside the scope of this book. For example, you have the option of turning auditing on or off. Most RDBMSs also have a configurable number of connections; each connection consumes system resources, but multiple connections can more efficiently overlay computation and reduce locking and I/O delays.

Furthermore, security is also important—granting permissions to users and defining views to control data. Often the best approach is to define privileges for prototypical users such as clerks, salespersons, managers, and system administrators. This is the way Sybase handles security. Each actual user then receives the permissions of a prototypical user. All these implementation details vary greatly among RDBMS products.

14.10 LESSONS LEARNED WITH THE PORTFOLIO MANAGER

We did not anticipate all operations during analysis and design and some operations became apparent during implementation. Nevertheless, we were able to extend the portfolio manager easily to accommodate the new operations. This has been typical of our experience with

OMT models. If you construct a sound object model and thoroughly consider the major aspects of behavior, you can readily add minor operations as you encounter them.

We have not thoroughly tested the portfolio manager software, so it may still have a few bugs. After all, the portfolio manager is just a device to illustrate the principles of this book. However, when we have developed software for real applications, we have found that thorough modeling makes it straightforward to fix bugs.

For the case study we decided to store derived data—portfolio value and ROI—to provide a better user interface. In our testing we have been pleased with the resulting software; it is indeed helpful to save past computations of portfolio value and ROI and be able to browse them. Nevertheless, the derived values complicated implementation more than we had expected. It is easy to overlook an update that can affect the derived data. We recommend that you carefully reconsider any decisions to use derived data in your applications.

Initially we had a problem with the accuracy of calculations. For example, calculations of portfolio value for simple portfolios had more errors than we expected (portfolios with five simple assets had errors as high as several tenths of a cent). When we changed all quantity attributes from single precision to double precision, the problem went away.

We could improve the portfolio manager in many ways. For example, we could refine the user interface by customizing the MS-Access menus and consulting a user interface expert. For a polished product we would certainly need to add context-sensitive help and interfaces to electronic sources of data (asset values and transaction data especially) to reduce data entry. We could also extend functionality for currency conversion and correct for the approximate effects of inflation and taxes.

14.11 CHAPTER SUMMARY

You should rigorously use referential integrity to avoid dangling references for associations and generalizations. You should declare primary and candidate keys; if the RDBMS does not create the corresponding unique indexes, you should explicitly do so. Furthermore, you should define indexes for all foreign keys so traversal of associations will be fast in your implementation, as the object model implies.

Indexes are important for performance tuning. But do not overlook the possibilities for manipulating the RDBMS query optimizer. Occasionally, you will have to define intermediate views to be able to express complex queries. Sometimes you will also find it helpful to define intermediate tables to guide the query optimizer. Sometimes it is helpful to use tables to implement portions of the functional model. For example, you can use tables to implement decision tables and discrete functions.

We presented an implementation of the portfolio manager using MS-Access 2.0. Our implementation is prototype software intended to illustrate the eminent practicality of taking object models and driving them through to the actual implementation.

Figure 14.8 lists the key concepts for this chapter.

allocating storage	dynamic SQL	performance tuning	static SQL
candidate key	foreign key	primary key	
constraint	index	referential integrity	

Figure 14.8 Key concepts for Chapter 14

BIBLIOGRAPHIC NOTES

[Celko-95] is a good source of information for advanced SQL queries. Celko presents many useful canonical queries that finesse SQL, such as queries for sequences, gaps, statistics, sets, trees, and graphs.

[Shasha-92] provides much useful advice for tuning databases and complements our coverage. Most of Shasha's advice transcends products, though he also discusses some product-specific details. His primary focus is RDBMSs, although he also discusses OO-DBMSs.

REFERENCES

[Celko-95] Joe Celko. *SQL for Smarties: Advanced SQL Programming*. San Francisco, California: Morgan Kaufmann, 1995.

[Melton-93] Jim Melton and Alan R. Simon. *Understanding the New SQL: A Complete Guide*. San Francisco, California: Morgan Kaufmann, 1993.

[Premerlani-90] WJ Premerlani, MR Blaha, and JE Rumbaugh. An object-oriented relational database. *Communications ACM 33*, 11 (November 1990), 99–109.

[Shasha-92] Dennis E Shasha. *Database Tuning: A Principled Approach*. Englewood Cliffs, New Jersey: Prentice Hall, 1992.

EXERCISES

14.1 For Exercises 13.1, 13.3, and 13.5 through 13.17, do the following. (The *Solutions Manual* answers parts a and b only.)

a. (7) List referential integrity actions for each association.

b. (7) Write SQL code for each table, defining candidate keys and a primary key. Also define indexes for each foreign key that is not subsumed by a primary key or candidate key constraint. Write SQL code for referential integrity actions.

c. (3) Execute the SQL code to create schema for an RDBMS.

d. (4) Execute the SQL code for the methods that you wrote for Chapter 13.

e. (8) Use a 4GL and quickly prototype the application.

14.2 Write SQL code to implement the pseudocode from your answers to the following exercises:

a. (5) Exercise 10.4, parts d, f, g, and h.

b. (5) Exercise 10.5, parts e, f, g, and i.

c. (7) Exercise 10.5, part h.

14.3 (9) Given pairs of parent and child parts, write enhanced pseudocode for an operation that generates a bill-of-material (BOM) in depth-first order. Such an operation is frequently needed for BOM applications. You should order sibling parts by ascending part number. Figure E14.1 shows sample input and output. This exercise illustrates the use of a DBMS as a computation engine.

Hint: Define *CrossReferenceTable (parentPart, childPart, quantity)* to hold the input. Define *PartTable (seqNumber, level, partNumber, quantity)* to hold the output. Assume that the maximum levels of a BOM hierarchy is 5 and the maximum number of siblings is 99. Then *seqNumber* requires two characters for each level for a total of 10 characters. (You should define *maxLevels* and *maxSiblings* as global constants to reduce the effort in extending the method to handle deeper and wider trees.) You should process the BOM hierarchy a level at a time using SQL and database cursors. You can then print a BOM in depth-first order by selecting from *PartTable* ordered by *seqNumber*.

Input: CrossReferenceTable

parentPart	childPart	quantity
LM16G	B16M	1
LM16G	E1	1
LM16G	W3	4
LM16G	D16	1
E1	G1	1
E1	F35	1
F35	P123	1
W3	V2	1

execute operation orderDepthFirst (treeRoot, inputTable, outputTable)

Output: PartTable

seqNumber	level	partNumber	quantity
0100000000	1	LM16G	1
0101000000	2	B16M	1
0102000000	2	D16	1
0103000000	2	E1	1
0104000000	2	W3	4
0103010000	3	F35	1
0103020000	3	G1	1
0104010000	3	V2	1
0103010100	4	P123	1

execute SQL statement . . . select level, partNumber, quantity from PartTable order by seqNumber;

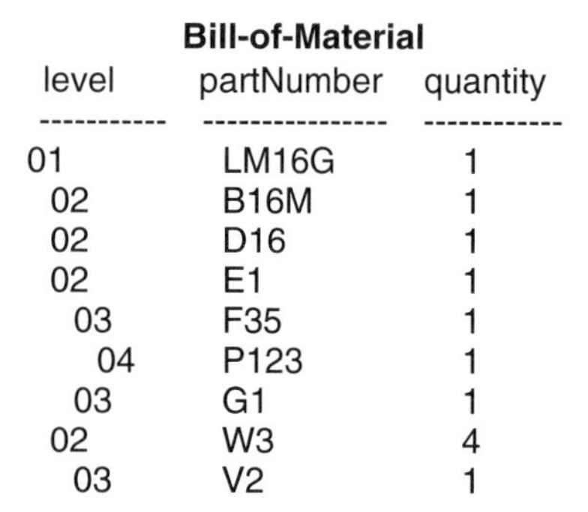

Bill-of-Material

level	partNumber	quantity
01	LM16G	1
02	B16M	1
02	D16	1
02	E1	1
03	F35	1
04	P123	1
03	G1	1
02	W3	4
03	V2	1

Figure E14.1 Computing a topological sort for a bill-of-material

15

Object-Oriented Databases: Basics

In the last two chapters we described how to take the optimized models from detailed design and implement a relational database. Now we describe how to take the same models and implement an object-oriented database. OMT models are easier to implement with an OO-DBMS, because the concepts and paradigm closely match. The RDBMS paradigm requires greater translation effort. We are not saying it is necessarily better to use an OO-DBMS; each type of DBMS has different strengths and weaknesses for a particular application. This chapter shows how to implement basic OMT constructs. Chapter 16 describes advanced issues.

To implement the portfolio manager, we used the ObjectStore DBMS (release 4.0), a product of Object Design Inc. OO-DBMSs vary widely so no OO-DBMS is truly representative. We chose ObjectStore because it is a radical departure from relational DBMSs and a dominant OO-DBMS product. Our purpose is not to evaluate OO-DBMSs and ObjectStore, but rather to focus on how to implement OMT models with ObjectStore. We believe much of our advice will transcend ObjectStore and apply to other OO-DBMS products.

We have only partially implemented our case study with ObjectStore; it would be too tedious to implement the user interface without a complete class library, which is beyond the scope of this book. Instead we show fragments of the case study and focus on the proper ways to use ObjectStore. All our examples are in C++, and we assume you are familiar with this language.

A detailed discussion of physical implementation issues is also outside the scope of this book because such issues tend to be product specific. Our goal is to provide advice that transcends products. Moreover, for large, complex applications, you must think abstractly, in terms of models, or you will become overwhelmed. After sound modeling, you then have the luxury of addressing physical details.

15.1 INTRODUCTION TO OBJECTSTORE

We now summarize important features for readers who are unfamiliar with ObjectStore. Unlike relational databases, the focus of ObjectStore applications is procedural code that manipulates objects. ObjectStore supports both C and C++, but most developers prefer to use an object-oriented language with an object-oriented database.

15.1.1 Support for the Object Model

ObjectStore supports all fundamental object modeling concepts:

- **Classes**. Objects and classes are the central concepts for ObjectStore. Objects have existence-based identity. You should encapsulate classes as much as possible and clearly define dependencies on other classes.
- **Attributes**. ObjectStore directly maps attributes to the underlying language constructs. Normally, you should not directly access attributes, but should instead use read and write methods. ObjectStore does not directly support derived attributes, so you must write explicit methods to compute them.
- **Methods**. ObjectStore provides language capabilities with minimal performance overhead and compiles methods into executable code.
- **Associations**. ObjectStore supports simple binary associations but does not support advanced concepts such as link attributes, qualifiers, and association classes. We suggest appropriate workarounds in Chapter 16. You should treat aggregation like an ordinary association.
- **Inheritance**. As you would expect, ObjectStore supports both single and multiple inheritance. A subclass inherits the attributes, methods, and associations of its ancestors. ObjectStore does not implement discriminators; they are not needed because you are dealing with coherent objects, not fragmented records, as with an RDBMS. ObjectStore supports parameterized classes, which obviates the need for inheritance in some situations.

15.1.2 Persistent and Transient Objects

Objects in ObjectStore applications may be persistent or transient. A ***persistent object*** is an object that is stored in the database and can span multiple application executions. A ***transient object*** is an object that exists only in memory and disappears when an application terminates execution. Thus a transient object is an ordinary programming object. The same class can have both transient and persistent instances.

ObjectStore databases relate objects with persistent pointers and ObjectStore references. An ObjectStore reference is not the same as a C++ reference. It is an ObjectStore parameterized class that can be used syntactically as a pointer but that lets you override default restrictions on referring across databases and transactions. References incur some extra cost; they are larger than pointers and dereferencing usually involves a table lookup.

You can access transient objects as you would any other programming object or you can use ObjectStore queries (Chapter 16). You can access persistent objects via navigation from other objects (Section 15.11), by using *database roots* (Section 15.12.1), or with queries. Database roots are named entry points into the database that ObjectStore remembers.

You can create transient objects with any of the available C++ mechanisms (in static memory, on the stack, or on the heap). You can create persistent objects by invoking the *new* operator with an argument of the appropriate database, segment, or cluster.

In general, all the properties of types apply equally to transient and persistent objects, including inheritance and method application. However, you must follow some rules for pointers, depending on whether the objects being referenced are transient or persistent in the same or different databases. The rules in Table 15.1 indicate when pointers are valid.

Kind of pointer	When pointer is valid
Transient pointer to transient object	Valid during program execution only
Transient pointer to persistent object	Valid for a single transaction or for the duration of an application, depending on the specified options
Persistent pointer to persistent object in the same database	Always valid
Persistent pointer to persistent object in a different database	Valid for a single transaction or multiple transactions, depending on the specified options
Persistent pointer to transient object	Valid for a single transaction. You must set these pointers to null or convert them to point to a persistent object before the commit of a transaction.

Table 15.1 Rules for transient and persistent pointers

15.1.3 Collection Classes

A class library is an important component of any object-oriented development environment and consists of carefully polished and reusable code. Collection classes are some of the most important classes in a class library.

A ***collection*** is an object that aggregates other objects; the other objects can be lesser collections and primitive objects. ObjectStore supports both heterogeneous collections (possibly different types) and homogeneous collections (the same type). ObjectStore includes a convenient protocol for iterating over elements of a collection; you can add, remove, or update elements during iteration. ObjectStore supports persistent and transient collections subject to the restriction that a persistent collection can contain only persistent elements.

Collections serve many purposes. For example, you can use collections to implement associations. ObjectStore allows queries on collections of objects. We use collections to implement portions of the Object Navigation Notation. For database applications, it is common to define an ***extent***, a collection with all the instances for a class. (See Chapter 16.) Collections are also useful for implementing multivalued attributes.

ObjectStore supports the following collection classes:

- A ***set*** is an unordered collection of objects without duplicates. Insertion causes an object to be added to a set if it is not already an element. Removal eliminates an object from a set. Iteration accesses each object in an arbitrary order.
- A ***bag*** is an unordered collection of objects with duplicates allowed. Insertion causes an object to be added to a bag. Removal eliminates one occurrence of the object in the bag. Iteration accesses each object in an arbitrary order.
- A ***list*** is an ordered collection of objects with duplicates allowed. Insertion and removal is performed at the specified location in the list. Iteration accesses each object in the specified order.
- An ***array*** is an indexed, ordered collection of objects with duplicates allowed. When an array is created, its elements are initialized to null. The basic operations, read and write, are done at a particular location in the array. The array size does not change as individual objects are accessed, but is controlled separately. Objects in an array may be accessed through iteration or individually by location.
- A ***dictionary*** is an unordered collection of objects with duplicates allowed. A dictionary is different from the previous collection classes because it associates a key with each element. The key can be a value of any C++ fundamental type or pointer type. When you insert an element into a dictionary, you specify the key along with the element. You can retrieve an element with a given key or retrieve elements whose keys falls within a given range. You can delete the combination of an element and its key or you can delete all elements for a given key. Iteration accesses each object in an arbitrary order.

ObjectStore supports set-theoretic operations on collections including intersect, union, subtract, subset, and compute cardinality.

15.1.4 Database Features

ObjectStore supports both traditional and advanced database features.

- **Concurrency**. ObjectStore locks each disc page that it reads and writes (pessimistic locking). Page locking provides excellent performance. However, for some applications page locking provides less concurrency than individual object locking.
- **Crash recovery**. ObjectStore has the standard capabilities for logging data and recovering from software, computer, and disc crashes.
- **Security**. As with many OO-DBMSs, ObjectStore enforces security with the underlying file system. In contrast, RDBMSs provide tighter security with grant and revoke commands and the view mechanism.
- **Data dictionary**. The ObjectStore data dictionary is available, but is difficult for programs to access at run-time. This difficulty is due largely to C++, which compiles away the metadata stored in a data dictionary. If you have an advanced application, ObjectStore has features that let you circumvent this C++ limitation.

- **Transactions**. Transactions can involve any combination of persistent and transient objects and more than one database. For persistent objects, either the entire transaction is written to the database(s) or nothing is written. Transactions maintain database consistency despite concurrent access and occasional hardware and software failures. ObjectStore requires that all manipulations of persistent objects occur within a transaction. It supports both conventional and nested transactions.

  A ***nested transaction*** is a transaction placed within another transaction. A nested transaction can succeed or fail or at local level, but when grouped with additional DBMS commands, it can still fail within a broader context. This gives you finer control and lets you rollback to intermediate states. For example, you may decide to abort a hotel reservation if you cannot make a complete set of reservations for a trip.

- **Indexes**. ObjectStore indexes can speed the response to certain queries. You can create one or more indexes on a collection. ObjectStore transparently maintains indexes and uses them to improve the performance of navigation and queries. You can create or drop an index anytime during the execution of an application.

  When you define an index, you can use ObjectStore's *path* mechanism.* An ObjectStore path is a sequence of traversals from an attribute of the objects in the collection to an attribute in the index. For example, a *person* object could have a pointer to an *address* object with attributes *street*, *city*, and *state*. You could create an index on a collection of persons using a path to state.

  We recommend against using multiple-attribute indexes to improve performance with ObjectStore. They are complex and provide little benefit over using multiple single-attribute indexes.

  Furthermore, ObjectStore lets you define unique and nonunique indexes. It uses an exception handling mechanism to deal with violations of unique indexes. In Chapter 16 (Section 16.1.2) we present an approach to enforcing uniqueness that is especially helpful for multiple-attribute candidate keys.

- **Queries**. ObjectStore provides queries for retrieving objects. A ***query*** retrieves a collection of objects by specifying predicates on the attribute values of the desired objects. Essentially, you describe the conditions to be satisfied and leave the choice of algorithms to the query optimizer. ObjectStore allocates memory from the heap for the returned collection, so you must delete these returned collections when they are no longer needed to avoid memory leaks. Section 16.5 presents some examples of queries.

- **Navigation**. You can also retrieve ObjectStore objects with navigation, by traversing a chain of pointers such as *(obj1->obj2->obj3->obj4)*. With a navigational query you specify the actual computation steps.

- **Schema evolution**. Schema evolution refers to migration of data across changes in data structure. Developers must simultaneously preserve past data while extending the data

---

* An ObjectStore path is not the same as an OMT path (see Chapter 3). For an ObjectStore path each traversal must yield at most one object and cannot yield a set.

structure for new requirements. Schema evolution is a difficult theoretical problem that is only partially understood.

ObjectStore provides some support for schema evolution by maintaining separate metadata for both the database schema and the application schema. ObjectStore issues warnings during compilation when an application redefines the database schema. It migrates the data to a new schema in two phases. First, it modifies pointers and references for data that has moved and marks affected indexes and queries as obsolete. Second, it executes custom code (which you have written) to resolve any remaining problems.

- **Distribution**. Persistent objects are stored in databases, with each object in exactly one database. ObjectStore does not support replication. If you need to replicate your data, you must write extra application code.

  ObjectStore has been optimized for the client-server architecture. Each database is managed by a single server, and a server can manage multiple databases. An application can have multiple servers, clients, and databases.

- **Client-centric computing**. ObjectStore adopts a client-centric approach to client-server computing and strives to place the computing burden on the client rather than on the server. In contrast, RDBMSs provide a server-centric approach, with the database processing burden on the server. A client-centric DBMS can support more clients and scale up more gracefully than a server-centric DBMS.

15.2 IMPLEMENTING THE OBJECT MODEL

The first step in implementing an application is to map the object model to ObjectStore constructs. Mapping consists of the following tasks. (The numbers in brackets refer to sections that explain these tasks in detail.)

- **Choose persistent objects**. Decide which objects need to be persistent and appropriately declare them.
- **Implement domains**. Map domains to C++ type declarations. Write methods for domains. Place the type declarations in a header file to be included by any file referencing the domain. [15.3]
- **Implement classes and associations**. Prepare C++ declarations for each class and advanced association in the model. Declare all attributes and methods for a class, including the implementation of associations. We recommend a separate file for each class declaration. [15.4–15.7]
- **Create indexes**. Create single-attribute indexes to reduce performance bottlenecks. Chapter 16 presents a uniform technique for implementing single-attribute and multiple-attribute candidate keys.
- **Define extents**. Finally, you can define the appropriate extents. We present an approach to extents that works for both persistent and transient objects. [Chapter 16]

15.3 IMPLEMENTING DOMAINS

Each attribute and method variable should have an assigned domain from detailed design (see Chapter 10). ObjectStore does not directly support domains, but it can readily implement them. Domain implementation has four tasks.

First, you should define a datatype for each domain and give it the same name as the domain. For example, you could map domain *money* to typedef *long int money* and domain *longString* to typedef *char longString*.

The next task is to choose one of four techniques to access each attribute or method variable:

- **Direct access**. If you declare *datatypeName attributeName*, the compiler allocates memory for the attribute. You can use the C++ syntax *anObject.anAttribute* to access the attribute value.

  This technique is useful for accessing attributes within objects and variables that refer to transient objects. You may not use direct access for variables that refer to persistent objects, because ObjectStore requires that all memory for persistent objects be allocated at run-time.

- **Pointer**. If you declare *datatypeName * attributeName*, the compiler allocates memory for a pointer. The compiler does not allocate memory for the attribute value itself. The C++ syntax for accessing the attribute value is **anObject.anAttributePntr*.

- **ObjectStore reference**. If you declare *os_Reference<datatypeName> attributeName*, the compiler allocates an ObjectStore reference. An ObjectStore reference is a pointer that is valid across transactions or databases. The C++ syntax for accessing the attribute value is **anObject.anAttributePntr*.

- **C++ reference**. If you declare *datatypeName & attributeName = initializer*, the compiler allocates memory for a pointer. The compiler does not allocate memory for the attribute value itself. The C++ syntax for accessing the attribute value is *anObject.anAttributeReference*.

The third task is to decide if the datatype should be read-only. For example, you can improve the efficiency of some methods by returning a pointer to a private collection of objects. You can include the keyword *const* in the declaration of the method datatype to prevent the caller of the method from accidently modifying the collection. The compiler catches any attempts to modify read-only datatypes.

The last task is to write domain methods. For example, the portfolio manager implementation requires arithmetic and comparison methods for the *money* and *date* domains. Often it is helpful to write print methods that add units of measure or decode abbreviated values.

We now discuss several special kinds of domains.

15.3.1 Identifier Domain

ObjectStore has a built-in mechanism for identifying objects. You simply define the type of a variable or an attribute that identifies an object to be that object's class, and ObjectStore can then automatically identify the object. Similar to the four ways for accessing datatypes, there are four ways of identifying an object with ObjectStore in C++:

- **Direct access**. Simply identify an object by its name. Then, for example, you would access a direct attribute with *anObject.anAttribute* and a pointer attribute with **anObject.anAttributePntr*.
- **Pointer**. Identify an object with a pointer. Then you would access a direct attribute with *anObjectPntr->anAttribute* and a pointer attribute with **anObjectPntr->anAttributePntr*.
- **ObjectStore reference.** Use an ObjectStore reference. The access syntax is the same as for pointers.
- **C++ reference**. Identify an object with a C++ reference. Then you would access a direct attribute with *anObjectReference.anAttribute*.

15.3.2 Enumeration Domains

Enumerations are essentially a C++ issue. You can implement them using techniques similar to those for RDBMSs (see Chapter 13):

- **Enumeration string**. Store the enumeration attribute as a string.
- **Enumeration encoding**. Encode each enumeration value. For example, you could implement *portfolioType* with *typedef enum { normal, 401K, IRA, otherPort } portfolioTypeEnum*.

We recommend that you encode enumerations with ObjectStore. The C++ compiler can automatically convert enumeration values into codes and store them. Methods that access the attributes can then change enumeration codes into readable strings. OO databases tend to have little adhoc browsing, and their data is normally accessed via methods. Thus the disadvantages of enumeration encoding discussed for relational databases in Chapter 13 are less relevant for ObjectStore.

15.3.3 Structured Domains

ObjectStore supports two techniques for implementing structured domains. Once again, these approaches are essentially straightforward C++.

- **Embedded structure**. If you declare the attribute type to be a class, ObjectStore includes the structure of the attribute class in the parent class. Thus for a person's address, you could embed an *Address* class in the *Person* class. Do not confuse attribute structuring with inheritance. In attribute structuring, you embed classes; in inheritance, you share classes.
- **Pointer**. You may promote a structured domain to a class and then point to the appropriate instance. For example, each *Person* object could point to the appropriate *Address* object.

15.3.4 Multivalued Domains

Because it is based on an object-oriented language, ObjectStore can readily implement multivalued domains with a set of pointers to values.

15.4 IMPLEMENTING CLASSES

We use the excerpt of the portfolio manager model in Figure 15.1 for examples in the rest of this chapter and the next.

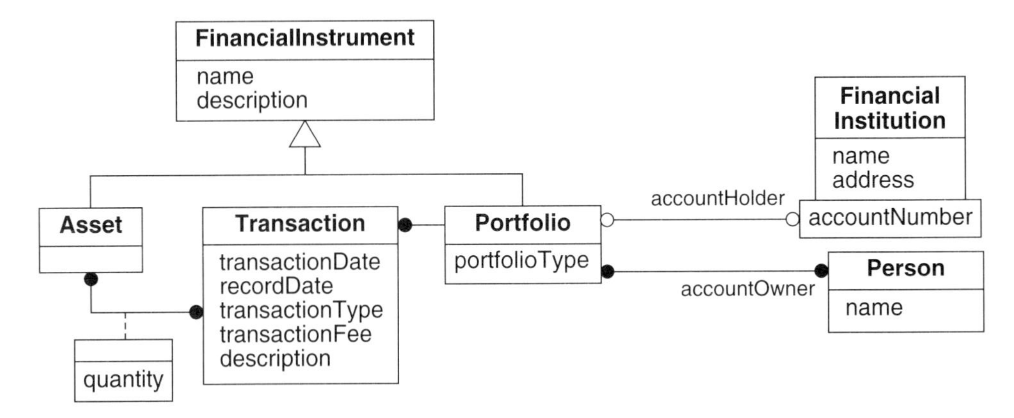

Figure 15.1 Excerpt of the portfolio manager model used for examples

A class in the object model maps to a C++ class. This implementation has three parts:

- **Declaration**. The C++ header files declare attributes from the object model, attributes for the implementation of associations, and methods.
- **Methods implied by the object model**. You must implement methods that are implicit in the object model, such as creating objects, destroying objects, accessing attribute values, and accessing association links.
- **Methods specified by the functional model**. You must implement methods that are specified by the functional model. This selection and implementation of algorithms is often the bulk of the work. We defer consideration of this step until we discuss the functional model later in this chapter.

15.4.1 Declaration for Classes

Classes in the object model map directly to C++ class declarations. You need to add attributes and methods for implementing associations, as we discuss in Section 15.6.

ObjectStore exposes the underlying C++ capability for controlling access. You may define attributes and methods to be *public*, *private*, or *protected*. A ***public*** attribute or method is globally visible and can be directly accessed by any method. A ***private*** attribute or method can be accessed only by methods of the containing class. A ***protected*** attribute or method can be accessed by methods of the containing class or any of its subclasses.

You should normally hide attributes by defining them as *protected* and accessing them via methods. For example, you should not directly modify attribute *description* of *FinancialInstrument*, but rather you should use the *getDescription* and *setDescription* methods. In contrast to attributes, most applications have all three kinds of methods: private, protected, and public.

As an example of declaration, we provide the standard C++ class declaration and method implementations for class *Person*:

```
class Person
{
public:

protected:
    name * namePntr ;
public:
    Person( name * newNamePntr = "" ) ;
    ~Person();
    const name * getName();
    void setName( name * newNamePntr );
};
```

In this code, *namePntr* is a *protected* pointer to attribute *name*. Declarations for four public methods follow. *Person* and *~Person* declare the constructor and destructor for the class. The constructor takes a single argument that is a pointer to type *name;* the default value of the argument is a zero length string. The last two method declarations access the *name* attribute. Method *getName* returns a pointer to a constant of type *name*.

15.4.2 Methods Implied by the Object Model

Methods may access transient and/or persistent data. You can implement derived attributes with a method that simply returns the value.

The implementation of the constructor, destructor, and access methods for class *Person* in the *person.cc* file uses standard C++ notation. The constructor sets the name attribute.

```
Person::Person( name * newNamePntr )
{
    this->setName( newNamePntr );
}
```

The destructor releases the memory used to store the name.

```
Person::~Person(){
    if(namePntr) delete namePntr;
}
```

Try to avoid dangling pointers when you delete objects. If you delete an object, also delete it from any collections to which it belongs; you can do this most easily by adding appropriate code to the destructor. You can often use ObjectStore inverse members to resolve dangling pointers from associations (see Section 15.6).

The *setName* method sets the *name* attribute of *Person* using a function that copies strings. (We wrote *strnewcpy*; it is not a library function.) The first argument to *strnewcpy* is a pointer to the target, the second is a pointer to the source, and the third specifies the database—a slight departure from standard C++. In this case, *os_database::of(this)* is the database that holds the instance of *Person*.

```
void Person::setName( name * newNamePntr )
{
   strnewcpy( namePntr,newNamePntr, os_database::of(this));
}
```

Finally, the *getName* method returns a read-only pointer to *name*.

```
const name * Person::getName()
{
   return namePntr ;
}
```

15.5 IMPLEMENTING GENERALIZATIONS

You can readily implement both single and multiple inheritance with ObjectStore. OO languages and OO-DBMSs are explicitly architected to provide thorough support for generalization. ObjectStore permits both abstract and concrete superclasses, as well as inheritance for both transient and persistent data. Furthermore, you can define abstract methods (C++ *virtual functions*) in which the superclass defines the protocol and the subclasses provide the implementation.

You may want to compose constructors and destructors across a generalization hierarchy. For example, the constructor for the subclass invokes the constructor for the superclass. Remember that a subclass inherits attributes and methods (and thus the associations) of a superclass.

15.6 IMPLEMENTING SIMPLE ASSOCIATIONS

ObjectStore makes it easy to implement simple binary associations, including ordered and symmetric associations. You can embed pointers or collections of pointers in associated classes. You may use bidirectional (put pointers in both classes) and unidirectional (put pointers in only one class) implementations.

We normally recommend a bidirectional implementation for two reasons:

- **Performance**. A bidirectional implementation is fast when queries can traverse an association in both directions. Bidirectional implementation can add overhead for inserts and deletes, but it is seldom significant.
- **Extensibility**. A bidirectional implementation is easier to extend than a unidirectional implementation.

ObjectStore provides *inverse members*, pairs of mutually consistent pointers, to facilitate bidirectional implementation of associations. ObjectStore maintains the integrity of inverse members; when one member is updated, ObjectStore transparently updates the other. Similarly, when an object is deleted, ObjectStore automatically updates any inverse members to prevent dangling pointers. When you use inverse members, be sure to observe the restrictions on pointers between persistent and transient objects specified by Table 15.1.

You should consider three aspects in implementing an association:

- **Implementation strategy**. Choose an approach for implementing the association.
- **Declaration**. Prepare the corresponding C++ attribute and method declarations.
- **Methods implied by the object model**. Implement methods that are implicit in the object model.

15.6.1 Implementation Strategy

The implementation strategy for simple associations depends on the multiplicity.

- **One-to-one associations**. As Figure 15.2 shows, you can put pointers in the associated classes or include one class in another. We recommend that you normally use pointers. Inclusion damages extensibility because you mix information for two separate and distinguishable classes. Furthermore, inclusion is not possible for a zero-or-one to zero-or-one association.

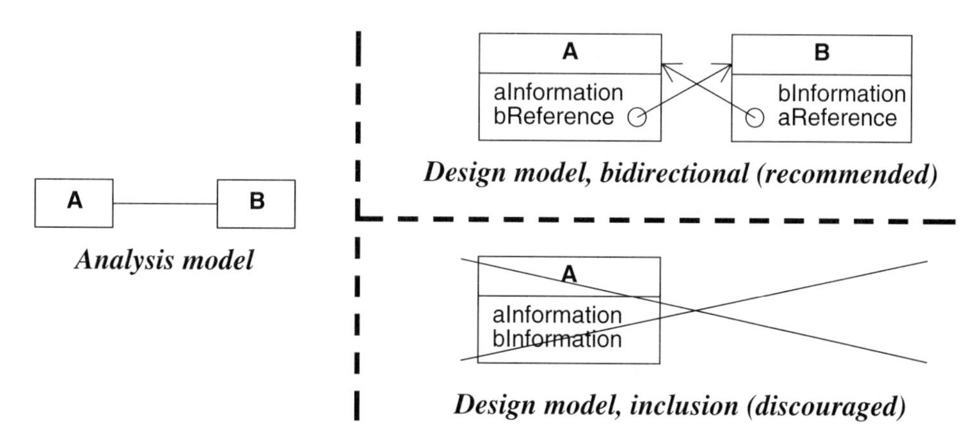

Figure 15.2 Implementation strategies: one-to-one associations

- **One-to-many associations**. You should put a pointer in the "many" class and a collection of pointers in the "one" class. In Figure 15.3 we arbitrarily show three elements for the collection of pointers. The collection is a set if the role is not ordered, and a list if it is ordered.

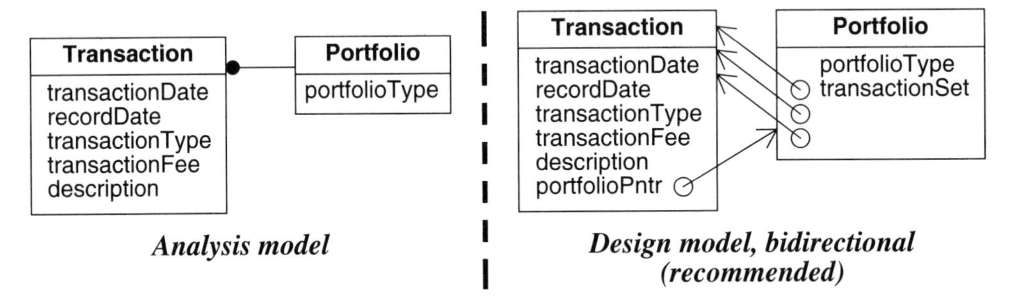

Figure 15.3 Implementation strategy: one-to-many associations

- **Many-to-many associations**. You should put collections of pointers in each of the associated classes. In Figure 15.4 we arbitrarily show three elements for the collections. Each collection is a set if the role is not ordered, and a list if it is ordered.

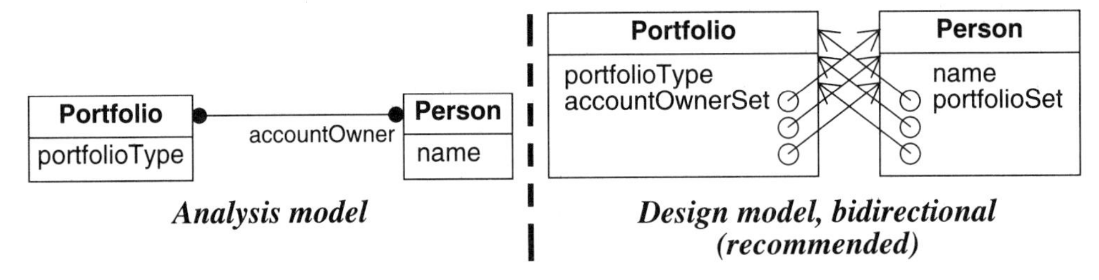

Figure 15.4 Implementation strategy: many-to-many associations

15.6.2 Declaration for Associations

You need four macro instantiations to implement a simple association. The first macro generates declarations for forward traversal of the association; the second generates declarations for backward traversal. The third macro implements forward traversal of the association, and the fourth implements backward traversal. We describe the declaration macros in this section and the implementation macros in Section 15.6.3.

Logically, there is redundancy between these four instantiations. It would be more concise to use a single macro. However, C++ makes it difficult to do this. In any case, the C++ compiler checks the four instantiations.

As Figure 15.5 shows, we place the names of the association implementation attributes as role names on the object model. Thus a bidirectional implementation will have two role names, and a unidirectional implementation will have one role name. You should use a consistent naming protocol for the attributes. Our convention is that a suffix indicates whether the attribute is an object (no suffix), a pointer to an object (*Pntr* suffix), or a set of pointers to objects (*Set* suffix).

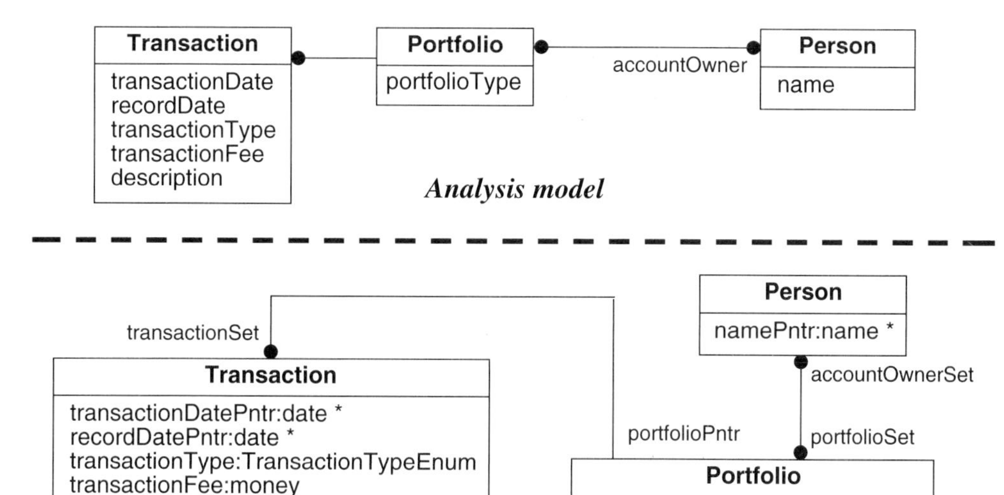

Figure 15.5 Design model with roles that implement associations

We included the following two instantiations of ObjectStore's inverse member macros in the declaration of the public attributes of the *Portfolio* class. You can use ObjectStore inverse members for public, protected, and private attributes.

```
os_relationship_m_m(Portfolio,accountOwnerSet,Person,
   portfolioSet,os_Set<Person*>) accountOwnerSet;
os_relationship_m_1(Portfolio,transactionSet,Transaction,
   portfolioPntr,os_Set<Transaction*>) transactionSet;
```

The first macro declares *accountOwnerSet* to be an attribute of the class *Portfolio* and the attribute *portfolioSet* in the class *Person* to be the inverse member. The type of *accountOwnerSet* is an *os_Set* of pointers to objects of type *Person*. (*os_Set* is the name of the ObjectStore set.)

Similarly, the second macro declares *transactionSet* to be an attribute of the class *Portfolio* and the attribute *portfolioPntr* in the class *Transaction* to be the inverse member. The type of *transactionSet* is an *os_Set* of pointers to objects of type *Transaction*. The multiplicity "m" in the name of the macro refers to *Transaction*; the "1" refers to *Portfolio*.

We included the following code in the declaration of the public attributes of the *Person* class:

```
os_relationship_m_m(Person,portfolioSet,Portfolio,
    accountOwnerSet,os_Set<Portfolio*>) portfolioSet ;
```

This declares *portfolioSet* to be an attribute of the class *Person*, with the attribute *accountOwnerSet* in the class *Portfolio* as the inverse member. The type of *portfolioSet* is an *os_Set* of pointers to objects of type *Portfolio*. Taken together with the previous declaration of *accountOwnerSet* in *Portfolio*, this completes the declarations needed for the association between *Portfolio* and *Person*.

To complete the inverse declarations for the association between *Transaction* and *Portfolio*, we included the following code in the declaration of the attributes of the *Transaction* class:

```
os_relationship_1_m(Transaction,portfolioPntr,Portfolio,
    transactionSet,Portfolio*) portfolioPntr;
```

15.6.3 Methods Implied by the Object Model

After establishing the declarations, you can implement the associations. ObjectStore macros can generate the implementation code. Code is generated for object constructors; the constructors initialize each pointer to null and each collection to empty. Also code is generated for synchronizing bidirectional implementations. This code maintains the integrity of inverse members across insertions, updates, and deletions.

For Figure 15.5 we can implement the associations with macros. The arguments to the implementation macros are the same as the first four arguments of the declaration macros. For the association between *Portfolio* and *Person* the macros are

```
os_rel_m_m_body(Portfolio,accountOwnerSet,Person,
    portfolioSet);
os_rel_m_m_body(Person,portfolioSet,Portfolio,
    accountOwnerSet);
```

For the association between *Transaction* and *Portfolio* the macros are

```
os_rel_m_1_body(Portfolio,transactionSet,Transaction,
    portfolioPntr);
os_rel_1_m_body(Transaction,portfolioPntr,Portfolio,
    transactionSet);
```

15.7 IMPLEMENTING ADVANCED ASSOCIATIONS

As we mentioned earlier, ObjectStore does not directly support advanced associations. Inverse members are insufficient for implementing link attributes, association classes, qualifiers, and ternary associations, so you require additional code. We summarize the strategy for the various kinds of advanced associations here and elaborate in Chapter 16.

One strategy you can always consider is promotion—treating an association as if it were a class. Promotion is one of the transformations (*T20*) we described in Chapter 10.

- **Link attributes and association classes**. The approach depends on the multiplicity.
  For a one-to-one or one-to-many association, you can simply migrate the link attributes to a related class. (Apply transformation *T19* in Chapter 10.) You must also provide access methods to implement the behavior of the link attributes. Alternatively, you can promote the association to a class.
  For a many-to-many association, you must promote the association to a class and add the link attributes to the promoted class. You must also add methods to navigate the resulting model and to ensure that the creation and deletion of promoted objects are consistent with those of related objects.
- **Qualified associations**. The approach depends on the multiplicity.
  For an association that is one-to-many with the qualifier removed, you can simply transform the qualifier to an attribute in the class opposite the qualifier. (Apply transformations *T18* and *T19* in Chapter 10.) You must also provide access methods to implement the behavior of the qualifier. Alternatively, you can promote the qualified association to a class.
  For a qualified many-to-one or many-to-many association, you must first promote the association.
- **Cascaded qualified associations**. Although cascaded qualified associations are advanced associations, they do not cause any special problems for ObjectStore because identity is existence based. You can implement each qualified association separately.
- **Ternary associations**. You should promote the association to a class and reduce the problem to implementing multiple binary associations. You must also implement methods for inserting, deleting, and navigating association links and maintain integrity constraints in the constructors and destructors of the participating classes.
- **Aggregation**. Aggregation is a kind of association, so you can implement aggregations as you would associations.

15.8 SUMMARY OF OBJECT MODEL MAPPING RULES

Table 15.2 summarizes the mappings for implementing the object model with ObjectStore.

15.9 IMPLEMENTING THE DYNAMIC MODEL

The dynamic model is seldom important for database applications—other than for development of user interfaces, which is outside the scope of this book. If you do encounter an application that involves the dynamic model, we offer advice on two issues:

- **External events**. Be careful not to service external events in ObjectStore transactions. Otherwise, the resulting locks may be held for an unpredictable time until the external event occurs. For example, do not ask for a user response within a transaction.

Concept	Object model construct	ObjectStore or C++ construct
Domain	Simple domain	C++ datatype
	Enumeration	C++ encoding
	Structured	C++ typedef or embed one object in another
	Multivalued	ObjectStore set of pointers to values
Class	Class	Map each class directly to a C++ class
Generalization	Single inheritance	Normal C++
	Multiple inheritance	Normal C++
Association	Simple binary association	ObjectStore inverse members with bidirectional pointers or sets of pointers
	Ordered association	Same as simple binary association, except use a list instead of a set
	Symmetric association	Same as simple binary association
	Link attribute	Promote the association or degrade it by putting the link attribute inside a class
	Association class	Promote the association to a class
	Qualified association	Promote the association or degrade it by putting the qualifier inside a class
	Ternary association	Promote the association to a class
	Aggregation	Same as association

Table 15.2 Summary of object model mapping rules for ObjectStore

- **Transaction modeling**. Some database design books recommend that you do transaction modeling, in which you characterize the queries and updates to a database, as well as their expected frequency. (See Chapter 10.) We do not agree with this recommendation. We have found that transaction modeling is unnecessary for most applications. Most database interaction concerns the navigation of the object model. If you implement the object and functional models as we recommend, navigation will be fast and your methods will perform well for most applications.

15.10 IMPLEMENTING THE FUNCTIONAL MODEL

You should prepare the bodies of the methods listed in the class declarations, as well as the main parts of an application. You should also write methods to create and initialize the database. The methods involve the bulk of the work for implementing an application. The functional model specifies many of the methods.

You can apply any method to transient or persistent objects. The application of methods to transient objects involves only language issues, but you must be aware of some special considerations for persistent objects. First, ObjectStore lets you access persistent objects only within the context of a transaction. Whenever you create a persistent object, you must supply a database as an argument to the constructor for the class. You must also observe the rules for the scope of pointer validity (see Section 15.1.2).

Implementing the functional model consists of three tasks:

- **Map the Object Navigation Notation to ObjectStore code**. The navigation code will provide the bulk of the logic for methods. [15.11]
- **Implement other methods**. Flesh out the remainder of the methods and translate pseudocode to ObjectStore code. Most of the work involves C++ programming. [15.12]
- **Write queries**. ObjectStore queries can help you find entry points for navigating the database. Also, sometimes you can reduce effort by substituting an ObjectStore query for programming code. We defer discussion of this task to Chapter 16 because it involves more advanced concepts.

15.11 MAPPING ONN CONSTRUCTS

Ideally, we would like to map each ONN construct to an ObjectStore expression. However, for some constructs we must write custom methods or promote an association to a class. Our examples use pointers to access objects.

15.11.1 Traversal of Simple Binary Association

The *objectOrSet.targetRole* construct traverses an association to a target role. The *objectOrSet.~sourceRole* construct traverses an association from a source role. You can start traversal with a single object or a set of objects (or null, which can arise for a composition of constructs). Both expressions may yield a single object, a set of objects, or null.

You can readily implement *object.targetRole* by using the association access method that returns an instance or set of instances of the class of *targetRole*. For example, the ONN expression *aTransaction.portfolio* maps to

```
aTransactionPntr->portfolioPntr.getvalue()
```

As another example, *aPortfolio.Transaction* maps to

```
aPortfolioPntr->transactionSet.getvalue()
```

We call a method for the advanced association, *aTransaction.asset*, (see Chapter 16)

```
aTransactionPntr->findAllAssets()
```

We must have already defined the method *findAllAssets* with the following signature:

```
const os_Set<Asset *> * Transaction::findAllAssets()
```

For *objectSet.targetRole* you can write a function to scan the *objectSet*, performing a union as you go. It would be nice to use a method instead of a function, but extending ObjectStore's parameterized collection classes is not easy. The ONN expression *aTransactionSet.asset* maps to

```
findAllAssets( aTransactionSetPntr )
```

where the following function implements *findAllAssets*:

```
const os_Set<Asset *> * findAllAssets( os_Set<Transaction *>
    * transactionSetPntr )
{
    Transaction * transactionPntr ;
    static os_Set<Asset *> * result = new os_Set<Asset *> ;
    result->clear();
    os_Cursor<Transaction*>
             transactionCursor(* transactionSetPntr);
    for(  transactionPntr = transactionCursor.first();
          transactionCursor.more();
          transactionPntr = transactionCursor.next())
       {
          * result |= * ( transactionPntr->findAllAssets() );
       }
    return result ;
}
```

We are using the overloading of |= provided in ObjectStore, which performs set union. If you are traversing a promoted association, you may need to create and maintain some auxiliary data structures or indexes to obtain adequate performance. We describe this in Chapter 16.

You can implement *object.~sourceRole* and *objectSet.~sourceRole* by translating to the corresponding target role in the implementation object model. (In Section 15.6.2 we suggested that you name all roles for the implementation object model.) For example, the ONN expression *aPerson.~accountOwner* maps to the code *aPersonPntr->portfolioSet.getvalue()*.

15.11.2 Traversal of Qualified Association

You can always traverse a qualified association without specifying the qualifier. You should implement this the same as simple binary associations described in Section 15.11.1.

Or you can use a qualifier in an ONN expression (*objectOrSet.role[qualifier = value]*). If traversal starts with an object, you can invoke an access method with the qualifier as an argument. If traversal starts with a set of objects, you should use code similar to that shown for *objectSet.role* in Section 15.11.1.

15.11.3 Traversal of Generalization

Both *objectOrSet:subclass* and *objectOrSet:superclass* are directly supported by the C and C++ casting operator. For example, the ONN expression *aFinancialInstrument:Asset* maps to *(Asset *) aFinancialInstrumentPntr.*

15.11.4 Traversal from Link to Object

For *link.role* you should promote the association to a class using the techniques described in Chapter 16. You can trivially implement this ONN construct by a method on the link object pointer that returns the related object pointer. For example, *anAssetTransaction.asset* maps to *anAssetTransactionPntr->getAsset()*.

For *linkSet.role* you should write a function to do the scanning. For example, the ONN expression *anAssetTransactionSet.asset* maps to *findAllAssets(anAssetTransactionSetPntr)* implemented by the following code:

```
const os_Set<Asset *> * findAllAssets
   ( os_Set<AssetTransaction *> * assetTransactionSetPntr )
{
   AssetTransaction * assetTransactionPntr ;
   static os_Set<Asset *> * result = new os_Set<Asset *> ;
   result->clear() ;
   os_Cursor<AssetTransaction*>
         assetTransactionCursor(* assetTransactionSetPntr);
   for(  assetTransactionPntr =
         assetTransactionCursor.first();
      assetTransactionCursor.more() ;
      assetTransactionPntr =
         assetTransactionCursor.next() )
      {
         result->insert(assetTransactionPntr->getAsset()) ;
      }
   return result ;
}
```

15.11.5 Traversal from Object to Link

The ONN provides two ways to retrieve links.

The construct *{ role1 = ObjectOrSet1 , role2 = ObjectOrSet2 , . . . , associationName }* finds link(s) between objects by using multiple roles. You must include the parameter of *associationName* when the role names are not sufficient to identify a link. You can implement this construct by promoting the association and invoking the appropriate methods. For example, the ONN expression *{ Asset = anAsset , Transaction = aTransaction }* maps to *anAssetPntr->findAssetTransaction(aTransactionPntr)* or *aTransactionPntr->findAssetTransaction(anAssetPntr)*.

The construct *objectOrSet@role* lets you retrieve links by specifying only one role. You can implement *object@role* by applying the appropriate association access method. For example, we can implement *anAsset@Transaction* by applying the *findAllAssetTransactions* method to *anAsset*. You can implement *objectSet@role* in a way similar to that for *objectSet.role*; write code that collects the links for each object in *objectSet*.

15.11.6 Filtering

The construct *objectOrSet[filter]* lets you specify a general expression to winnow a set of objects. For a single object you can just test the filter with a simple method that invokes an access method. Or you can apply the filter as you scan a set of objects. As a third option, you can use a query on an extent (see Chapter 16).

15.11.7 Traversal from Object to Value

You can access an attribute value of an object or a set of attribute values for a set of objects with *objectOrSet.attribute*. Normally, you should preserve encapsulation by making attributes protected and accessing the values with read and write methods.

You can implement *object.attribute* with a method that returns the attribute. The appropriate syntax depends on whether implementation of *object* is direct, a pointer, an ObjectStore reference, or a C++ reference (see Section 15.3.1). The method could return an attribute, a pointer to an attribute, or an attribute reference. For example, *aTransaction.transactionDate* maps to *aTransactionPntr-> getTransactionDate().*

You can implement *objectSet.attribute* by writing a function to scan the set of objects. You should then return a pointer to a set of attribute pointers. For example, *aTransactionSet.transactionDate* maps to *getTransactionDates(aTransactionSetPntr).* The following code implements *getTransactionDates*:

```
const os_Set<const date *> * getTransactionDates( const
   os_Set<Transaction *> * transactionSetPntr )
{
   Transaction * transactionPntr ;
   static os_Set<const date *> * result =
      new os_Set<const date*> ;
   result->clear() ;
   os_Cursor<Transaction *>
      transactionCursor(* transactionSetPntr);
   for(  transactionPntr = transactionCursor.first() ;
         transactionCursor.more();
         transactionPntr = transactionCursor.next())
      {
         result->insert(
            transactionPntr->getTransactionDate());
      }
   return result ;
}
```

15.11.8 Traversal from Link to Value

The construct *linkOrSet.attribute* is similar to *objectOrSet.attribute*, except the values are scanned from links. ObjectStore requires that you implement link attributes by promoting an association to a class. Normally, you should preserve encapsulation by making attributes protected and accessing the values with public read and write methods.

You can implement *link.attribute* with a method that returns the link attribute. For example, *anAssetTransactionLink.quantity* maps to *anAssetTransactionLinkPntr->getQuantity()*.

You can implement *linkSet.attribute* by writing a function to scan the set. For example, the expression *anAssetTransactionLinkSet.quantity* maps to *getQuantities(anAssetTransactionLinkSetPntr)*, where the following function scans the set of links:

```
const os_Set<const int *> * getQuantities(
   os_Set<AssetTransaction *> * assetTransactionSetPntr )
{
   AssetTransaction * assetTransactionPntr ;
   static os_Set<const int *> * result =
      new os_Set<const int *> ;
   result->clear() ;
   os_Cursor<AssetTransaction *>
      assetTransactionCursor(* assetTransactionSetPntr);
   for(  assetTransactionPntr =
            assetTransactionCursor.first();
         assetTransactionCursor.more();
         assetTransactionPntr =
            assetTransactionCursor.next())
      {
         result->insert(
            assetTransactionPntr->getQuantityPntr() ) ;
      }
   return result ;
}
```

15.11.9 Combining ONN Constructs

The ONN is an *analysis* specification that by definition cannot violate the *design* principle of encapsulation. An ObjectStore implementation of an ONN expression, however, should observe encapsulation. All methods should honor the law of Demeter: An object should access only its own attribute values or objects that are directly related. Indirectly related objects should be accessed indirectly via the methods of intervening objects. Our mapping rules for the primitive ONN constructs honor encapsulation.

You must carefully implement ONN expressions that combine traversals. For example, we could implement the ONN expression *aPerson.Portfolio.FinancialInstitution* by invoking the *findAccountHolders* function with an argument of the collection of portfolios for a person:

```
findAccountHolders (aPersonPntr->portfolioSet.getvalue())
```

The following code implements the findAccountHolders function:

```
const os_Set<FinancialInstitution*> & findAccountHolders(
      const os_Set<Portfolio*> & portfolioSet )
{
   Portfolio * portfolioPntr ;
   static os_Set<FinancialInstitution*> * result =
      new os_Set<FinancialInstitution*> ;
   result->clear();
   os_Cursor<Portfolio*>
            portfolioCursor(portfolioSet);
   for(  portfolioPntr = portfolioCursor.first();
         portfolioCursor.more();
         portfolioPntr = portfolioCursor.next())
      {
         result->insert(
            portfolioPntr->
               financialInstitutionPntr.getvalue());
      }
   return * result;
}
```

In general, you can implement a long ONN expression by using a sequence of nested function calls. Each function traverses one association in the expression. The functions are nested in the opposite order to the sequence in the ONN expression. Thus the outermost function traverses the last association in the expression.

15.11.10 Summary of ONN Mapping Rules

Table 15.3 summarizes the mapping rules for the Object Navigation Notation.

15.12 CREATION AND DELETION METHODS

You implement methods by writing code that meets the functional specification. You should declare each method in the header file for the class and implement each method in the corresponding source file. The mapping of pseudocode to C++ is largely straightforward, so we do not discuss it here. However, we do discuss methods for creating and deleting objects and links because they are more complex.

15.12.1 Creating Objects

You can create both transient and persistent objects. You can manipulate transient objects with standard C++ techniques, so we do not discuss this implementation further.

Concept	ONN construct	ObjectStore or C++ construct
Traverse binary association	object.role	Access method
	objectSet.role	Scan the set of objects and apply access method.
Traverse qualified association	objectOrSet.role [qualifier=value]	Same as binary association but can specify a qualifier value as an argument to access method.
Traverse generalization	objectOrSet:superclass objectOrSet:subclass	C and C++ casting operator
Traverse from link to object	link.role	Promote association to class, then apply access method.
	linkSet.role	Promote association to class, then scan the set of objects and apply access method.
Traverse from object to link	{role1=objectOrSet1, role2=objectOrSet2, . . . , associationName}	Promote association to class. If each argument is an object, then apply access method. If at least one argument is a set, then scan the set of objects and apply access method.
	object@role	Access method
	objectSet@role	Scan the set of objects and apply access method.
Filter objects	object[filter]	Apply filter within access method.
	objectSet[filter]	Scan the set of objects and apply filter within access method.
Filter links	link[filter]	Promote association to class, then apply filter within access method.
	linkSet[filter]	Promote association to class, then scan the set of objects and apply filter within access method.
Traverse object to value	object.attribute	Access method
	objectSet.attribute	Scan the set of objects and apply access method.
Traverse link to value	link.attribute	Promote association to class, then apply access method.
	linkSet.attribute	Promote association to class, then scan the set of objects and apply access method.

Table 15.3 Summary of ONN mapping rules for ObjectStore

ObjectStore overloads the *new* operator for persistent objects. Generally, objects are accessed through pointers, so you must include a pointer declaration. For example, the following code declares a pointer to an object:

```
Person * companyPresidentPntr ;
```

Also, the *new* operator requires an ObjectStore typespec. You should provide the typespec for a class by including the following static member in each class:

```
static os_typespec * get_os_typespec();
```

You can then create a persistent *Person* object for *Sue* with the following code. Argument *a_database* specifies the ObjectStore database that holds the persistent object:

```
companyPresidentPntr =
   new( a_database , Person::get_os_typespec())
   Person( "Suc") ;
```

To retrieve a persistent object, you must provide at least one navigation path to it. There are three possibilities: Navigate from associated objects, navigate from the class extent (see Chapter 16) if one exists, or look up the object using a *database root.* A database root is an ObjectStore feature that establishes a lookup to an object with a name that you provide. For example, the following code creates a database root *companyPresident*:

```
os_database_root * companyPresidentRoot =
   a_database->create_root( "companyPresident" )
```

You can then link the person object pointed to by *companyPresidentPntr* to the root:

```
companyPresidentRoot->setvalue( companyPresidentPntr )
```

You can later retrieve the root bound to *companyPresident.*

```
os_database_root * companyPresidentRoot =
   a_database->find_root("companyPresident")
```

Finally, you can retrieve the person object from the database root:

```
companyPresidentPntr = ( Person * )
   companyPresidentRoot->getvalue() ;
```

During object creation you must enforce any candidate keys and update the corresponding extents (see Chapter 16.) Check the uniqueness of all candidate keys before creating a new object. If you find a duplicate key, take some action appropriate to the application.

15.12.2 Deleting Objects

To delete an object (transient or persistent), you can apply the *delete* operator to a pointer to the object:

```
delete companyPresidentPntr;
```

15.12.3 Creating Links

The creation of links is mainly a C++ programming issue. Unlike object creation, you do not specify a database. Links are just formed between the specified objects and may span databases. However, the resulting pointers are subject to the validity rules in Section 15.1.2.

ObjectStore provides three approaches for accessing inverse members: *simple data member*, *relationship*, and *functional interface*. The *simple data member* approach operates directly on the attribute that implements a link, setting it for a multiplicity of "one" and inserting into it for a multiplicity of "many." The *relationship* approach applies a method to the attribute. The *functional interface* approach applies a method to the object.

We recommend against using the *simple data member* approach because it violates encapsulation. Instead, read and write attributes indirectly via methods. The choice between the *relationship* and *functional interface* approaches depends on whether an association is simple or advanced.

Simple Associations

We recommend that you use the *relationship* approach for simple associations (Section 15.6) because ObjectStore directly supports it. The *relationship* approach uses methods such as *getvalue()* to return a pointer or collection of pointers, *setvalue()* to set a pointer, *insert()* to add a pointer to a collection, and *remove()* to remove a pointer from a collection. The resulting code (a couple of declarations) is compact and easy to understand.

The following code adds a many-to-many link. The *aPersonPntr* is a pointer to a person, *portfolioSet* is the set of portfolios, *getvalue* returns the set, and *insert* inserts the portfolio pointer.

```
aPersonPntr->portfolioSet.getvalue().
    insert(aPortfolioPntr);
```

In the preceding code *portfolioSet* is not a pointer. If it had been a pointer the syntax would be

```
aPersonPntr->portfolioSetPntr.getvalue()->
    insert(aPortfolioPntr);
```

ObjectStore automatically updates the other end of the association. (This is a property of inverse members.) We could have formed a link from the other role.

```
aPortfolioPntr->accountOwnerSet.getvalue().
    insert(aPersonPntr);
```

A *Transaction* object refers to one *Portfolio* object, so we use *setvalue* instead of *insert* in the following code:

```
aTransactionPntr->portfolioPntr.setvalue(aPortfolioPntr);
```

Advanced Associations

We recommend that you use the *functional interface* approach for advanced associations (see Section 15.7 for a list of advanced associations).

ObjectStore does not directly support qualified associations such as between *Financial-Institution* and *Portfolio*. To form a link, we must supply a qualifier value to the method. (Chapter 16 explains the details.)

```
aFinancialInstitution->
    linkPortfolio(aPortfolioPntr, anAccountNumber);
```

We also return the count of the links created for a promoted association:

```
linkCount = aTransactionPntr->linkAsset(anAssetPointer);
```

For one-to-one and one-to-many associations, adding a link to an object on the "one" end that is already linked causes the existing link to be deleted (overwritten).

15.12.4 Deleting Links

Deleting a link is similar to inserting one. If you are using inverse members, you can remove a link by breaking it on one end; ObjectStore automatically takes care of the other end. For example, on a "one" end you should simply set the pointer to zero.

```
aTransactionPntr->portfolioPntr.setvalue(0);
```

Note the subtle C++ distinction between "->" and "." in the preceding code. At first glance, the code might appear wrong, because *portfolioPntr* is a pointer to a portfolio, and you might think the *setvalue* method should be applied with "->". However, in this case the method is being applied to the pointer, not to the object being pointed to. The pointer itself is being set to zero.

On a "many" end, you can remove the pointer from the collection. ObjectStore automatically adjusts the inverse member.

```
aPortfolioPntr->accountOwnerSet.getvalue().
    remove(aPersonPntr);
```

You can also delete all links to an object for a many-to-one or many-to-many association by applying the ObjectStore *clear* operation to the collection class on the many end. (If you set the pointer to the collection object to zero, be sure to reclaim the memory to avoid memory leaks.)

For qualified associations or association classes, you should apply the appropriate method. We suggest that you return a count of the deleted links, as in the following code:

```
linksDeleted = aFinancialInstitution
    ->unlinkPortfolio(aPortfolioPntr);
```

15.13 CHAPTER SUMMARY

This chapter describes how to take the optimized object model from detailed design and develop an implementation for an OO-DBMS. Although the focus is on ObjectStore, we believe much of our advice will apply to other OO-DBMSs as well. ObjectStore preserves the flavor of native C++; part of the philosophy of ObjectStore is to make language capabilities available with minimal overhead.

As you would expect, ObjectStore has robust support for objects, classes, attributes, and inheritance. It also supports simple associations with the inverse member mechanism. Essentially, when you use an OO-DBMS you gain some features and lose others relative to relational DBMSs; the appropriate trade-off depends on the application.

The first implementation step is to map OMT domains to ObjectStore constructs. You should define a datatype for each domain. You may access each attribute or method variable directly, by pointer, by ObjectStore reference, or by C++ reference.

Then you must map classes from the object model to C++ classes. Ordinarily, you should observe the principle of information hiding and define all attributes as *protected.* By accessing attributes indirectly via read and write methods, you preserve the flexibility of changing the implementation at a later date. As you define methods, you must decide which methods are *private*, *protected*, and *public*. Implementation of inheritance is straightforward with ObjectStore and involves the same issues as with native C++.

You can implement simple associations by using the inverse members provided by ObjectStore. ObjectStore does not directly support advanced associations, so you must promote them to a class. We outlined the approach for advanced associations, deferring the details until the next chapter.

We presented each construct from the Object Navigation Notation and showed how to implement it with ObjectStore. You can implement the simple constructs by invoking an access method. Other constructs require that you scan a set and accumulate the result. For a few complex constructs, you must first promote the association to a class. The remainder of the functional model implementation consists of writing methods and writing queries.

Figure 15.6 lists the key concepts for this chapter.

association, bidirectional implementation	object model, implementing associations
association, unidirectional implementation	object model, implementing classes
datatype access, direct access	object model, implementing domains
datatype access, pointer	object model, implementing generalizations
datatype access, C++ reference	ObjectStore
datatype access, ObjectStore reference	persistence
functional model, mapping the ONN	transience
inverse member	

Figure 15.6 Key concepts for Chapter 15

BIBLIOGRAPHIC NOTES

We have not surveyed OO-DBMSs in this chapter, but have instead focused on ObjectStore [ObjectDesign-96]. Many articles and books discuss general OO-DBMS concepts and features, such as [Cattell-91], [Loomis-95], [Ahmed-92], and [Bertino-91]. Mary Loomis is a regular contributor to the theory and application of object-oriented databases and writes in both the *Journal of Object-Oriented Programming* and *Object Magazine*. [Cattell-96] discusses the efforts of the *Object Data Management Group* to standardize object DBMS features across vendors.

The October 1991 issue of the *Communications of the ACM* was a special issue on object-oriented databases and discussed several commercial products. Douglas Barry regularly publishes comparisons of OO-DBMS products. See, for example, [Barry-95].

Much work ([Loomis-94], [Czejdo-90], [Bertino-92]) is underway on object-oriented queries and navigation.

Most of the issues in using ObjectStore are C++ programming issues. Standard C++ texts include [Lippman-91] and [Stroustrup-91].

REFERENCES

[Ahmed-92] Shamim Ahmed, Albert Wong, Duvvuru Sriram, and Robert Logcher. Object-oriented database management systems for engineering: A comparison. *Journal of Object-Oriented Programming 5*, 3 (June/July 1992), 27–44.

[Barry-95] Douglas K. Barry. ODBMS—Feature coverage in the current market. *Object Magazine* (July/August 1995).

[Bertino-91] Elisa Bertino and Lorenzo Martino. Object-oriented database management systems: concepts and issues. *Computer (*April 1991), 33–47.

[Bertino-92] Elisa Bertino, M Negri, G Pelagatti, and L Sbattella. Object-oriented query languages: the notion and the issues. *IEEE Transactions on Knowledge and Data Engineering 4*, 3 (June 1992), 223–237.

[Cattell-91] RGG Cattell. *Object Data Management: Object-Oriented and Extended Relational Database Systems*. Reading, Massachusetts: Addison-Wesley, 1991.

[Cattell-96] RGG Cattell, editor. *The Object Database Standard: ODMG-93, Release 1.2*. San Francisco, California: Morgan Kaufmann, 1996.

[Czejdo-90] Bogdan Czejdo, Ramez Elmasri, Marek Rusinkiewicz, and David W. Embley. A graphical data manipulation language for an extended entity-relationship model. *Computer* (March 1990).

[Lippman-91] Stanley B. Lippman. *C++ Primer,* 2nd edition. Reading, Massachusetts: Addison-Wesley, 1991.

[Loomis-94] Mary E S Loomis. ODBMS—Querying object databases. *Journal of Object-Oriented Programming 7*, 3 (June/July 1994).

[Loomis-95] Mary E S Loomis. *Object Databases: The Essentials*. Reading, Massachusetts: Addison-Wesley, 1995.

[ObjectDesign-96] Object Design, Inc. *ObjectStore C++ Release 4.0.2 Documentation*. Burlington, Massachusetts: ObjectDesign, Inc., 1996.

[Stroustrup-91] Bjarne Stroustrup. *The C++ Programming Language*, 2nd edition. Reading, Massachusetts: Addison-Wesley, 1991.

EXERCISES

15.1 Figure E15.1 adds domains to the airline flight reservation object model in Figure 2.23. (We have also added the three minor extensions mentioned in Section 5.4.2.)

a. (5) Implement each domain and define datatypes, choose access techniques, and decide if read-only. *AirportCode* is a three-letter code that uniquely identifies an airport. A *name* is an arbitrary alphanumeric string. Keep in mind that some of the domains are enumerations.

b. (3) Prepare ObjectStore C++ declarations for the classes. Include declarations for classes and attributes as well as for methods for reading and writing attributes and constructors and destructors. You do not need to write code for implementing any methods, and you may ignore the associations. (Note to the instructor: You may wish to assign specific classes.)

c. (4) Implement the following methods for the *FlightDescription* class: constructor, destructor, reading attributes, and writing attributes.

d. Extend the class declarations for the following associations. Use a bidirectional implementation and the recommended object model mapping rules. You will need a total of four macro instantiations to implement each association.

- (2) Flight@FlightReservation
- (2) FlightDescription@destination
- (2) Airport@City
- (2) FlightReservation@Seat
- (2) TripReservation@Ticket
- (2) TripReservation@FlightReservation. Keep in mind that this is an ordered association.

e. Implement the following ONN expressions with ObjectStore C++ code. Assume that methods required to traverse associations are available and do not write them. Section 5.4.1 explains the expressions. Make sure that you honor encapsulation (Section 15.11.9).

- (5) theStLouisAirport.~origin.destination
- (6) theStLouisAirport.~origin.destination.~origin.destination
- (7) aFlight.FlightDescription.AircraftDescription.modelNumber
- (7) aTripReservation.FlightReservation.Flight.FlightDescription.Airline

f. Declare and implement the following methods described in Section 5.4.2. Assume that methods required to traverse associations are available and do not write them.

- (5) *Airport::findZeroOneStops()*
- (6) *TripReservation::hasOnlyAisleSeats()*
- (6) *TravelAgency::calcMonthlySales(month,year)*
- (5) *TripReservation::setFrequentFlyerAccount(aFrequentFlyerAccount)*

15.2 Consider the prescription drug information system described in Exercises 9.3 and 10.4.

a. Repeat Exercise 15.1a for the domains of the following attributes:

- (1) Allergy.description
- (1) Drug@Compound.strength
- (1) ControlledSubstanceClassification.schedule
- (1) Compound.administration

b. (3) Repeat Exercise 15.1b for the classes in Figure E9.2.

c. Extend the class declarations for the following associations. Use a bidirectional implementation and the recommended object model mapping rules. You will need a total of four macro instantiations to implement each association.

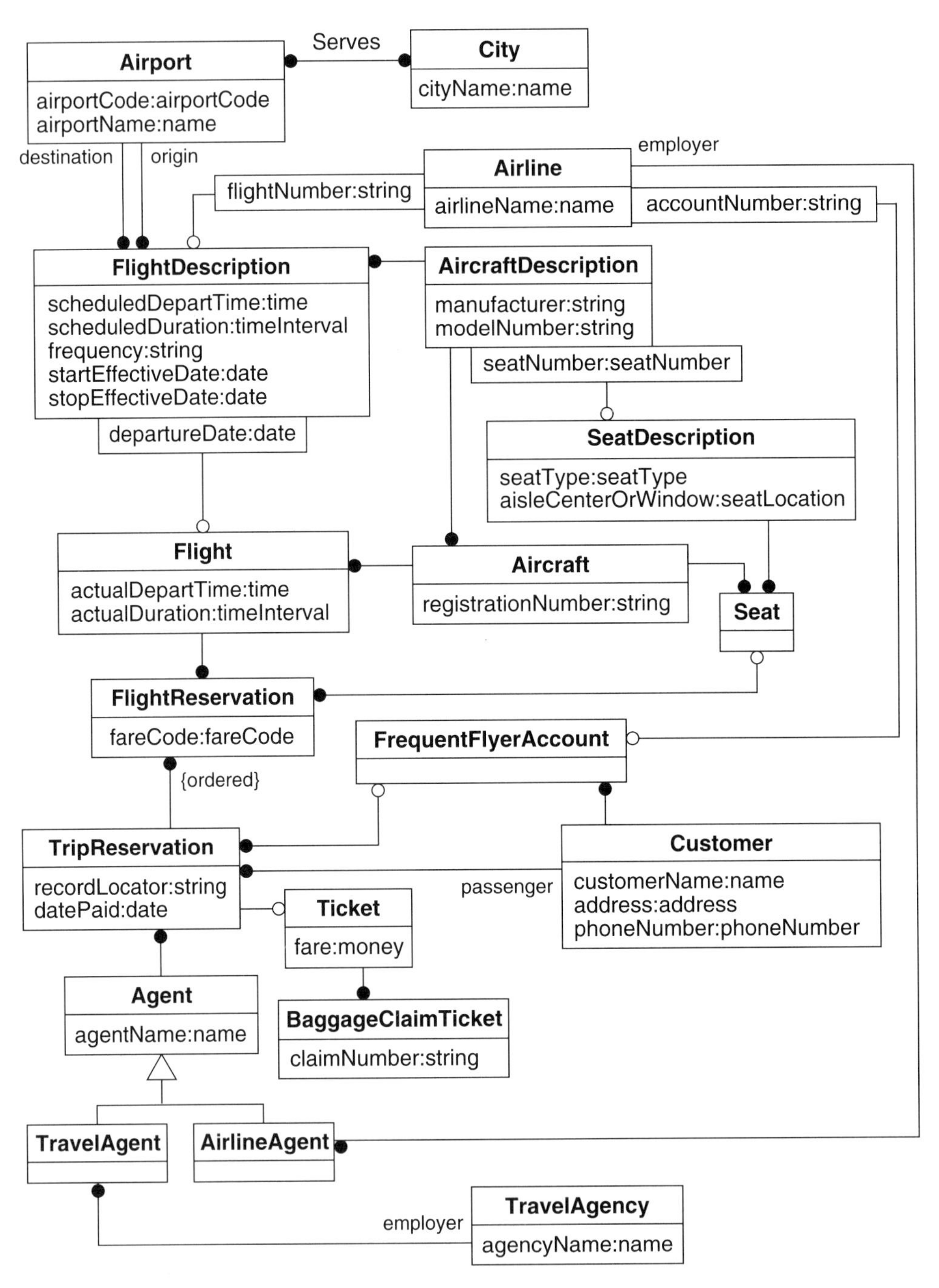

Figure E15.1 Object model for an airline flight reservation system with domains

- (2) Drug@ControlledSubstanceClassification
- (2) Drug@DrugInteraction
- (2) Compound@distributor
- (2) Compound@manufacturer

d. Implement the following ONN expressions with ObjectStore C++ code. Assume that methods required to traverse associations are available and do not write them. Ignore any link attributes. Make sure that you honor encapsulation.

- (2) aDrug.Compound.Form
- (2) aDrug.Compound.manufacturer
- (2) aDrug.Compound.distributor
- (3) aCompound.Drug.Allergy

e. Declare and implement the following methods. Assume that methods required to traverse associations are available and do not write them. You may ignore the link attribute.

- (4) *Drug::findCompounds (anotherDrug)*—Find all compounds that contain a given pair of drugs. (You prepared pseudocode for this method in Exercise 10.4g.)
- (4) *Company::findCompounds (aDrug)*—Find all compounds from a given manufacturer, in any form, that contain a drug. (You prepared pseudocode for this method in Exercise 10.4h.)

f. Declare and implement the methods for the following enhanced pseudocode. Assume that any methods that you might need to access associations have already been written.

- (4) *Drug::producesAllergy(allergies)*—Test for allergies. Determine whether a given drug may produce an allergic reaction in patients who have the allergies specified by *allergies*. Returns a collection of *allergies* that the drug matches.

```
Drug::producesAllergy(allergies) returns set of Allergy
   return self.Allergy intersect allergies;
```

- (5) *Drug::findCompounds(aForm)*—Return a collection of *Compound* instances for a given *Drug* in a given instance of *Form*. Do not use an ObjectStore query.

```
Drug::findCompounds(aForm) returns set of Compound
   return self.Compound[self.Compound.Form==aForm] ;
```

15.3 Consider the prescription refill software described in Exercises 9.4 and 10.5.

a. Repeat Exercise 15.1a for the domains of the following attributes:

- (1) Prescription.quantity
- (1) Prescription.refills
- (1) Prescription.fillGenerically

b. (3) Repeat Exercise 15.1b for the additional classes in Figure E9.5.

c. Extend the class declarations for the following associations. Use a bidirectional implementation and the recommended object model mapping rules. You will need a total of four macro instantiations to implement each association.

- (2) Customer@Allergy
- (2) Customer@Prescription

d. Implement the following ONN expressions with ObjectStore C++ code. Assume that methods required to traverse associations are available and do not write them. Ignore any link attributes. Make sure that you honor encapsulation.

- (2) aPrescription.Customer.telephone
- (3) aCustomer.Allergy.Drug.name
- (4) aCompound.Prescription.Customer.name
- (4) aPrescription.Customer.Prescription.prescriptionNumber

e. Declare and implement the following methods using the given pseudocode. Assume that methods required to traverse associations are available and do not write them.

- (4) *Prescription::computeRemainingRefills()*—Compute the number of remaining refills. If the prescription has expired (the prescription is older than one year), the number of remaining refills is zero. Partial refills are not allowed. Assume that there is a library function to get today's date.

```
Prescription::computeRemainingRefills() returns integer
  if getTodaysDate() > self.date + ONEYEAR then return 0 ;
  else return self.numberOfRefills - cardinality (self.Refill);
  end if
```

- (4) *Customer::determinePotentialAllergies(aDrug)*—Detect and return any possible allergic reactions that the customer might have to a drug.

```
Customer::determinePotentialAllergies(aDrug)
      returns set of Allergy
  return self.Allergy intersect aDrug.Allergy ;
```

- (3) Repeat the implementation of the previous method for the following pseudocode. (Note that we are reusing a method from Exercise 15.2f.)

```
Customer::determinePotentialAllergies(aDrug)
      returns set of Allergy
  return aDrug#producesAllergy( self.Allergy ) ;
```

- (6) *Customer::detectPotentialInteractions(aPrescription)*—Detect possible interactions of the drugs in a new prescription with old prescriptions, including expired prescriptions. Return the drugs in the old prescriptions that may interact. Assume that old prescriptions have already been checked.

```
Customer::detectPotentialInteractions(aPrescription)
    returns set of Drug
  oldDrugs := self.Prescription.Compound.Drug ;
  newDrugs := aPrescription.Compound.Drug ;
  interactions := oldDrugs.DrugInteraction intersect
    newDrugs.Druginteraction ;
  return interactions.Drug intersect oldDrugs ;
```

15.4 Consider the insurance claim software described in Exercises 9.5 and 10.6.

a. (3) Repeat Exercise 15.1b for the additional classes in Figure E9.8.

b. Extend the class declarations for the following associations. Use a bidirectional implementation and the recommended object model mapping rules. You will need a total of four macro instantiations to implement each association.

- (2) Customer@primaryCarrier
- (2) insured@dependent
- (2) InsuranceTransaction@Pharmacy

c. Implement the following ONN expressions with ObjectStore C++ code. Assume that methods required to traverse associations are available and do not write them. Ignore any link attributes.

- (4) aRefill.Prescription.Customer.primaryCarrier.name
- (3) aRefill.InsuranceTransaction.approvalCode
- (3) aCustomer.dependent.Prescription.prescriptionNumber

d. Declare and implement the following methods using the given pseudocode. Assume that methods required to traverse associations are available and do not write them.

- (4) *Customer::getFamilyAllergies()*—Return all allergies of a customer and the customer's dependents.

```
Customer::getFamilyAllergies() returns set of Allergy
  return self.Allergy union self.dependent.Allergy ;
```

- (4) *Refill::getInsuranceCarrierTransactions(anInsuranceCarrier)*—Return all insurance transactions with the given insurance carrier concerning the given refill.

```
Refill::getInsuranceCarrierTransactions(anInsuranceCarrier)
   returns set of InsuranceTransaction
   return self.InsuranceTransaction intersect
      anInsuranceCarrier.InsuranceTransaction ;
```

- (4) *Refill::getPrimaryInsuranceTransactions()*—Return all insurance transactions with the primary insurance carrier of the customer refilling a prescription. The previous bullet implements method *getInsuranceCarrierTransactions*. You do not need to write method *getPrimaryCarrier*. Implement the following pseudocode:

```
Refill::getPrimaryInsuranceTransactions()
         returns set of InsuranceTransaction
   aPrimaryCarrier :=
      self.Prescription.Customer#getPrimaryCarrier();
   return      self#getInsuranceCarrierTransactions(aPrimaryCarrier)
;
```

15.5 (2) Classes *Physician*, *Company*, *Pharmacy*, *InsuranceCarrier*, and *Pharmacist* have similar attributes. Redo their class declarations to reflect the inheritance shown in Figure E15.2 and discuss the relative merits of your new implementation.

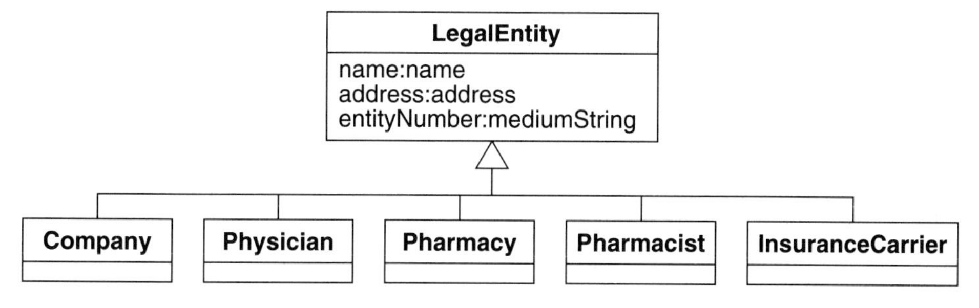

Figure E15.2 Object model for an inheritance relationship in the insurance claim software

16

Object-Oriented Databases: Advanced

This chapter explains advanced aspects of implementing the OMT models with the ObjectStore OO-DBMS, building on the basic material in Chapter 15. We discuss keys, extents, and more about advanced associations (associations with link attributes, association classes, ternary associations, and qualified associations). We also address how to use queries and provide guidelines in dealing with software engineering issues.

The ObjectStore methods for inverse members support only simple binary associations; they do not address the semantics of advanced associations. We present two workarounds: folding link attributes and qualifiers into a related class or promoting the association to a class. Alternatively, you could use the ObjectStore dictionary class to implement link attributes and qualifiers, but we do not describe how to do this.

16.1 KEYS

16.1.1 Foreign Keys

ObjectStore pointers obviate the foreign keys that must be used with RDBMSs. Our mapping rules maintain the integrity of ObjectStore pointers during database activities.

16.1.2 Candidate Keys

You can use ObjectStore indexes to implement single-attribute candidate keys. However, another approach can also handle multiple-attribute candidate keys.

- **Implement extent**. For each class with a candidate key, implement an extent using the techniques discussed in Section 16.2.
- **Define a search method**. Define a method for each candidate key that searches the class extent for a match on the key. The arguments of the method are the values of the key's

attributes. The method is simple to implement, because it merely queries the extent for candidate key attributes that match the specified argument and returns the object, if any, that matches the key.

- **Check for a duplicate key**. Write a *create* method for each class with a candidate key. The *create* method executes the methods for searching extents preventing the creation of an instance with a duplicate key and then invokes the constructor.

16.1.3 Primary Keys

ObjectStore uses existence-based identity, creating an artificial number as the primary key of each object. The pointers that implement associations refer to these object identifiers.

16.2 EXTENTS

An extent (the set of objects for a class) can have several purposes. In Section 16.1.2 we enforced uniqueness by searching the extent for a class. You can also initiate queries with extents (see Section 16.5). Sometimes extents can make it easier to implement algorithms for methods.

Because a class may have both transient and persistent objects, you must carefully implement extents. Furthermore, a persistent extent cannot refer to transient objects when you actually write the persistent extent to a database. Our solution is to define two sets that are static members for each class with an extent; one contains persistent objects and the other contains transient objects. We have implemented our methods for extents so that they transparently access the appropriate set.

The static members that implement extents correspond to class attributes in an object model. In Section 3.1.2 we discouraged use of class attributes because they often lead to an inferior model. The use of class attributes to implement extents is an exception to this general rule.

16.2.1 Declarations

You should define two sets that are static members for each class with an extent. For example, if class *Person* has an extent, you should declare the following attributes:

```
static os_Set<Person*> * persistentExtent ;
static os_Set<Person*> * transientExtent ;
```

The following code declares static accessor methods for class *Person*. The methods retrieve the extent and perform any required initialization.

```
static os_Set<Person*> & getPersistentExtent() ;
static os_Set<Person*> & getTransientExtent() ;
```

Also you must declare a method that reinitializes a persistent extent between transactions or during a database switch. You must reinitialize a persistent extent at the start of a transaction, because other concurrent transactions may have invalidated the copy of a persistent extent that may remain from a prior transaction.

```
static void resetPersistentExtent() ;
```

16.2.2 Implementation

When you create an object, you must insert the object into its transient or persistent extent. For example, you should include the following code in the constructor for *Person*:

```
if( objectstore::is_persistent( this ) )
{
   getPersistentExtent().insert( this ) ;
}
else
{
   getTransientExtent().insert( this ) ;
}
```

Similarly, when you delete an object, you must delete the object from its transient or persistent extent. For example, you should include the following code in the destructor for *Person*:

```
if( objectstore::is_persistent( this ) )
{
   getPersistentExtent().remove( this ) ;
}
else
{
   getTransientExtent().remove( this ) ;
}
```

You must then implement the method to get the persistent extent. If the extent has already been retrieved, just return it. Otherwise, try to retrieve the database root. If you cannot retrieve the root, create a new extent and root. The following code implements *getPersistentExtent* for *Person*:

```
os_Set<Person*> & Person::getPersistentExtent()
{
   if ( persistentExtent )
   { return * persistentExtent ; }
   else
   {
      os_database_root * persistentExtentRoot =
         aDatabase->find_root("personPersistentExtent");
      if ( persistentExtentRoot )
      {
         persistentExtent = ( os_Set<Person*> * )
            (persistentExtentRoot->get_value()) ;
```

```
        }
        else
        {
           persistentExtent =
              & os_Set<Person*>::create(aDatabase) ;
           persistentExtent->change_behavior(
              os_collection::pick_from_empty_returns_null );
           os_database_root * personPersistentExtentRoot =
            aDatabase->create_root("personPersistentExtent");
           personPersistentExtentRoot->set_value(
              persistentExtent ) ;
        }
        return * persistentExtent ;
     }
  }
```

You can use a simpler process to get the transient extent. If the extent has been already retrieved, just return it. Otherwise, create a new transient extent. The following code implements *getTransientExtent* for *Person*:

```
  os_Set<Person*> & Person::getTransientExtent()
  {
     if ( transientExtent )
        { return * transientExtent ; }
     else
        {
           transientExtent = & os_Set<Person*>::create(
              os_database::get_transient_database()) ;
           return * transientExtent ;
        }
  }
```

Finally, the following code implements the code to reinitialize a persistent extent. This code discards the *persistentExtent* that may remain in memory from a prior transaction. The *getPersistentExtent* method will then retrieve the *persistentExtent* from the database at the start of a transaction.

```
  void Person::resetPersistentExtent()
  {
     persistentExtent = 0 ;
  }
```

16.2.3 Memory Allocation and Initialization

You allocate memory by including a declaration in the source code for a class. The following code also sets the pointers to zero, so the extent will automatically initialize the first time an attempt is made to access it. However, you must manually invoke the *resetPersistentExtent* method between transactions or when switching databases.

```
os_Set<Person*> * Person::transientExtent = 0 ;
os_Set<Person*> * Person::persistentExtent = 0 ;
```

16.2.4 Use of Extents

Extents are easy to use because the constructor and destructor automatically maintain them, inserting and deleting objects from the persistent or transient extent, as appropriate. You should use the following code to access the transient and persistent extents for the *Person* class:

```
Person::getPersistentExtent()
Person::getTransientExtent()
```

16.3 FOLDING ATTRIBUTES INTO A RELATED CLASS

We now discuss two workarounds for implementing advanced associations with ObjectStore. The first is folding attributes into a related class. The second, which we describe in Section 16.4, is to promote the association to a class.

You can simplify one-to-one and one-to-many binary associations by folding their link attributes and qualifiers into a related class. Essentially, you can apply one or more transformations (see Chapter 10) to degrade an advanced association into a simple association. You are left with a simple association that you can directly implement.

16.3.1 Folding Link Attributes into a Class

In Figure 16.1 a person works for a company as an employee and has a current salary. You cannot directly implement this association with ObjectStore, but you can apply transformation *T19* from left to right and fold the link attribute into the *Person* class. You can then implement the resulting association according to Chapter 15. You can fold link attributes for associations that are one-to-one or one-to-many.

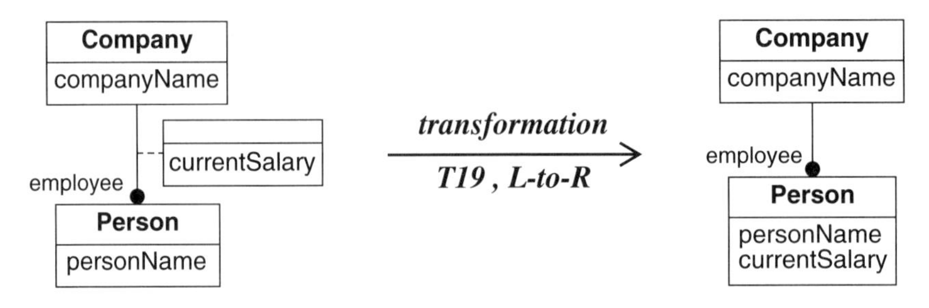

Figure 16.1 Folding a link attribute into a related class

Remember that you are degrading the association when you migrate the link attribute to a related class. If you want to extend the model and allow an employee to work for multiple companies, once again you would have to make *currentSalary* a link attribute (and then implement it by promoting the association, as we describe in Section 16.4). More important, you must document the logic of the original analysis models; otherwise someone might not see that *currentSalary* is really a link attribute and might incorrectly extend the corresponding application.

16.3.2 Folding Qualifiers into a Class

Similarly, you can apply transformations *T18* and *T19* from left to right to bury a qualifier in a related class. In Figure 16.2 we bury the *accountNumber* qualifier in the *Portfolio* class. You can use this workaround for qualified associations that are one-to-many with the qualifier removed.

Figure 16.2 Folding a qualifier into a related class

Once again the workaround degrades the model. In reality, *accountNumber* is not a property of a portfolio, but a property of the relationship because it is assigned by a financial institution. (Different financial institutions may assign the same account number to different accounts.) You can choose to accept this degradation during implementation, but you must understand the semantic implications and remember the original intent.

Figure 16.3 shows the implementation model for the buried qualifier, in which we have added role names (*accountHolderPntr, portfolioSet*) for the inverse data members and methods to implement the qualified association. For clarity, we omit the other attributes of *FinancialInstitution* and *Portfolio*. We use ObjectStore inverse members to maintain the integrity of dual association pointers, as we did with simple binary associations in Chapter 15. You can use the built-in ObjectStore methods to implement simple binary associations, as we describe in Chapter 15, but you must provide additional methods to implement a qualifier.

If a portfolio is not an account in a financial institution, we set its account number to zero. The *getAccountNumber* and *setAccountNumber* methods read and write the account number.

The *linkPortfolio* and *linkFinancialInstitution* methods create a link between a *Portfolio* instance and a *FinancialInstitution* instance if the link does not already exist. The methods return the number of links created. We show the code for the *linkPortfolio* method. The *linkFinancialInstitution* method simply invokes the *linkPortfolio* method.

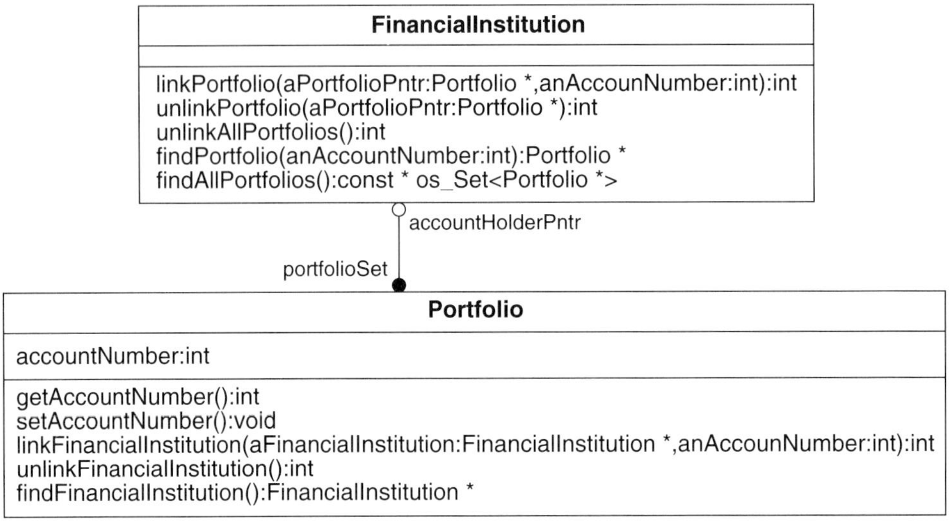

FinancialInstitution

linkPortfolio(aPortfolioPntr:Portfolio *,anAccounNumber:int):int
unlinkPortfolio(aPortfolioPntr:Portfolio *):int
unlinkAllPortfolios():int
findPortfolio(anAccountNumber:int):Portfolio *
findAllPortfolios():const * os_Set<Portfolio *>

accountHolderPntr

portfolioSet

Portfolio

accountNumber:int

getAccountNumber():int
setAccountNumber():void
linkFinancialInstitution(aFinancialInstitution:FinancialInstitution *,anAccounNumber:int):int
unlinkFinancialInstitution():int
findFinancialInstitution():FinancialInstitution *

Figure 16.3 Implementation model for qualifier folded into a class

```
int FinancialInstitution::linkPortfolio( Portfolio *
   aPortfolioPntr , int anAccountNumber )
{
   Portfolio * existingPortfolioPntr ;
   if ( this != 0 && aPortfolioPntr != 0 )
   {
      existingPortfolioPntr = this->findPortfolio
         ( anAccountNumber );
      if ( existingPortfolioPntr )
      {
         return 0;
      }
      else
      {
         aPortfolioPntr->setAccountNumber(anAccountNumber);
         this->portfolioSet.getvalue().
            insert(aPortfolioPntr);
         return  1;
      }
   }
   else return 0;
}
```

The *unlinkPortfolio* and *unlinkFinancialInstitution* methods remove a *Portfolio* from the *portfolioSet* of a *FinancialInstitution*, using ObjectStore's *remove* method. The methods return the change in the number of links and, if necessary, set the *accountNumber* to zero. The following code implements the *unlinkPortfolio* method:

```
int FinancialInstitution::unlinkPortfolio( Portfolio *
   aPortfolioPntr)
{
   int count ;
   if ( this != 0 && aPortfolioPntr != 0 )
   {
      count = this->portfolioSet.getvalue().cardinality();
      this->portfolioSet.getvalue().remove(aPortfolioPntr);
      count = this->portfolioSet.getvalue().cardinality() -
         count ;
      if ( count == 0 )
      {
         return 0 ;
      }
      else
      {
         aPortfolioPntr->setAccountNumber( 0 ) ;
         return -1 ;
      }
   }
   else return 0 ;
}
```

The *unlinkAllPortfolios* method removes all *Portfolios* from a *FinancialInstitution* by setting the account number to 0 and clearing *portfolioSet*:

```
int FinancialInstitution::unlinkAllPortfolios( )
{
   int count = 0 ;
   Portfolio * portfolioPntr;
   if ( this != 0 )
   {
      os_Cursor<Portfolio*> portfolioCursor(
         portfolioSet.getvalue() );
      for( aPortfolioPntr = portfolioCursor.first() ;
         portfolioCursor.more() ;
         aPortfolioPntr = portfolioCursor.next() )
         {
            aportfolioPntr->setAccountNumber( 0 );
            count--;
         }
      portfolioSet.getvalue().clear();
```

```
    }
    return count ;
}
```

The *findAllPortfolios* method simply returns a pointer to *portfolioSet*.

```
const os_Set<Portfolio*> *
    FinancialInstitution::findAllPortfolios()
{
    return & portfolioSet.getvalue() ;
}
```

The *findPortfolio* method uses a query (see Section 16.5). It returns a pointer to the instance of *Portfolio* that is linked to a given *FinancialInstitution* and *accountNumber*.

```
Portfolio * FinancialInstitution::findPortfolio( int
    anAccountNumber )
{
    static Portfolio * portfolioPntr;
    portfolioPntr = portfolioSet.getvalue().query_pick(
        "Portfolio*",
        "anAccountNumber == this->accountNumber",
        os_database::of(this));
    return portfolioPntr;
}
```

The following *findFinancialInstitution* method returns a pointer to the instance of *FinancialInstitution* linked to the specified instance of *Portfolio* with *accountHolderPntr*.

```
FinancialInstitution *
    Portfolio::findFinancialInstitution()
{
    static FinancialInstitution * result ;
    result = this->accountHolderPntr.getvalue() ;
    return result ;
}
```

16.4 PROMOTING ASSOCIATIONS

The other technique for implementing advanced associations is promotion to a class. Figure 16.4 gives an example. The earlier technique of folding attributes works only for some associations with link attributes and qualifiers; promotion works for any advanced association (associations with link attributes, association classes, ternary associations, and qualified associations). Promotion involves the following steps:

- **Replace the promoted association with a class**. The new class includes any link attributes and qualifiers of the promoted association. Instances of the new class correspond to links of the original association.

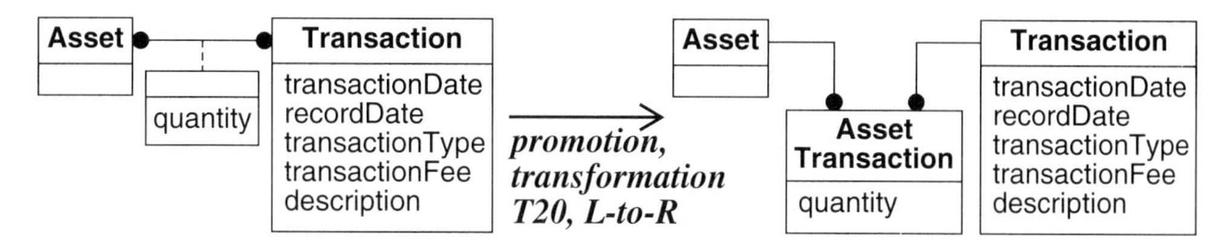

Figure 16.4 An example of association promotion

- **Add associations**. Add binary associations between the new class and the classes for the original association. You need one association for each role of the original association.
- **Implement methods**. Define and implement methods for traversing the association, as well as for creating and deleting links.

ObjectStore does not automatically recognize and enforce the dependencies that result from promotion, so you must write additional methods to enforce integrity constraints. For example, upon deletion of an object, you must delete any instances of promoted associations. For insertion into a promoted association, you must check for prior existence of the link, create an instance of the promotion class, and finally update the new primitive associations.

You must write declarations and code for each promoted association. If you have many associations to promote, you will find yourself writing many similar methods, and you may want to try one of the following advanced techniques to eliminate duplication:

- **Code generation**. Write your own code generator or use the code-generation facilities offered by some OO tools.
- **Macros**. Write your own promotion macros, similar to the ones that ObjectStore has for inverse members.

16.4.1 Promoting a Simple Association

In this section, we promote a simple association (Figure 16.5) and then implement the promoted association to explain the basic mechanics. You can use a similar strategy for simple ternary associations. Sections 16.4.2 and 16.4.3 discuss promotion for link attributes, association classes, and qualifiers.

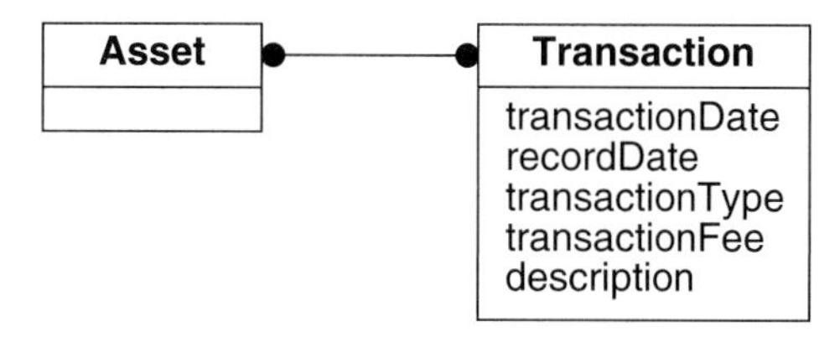

Figure 16.5 A simple binary association

Figure 16.6 shows the implementation model for the promoted *AssetTransaction* association. We have added role names for the inverse data members and methods to implement the promoted association. The role names *assetPntr* and *transactionPntr* indicate that an *AssetTransaction* instance has a pointer to an *Asset* instance and a pointer to a *Transaction* instance. The role name *assetTransactionSet* (used twice) indicates that an instance of *Asset* or *Transaction* has a set of pointers to *AssetTransaction* instances. For clarity, we show only the attributes and methods relevant to the promotion. We have omitted other attributes and methods, such as methods for accessing attributes and declarations for the implementations of classes *Asset* and *Transaction*.

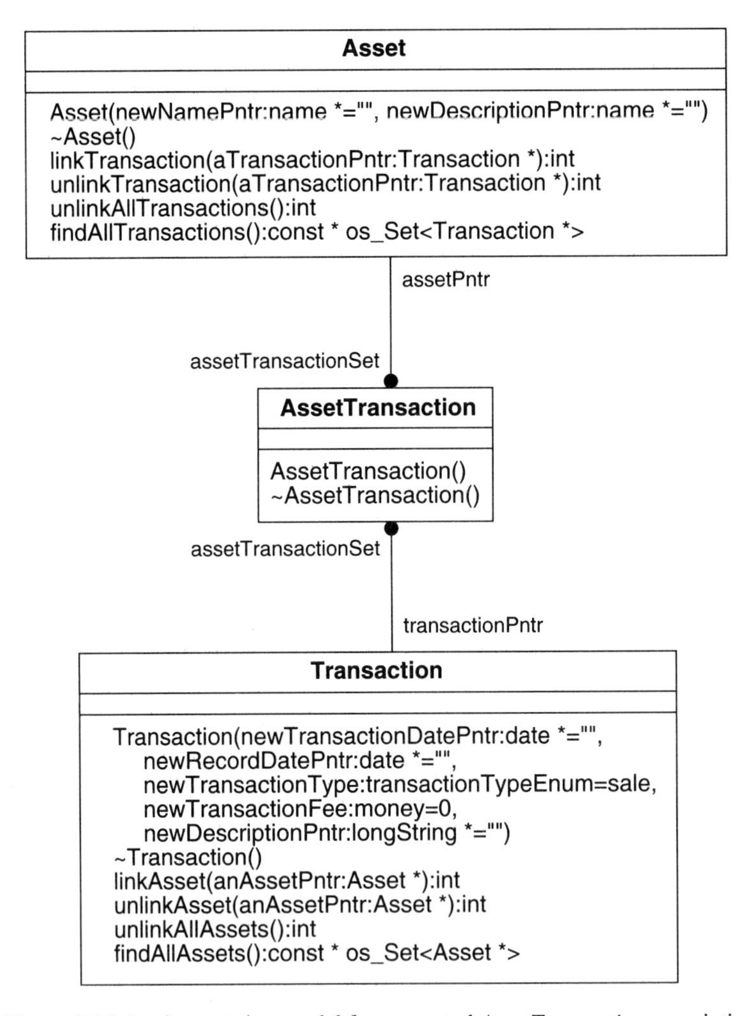

Figure 16.6 Implementation model for promoted *AssetTransaction* association

We do not include any methods on *Asset* or *Transaction* for returning the actual links of the association, because they are not needed. The code for *AssetTransaction* implements the methods for accessing the association. We now provide declarations and code for Figure 16.6.

Declarations

The following code declares *assetTransactionSet* in *Asset* to be a set of pointers to *AssetTransaction* instances; *assetPntr* is the inverse member in *AssetTransaction*. We would include a similar declaration in the public attributes of *Transaction* for the association between *AssetTransaction* and *Transaction*.

```
os_relationship_m_1(
   Asset,assetTransactionSet,AssetTransaction,assetPntr,
   os_Set<AssetTransaction *>) assetTransactionSet ;
```

The following code declares inverse members for *AssetTransaction*:

```
os_relationship_1_m(
   AssetTransaction,assetPntr,Asset,assetTransactionSet,
   Asset *) assetPntr ;
os_relationship_1_m(
   AssetTransaction,transactionPntr,
   Transaction,assetTransactionSet,
   Transaction *) transactionPntr ;
```

C++ is a strongly typed language. The previous macros for implementing inverse members must know that the classes *Asset* and *Transaction* exist. To do this, be sure to put the following at the head of the declarations for *AssetTransaction*:

```
class Asset ;
class Transaction ;
```

The following declaration, included in the protected attributes of *Asset*, allocates memory for the set that is used by the *findAllTransactions* method. It declares *allTransactions* to be a set of pointers to *Transaction* instances.

```
os_Set<Transaction *> allTransactions ;
```

Finally, we implement the inverse data members for *AssetTransaction* with the following macro instantiations:

```
os_rel_m_1_body(Asset,assetTransactionSet,AssetTransaction,
   assetPntr);
os_rel_m_1_body(Transaction,assetTransactionSet,
   AssetTransaction,transactionPntr);
os_rel_1_m_body(AssetTransaction,assetPntr,Asset,
   assetTransactionSet);
os_rel_1_m_body(AssetTransaction,transactionPntr,
   Transaction,assetTransactionSet);
```

Link Methods

Both the *linkTransaction* and *linkAsset* methods create a link between an *Asset* instance and a *Transaction* instance. Because the two methods do exactly the same thing, you need implement only one of them and have the other simply call the first method. The methods return a count of the change in the number of links: +1 when a link is added and 0 if the link already exists.

The following code implements the *linkTransaction* method. We first determine if the link exists, using the *findAssetTransaction* method, whose implementation is shown in Section 16.4.2. If the link does not exist, we create a new *AssetTransaction* instance in the same database and set its pointers. The inverse member facilities automatically update the sets of *AssetTransactions*.

```
int Asset::linkTransaction( Transaction *
   this_transaction_pntr )
{
   AssetTransaction * asset_transaction_pntr ;
   asset_transaction_pntr =
      this->findAssetTransaction( this_transaction_pntr );
   if( asset_transaction_pntr)
   {
      return 0;
   }
   else
   {
      asset_transaction_pntr = new(os_database::of(this))
         AssetTransaction( );
      asset_transaction_pntr->setAssetPntr(this);
      asset_transaction_pntr->setTransactionPntr(
         this_transaction_pntr);
      return 1;
   }
}
```

We can implement the *linkAsset* method by swapping the target and the argument and calling *linkTransaction*.

```
int Transaction::linkAsset( Asset * this_asset_pntr )
{
   return ( this_asset_pntr->linkTransaction( this ) );
}
```

Unlink Methods

The *unlinkTransaction* and un*linkAsset* methods delete a link between an *Asset* instance and a *Transaction* instance. The methods return a count of the change in the number of links: –1 when a link is deleted and 0 if the link does not exist.

The following code implements the *unlinkAsset* method. We first check to ensure that the link exists. If it does, we simply delete the *AssetTransaction* instance. ObjectStore's inverse members will automatically remove the specified link from the *assetTransactionSets* in *Asset* and *Transaction*.

```
int Transaction::unlinkAsset( Asset * this_asset_pntr)
{
   AssetTransaction * asset_transaction_pntr ;
   asset_transaction_pntr =
      this->findAssetTransaction( this_asset_pntr );
   if( asset_transaction_pntr )
   {
      delete asset_transaction_pntr;
      return -1;
   }
   else
   {
      return 0;
   }
}
```

We can implement *unlinkTransaction* by swapping the target and the argument and calling *unlinkAsset*.

The *unlinkAllTransactions* method drops all links to a given *Asset* instance and returns a count of the number of links dropped. Similarly, *unlinkAllAssets* drops all links to a given *Transaction* instance.

The next code segment implements *unlinkAllAssets*; *unlinkAllTransactions* has similar code. We use a cursor to scan all members of *assetTransactionSet*, deleting as the code proceeds. The *os_cursor::safe* option lets us modify the scanned set. We must use the *getvalue()* method to access the set because we are using inverse members. However, because we are using inverse members, ObjectStore updates everything else automatically.

```
int Transaction::unlinkAllAssets()
{
   int result = 0 ;
   AssetTransaction * asset_transaction_pntr ;
   os_Cursor<AssetTransaction*>
      assetTransactionCursor(
         assetTransactionSet.getvalue(),
         os_cursor::safe) ;
   for(  assetTransactionPntr =
            assetTransactionCursor.first();
      assetTransactionCursor.more() ;
      assetTransactionPntr =
         assetTransactionCursor.next() )
      {
```

```
            result--;
            delete asset_transaction_pntr ;
        }
    return result;
}
```

Find Methods

The *findAllTransactions* method returns a read-only set of pointers to *Transaction* instances associated with a given *Asset* instance. Similarly, *findAllAssets* returns a read-only set of pointers to *Asset* instances associated with a *Transaction* instance. The implementations of these methods are similar; we illustrate the logic with the *findAllAssets* method. With this method, we first clear the set *allAssets* that returns the result. (We do not need to create *allAssets* explicitly, because ObjectStore transparently creates it as a consequence of creating an object that contains a collection attribute.) We then scan the set of *AssetTransactions*, inserting the associated *Transaction* instances and returning the result as read-only.

```
const os_Set<Asset *> * Transaction::findAllAssets()
{
    allAssets.clear() ;
    Transaction * this_transaction_pntr = this ;
    AssetTransaction * asset_transaction_pntr ;
    os_Cursor<AssetTransaction*>
        assetTransactionCursor(
            this_transaction_pntr->assetTransactionSet.
            getvalue()) ;
    for(    assetTransactionPntr =
                assetTransactionCursor.first();
            assetTransactionCursor.more() ;
            assetTransactionPntr =
                assetTransactionCursor.next() )
        {
            allAssets.insert(
                asset_transaction_pntr->
                getAssetPntr().getvalue());
        }
    return & allAssets ;
}
```

Constructors and Destructors

Promoting the association affects the constructor and destructor methods for *Asset* and *Transaction*. We illustrate some code that must be included in the constructor and destructor of *Asset*. *Transaction* would require similar code.

We include the following code in the constructor for *Asset* to change the behavior of *assetTransactionSet* from its default behavior:

```
assetTransactionSet.change_behavior(
   assetTransactionSet.get_behavior()
   | os_collection::maintain_cursors
   | os_collection::pick_from_empty_returns_null );
```

In this example, *change_behavior* establishes the desired behavior, *get_behavior* gets the existing behavior, *maintain_cursors* makes it possible for create methods to modify the set during iteration, and *pick_from_empty_returns_null* lets iterators and queries operate on an empty set without generating an exception error.

We include the following code in the destructor for *Asset* to avoid dangling association links:

```
this->unlinkAllTransactions() ;
```

16.4.2 Promoting Link Attributes and Association Classes

You can implement link attributes and association classes by extending the code for implementing a simple promoted association, as we described in Section 16.4.1. You must implement the following additional methods:

- Return the link, if any, between two objects. (For a ternary association, you must have a similar method that returns the link between three objects.)
- Return all links to an object in a given association.
- Get and set attribute values for link attributes.
- Return the objects connected to a link.

As an example, we use the association between *Asset* and *Transaction* in Figure 16.7. Figure 16.8 shows the corresponding implementation.

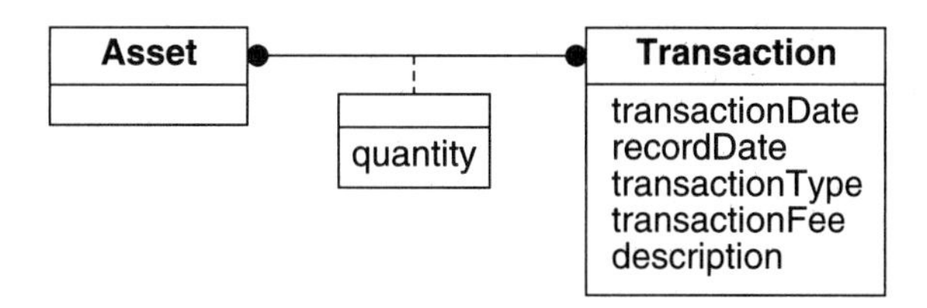

Figure 16.7 An association with a link attribute

We have promoted the *AssetTransaction* association, making the link attribute *quantity* an object attribute of the new class. Figure 16.8 shows the attributes and methods required in addition to those in Figure 16.6. The role names are the same as those in Figure 16.6 and indicate the names of the variables that implement the association. The *getQuantity* and *setQuantity* methods for the *AssetTransaction* class are similar to those for any other attribute of a class.

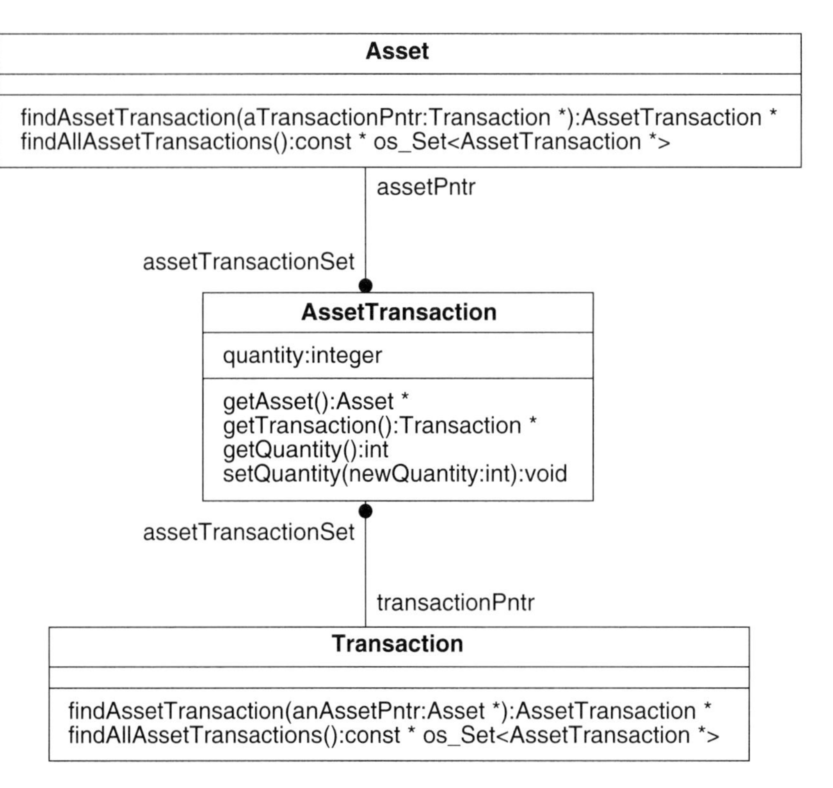

Figure 16.8 Implementation model with additional methods for a promoted link attribute

```
int AssetTransaction::getQuantity()
{
   return quantity;
}
void AssetTransaction::setQuantity( int newQuantity )
{
   quantity = newQuantity ;
}
```

The methods *linkTransaction* and *linkAsset*, shown in Figure 16.6, now add an argument for the initial value of *quantity*. The methods pass the argument to the *AssetTransaction* constructor, which then initializes the value of *quantity*. The *getAsset* method simply returns the *assetPntr* attribute of *AssetTransaction*. Similarly, the *getTransaction* method returns the *transactionPntr*. The additional methods for *Transaction* are similar to the additional methods for *Asset*, so we give only one example of each.

The following code implements the *Transaction::findAssetTransaction* method. We first find the intersection of *assetTransactionSet* in *Asset* and *assetTransactionSet* in *Transaction* and then pick one element from the intersection.

```
AssetTransaction * Transaction::findAssetTransaction
   ( Asset * this_asset_pntr )
{
   os_Collection<AssetTransaction *> *
      intersectAssetTransactionSet ;
   AssetTransaction * result ;
   Transaction * this_transaction_pntr = this;
   intersectAssetTransactionSet =
      & /* & is address operator here */
      ( this_transaction_pntr->assetTransactionSet.
      getvalue() & /* & indicates set intersection here */
      this_asset_pntr->assetTransactionSet.getvalue() ) ;
   if ( intersectAssetTransactionSet->cardinality() == 0 )
   {
      result = 0 ;
   }
   else
   {
      result = intersectAssetTransactionSet->pick() ;
   }
   delete intersectAssetTransactionSet ;
   return result ;
}
```

The following method, *findAllAssetTransactions*, simply returns *assetTransactionSet*.

```
const os_Set<AssetTransaction *> *
   Transaction::findAllAssetTransactions()
{
   return  & assetTransactionSet.getvalue();
}
```

16.4.3 Promoting Qualified Associations

You must promote an association that is many-to-many with the qualifier removed. The class declarations and inverse data members are similar to those described in Section 16.4.1. As we demonstrated with direct implementation of qualified associations (Section 16.3.2), you must have additional methods to implement the semantics of the qualifier.

16.5 USING OBJECTSTORE QUERIES

The query language of ObjectStore is much different from the SQL language of RDBMSs. SQL is more expressive than the ObjectStore query language. However, ObjectStore cleanly integrates queries with the programming language, in contrast to the awkward amalgam of

SQL and programming code. For that reason, we emphasize programming more and queries less with ObjectStore than with RDBMSs. Nevertheless, you may want to use ObjectStore queries, for example, to find a value in an extent and enforce the uniqueness of a candidate key. You may also use queries to initiate navigation by finding an object in an extent. We give a few examples of ObjectStore queries to illustrate their power and style.

An ObjectStore query lets you specify a predicate on a collection of pointers to objects. ObjectStore has two functions for performing queries. The *query* function returns the collection of pointers to objects that satisfy the predicate. The *query_pick* function returns a pointer for a single object—the first object found that satisfies the predicate. ObjectStore may implement a query by scanning the collection or by using an index to search only the most promising objects. Whether or not ObjectStore uses an index is immaterial to your application, aside from the performance impact.

For example, you can query the persistent extent of the *Person* class to retrieve a pointer to the *Person* object with a given name. The *getPersistentExtent* method retrieves the extent if it has not already been brought into memory. As the collection of pointers is processed, *this* points to each object in the collection. The C++ function *strcmp* determines if two strings have the same value; the *getName* function retrieves the *name* attribute for a *Person* object. ObjectStore requires you to "register" the *strcmp* and *getName* functions before using them in a query. Registration declares the functions to the ObjectStore query processor. The overall query returns a pointer of 0 if there is no match. Otherwise, it returns a pointer to the first *Person* object found with the given name.

```
aPersonPntr = Person::getPersistentExtent().query_pick(
    "Person*", "strcmp( this->getName() , aNamePntr ) == 0",
    os_database::of(this));
```

As another example, the following query finds the collection of portfolios with at least one transaction. We apply the query to the persistent extent of *Portfolio,* and *transactionSet* is the set of transactions for a given portfolio. ObjectStore supplies the method *empty()*, which returns a value of 0 if the collection is not empty.

```
portfolios = Portfolio::getPersistentExtent().query(
    "Portfolio*",
    "this->transactionSet.getvalue().empty() == 0",
    os_database::of(this));
```

For a more complex example, we can find and print all portfolios for a given person with at least one transaction. First, we find the person. If the person is found, we then query the portfolios and transactions, as we did in the previous example.

```
aPersonPntr = Person::getPersistentExtent().query_pick(
    "Person*", "strcmp( this->getName() , aNamePntr ) == 0",
    os_database::of(this));
if ( aPersonPntr )
{
    portfolios =
        aPersonPntr->portfolioSet.getvalue().query(
        "Portfolio*",
```

```
        "this->transactionSet.getvalue().empty()==0",
        os_database::of(this));
    os_Cursor<Portfolio*> portfolioCursor( portfolios );
    for( aPortfolioPntr = portfolioCursor.first() ;
        portfolioCursor.more() ;
        aPortfolioPntr = portfolioCursor.next() )
        {
            aPortfolioPntr->print () ;
        }
}
else
{
    cout << endl << "Not found." << endl ;
}
```

16.6 SOFTWARE ENGINEERING ISSUES

As you complete your OO-DBMS implementation, you should consider a number of software engineering issues:

- Dependencies
- Performance
- Style
- Memory management
- Testing and debugging
- Large system issues.

16.6.1 Dependencies

A dependency is a directed relationship between items of data, including modeling constructs, source code, compiled code, and actual application data. For example, related applications may share files with declarations, code, and initializations. On update you may have to perform some actions to maintain consistency between the source data and dependent target data. Dependencies are not limited to compile-time activities. For example, you may need to reload a database after restructuring an application.

You can minimize dependencies by carefully structuring declarations, code, initializations, makefiles, and other information. You can further limit the effect of change if each application includes only the classes actually used. The following additional strategies will simplify dependency handling:

- **Put declarations in a separate header file for each class or promoted association.** The header file should declare attributes, methods, association roles, methods for association role access, and class extents.

- **Build a separate source file for each class with method implementation and extent initialization**. This file should include the corresponding header files for the class and direct superclasses.
- **Put the implementation of each association and extent initialization in a separate source file**. For unpromoted associations, this file should simply include the header files for the associated classes and the macros for implementing the association. For promoted associations, it should also contain the implementation of the association and the corresponding header file.
- **Avoid repeated inclusion**. You should have the header and source files use *ifndefs* to avoid repeated inclusion. For example, you can use *ifndef* to define *FILE_NAME_HH* and ensure the header file *FILE_NAME.HH* is included at most once. The first line tests the existence of a definition of *FILE_NAME_HH*. If it is not defined, the next two lines are executed, which define *FILE_NAME_HH* and include the header file. The *endif* statement terminates the *ifndef* statement.

```
#ifndef FILE_NAME_HH
#def FILE_NAME_HH
#include FILE_NAME.HH
#endif
```

16.6.2 Performance

Object Design, Inc. has dedicated an entire manual to tuning ObjectStore. Their advice is important, but is specific to ObjectStore, so we do not repeat it here. Instead, we offer tuning techniques that apply to a range of OO-DBMSs. First we emphasize that you carefully conceive and implement your models. Only then can you receive the full benefits of physical optimization. Second, we recommend some conceptual techniques for performance tuning.

- **Indexes**. Indexes can improve the performance of some queries. You can define indexes for a single attribute of type integer, string, or pointer. ObjectStore can also create indexes on user-defined classes via user-defined comparison and hashing functions, but we recommend against using this feature because of its complexity. Instead, you can write your own mapping function from the domain of interest (such as a pair of small integers) to an integer or string.
- **Clustering**. You can use clustering physically to collocate objects that are accessed together. Clustering can increase reference locality and improve data access times.
- **Unidirectional associations**. Unidirectional implementation can improve performance for an association if most of the activity is to create and delete links. Unidirectional associations damage extensibility, so use them only when the performance benefit is compelling.
- **Preanalyzed queries**. You can use preanalyzed queries to shift query parsing overhead from run-time to compile-time. Preanalysis is especially useful for frequently executed queries.

- **Derived attributes**. You can choose between eager evaluation (compute in advance and cache the results) and lazy evaluation (compute results on demand) for derived attributes. The decision is transparent to clients of the class when you use information hiding and access attributes through methods.
- **Collection attributes**. You can declare attributes that are collection classes either directly (*os_Set<>*) or as pointers to collection classes (*os_Set<>* *). You should generally use direct implementation, with pointer implementation used only for performance tuning. We have several reasons for this recommendation. With direct implementation, the collection class is automatically initialized when the enclosing object is instantiated. Also, the direct syntax is simpler than the pointer syntax.

16.6.3 Style

You should observe proper programming practices in your implementation.

- **Naming protocol**. Use a naming protocol for variables and implementation attributes to avoid name collisions. The naming protocol should preserve the names in the object model from analysis. Distinguish between variables that are objects and those that are pointers to objects, as we have done in this chapter.
- **Attribute access methods**. Normally, make attributes protected and access their values only via read and write methods.
- **Direct access, pointer, ObjectStore reference, or C++ reference**. See Section 15.3. For transient objects, ObjectStore lets you use any of the four techniques for accessing attributes and method variables. For persistent objects, use pointers, references, or direct access for attributes within objects. Do not use direct access for variables that refer to persistent objects, because all memory for persistent objects must be allocated at run-time.

  Try to use a uniform protocol. We recommend that you use mainly pointers or C++ references, with ObjectStore references as required.

16.6.4 Memory Management

Memory management is an issue because memory is allocated as a side effect of some operations. This is an issue for both transient and persistent memory. Mismanaging transient memory causes a "memory leak." Mismanaging persistent memory causes a "database leak." Unfortunately, persistence complicates the use of software for detecting memory leaks. We recommend that each method that allocates memory be responsible for recovering memory when it is no longer needed.

For transient objects, you can allocate memory from either the heap or the stack. (Persistent objects are always allocated from the heap.) This is strictly a C++ issue. The stack is managed automatically; you must explicitly manage memory allocated from the heap. The stack is normally used for objects known at compile-time, such as fixed-length buffers; the heap is normally used for objects created during the execution of an application.

Some methods for traversing associations return a collection of pointers. You can simplify memory management by returning a pointer to a constant collection and making the method responsible for memory management. If the calling method must modify the collection, it should create a copy and then manage memory for the copy.

You should also check for out-of-memory on every *new* operation. It is better to have the execution overhead in your code than risk failure. For brevity we have omitted the code that deals with memory management in our examples.

16.6.5 Testing and Debugging

Testing and debugging for an ObjectStore application is similar to an ordinary C++ application. Strong typing will catch many errors at compile-time. You can make testing and debugging easier by putting assertions in the code. For methods with critical performance needs, you can surround assertions with macro code that include or omit the assertions at compile-time, depending on a switch.

ObjectStore supports both heterogeneous collections (possibly different types) and homogeneous collections (the same type). When possible, we prefer to use homogeneous collections so the compiler can perform type checking. We were able to catch many errors in our ObjectStore code by using homogeneous collections.

16.6.6 Large System Issues

Large systems give rise to additional implementation issues:

- **Configuration management**. Configuration management is especially important when a large team works on an application. A configuration management system can track versions of applications and control checkouts. Configuration management is more difficult for database applications because you must coordinate the schema with the programming code. One solution is to integrate configuration management and schema evolution into the build procedure.
- **Project libraries**. You may want to develop your own class libraries for large projects. You could consider gaps in the ObjectStore library as well as classes peculiar to your applications.
- **Third-party libraries**. In addition to building your own project libraries, you may want to purchase third-party class libraries, such as user interface, communications, and digital signal processing classes.

16.7 CHAPTER SUMMARY

This chapter discusses advanced aspects of implementing the OMT models with the ObjectStore OO-DBMS.

ObjectStore provides robust indexes to handle single-attribute candidate keys. We present a workaround for multiple-attribute candidate keys. To enforce the uniqueness of candidate keys, you can define an extent for the class and search the extent before creating a new value. Our technique for implementing extents supports both transient and persistent objects.

We also covered implementation of advanced associations—associations with link attributes, association classes, ternary associations, and qualified associations. Because ObjectStore's inverse data members do not directly support these advanced associations, you must write your own implementation code. We discussed two workarounds: folding link attributes and qualifiers into a related class and promoting the association to a class.

For some associations you can apply one or more transformations to fold link attributes and qualifiers into a related class. You can then directly implement the simplified association. For qualifiers you will also have to extend the resulting methods to enforce the uniqueness and navigation behavior that is normally implied by a qualifier.

The second workaround always applies: You can always promote an advanced association and directly implement the resulting model with ObjectStore. Promotion replaces an advanced association with a class; each link of the association becomes an object of the new class. You must define and implement your own methods to enforce integrity constraints and association traversal.

We briefly discussed ObjectStore queries and illustrated their use. You can use an ObjectStore query to find a value in an extent and enforce the uniqueness of a candidate key. You may also find an object in an extent to initiate navigation.

We concluded the chapter with a discussion of various software engineering issues. Figure 16.9 lists the key concepts for this chapter.

association promotion
candidate key
extent
find methods
folding attributes into a related class
implementation of association class
implementation of link attributes
implementation of qualified association
implementation of ternary association
index
integrity constraints
link methods
memory management
performance tuning
query
unlink methods

Figure 16.9 Key concepts for Chapter 16

REFERENCES

[Lippman-91] Stanley B Lippman. *C++ Primer*, 2nd edition. Reading, Massachusetts: Addison-Wesley, 1991.

[Loomis-95] Mary E S Loomis. *Object Databases: The Essentials*. Reading, Massachusetts: Addison-Wesley, 1995.

[ObjectDesign-96] Object Design, Inc. *ObjectStore C++ Release 4.0.2 Documentation*. Burlington, Massachusetts: Object Design, Inc., 1996.

[Stroustrup-91] Bjarne Stroustrup. *The C++ Programming Language*, 2nd edition. Reading, Massachusetts: Addison-Wesley, 1991.

EXERCISES

16.1 Consider the airline flight reservation application discussed in Exercise 15.1.

a. (3) Attribute *Airport.airportCode* is a candidate key. Implement it. You do not need to write code for the extent and can assume it has already been written.

b. Write ObjectStore C++ declarations and methods to fold the qualifier into a class for the following associations:

- (8) FlightDescription@Flight
- (8) Airline@FrequentFlyerAccount
- (8) AircraftDescription@SeatDescription

c. (8) Write ObjectStore C++ declarations and methods to promote the association *Airport@City*.

d. For the following classes, write ObjectStore C++ code for declaring and implementing transient and persistent extents:

- (4) TripReservation
- (4) Customer
- (4) FrequentFlyerAccount

16.2 Consider the prescription drug information system described in Exercises 9.3, 10.4, and 15.2.

a. The following attributes are candidate keys. Implement each of them. You do not need to write code for the extents and can assume they have already been written.

- (3) ControlledSubstanceClassification.schedule
- (3) Allergy.allergyCode
- (3) Company.companyNumber
- (3) Compound.nationalDrugCode

b. (8) Write ObjectStore C++ declarations and methods to promote the association *Distributor@Compound.*

c. (10) Write ObjectStore C++ declarations and methods to promote the association *Drug@Compound* and to implement the link attribute *strength.*

d. Declare and implement the following methods:

- (6) *DrugInteraction::getInteractions(aCollectionOfDrugs)*—Test a collection of *Drug* instances and return a (possibly empty) collection of *DrugInteraction* instances that can be produced by some combination of the drugs. Note that this is a class method.

```
DrugInteraction::getInteractions(aCollectionOfDrugs)
    returns set of DrugInteraction
  result := emptySet of DrugInteraction ;
  possibleInteractions :=
    aCollectionOfDrugs.DrugInteraction ;
  for each aDrugInteraction in possibleInteractions
    if aDrugInteraction.Drug is
      contained in aCollectionOfDrugs
      insert aDrugInteraction into result ;
    end if
  end for each
  return result ;
```

- (5) *Drug::findCompounds(aForm)*—Return a set of *Compound* instances for a given *Drug* in a given *Form*. Use a query to implement the filter in the following pseudocode:

```
Drug::findCompounds(aForm) returns set of Compound
  return self.Compound[self.Compound.Form==aForm];
```

- (6) *Drug::create (name, action, chemicalFormula, schedule, functionalClass)*—Insert a new drug into the database. Note that this is a class method. You prepared pseudocode for it in Exercise 10.4. Assume that any methods that you might need to access associations have already been written.

16.3 Consider the prescription refill software described in Exercises 9.4, 10.5, and 15.3.

a. (3) Attribute *Pharmacist.licenseNumber* is a candidate key. Implement it. You do not need to write code for the extent and can assume it has already been written.

b. (8) Write ObjectStore C++ declarations and methods to fold the qualifier into a class for the association *Prescription@Refill*.

c. For the following classes, write ObjectStore C++ code for declaring and implementing transient and persistent extents:

- (4) Customer
- (4) Prescription
- (4) Compound

d. Declare and implement the following methods:

- (6) *LatinTranslation::translateToEnglish(aLatinPhrase)*—Translate a Latin phrase and return the corresponding English phrase. The phrase must match exactly. Note that this is a class method.

```
LatinTranslation::translateToEnglish(aLatinPhrase)
    returns mediumString
  dictionaryEntry := query extent of
      LatinTranslation[latinPhrase == aLatinPhrase]
  if dictionaryEntry return dictionaryEntry.englishPhrase ;
  else return NULL ;
  end if
```

- (5) *Prescription::locatePrescription(aPrescriptionNumber)*—Locate a prescription by prescription number. Note that this is a class method.

```
Prescription::locatePrescription(aPrescriptionNumber)
    returns Prescription
  return query extent of Prescription[prescriptionNumber==
      aPrescriptionNumber];
```

- (5) *Customer::create (name, address, taxNumber, dateOfBirth, telephone)*—Insert a new customer into the database. Note that this is a class method. You prepared enhanced pseudocode for it in Exercise 10.5. Assume that any methods that you might need to access associations have already been written.

16.4 (2) Write ObjectStore C++ declarations and methods needed to fold the link attribute *currentSalary* shown in Figure 16.1 into the class *Person*.

16.5 (7) Write ObjectStore C++ declarations and methods to fold the qualifier into a class for the association *Bank@Account* in Figure 2.20.

17

Implementation Review

This chapter completes our presentation of implementation for data management applications. First we summarize the similarities and differences among implementations for files, relational databases, and object-oriented databases. We then preview the treatment of large systems issues, which we cover in Part 4.

17.1 IMPLEMENTING THE OBJECT MODEL

We have presented a process and much advice for implementing the object model with files, RDBMSs, and ObjectStore. We believe many of our recommendations for ObjectStore will apply to other OO-DBMS products as well. The OMT methodology helps you develop an abstract understanding of an application and then implement the resulting models with various implementation platforms.

17.1.1 Files

You can implement an object model with files by performing the following steps:

- **Organize data into files**. Decide which classes and associations must be persistent and group them into file types. For each file type decide how to apportion application data among one or more files.
- **Select file approach(es)**. For each file type, choose a file approach. Most often you will use sequential-access ASCII files.
- **Implement identity**. Implement the approach to identity that you chose during system design.
- **Select mapping strategies**. Implement the domains, classes, associations, and generalizations for the file types.

17.1.2 Relational Databases

You can implement an object model with an RDBMS by performing the following steps:

- **Implement identity**. During system design you chose a strategy for identity—existence based or value based. During implementation you follow through and adjust your object model accordingly.
- **Implement domains**. Determine how to implement the domains that you added during detailed design. For simple domains, merely substitute the corresponding data type and size. Complex domains require more effort. For each attribute that uses a domain, add an SQL *check* clause for each domain constraint.
- **Define tables**. Map the constructs in your object model to tables.
- **Define constraints on tables**. You can use RDBMS constraints to enforce the structure of the object model. Define primary and candidate keys, as well as referential integrity for foreign keys.
- **Create indexes**. Tune the database so that navigation is fast, as the object model implies. Indexes are the primary means for tuning a relational database.
- **Allocate storage**. Most RDBMSs let you allocate space for each table and index.

17.1.3 Object-Oriented Databases

You can implement an object model with ObjectStore by performing the following steps:

- **Choose persistent variables**. Decide which variables need to be persistent and appropriately declare them.
- **Implement domains**. Map domains to C++ type declarations. Write methods for domains. Place the type declarations in a header file to be included by any file referencing the domain.
- **Implement classes and associations**. Prepare C++ declarations for each class and advanced association in the model. Declare all attributes and methods for a class, including the implementation of associations.
- **Create indexes**. Create single-attribute indexes to reduce performance bottlenecks. Indexes can also enforce single-attribute candidate keys.
- **Define extents**. Define the appropriate extents. You can use extents to implement multiple-attribute candidate keys.

17.1.4 Summary of Object Model Mapping Rules

Table 17.1 summarizes our recommended object model mapping rules for files, RDBMSs, and OO-DBMSs. This table compiles the earlier advice from Table 12.2, Table 13.2, and Table 15.2.

Concept	Object model construct	Files	RDBMS	ObjectStore
Domain	Simple	Map to a logical data type, then to a physical data type	A data type and size	C++ data type
	Identifier	Write code for existence-based identity	Use available RDBMS-specific features	Built into ObjectStore
	Enumeration	Enumeration string	Typically enumeration string	C++ encoding
	Structured	Hierarchical production rules	Concatenation, multiple columns, or additional table	C++ typedef or embedded object
	Multivalued	Put asterisk after attribute token		ObjectStore set of pointers to values
Class	Class	Multiple production rules	Map each class to a table	Map each class directly to a C++ class
Association	Simple binary	Explicit links	Buried association or distinct table—depending on the multiplicity	Inverse members with bidirectional pointers or sets of pointers
	Qualified			Fold attribute or promotion
	Link attribute			
	Association class			Promotion
	Ternary			
	Ordered		Use a sequence number	Same as simple binary, but use list instead of set
	Symmetric		Try to break the symmetry	Same as simple binary
	Aggregation		Same as corresponding association	
Generalization	Single inheritance	Reference superclass attributes for the subclasses	Separate superclass and subclass tables	Normal C++
	Multiple inheritance			

Table 17.1 Summary of recommended object model mapping rules

17.2 IMPLEMENTING THE FUNCTIONAL MODEL

In Part 3 we also described how to implement the functional model. Our primary focus was the implementation of the Object Navigation Notation (ONN).

17.2.1 Files

Grammars for files do not involve much functional behavior. Essentially, applications load file(s), process the data, and then save file(s). We did not show an implementation of the ONN for files because this is a matter of programming, not file structure.

17.2.2 Relational Databases

We systematically mapped ONN expressions to SQL code and then addressed miscellaneous functional modeling issues that are important for relational databases.

17.2.3 Object-Oriented Databases

Once again, our focus was on implementing ONN expressions. We also discussed creation and deletion methods, as well as ObjectStore queries.

17.2.4 Summary of Object Navigation Notation Mapping Rules

Table 17.2 summarizes the mapping rules for the ONN for relational databases and object-oriented databases. This table compiles the earlier advice in Table 13.3 and Table 15.3. In the table *A* and *B* denote tables for classes *A* and *B*. *AB* denotes a distinct table that implements an association between classes *A* and *B*.

17.3 OVERVIEW OF LARGE SYSTEM ISSUES

We conclude the book with a discussion of several large system issues. Until now, our emphasis has been on grass-roots applications. We have described concepts and notation (Part 1), developed application models (Part 2), and driven application models forward to various implementation targets (Part 3). With Part 4 we broaden our focus and consider several topics of deeper complexity. Our treatment in Part 4 is not complete, but it gives a flavor of several advanced technologies:

- **Distributed databases** (Chapter 18). You can disperse a database across several physical locations. A distributed DBMS provides the illusion of a single database.
- **Integration of applications** (Chapter 19). Organizations typically use many applications, all of which must be coordinated to form a coherent system.
- **Reverse engineering** (Chapter 20). Few applications are truly grass roots. Usually there is a predecessor application that can seed the new application with both data and ideas.

Concept	ONN construct	ObjectStore ONN implementation	RDBMS Object model implementation	RDBMS ONN implementation
Traverse binary association	objectOrSet. role	Access method -or- scan + access method	Buried association	Select from table A, B on join condition
			Distinct association table	Select from table A, B, AB on 2 join conditions
Traverse qualified association	objectOrSet. role[qualifier =value]	Same as binary assoc, but can specify qualifier argument		Similar to binary association except can specify qualifier value
Traverse generali-zation	objectOrSet: subclass -and- objectOrSet: superclass	C and C++ casting operator	Separate tables	Join subclass table to superclass table
			Push down superclass attributes	Query appropriate subclass table
			Push up subclass attribs	Query superclass table
Traverse from link to object	linkOrSet. role	Promotion, then same as binary association	Buried association	Select from table
			Distinct association table	Select from table B, AB on join condition
Traverse from object to link	{role1=object OrSet1, . . . , association Name}	If advanced association, then promotion. Otherwise, same as binary association	Distinct association table	Select from table
	objectOrSet @role		Buried association	Select from table
			Distinct association table	Select from table B, AB on join condition
Filter objects	objectOrSet [filter]	Apply filter within access method		Specify filter in where clause
Filter links	linkOrSet [filter]			
Traverse object to value	objectOrSet. attribute	Access method -or- scan + access method		Specify desired attribute in SQL select list
Traverse link to value	linkOrSet. attribute	Promotion, then same as object attribute		

Table 17.2 Summary of Object Navigation Notation mapping rules

PART 4: LARGE SYSTEM ISSUES

18

Distributed Databases

As part of architecting a system, you must decide *where* the data is stored (distribution) as well as *how* it is stored (persistence). You can encounter distribution with in-memory data structures, files, relational databases, object-oriented databases, and other data management paradigms. Applications that require fast performance may use an in-memory database distributed over a local area network. Modern operating systems transparently distribute file directories across networks. The most complex and robust support for distribution arises with relational databases and object-oriented databases.

This chapter addresses implementation issues for distributed relational databases and distributed object-oriented databases. We do not consider distributed computation (parallel processing) because it is outside the scope of this book. To a large extent, a distributed DBMS hides data location, and applications can be written as if against a centralized DBMS.

We distinguish between *distributed* databases and *heterogeneous* databases. A distributed database concerns one database paradigm. In contrast, a heterogeneous database may involve multiple paradigms. Consequently, software development is much different for distributed databases than for heterogeneous databases. You should design a distributed database top-down, designing the global schema before the local schema. In contrast, heterogeneous databases require a bottom-up approach because you must integrate separate applications that already exist. Chapter 19 discusses the integration technology that is appropriate for heterogeneous databases.

18.1 INTRODUCTION TO DISTRIBUTED DATABASES

A ***distributed database*** is a database that is built on top of a computer network rather than on a single computer. Data may be distributed over a local area network (LAN) or a wide area network (WAN). Data is stored at different network sites; DBMS processing can occur at individual sites or combinations of sites. A distributed database has a single global logical schema that is implemented with multiple physical schema, one for each site.

A ***distributed DBMS*** is the software for managing access to a distributed database. A distributed DBMS lets application software be written as if it were accessing a centralized DBMS, by isolating an application from distribution details.

A distributed DBMS is more complex than a centralized DBMS because a distributed DBMS must balance communication costs in addition to CPU and I/O costs. In practice, the CPU is fast, so the emphasis is on balancing communications and I/O. The trade-off depends on the geographic dispersal of sites—whether the component DBMSs are close and connected via a LAN or far apart and connected by a WAN. The relative costs of communications and I/O influence the optimization of DBMS processing, as well as the design and implementation of the distributed database.

Distributed databases have several advantages:

- **Managed distribution of data**. Data can be placed at different sites. Local processing suffers little degradation yet the entire database can be queried as a coherent whole.
- **Modular growth**. You can incrementally add data management capacity, transparent to applications.
- **Fault tolerance**. With judicious replication, data can remain available despite failure of individual sites (component databases). There is graceful degradation of processing capacity when sites fail. Multiple copies of data protect against catastrophe.
- **Greater capacity**. You can apportion computation among multiple DBMSs in a network. This lets you deploy resources more effectively than a centralized DBMS.
- **Cheaper computing**. Distributed data management offers further flexibility for configuring computing resources.

However, distributed database technology is complex and problems can arise:

- **Predictability of performance**. Distributed DBMSs are not only subject to variations in performance from I/O competition but are also vulnerable to changes in network loading. The performance fluctuations can confuse users and complicate database administration.
- **Security**. Distributed DBMSs have conflicting impacts on security. A beneficial effect is that local sites can provide protection independent of the overall database. However, communication networks can be prone to security breaches.
- **Integrity**. You must carefully maintain referential integrity across site boundaries. Referential integrity issues arise through foreign keys with relational databases and associations (or pointers) between classes for object-oriented databases. Configurations of versions can involve data at different sites and must also be kept consistent.
- **Database administration**. Distributed databases are more difficult to administer than centralized DBMSs. The database administrator must cope with additional tuning parameters and subtle interactions. Crash recovery becomes more complex because there is another failure mode (network failure) to consider.

Centralized DBMSs promote data independence, so the developer can focus on logical aspects of data and defer implementation details. With logical data independence an application is made aware only of relevant portions of the database; this lets you add new applications to the database without disrupting existing applications. Physical data independence insulates applications from changes in tuning mechanisms. Distributed DBMSs extend the notion of data independence.

- **Location transparence**. You can write applications independently of data location. To a large extent, applications simply request data; the distributed DBMS tracks data location and accesses data in the most appropriate manner.
- **Migration transparence**. You can freely migrate data to reduce response time, improve resource utilization, and change responsible site(s). The primary key of an object need not be changed upon physical relocation.
- **Replication transparence**. Data may be replicated at multiple sites, and the distributed DBMS automatically maintains consistency across updates. Ideally, replicated copies of data should behave like a single copy. In practice, this is difficult to accomplish fully due to equipment failures and network lag.

There are two dimensions to distribution for database applications:

- **Client-server computing**. Client-server computing ostensibly concerns distribution of computing, but is often encountered with database applications. You can manage databases on a machine other than the machine that executes the user interface. Section 18.2 compares two different approaches to client-server computing.
- **Distributing data**. A distributed DBMS lets you spread a logical database across multiple machines. Section 18.3 presents a simple process for allocating data.

18.2 CLIENT-SERVER COMPUTING

Client-server computing is an architecture for which a resource, the ***server***, provides computation for multiple components, the ***clients***. As Figure 18.1 shows, a client can access one server or multiple servers. Clients and servers may run on the same or different machines, be written in various languages, and use different operating systems. Client-server computing lets you distribute workload across machines and coordinate the results. We will discuss the aspects of client-server computing that are relevant to database applications.

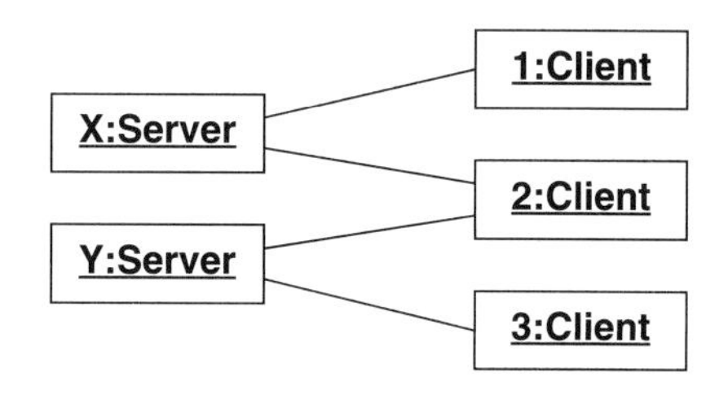

Figure 18.1 Clients and servers have a many-to-many relationship

Client-server computing offers many benefits for database applications:

- **More efficient hardware deployment**. You can use inexpensive machines (such as PCs) for clients and more robust machines (such as UNIX workstations) for servers. You can tightly control and coordinate data on servers, yet quickly respond to individual user requests with clients.
- **Familiar environment**. You can choose client hardware that is easy to use and familiar to users. For example, you can use a PC to run applications on mainframes, workstations, and other PCs.
- **Performance**. A dedicated PC can quickly respond to a user. Contention with other users and applications occurs only upon substantive database activity.
- **Extensibility**. The database that stores data is much more stable than the interface that accesses data. By separating the user interface from data management, you can readily update a user interface without disrupting the schema and data for an application.

As with any technology, there are also disadvantages:

- **Complexity**. With client-server computing you must make more decisions. You must apportion capability between clients and servers.
- **Dependencies**. There is a mutual dependency between the clients and the servers. You can mitigate such dependencies by carefully designing interfaces. You must carefully coordinate updates that affect interfaces.

18.2.1 Three-Tier Architecture

When possible, you should use a three-tier client-server architecture. We wanted to use such an architecture for our full implementation of the portfolio manager, but were unable to do so because of the limitations of MS-Access. (See Chapter 13.) As Figure 18.2 shows, the three-tier architecture cleanly separates data management, application functionality, and the user interface. The data management layer holds the database schema and data. The applica-

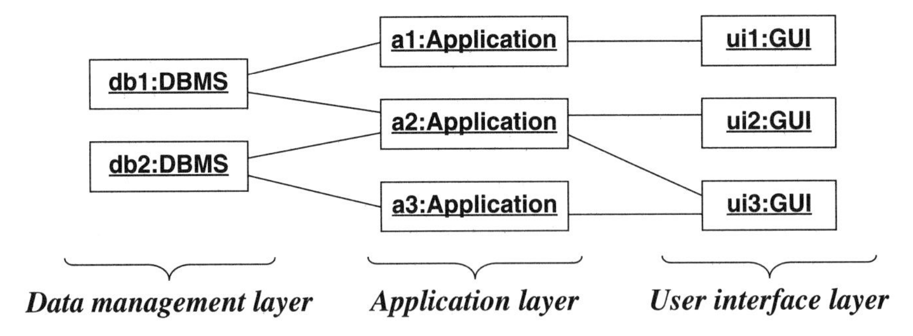

Figure 18.2 Try to use a three-tier client-server architecture for database applications

tion layer holds the methods that embody the application logic. The user interface layer manages the forms and reports that are presented to the user.

The three-tier architecture complements client-server computing. The application layer is a client with regard to the data management layer. The user interface layer is a client with regard to the application layer.

Similarly, object-oriented technology complements client-server computing. The OMT models describe the application layer. This book shows how to drive the OMT models into an implementation of the database layer. Chapter 14 briefly describes the user interface layer; a more thorough treatment of user interfaces is outside our scope.

The exact boundaries between the layers are somewhat subjective. For example, you can place application logic within the data management layer with stored procedures. Some structural constraints are difficult to represent with the data management layer; you can enforce them with the application and user interface layers. Nevertheless, you should try to adhere to the three-tier ideal as much as possible—to increase the flexibility and extensibility of your systems.

You can apply the three-tier architecture to both RDBMS and OO-DBMS applications. For RDBMS applications there is a natural partition between the data that resides in the database and the logic in the application methods. Such a partition is less apparent for an OO-DBMS, because objects combine data and functionality. Nevertheless, you can still use a variation of the three-tier architecture. For an OO-DBMS the data management layer consists of objects that apply to multiple applications and are likely to be reused. The application layer holds objects that are peculiar to an application.

You should design applications so that all layers ensure data integrity. The user interface should check data that is entered into forms and make it easy for users to fix mistakes. The application methods normally perform the detailed checking. The database layer must ultimately safeguard data despite access by multiple users and applications.

You also must carefully consider the scope of a query and the ensuing quantity of data passed between layers. You can reduce network traffic, increase concurrency, and improve responsiveness if the user does not browse large collections of data. For example, for a database of phone numbers you would probably not let the user browse an entire country. Instead you could require that the user be more specific and provide part of a name or a category of interest.

Client-server applications require the use of communications software for connecting clients and servers. The remainder of Section 18.2 discusses two major communications approaches: CORBA* and client-server SQL. CORBA connects clients to servers via explicit calls to methods; clients invoke methods that CORBA dispatches to the appropriate servers. In contrast, RDBMSs use SQL to connect clients to servers implicitly via references to data; clients execute application code that may reference data on servers.

---

* Our purpose is to contrast the RDBMS approach with an object-oriented approach to client-server computing. We present CORBA as representative of an object-oriented approach. Microsoft's OLE/DCOM and the Java class libraries also provide object-oriented approaches to client-server computing.

18.2.2 CORBA

CORBA, the Common Object Request Broker Architecture, is a standard object-oriented architecture for applications. The CORBA object request broker separates clients and servers and mediates their communication. The Object Management Group (OMG)[†] invented CORBA for the purpose of integrating applications and promoting interoperability. The OMG is a nonprofit organization with more than 500 members that was established to foster object-oriented standards. CORBA combines object-oriented computing with distributed computing using the client-server model.

As Figure 18.3 shows, clients and servers communicate through CORBA rather than directly with each other. CORBA decouples clients from servers and lets them evolve separately. Clients and servers must agree only on the request format; the details of internal computations are encapsulated. A portion of the object request broker resides on each client and server. The internal coupling between the server brokers and client brokers accomplishes communications and the processing of requests. The server brokers and client brokers could be from different vendors and still be able to communicate.

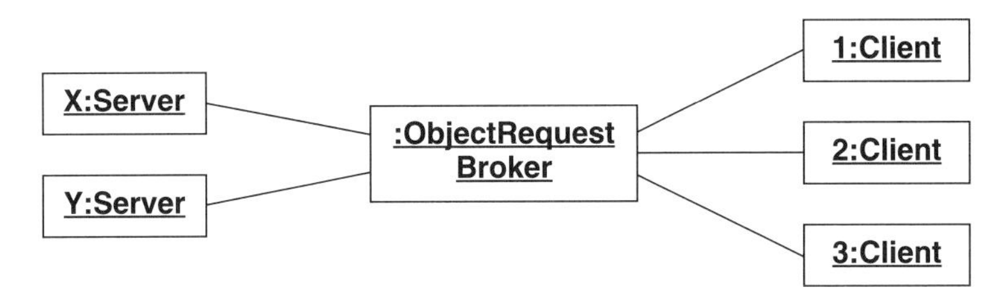

Figure 18.3 Client-server computing with an object request broker

With CORBA you can add new servers without changing client code, or add new clients for existing servers. Furthermore, CORBA clients can dynamically couple to servers. CORBA lets the same software act as a client for one request and as a server for the next request. Another major benefit of CORBA is location transparency; CORBA hides functions such as finding if there is a server, locating the server, and determining how to access the server.

CORBA has benefits beyond distributed computing. CORBA lets your application access code written in another language. CORBA also provides a bridge across operating systems and hardware platforms. For example, you could have a PC client access a UNIX server. The PC client may be written in Smalltalk and the UNIX server in C++.

CORBA supports the three-tier architecture for client-server computing. You can use one CORBA instance to mediate communication between a data manager and application methods. Another CORBA instance can couple application methods to a user interface.

[†] Do not confuse the *OMG* with the *ODMG*. The OMG and the ODMG are distinct standards bodies. The purpose of the OMG is to foster standards for OO computing in general. In contrast, the focus of the ODMG is more specific; the ODMG promotes standards for OO-DBMSs. The ODMG coordinates its efforts with those of the OMG.

18.2.3 Client-Server SQL

RDBMSs offer another approach to client-server computing; they use SQL as a generic language for connecting clients to servers. Portions of the RDBMS software reside on both the client and the server computers, and the RDBMS internally manages communications. You may use the same RDBMS product for the client and server. Or because of the SQL standard, you can sometimes use different products. (See Chapter 13 for a discussion of the SQL standard.)

As Figure 18.4 shows, client-server SQL implementations often assume a two-tier architecture, linking a user interface client to a data management server. Application logic is dispersed between client and server. The precise placement of application methods depends on the nature of the application, the RDBMS product, and network constraints.

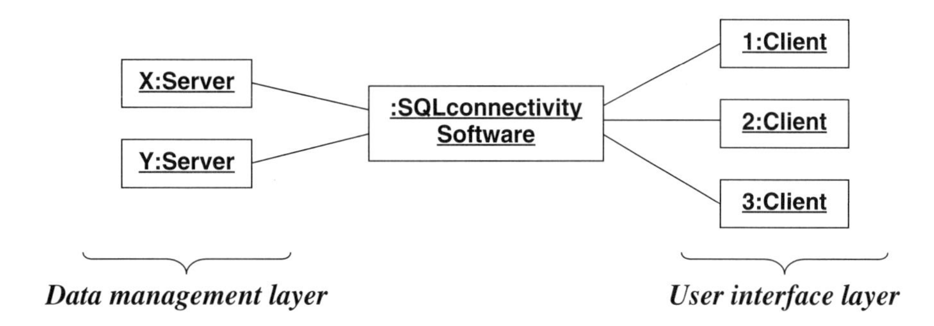

Figure 18.4 Client-server computing with SQL

Oracle's SQL*Net product is an example of a two-tier client-server SQL software. SQL*Net is a heterogeneous network interface that sends SQL commands from the client to the server and returns data. The client provides the user interface and application programming. The server parses and executes SQL statements. The Sybase approach to client-server computing is similar to Oracle, but more of the application logic is located on the server as stored procedures.

Microsoft's MS-Access is a popular product that supports a two-tier architecture. For the portfolio manager case study we defined two MS-Access databases. The server database holds tables, indexes, referential integrity, and other declarative constraints. The client database holds views, queries, macros, methods, forms, and reports. We used MS-Access attachments to communicate transparently between the databases.

18.2.4 Comparison of CORBA and Client-Server SQL

Table 18.1 compares the object-oriented and RDBMS approaches to client-server computing. The choice of the better technique depends on the application problem. For example, CORBA is helpful for complex problems, such as software frameworks. SQL is often appropriate for delivering routine business information systems.

CORBA (object-oriented)	Client-Server SQL (RDBMSs)
Explicitly connects clients to servers via calls to methods	Implicitly connects clients to servers via references to data
More flexible	Less flexible
More effort to use	Less effort to use
No replication	Replication via distributed RDBMS
Imperative language	Declarative language
Cleanly integrates with programming languages	Awkwardly grafts on programming languages

Table 18.1 Different approaches to client-server computing

There is a major difference in style between CORBA and client-server SQL. CORBA adopts the imperative style of object-oriented programming languages; you must reduce application logic to some mix of algorithms and write programming code. In contrast, SQL adopts a declarative paradigm; you state the desired properties of data and the RDBMS chooses algorithms that achieve your stated request. Thus CORBA is organized about operations, while client-server SQL is organized about data.

[Otte-96] presents an electronic mail system as an example of CORBA. A file viewer can act as a server and display files for an electronic mail client. Then the file viewer can act as a client when invoking the print server. SQL cannot express this kind of arbitrary behavior and is not intended to do so. Instead, SQL has been developed to manage data and perform certain common generic operations on data.

CORBA is a lower-level language than SQL. As a consequence, it is more powerful than SQL, but it is also more difficult to use. In contrast, SQL provides a clean approach as long as your problem can be represented with a relational database and your computation adequately expressed with SQL. Essentially, when choosing between CORBA and SQL, you trade the flexibility of a lower-level language for the abstraction of a higher-level language.

18.3 DISTRIBUTING DATA

You can not only distribute computation (Section 18.2), but you can also distribute data. You can implement a database design by spreading the data across several component databases. A distributed DBMS automatically locates the data and apportions computing.

To design a distributed database rigorously, you must perform transaction modeling and estimate the query mix (frequency of reads and writes for the expected queries) at each site and characterize the data. This would allow you to determine the effects of data fragmentation and replication on processing locality. Often it is impractical to obtain such detailed statistics. Our experience has been that you can often accomplish an adequate design with heuristics. We present some implementation guidelines for distributed relational and distributed object-oriented databases.

We suggest the following process for distributing data. (The numbers in brackets refer to sections that explain these tasks in detail.)

- Fragment the schema and data of the database. [18.3.1]
- Allocate fragments to distributed DBMS sites. [18.3.2]
- Replicate data. [18.3.3]
- Provide location transparency. [18.3.4]
- Implement schema for the local databases. [18.3.5]

18.3.1 Fragmenting a Database

The first step in implementing a distributed database is to partition the data and/or the model from detailed design. A distributed database has one conceptual model but may have multiple physical schema, one for each site. You may use schema fragmentation and data fragmentation separately or together to divide a distributed database.

- **Schema fragmentation**. You can partition the types themselves, ultimately assigning the different classes and associations to various distributed database sites.
- **Data fragmentation**. You can assign the actual instances to different database sites.

Schema Fragmentation

In general, you should choose schema fragments to minimize update transactions and referential integrity actions across sites. (Section 18.3.3 discusses the simpler situation in which you are merely reading across sites.) As Figure 18.5 shows, the best approach is to partition an object model across unrelated classes. (The heavy dotted lines are not object modeling notation, but informally denote separate sites.) If you must partition a model across related classes, the next best approach is to disrupt many-to-many associations without qualifiers and link attributes. These many-to-many associations are the simplest associations because they do not require enforcement of maximum multiplicity or existence dependency constraints.

You should try to avoid schema fragmentation across any other associations (that is, associations with qualifiers, link attributes, or non-many-to-many multiplicity). You also should not fragment aggregations, because the aggregation semantics imply a close correspondence between the assembly and the components that will probably be reflected in database access.

As Figure 18.6 illustrates, you should not partition the attributes of a class. An object-oriented approach deals with coherent objects that are described by attributes. You should not compromise the semantics of object encapsulation and object identity by dispersing the contents of an object across multiple sites. Otherwise users may become confused by behavior quirks caused by site failures and network timing delays.

Unrelated classes (the best approach):

A B fragment Site 1 A Site 2 B

Simple many-to-many association (another approach):

A B fragment Site 1 A Site 2 B

Figure 18.5 Recommended approaches for schema fragmentation

Do not partition the attributes of a class:

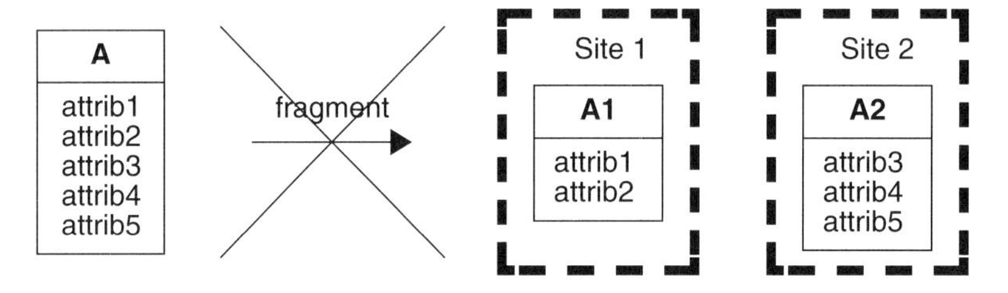

Do not disperse generalization levels across sites:

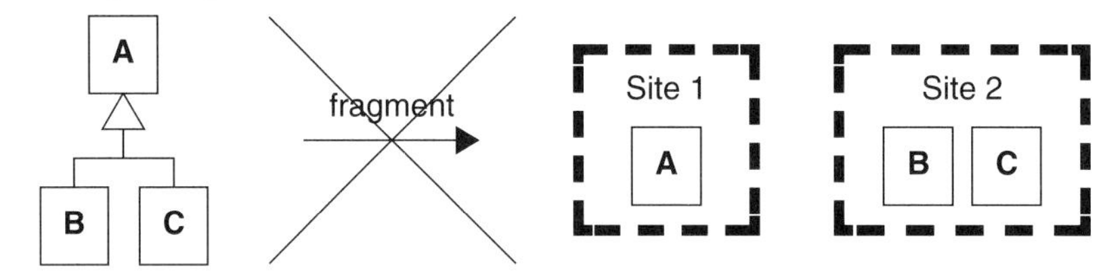

Figure 18.6 Discouraged approaches for schema fragmentation

For similar reasons, you should not distribute across generalization levels. Each generalization level describes different aspects of an object. You should place the superclasses at all sites where subclasses may be found. This advice applies to both single inheritance and multiple inheritance.

Do not confuse fragmentation of object models with the vertical fragmentation of tables discussed in the relational database literature. As Figure 18.7 shows, you can vertically partition columns among multiple tables. With the "old" attribute-driven approach to relational database design (see Section 13.1.3), attributes are empirically grouped into tables based on functional dependencies and normal forms. Since the resulting tables can combine dissimilar things, vertical fragmentation is a reasonable option.

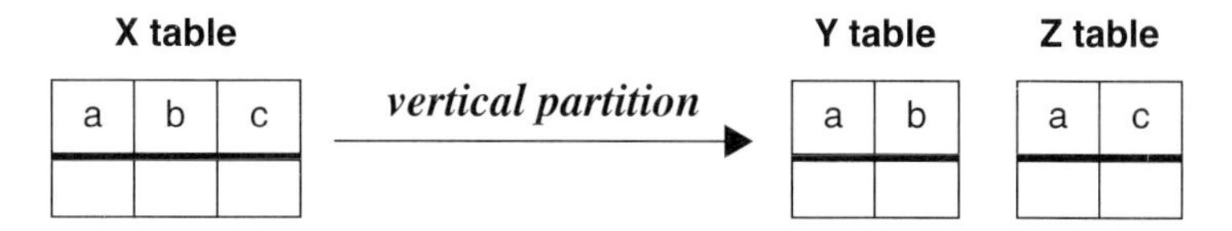

Figure 18.7 Vertical partitioning of tables is different from fragmentation of objects

In contrast, vertical fragmentation is inappropriate for object models. Objects are coherent, atomic things. All the attributes of a class directly describe the objects of the class, and attributes are not juxtaposed in accidental combinations. You should not vertically fragment objects. Otherwise, the vagaries of network traffic and equipment failures may disrupt the intended atomicity of objects and confuse the user.

Data Fragmentation

You can not only distribute a database by fragmenting schema, but you can also fragment data and assign instances to different sites. You could organize instances about a theme that spans classes. For example, we could partition the simple payroll database modeled in Figure 18.8 on the basis of department. Such a partition would be especially helpful if departments were at different locations and had separate computers. Time, location, and important enumeration attributes are common bases for data fragmentation.

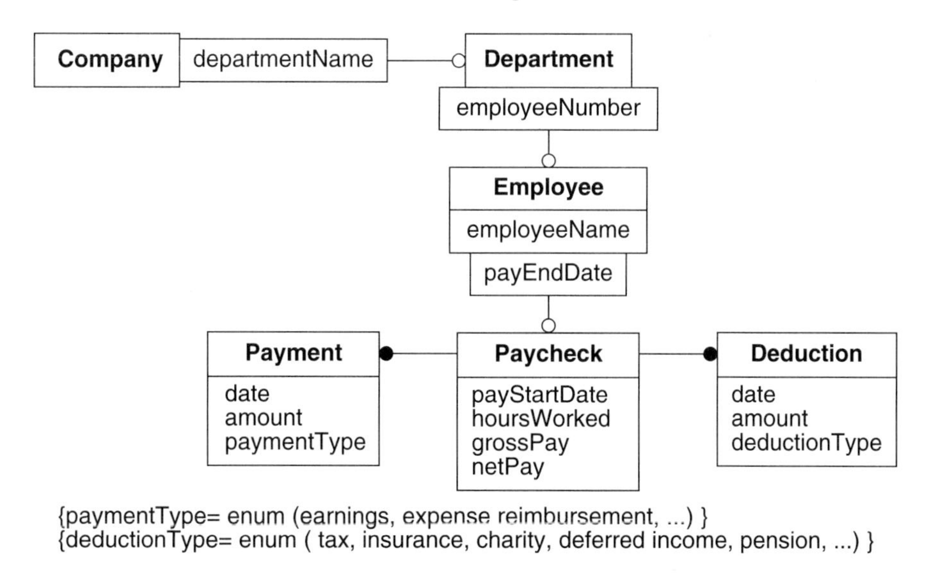

Figure 18.8 An object model for a simple payroll database

You can also use generalization as a basis for spreading objects across distributed sites. In Figure 18.9 subclass schema are assigned to sites; superclass schema are replicated at subclass sites.

Generalization provides a basis for apportioning objects:

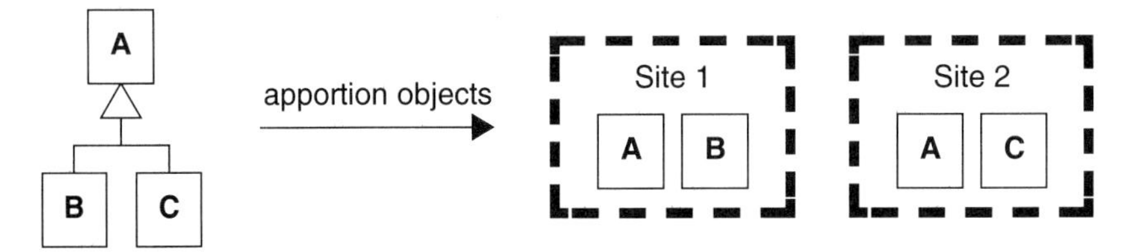

Figure 18.9 An acceptable approach for data fragmentation

You should not fragment versions of an object because it would violate the semantics of object encapsulation and object identity. Furthermore, you should try to avoid fragmenting configurations of versions across sites. (A ***configuration*** is a set of mutually consistent versions.) The semantics of a configuration imply that the constituent versions will frequently be accessed together. When there are many cross dependencies on the same versions, as Figure 18.10 illustrates, fragmentation may be difficult to avoid.

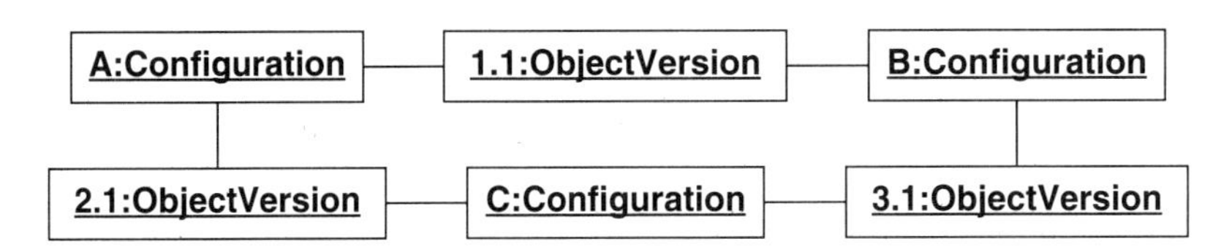

Figure 18.10 Configurations of versions with cross dependencies

Distribution of data can complicate enforcement of candidate keys. A distributed candidate key can occur when instances of a class are located at multiple sites. An application must enforce distributed candidate keys if the DBMS does not. Furthermore, the performance of your application could be impaired by the network traffic from frequent checking. You can resolve this problem by including a site prefix—either explicitly as a separate attribute or implicitly within the substructure of a candidate key attribute. For example, most distributed DBMSs incorporate the prefix of the creation site in object identifiers.

18.3.2 Allocating Fragments to Sites

The next step is to allocate each fragment to a distributed DBMS site. (Some replication policies let you allocate a fragment to more than one site. The next section discusses replication.) You should try to maximize locality by placing data close to the applications that use it. The goal is to achieve fast response time and high availability with minimal consumption of resources. You must consider the capabilities of each site, other (nondistributed DBMS) computing demands, the mix of read and write queries for each site, network capacity, security needs, and effects on integrity. The initial allocation of data to sites will be less critical if the distributed DBMS supports migration transparency.

For example, consider our object model of a simple payroll database in Figure 18.8. If each department has a computer, we could put the payroll information for each department on its computer. Then most accesses could be performed locally; only the occasional query would involve other departments. The multiple payroll sites would tend to generate little network traffic.

18.3.3 Replicating Data

Now you can replicate data at additional sites to improve the overall performance. In general, replication reduces communication costs for reading and complicates updates. When an update occurs, copies of data must also be updated or marked as invalid. The desirability of replication depends on the efficiency and robustness of the distributed DBMS update algorithms. Most distributed DBMSs ensure the correctness of updates—that updates behave the same as for a database without replication. However, a distributed DBMS may be more lax for read queries and allow occasional access to old data.

Replication can improve database availability because data can be found at an alternate site following a site failure. However, redundancy can also reduce availability, depending on the concurrency control mechanism of the distributed DBMS. For example, if a transaction locks all replicates of a data item, then a site failure can block subsequent transactions with the data item. Alternate locking mechanisms (such as primary copy and majority voting; see [Son-88]) can mitigate the effect of failure.

Replication can also improve the reliability of a database—the ability of a system to recover from software crashes, machine crashes, disc failures, and network failures. Reliability is enhanced by storing multiple copies of the same information, as long as different database sites have independent failure modes.

You can adopt various replication policies. For example, you can choose complete replication, allocating a full copy of the database to each network site. A more practical option is selective replication, replicating data that is often read but seldom updated. You should devise a broad policy, because it is unwieldy to choose a replication policy for each individual data item.

In Section 18.3.1 we cautioned that fragmentation of a distributed database must honor the semantics of object models. You do not want to fragment an object model unnaturally and encounter troublesome updates across sites. The fragmentation restrictions do not apply for replicate data that is *only intended for reading*. Partial and summary data is often used for decision support and populating data warehouses.

For example, consider the equipment model in Figure 18.11. The full database contains manufacturing data. A replicated database may just contain *dateOfPurchase* and *cost*, so accounting can compute depreciation for tax returns. It would be acceptable to replicate this summary information, as long as the replicate is used for reading and not writing.

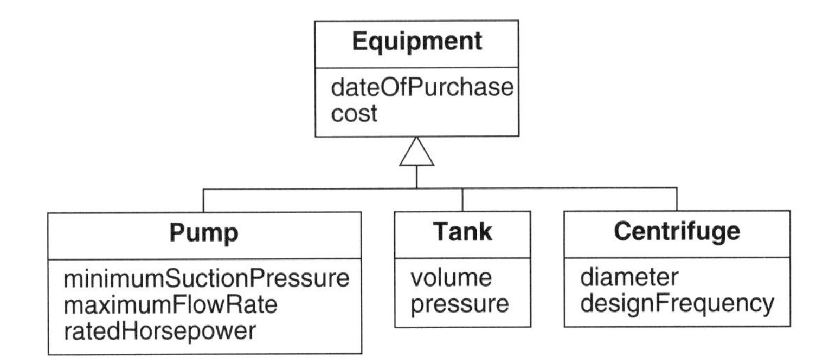

Figure 18.11 An object model for an equipment database

It is important that the initial replication of data be reasonable, but you can tolerate some flaws. The database administrator can refine replication strategy at run-time as performance is monitored and the query mix becomes better known.

18.3.4 Providing Location Transparency

You can achieve some location transparency by replicating data. Recall that we defined location transparency as the ability to write applications independently of data location. You can access a replicate at your local site like any other local data; the distributed DBMS handles the details of synchronization with other sites. However, there are additional mechanisms that you can use to achieve location transparency.

You can use synonyms to hide the location of data. You treat a synonym like any other name, but the DBMS automatically resolves a synonym into its site and base name. RDBMS views can also hide data location; a view defines a virtual table that is computed from other tables (both local and remote), views, and synonyms. RDBMS stored procedures and OO-DBMS methods provide a third mechanism for location transparency. Stored procedures and methods can access remote information without the user being explicitly aware.

The mechanisms for location transparency have different impacts on security. You can usually grant rights on views, stored procedures, and methods to other users. In contrast, synonyms just provide an alias and you cannot grant them. You should be careful to avoid circular references when you define views, stored procedures, and methods.

18.3.5 Implementing Schema for the Local Databases

Finally, you should implement the schema for each of the local databases by applying the techniques discussed in Chapters 13 through 16.

18.4 CHAPTER SUMMARY

As part of architecting a system, you must decide where the data is stored (distribution) as well as how it is stored (persistence). There are two dimensions to distribution for database applications: client-server computing and distributing data.

Client-server computing is an architecture in which a resource, the server, provides computation for multiple components, the clients. When possible, we recommend that you use a three-tier client-server architecture consisting of data management, application, and user interface layers. The data management layer holds the database schema and data. The application layer holds the methods that embody the application logic. The user interface layer manages the forms and reports that are presented to the user.

We discussed two kinds of communications software for connecting clients and servers. CORBA is a standard object-oriented architecture for connecting clients to servers; clients invoke methods that CORBA dispatches to the appropriate servers. In contrast, RDBMSs use SQL to connect clients to servers implicitly via references to data; clients execute application code that may reference data on servers. CORBA and SQL provide valuable, but very different, approaches to client-server computing.

You can not only distribute computation, but you can also distribute data. You can implement a database design by spreading the data across several component databases. You should begin by fragmenting the database—by partitioning the data, model, or both from detailed design. A distributed database has one conceptual model but may have multiple physical schema, one for each site. Our fragmentation guidelines are consistent with the semantics of object-oriented models. Then you must allocate fragments to physical database sites, attempting to achieve a fast response with minimal consumption of resources. The third step is to replicate the data selectively. You can disperse multiple copies of data, and the distributed DBMS keeps them consistent. Finally, by carefully defining synonyms, views, stored procedures, and methods, you can hide the precise location of data and increase the flexibility and maintainability of your applications.

Figure 18.12 lists the key concepts for this chapter.

client-server architecture	data replication	migration transparency
client-server SQL	distributed database	replication transparency
CORBA	distributed DBMS	schema fragmentation
data fragmentation	location transparency	three-tier architecture

Figure 18.12 Key concepts for Chapter 18

BIBLIOGRAPHIC NOTES

[Ceri-84], [Chin-91], and [Elmasri-94] thoroughly discuss implementation of distributed DBMSs and their external properties. [Son-88] summarizes different replication strategies that DBMSs use. Chapter 7 of [Loomis-95] has a good explanation of some of the factors that you must consider when devising a replication strategy.

[Otte-96] provides an articulate explanation of CORBA. [Brockschmidt-95] explains the historical evolution and capabilities of OLE, which is Microsoft's approach to distribution and object orientation. [Orfali-96] explains the many architectural nuances of client-server systems. We highly recommend these last three books for further reading.

REFERENCES

[Brockschmidt-95] Kraig Brockschmidt. *Inside OLE,* 2nd edition. Redmond, Washington: Microsoft Press, 1995.

[Ceri-84] Stefano Ceri and Giuseppe Pelagatti. *Distributed Databases: Principles and Systems*. New York: McGraw-Hill, 1984.

[Chin-91] Roger S Chin and Samuel T Chanson. Distributed object-based programming systems. *ACM Computing Surveys 23*, 1 (March 1991), 91–124.

[Elmasri-94] Ramez Elmasri and Shamkant B Navathe. *Fundamentals of Database Systems*. New York: Benjamin Cummings, 1994.

[Loomis-95] Mary E S Loomis. *Object Databases: The Essentials*. Reading, Massachusetts: Addison-Wesley, 1995.

[Orfali-96] Robert Orfali, Dan Harkey, and Jeri Edwards. *The Essential Client/Server Survival Guide,* 2nd edition. New York: Wiley, 1996.

[Otte-96] Randy Otte, Paul Patrick, and Mark Roy. *Understanding CORBA: The Common Object Request Broker Architecture*. Upper Saddle River, New Jersey: Prentice Hall PTR, 1996.

[Son-88] Sang Hyuk Son. Replicated data management in distributed database systems. *SIGMOD Record 17*, 4 (December 1988), 62–69.

19

Integration of Applications

Today, most corporate information systems consist of many "islands of automation." Most software is developed to service the needs of a single application with little regard for the effect on a corporation as a whole. The individual applications perform well alone, but do not collaborate for the overall benefit of a corporation. As a consequence, there is much tedious re-entry of data. It is difficult to analyze data that spans applications. And much of the potential benefits of automation are wasted. The purpose of integration technology is to facilitate information exchange between applications.

With the current state of the art, the focus is on integrating data. Thus this chapter emphasizes the object model. Each application has its own database, the logical content of the databases partially overlap, and the databases must be made to exchange common data. There may be multiple database paradigms (hierarchical, network, relational, and object-oriented) and multiple operating systems, though most applications to be integrated use a relational database. The legacy applications that use hierarchical and network DBMSs are often only partially understood, brittle, and too difficult to integrate. The advanced applications that use object-oriented DBMSs tend to be removed from the business mainstream and have less overlap.

Future technology will also address integration of behavior; integrated behavior can then be attached to integrated data. Integration of data causes applications to have consistent data. However, integration of behavior is required to have a common look and feel across applications. As more applications are built using class libraries, applications will become easier to customize and integration of behavior will become practical.

19.1 OVERVIEW

We recognize that there are significant advantages of the application-centric development that has caused the "islands of automation." Applications are justified by a business need. Application-centric development is focused on servicing the business need and quickly realizing the potential payback. With application-centric development, projects are more readily managed and it is easier to define and measure success.

Nevertheless, as information systems increase in size and complexity, many organizations are finding the limitations of application-centric development. There are many business motives for integrating applications.

- **Cost reduction**. It is wasteful for applications to acquire the same data repeatedly. Furthermore, it is costly to maintain a patchwork of interfaces. Integration technology can reduce multiple entry of data and provide a disciplined approach to connecting applications.
- **Data integrity**. It is relatively easy to store data. It is more difficult to store information—to ensure data is correct, consistent, and timely. By reconciling applications, integration can improve consistency, deepen the understanding of data, and improve the quality of the individual application databases.
- **Greater flexibility**. Information systems must be able to respond rapidly to business opportunities. Information systems should enable, not inhibit, business decisions. Applications are inflexible when they are balkanized and changes have an unknown effect on other applications. Integration technology can reduce the stasis of a network of applications, making changes easier to perform and the effects more predictable.
- **More functionality**. Applications require a substantial investment. It is desirable to exploit application synergy to derive additional business benefit. Data is a corporate asset transcending the needs of a single application. Integration can effectively span applications, enabling information mining and decision support.

By definition, integration must cope with heterogeneity. Applications come from various sources: legacy, new development, and purchases. Applications are developed at different points in time by different project teams using different DBMSs. Integration allows an evolutionary, rather than revolutionary, improvement to the existing, application-centric practice. You can gradually and incrementally assimilate applications into an integration architecture with little disruption to other applications.

We recommend the following process for achieving integration. Of course, you need not carry out these steps in strict sequence, and can perform some tasks in parallel. (The numbers in brackets refer to sections that explain these tasks in detail.)

- **Application modeling**. Develop analysis models for each individual application.
- **Enterprise modeling**. Integrate the application object models to form an enterprise model. An enterprise model spans applications and provides a focus for integration. [19.2]
- **Integration techniques**. Choose a combination of integration techniques for relating the application models to each other and an enterprise model. [19.3]
- **Integration architecture**. Forge a full integration architecture using the combination of integration techniques as the core. [19.4]
- **Data warehouse**. Use the integrated applications and enterprise model as the foundation for constructing a data warehouse. [19.5]

19.2 ENTERPRISE MODELING

The previous chapters have been concerned with models for individual applications. You can use models to build superior, extensible applications in less time than with an undisciplined approach. But models can serve purposes beyond application development; the models that improve application development are precisely the artifacts needed for integration. Object-oriented models are particularly appropriate because they are expressive, span database paradigms, and provide a path for evolving from the current emphasis on data to a future emphasis on objects. We now discuss the use of models for integration.

An ***enterprise model*** is a model that describes an entire organization or some major aspect of an organization. An enterprise model abstracts multiple applications, combining and reconciling their logical content. By constructing an enterprise model, you can acquire a high-level understanding of the relationship between applications and the processing that must occur to exchange data. You can also use an enterprise model to seed the development of new applications—to provide reuse of models.

There may be multiple tiers of enterprise models. For example, in Figure 19.1 finance, engineering, manufacturing, and sales may each have their own enterprise model that reconciles application models. You can then reconcile these intermediate enterprise models to form the overall enterprise model. In the remainder of this chapter for simplicity we only consider a single enterprise model.

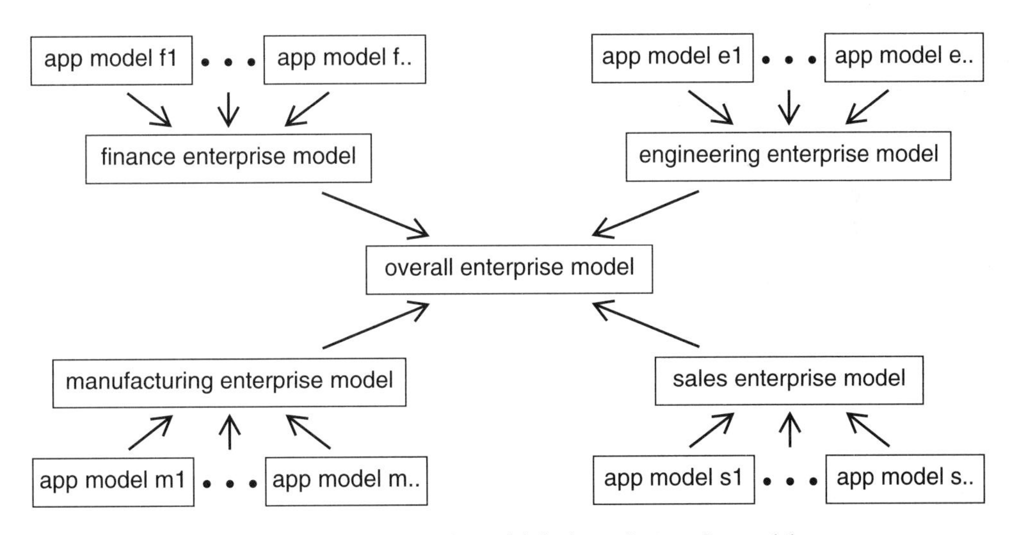

Figure 19.1 There may be multiple tiers of enterprise models

You can present an enterprise model at different levels of abstraction. Executive management may wish to view a simplified enterprise model summarizing the entire corporation. In contrast, the exchange of application data requires a detailed enterprise model.

You should construct an enterprise model by reconciling models rather than physical schema. It is just too difficult to construct an enterprise model directly from physical schema and simultaneously address the artifacts of implementation as well as different conceptualizations. If you lack models for applications, then you will first need to perform reverse engineering. (See Chapter 20.) It is more practical first to abstract physical schema into models and then reconcile the models.

You need not incorporate the entire contents of application models in the enterprise model. However, you should include modeling constructs that are currently shared between applications or likely to be shared in the future. Thus the enterprise model need not be the union of the application models, but must encompass the intersection. The reason for making the enterprise model the intersection is to make it smaller in size and easier to manage.

In principle, an enterprise model could encompass all three OMT models. However, given that the focus of this book is database applications, we will only consider the object model. In industrial practice, enterprise functional models are often constructed to obtain an overall understanding of the business and detect redundant and wasteful business practices. We have not seen the dynamic model used at the enterprise level.

The remainder of Section 19.2 presents a process for constructing an enterprise model consisting of the following steps:

- Choose a development approach—top-down versus bottom-up. [19.2.1]
- Choose a sequence for merging applications. [19.2.2]
- Compare application models and find discrepancies. [19.2.3]
- Resolve discrepancies. [19.2.5]
- Verify the enterprise model. [19.2.6]

19.2.1 Choosing a Development Approach

In principle, there are two approaches to developing an enterprise model—top-down or bottom-up. With the top-down approach you construct a high-level model of an organization and then elaborate the model until you have sufficient detail. In contrast, the bottom-up approach synthesizes an enterprise model by combining the models of relevant applications.

Top-down development has the advantage of yielding a coherent model. In contrast, the bottom-up approach must cope with disjointed application models built by different modelers over a period of time. Essentially, top-down modeling provides a clean start apart from the problems of the applications.

But top-down development has disadvantages. The most serious is that it can lack focus and become an obscure exercise without value to an organization. A top-down model may also lack the detail needed for application integration.

We recommend that you develop an enterprise model mostly bottom-up. The applications are important to an organization and have the business information, so you should use them to drive construction. You should successively rationalize applications with an enterprise model and add their information. We do not favor top-down construction, because there is too much risk that the modeling effort will become excessive and disconnected from reality. However, it is appropriate to guide construction of an enterprise model with a top-down vision.

19.2.2 Choosing a Sequence for Merging Applications

As Figure 19.2 shows, there are several different sequences you can use for merging application models [Batini-86]. We normally use the binary ladder merge.

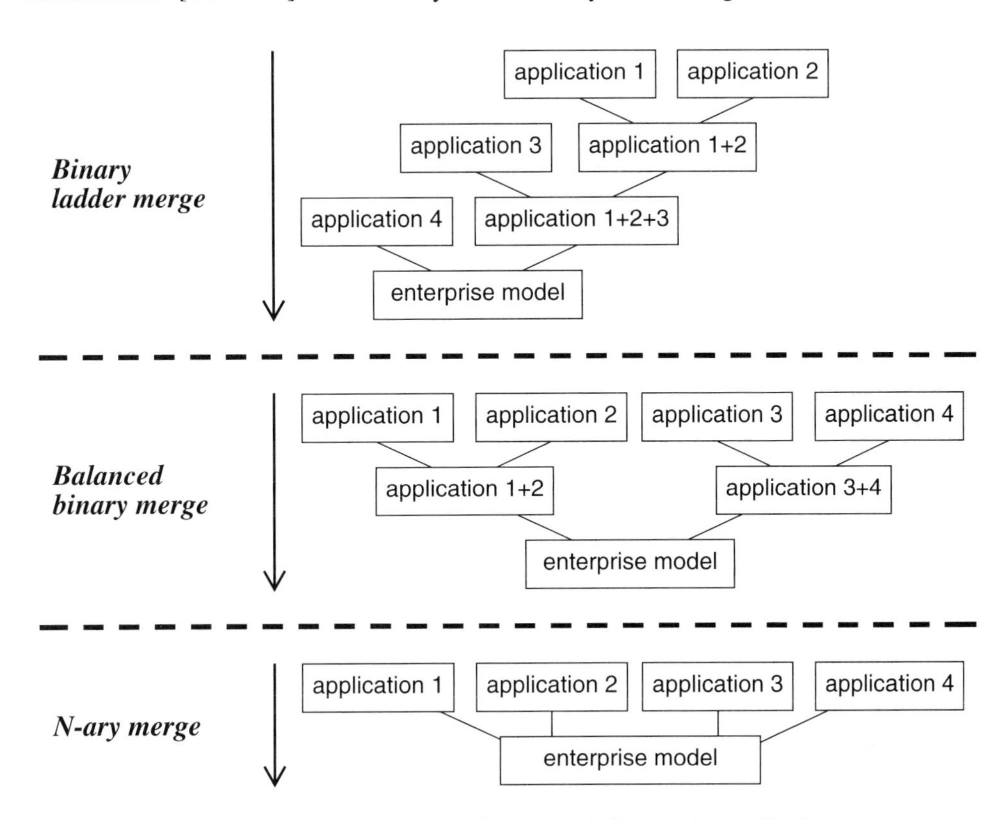

Figure 19.2 Use the binary ladder approach for merging applications

- **Binary ladder merge**. Merge two application models to form an intermediate model. Then add one application model at a time to the intermediate model until all applications are merged. The binary ladder merge has the advantage that you can integrate applications in decreasing order of relevance. So you can more easily bias an enterprise model toward the perspective of the more important applications.
- **Balanced binary merge**. Merge pairs of the application models to form *n*/2 intermediate models. Then merge the pairs to form *n*/4 intermediate models. Eventually only one model remains—the enterprise model.
- ***N*-ary merge**. Simultaneously merge multiple application models, all in one pass or in successive passes, as with the balanced binary merge.

19.2.3 Comparing Application Models

After you have chosen a sequence for merging applications, you compare application constructs and detect possible conflicts. We recommend that you begin by comparing classes. Corresponding classes will often have the same name or similar names and many of the attributes and relationships will match. You should compare definitions for similar classes to increase your confidence that they actually match.

Once you have found matching classes and classes without counterparts, you can proceed at a deeper level. Then you can compare associations, generalizations, and attributes. You should keep in mind that there may be some cross matches. Some classes will match associations, some classes will match attributes, and so on.

A model repository can facilitate comparisons. A repository can store the application models, the evolving enterprise model, the mapping between the application models and the enterprise model, and clarifying comments. The repository manages the accumulated data against accidental loss and provides summary queries and reports.

As you compare models, you will encounter several kinds of integration problems:

- **Naming conflicts**. The same concept may appear with different names (***synonyms***). For example, *customer* and *client* may refer to the same concept. Similarly, the same name may apply to different concepts (***homonyms***). For example, *account* may refer to a sales account or a general ledger account. You can often detect homonyms by noting classes with the same name that have inconsistent attributes, associations, or generalizations. Synonyms are more difficult to recognize and often require deep understanding of the models.

- **Domain conflicts**. You will encounter conflicts for attributes in data types, field length, and units of measure. You may encounter different formats for dates such as *dd/mm/yyyy* and *mm/dd/yyyy*. Applications may encode enumerations in different ways.

- **Type conflicts**. A class, association, or attribute could be used to model the same concept. For example, *city* may be a class in one application and an attribute in another application.

- **Multiplicity conflicts**. Attributes and roles may be optional (minimum multiplicity of zero) or mandatory (minimum multiplicity greater than zero). Similarly, attributes and roles may be single valued or multivalued.

- **Identity conflicts**. Different implementations of identity (see Chapter 9) may cause confusion. For example, in one application a class may have intrinsic identity; in another it may have derived identity. One application may model a concept as a class and another may show an association class.

- **Semantic conflicts**. All references to the same concept must observe the same constraints and have the same meaning.

19.2.4 An Example of Integration

Figure 19.3 presents a simple example—integrating a model of customer purchases with a model of data available from the post office. In the first application a customer has many locations that conduct business; we wish to track product purchases for each location. In the second application we have a cascade of qualifiers for identifying an address. *Missouri* is an example of a state name and *MO* is a state abbreviation.

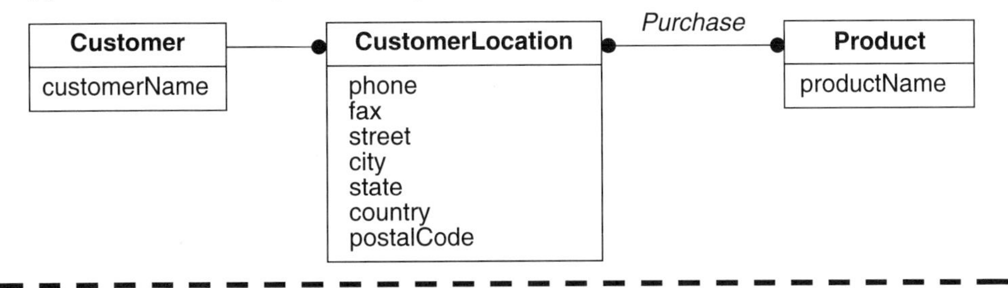

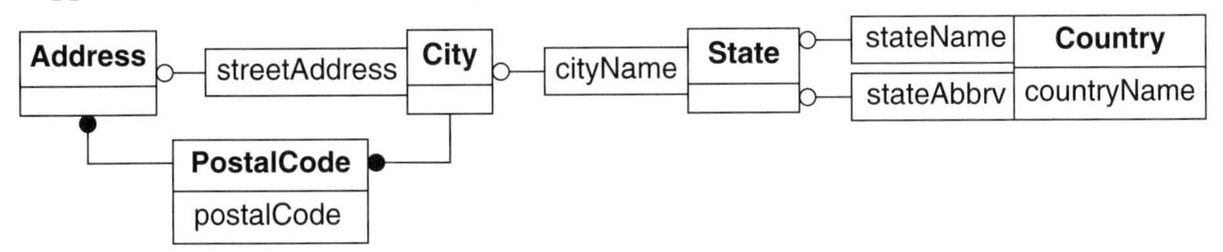

Figure 19.3 An integration example

A company may want to integrate these applications to increase the accuracy of customer data and to reduce data entry. If we preloaded postal data, a clerk could merely enter the street address and postal code for a customer purchase. Software could then automatically use the postal code to populate the city, state, and country.

As we compare these two models, it becomes apparent that there is substantial overlap, yet the models do not directly align. *CustomerLocation* overlaps the postal data and *Customer* and *Product* are disjoint from the postal data. All the postal classes overlap the purchase information.

This simple example illustrates many kinds of integration problems. There appear to be naming conflicts (such as *CustomerLocation* corresponds to *Address* and *city* corresponds to *cityName*). There is also a domain conflict; it is not clear whether *CustomerLocation.state* refers to a state abbreviation or a state name. The applications have type conflicts, as several *CustomerLocation* attributes correspond to qualifiers in the postal model. Furthermore, the purchase model is a weaker model than the postal model. In the purchase application postal code, city, state, and country are distinct attributes. In the postal model, postal code implies a unique city, state, and country.

Table 19.1 maps the attributes and roles of the two models, as might be stored in a model repository, to document our comparisons.

Customer purchase	Postal data
Customer.customerName	
CustomerLocation.phone	
CustomerLocation.fax	
CustomerLocation.street	Address@City.streetAddress
CustomerLocation.city	Address.City@State.cityName
CustomerLocation.state	Address.City.State@Country.stateAbbrv, Address.City.State@Country.stateName
CustomerLocation.country	Address.City.State.Country.countryName
CustomerLocation.postalCode	Address.PostalCode.postalCode
Product.productName	
CustomerLocation.Customer	
CustomerLocation.Product	
	PostalCode.City

Table 19.1 An attribute-to-attribute mapping for the integration example

19.2.5 Resolving Discrepancies

Next we must reconcile the application models. The mappings already couple corresponding constructs. You may have the following possible outcomes from comparing application models, and we list the appropriate response:

- **Identical constructs**. Do nothing.

  The *postalCode* in the purchase application exactly matches the *postalCode* in the postal data.

- **Conformable constructs**. You should note the transformations that convert the application model to match the enterprise model.

  In Figure 19.4 we transform address information in the purchase application to match the postal application; we wish to bias the enterprise model toward the postal model of address. First we apply transformation *T9a* from right to left (see Chapter 10) four times to fragment address information; these transformations are equivalence transformations because we have added no new information. The next steps convert several object attributes to link attributes and then convert the link attributes to qualifiers. Note

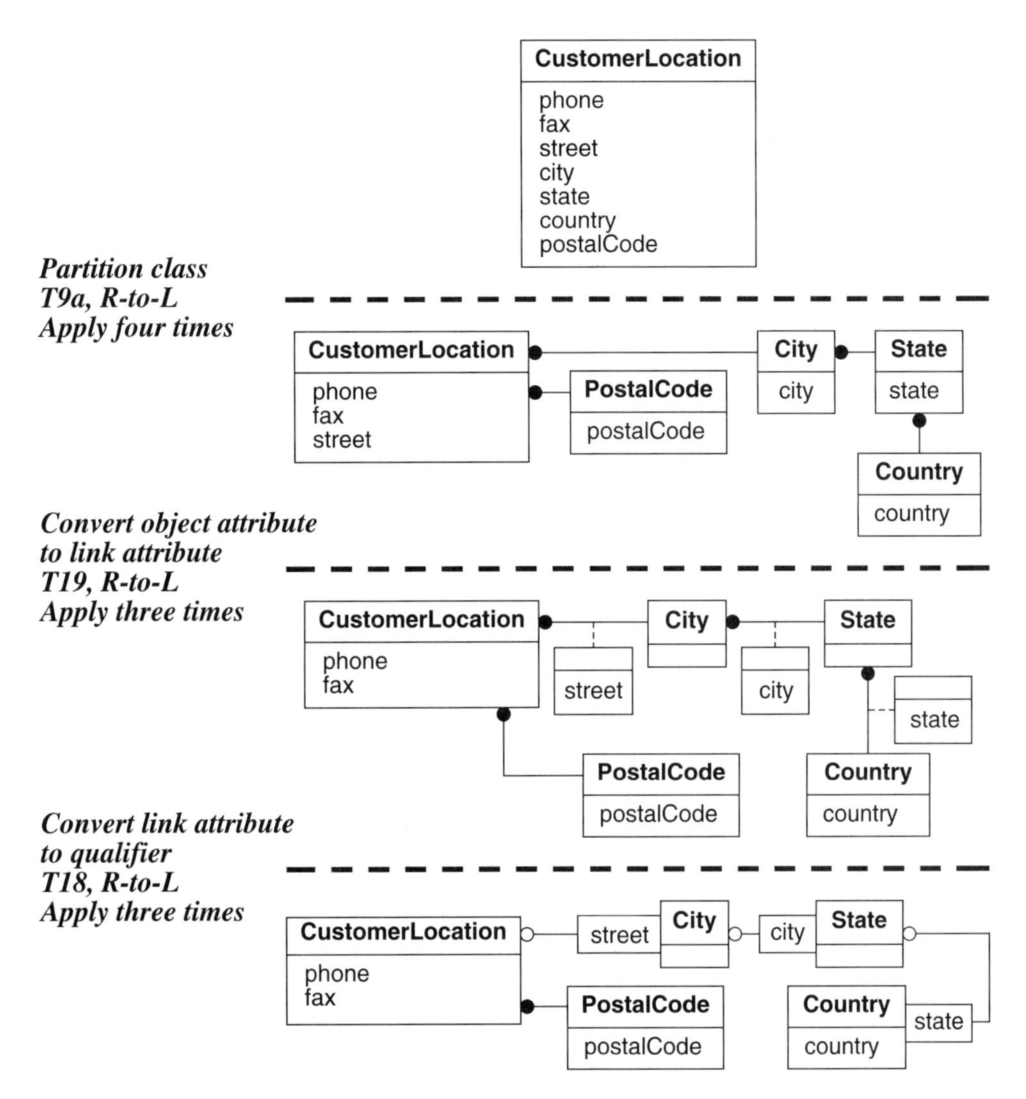

Figure 19.4 Transformations to reconcile applications for the integration example

that the combination of *City* and *street* can yield multiple *CustomerLocations*; this allows for multiple businesses being located in the same building.

- **Compatible constructs**. The enterprise model should combine the relevant information of the merged applications.

  The *Customer@CustomerLocation* and *Purchase* associations, as well as *fax* and *phone* attributes, only appear in the purchase application. The *City@PostalCode* association and an extra *state* attribute (*stateAbbrv* or *stateName*, whichever does not match *state* from the purchase application) only appear in the postal data.

- **Incompatible constructs**. Models can be contradictory because of different requirements or errors. You must fix any errors in the application models. If application requirements contradict, you must choose the requirements that apply to the enterprise. Often a difference in requirements is not deliberate, but is due to oversight.

  It is not clear whether *state* in the purchase application refers to *stateName* or *stateAbbrv* in the postal data. Fortunately, we can simply resolve this problem by including both *stateName* and *stateAbbrv* in the enterprise model. We would need further information to map state data more precisely.

Figure 19.5 shows the final enterprise model. We have included all the classes because we are assuming that they are all important for integration. Typically, when we construct an enterprise model that spans large applications, we only include constructs that are common to multiple applications or likely to be reused by future applications.

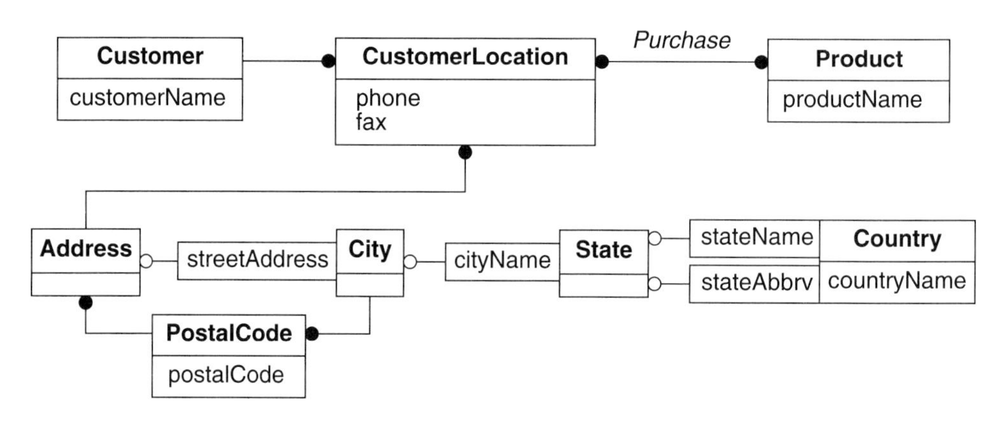

Figure 19.5 An enterprise model for customer purchases and postal data

19.2.6 Verifying the Enterprise Model

Finally, you should ensure the quality of your model by checking the following criteria:

- **Completeness**. All relevant application information must appear in the enterprise model. You may also need to add associations and generalizations to relate constructs from different application models. For example, in Figure 19.5 we added an association between *Address* and *CustomerLocation*.
- **Minimality**. Beware of redundancy in the enterprise model that accumulates from integrating successive waves of applications. You must deeply reconcile applications to minimize redundancy.
- **Correctness**. Make sure you are satisfied with the resolution of conflicting information. The enterprise model must correctly express the requirements from the perspective of the entire organization.

- **Understandability**. The enterprise model can become quite large when there are many applications. You should organize a large enterprise model into packages, as with any application model (see Chapter 3). In addition, you should prepare a narrative and data dictionary to document the enterprise model.

We are satisfied with the enterprise model for our simple example.

19.3 INTEGRATION TECHNIQUES

Now we discuss three techniques for connecting applications. In practice, you need not choose one technique to the exclusion of the others. In fact, we normally use a combination of the integration techniques with an emphasis on indirect integration.

19.3.1 Master Database

An obvious approach to integration is to require that all applications store data in a single enterprise database (called the master database) denoted by the solid line cylinder in Figure 19.6. (Multiple databases can still arise through distribution, but distribution is not pertinent to this discussion. We are focusing on logical databases.) Then each application operates on its own view of the enterprise database (shown as dotted line cylinders). For this approach the enterprise database subsumes and thoroughly integrates the individual applications.

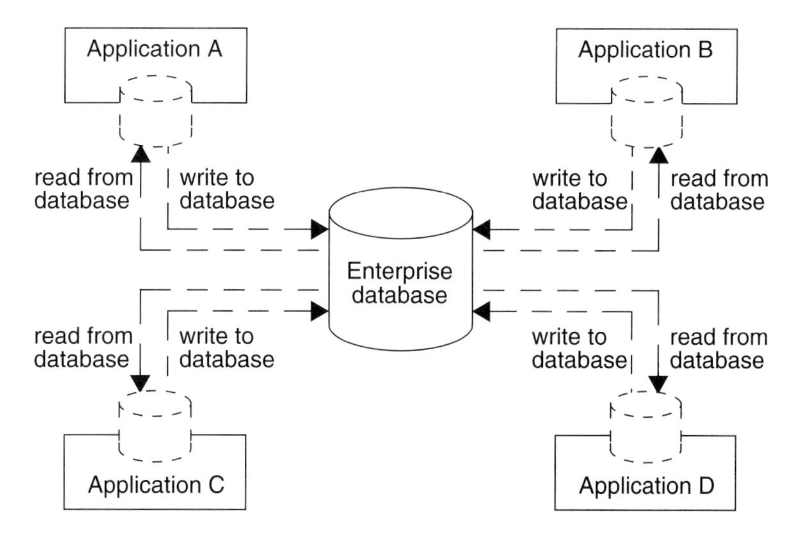

Figure 19.6 An integration technique: Master database

There are two ways of realizing the application databases. You can dynamically materialize the application databases as needed—with RDBMS views or OO-DBMS methods. Alternatively, you can physically materialize an application database with a long transaction.

An extract of the enterprise database is checked out and written to an application database, the application does its computation, and then the application data is checked in and written back to the enterprise database.

There are many advantages of integration with one database. The thorough integration is readily apparent to the end user; by definition, there are few inconsistencies between applications. The deep understanding that is needed to achieve a single database improves the extensibility and quality of applications. The master database technique is facilitated when a standards organization promotes acceptance of the master model.

Unfortunately, a master database can normally be accomplished only for small groups of applications. A master database is impractical for a large number of applications because developers must revise or rewrite too many applications. Thus in industrial settings we must cope with multiple databases, each with their own particular models. The master database technique is useful for enlarging the "islands of automation" and reducing the number of islands with which the other two integration approaches must deal.

The master database approach described here corresponds to the *three-schema architecture* that has been widely discussed in the literature [Tsichritzis-78]. In Figure 19.7 the conceptual schema is the same as the enterprise model and reconciles the partially overlapping applications. The external schema simplify the conceptual schema and correspond to the application models in Figure 19.6. The internal schema implement the conceptual schema for a particular DBMS and deal with product limitations and features.

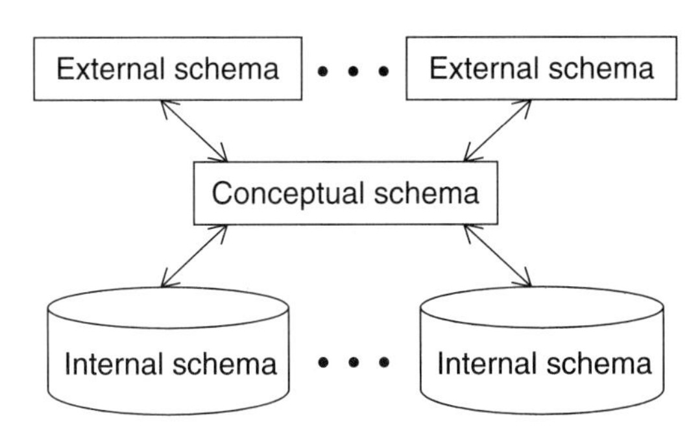

Figure 19.7 The three-schema architecture

19.3.2 Point-to-Point Interfaces

Another obvious integration approach is to connect pairs of applications directly. In Figure 19.8 there are two interfaces for each pair of applications that must exchange data. One interface reads from the first application and writes to the second; the other interface reads from the second application and writes to the first. The integration mechanism for point-to-point integration does not require an enterprise model, though an enterprise model might still be useful for deepening understanding and seeding new application models.

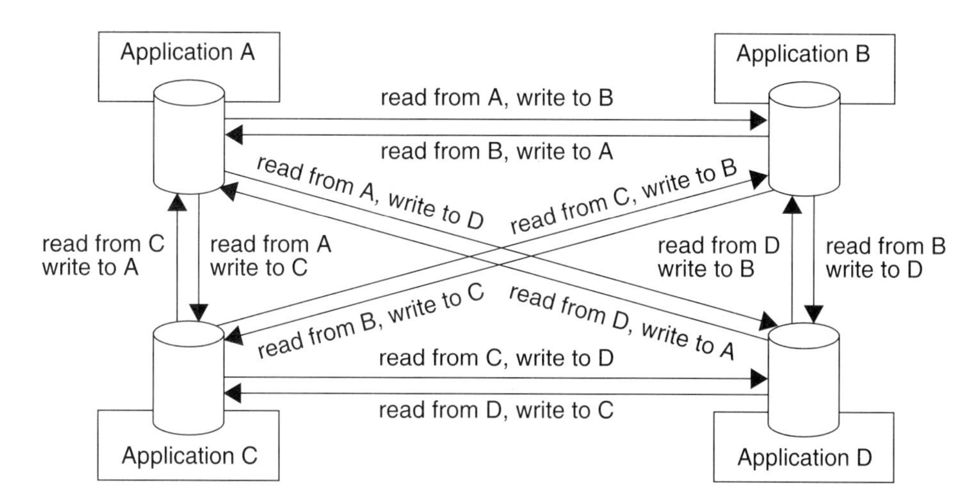

Figure 19.8 An integration technique: Point-to-point interfaces

As with the master database approach, there are different ways of realizing point-to-point interfaces. If the applications are batch programs that are run serially, the interfaces can be separate programs that wrap around application execution. If the execution of applications is interleaved, the interfaces could be a collection of RDBMS triggers and stored procedures or OO-DBMS methods that synchronize data as execution proceeds.

There are advantages to point-to-point integration. First, this approach is straightforward to understand and implement; the developer must deal with only two applications at a time. There is little disruption to ongoing business activities because each application can be run in isolation and then integrated when the interfaces become available. Point-to-point integration is often the preferred approach when there is little commonality between applications or there are few applications. Also, point-to-point integration can efficiently connect two applications with a high volume of traffic. Some vendors provide point-to-point interfaces, connecting their application to other applications.

However, there are severe drawbacks to broad use of point-to-point integration. The most serious problem is that the interfaces create a web of dependencies for a large number of applications. If n is the number of applications, then there could be as many as $n(n-1)$ dependencies. As the number of integrated applications grows, adding the next application becomes increasingly painful. Maintenance of the interfaces can become more difficult than the maintenance of the applications themselves. A few point-to-point interfaces are certainly acceptable. However, widespread use of point-to-point integration will frustrate the goals of integration (the ability to respond quickly and flexibly to changes in business processes).

In practice, every pair of applications will not exchange data. Thus n applications will not necessarily have $n(n-1)$ interfaces. However, it is often the case that enough applications will exchange data for point-to-point integration to be an $n \times n$ algorithm.

19.3.3 Indirect Integration

The third approach, indirect integration, is the most complex and provides the backbone of a robust, scalable integration strategy. In Figure 19.9 applications communicate indirectly via the enterprise database rather than directly with each other. The enterprise database holds metadata that maps read and write requests between the applications. Unlike the master database approach, indirect integration does not require application modifications and instead superimposes an integration layer.

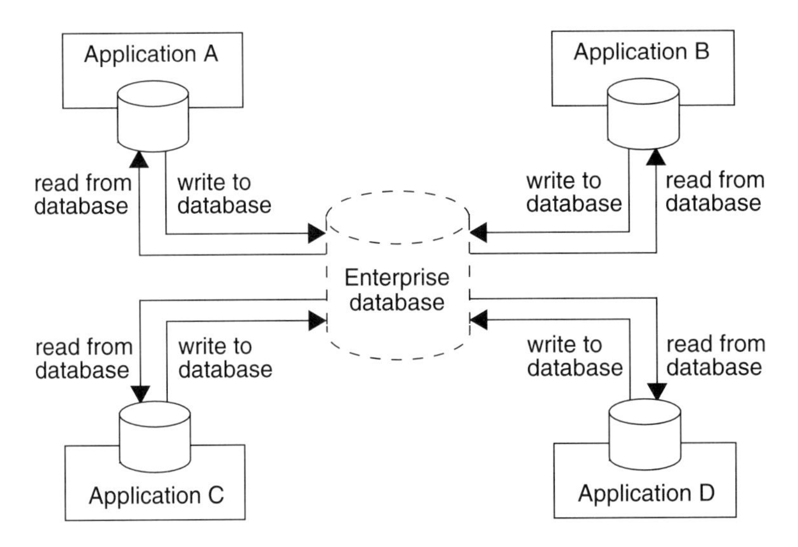

Figure 19.9 An integration technique: Indirect integration

You can implement indirect integration by hard-coding mappings between each application and the enterprise database, such as with RDBMS triggers and stored procedures or OO-DBMS methods. Or you can elevate mapping data to a metalevel and use generic code to effect the mappings. The enterprise database may permanently store data (redundant with application data), or it may simply serve as a transient staging area for copying data between applications. The decision to store data permanently or transiently is purely a design optimization that does not affect the complexity of indirect integration.

The most significant advantage of indirect integration is scalability; for each new application you need only add two interfaces for reading from and writing to the enterprise database. This contrasts starkly with point-to-point integration, where you must add *n* interfaces for each new application. Furthermore, with indirect integration you need not disrupt applications to achieve integration—in contrast to the master database approach. Indirect integration is clearly the preferred technique for coping with a large number of applications. You can use a master database and point-to-point integration within the context of indirect integration effectively to provide larger islands of automation.

The most severe disadvantage of indirect integration is complexity; metadata and generic software are difficult to implement and difficult to understand. However, you have little choice but to accept the complexity, if you desire an integration technique that gracefully scales with an increasing number of applications. Indirect integration becomes more difficult when application models and their implementations are greatly dissimilar.

Indirect integration corresponds to the *five-schema architecture* summarized in Figure 19.10 [Sheth-90]. Each application or group of applications integrated about a master database has its own internal schema and exports a schema to the federation. The federated schema corresponds to the enterprise model. The external schema then make portions of the applications and federated schema available for enterprisewide queries.

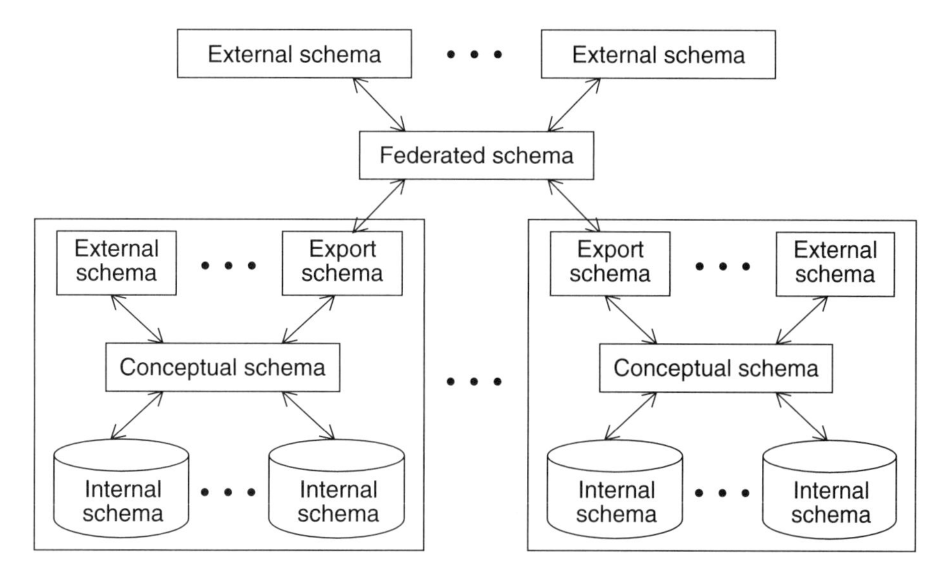

Figure 19.10 The five-schema architecture

19.4 INTEGRATION ARCHITECTURE

The core of an integration architecture is a blend of the three techniques from the previous section. However, you must consider several more issues for a complete integration architecture.

19.4.1 Identity Applications

Proper use of the integration techniques will yield an integration architecture that is flexible, robust, and linearly scalable with the number of applications. However, integration is still difficult. You have the enterprise-modeling problem of determining the precise correspondence of data and the operational problem of performing data exchange. You can further reduce the complexity of integration by establishing several identity applications.

An ***identity application*** is a dedicated application that enforces the identity of an important abstraction and serves as a resource for other applications. An identity application should service all requests to create a new instance of such an abstraction. It simplifies integration by providing an anchor point that cuts across applications and should lie at the core of the business and permeate many applications. For example, many applications refer to customers and parts; *Customer* and *Part* would be good candidates for identity applications for many companies.

Let us consider a *Customer* identity application. If each application can individually name customers, you can encounter many variations, such as *AT&T*, *A.T.&T.*, and *American Telephone and Telegraph*. With this kind of naming variation, it becomes difficult to combine data. In practice, integration does not fully occur and information is lost for some queries. In contrast, a *Customer* identity application would dictate the precise name for a company, and the name would be used uniformly across other applications. Then a compilation of *Customer* data will find all related data. Subsequent queries on other abstractions might still fail to match fully, but at least the complexity of integration has been reduced.

19.4.2 Isolating the Effects of Flawed Applications

In practice, many applications have errors in their schema. (See Chapter 20.) This leads to flaws in their data. For example, we have encountered applications with "primary keys" that have extraneous attributes. Then the DBMS enforcement mechanism (usually an index) cannot assure the uniqueness of the correct combination of attributes that really is the primary key. As another example, if the DBMS lacks foreign key definitions, you may encounter "dangling" foreign keys with a missing referent.

Ideally, it would be best to repair these flaws in schema and completely eliminate the corresponding flaws in data. However, it is not your job as integrator to correct and revise the participating applications. One reason is that you may fix some problems and create others if you are not fully aware of application nuances. Also, such work is time-consuming and will divert your focus from your primary task—integrating applications.

Another strategy is to let the applications alone and write separate software to check suspect databases. Typically, the errors in schema do not prevent the storing of correct data, they just let flawed data be stored. As Figure 19.11 shows, you can write queries that periodically inspect application databases for errors. When errors are found, you can then let application maintainers correct their database. This approach isolates the mistakes of applications and limits their impact on the integrated system.

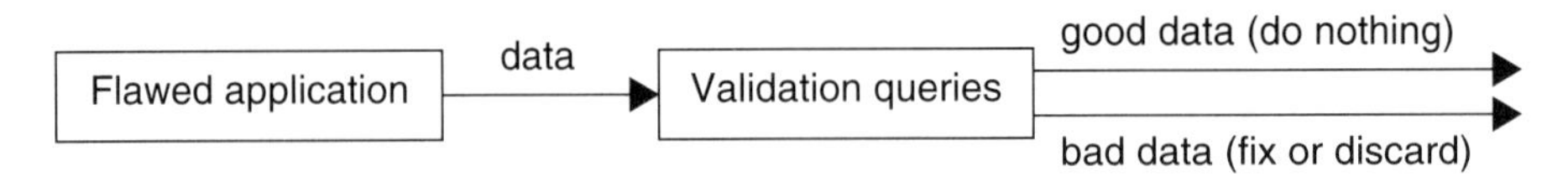

Figure 19.11 Isolate the mistakes of other systems

19.4.3 Integration for Distributed Applications

Chapter 18 discusses distributed databases. Conceptually, integration is different from distribution. Distributed databases involve multiple databases, but they all have the same schema. In contrast, integration addresses the rationalization of multiple logical schema, regardless of how they happen to be physically implemented. Thus with ordinary distributed databases, distribution and integration are different areas of technology that you can address separably.

However, the situation becomes more complex with mobile computing for which a user occasionally dials in with a modem to coordinate a client database with a server database. Users can operate autonomously with an optimistic concurrency algorithm—users are assumed to contend seldom for updating the same data. When contention occurs, different policies can resolve the problem. For example, the last person making a change may have his or her update posted in the server database. Or contending updates may fail and the users may be notified.

As Figure 19.12 shows, mobile clients impact integration via a server. The user may change some data for a mobile client and there may be a long delay before the update is posted to the server, conveyed to other applications on the server, and finally sent back to other client databases. Thus the combination of mobile distribution and integration can cause not only a delay in distributed synchronization, but also a delay in propagating integrated data.

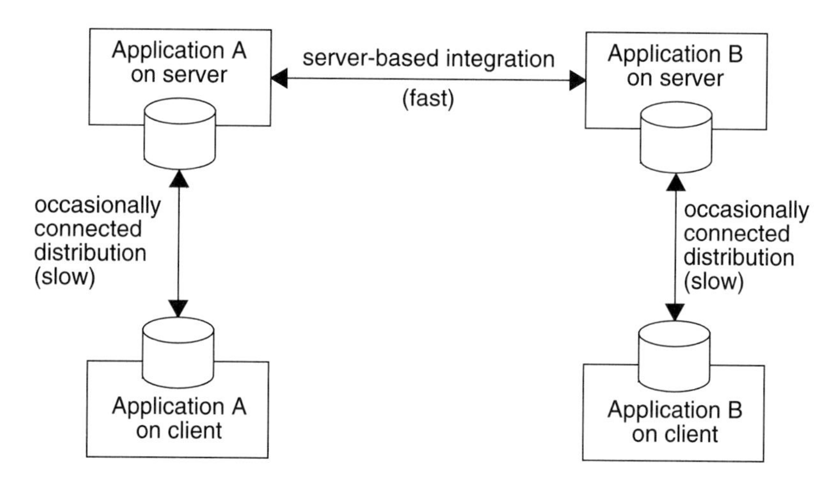

Figure 19.12 Integration for mobile client-server applications

An obvious solution to this problem would be to integrate directly on the client. However, this is not always possible. The server databases are often more robust than the client databases and provide richer mechanisms for integration. For example, most server databases involve a major commercial product, such as Oracle, Sybase, or DB2. Client databases vary widely and are often less robust products.

19.4.4 Legacy Data Conversion

Integration allows data to be moved between applications on an ongoing basis. However, sometimes it is only necessary to connect applications on a one-shot basis—to convert a batch of data from an old application to a new successor application. Once the new application is deployed, the old application can be discarded.

To migrate legacy data, you must model both the old and new applications. You must deeply understand the similarities and differences between the applications before you can attempt to move data. If you have an enterprise model, you will have already reconciled the applications and caused a uniform bias in your applications conducive to data conversion.

When you perform the actual data conversion, you should first stage the legacy data into mirror database structures. For example, if the old application used files and the new application uses a relational database, then you should have one staging table for each file. Each column in a file would map to one column in a table with the same data type and length. With staging database structures, you have the full power of database queries at your disposal to convert data from the old format to the new format.

Do not try to convert data by distorting the model of the new application to match that of the old application. Instead, you should model the new application without regard for legacy data and consider only the requirements. Essentially, you have a choice of complicating the structure of the new application or complicating the conversion code. We much prefer to complicate the one-shot conversion code than the long-lived structure of the new application.

19.5 DATA WAREHOUSE

Database applications can be classified into two categories: decision support and operational. Decision-support applications tend to emphasize complex queries that read large quantities of data and have few updates. The queries may combine data from several applications and execute for a long time, possibly for hours. In contrast, most operational applications are concerned with rapid transaction processing. The transactions tend to be simple, such as update of a few records, and execute in a fraction of a second. Operational transactions are often used to service forms on a screen for simple inquiries and data entry.

A ***data warehouse*** [Inmon-93] is a database that is dedicated to the needs of decision-support applications. A data warehouse is distinct from operational databases and combines disparate data sources, providing one location for decision-support data. Furthermore, the data is placed on a common basis—for the same period of time, for the same portion of a corporation, and for the same currency (such as dollars, yen, marks). A business uses a data warehouse to monitor overall performance and develop strategies. Normally, the data warehouse is placed on a different machine from operational systems, so the exploration of an analyst does not contend with operational activities.

Most operational databases store only the current snapshot of data and an update overwrites the prior value. In contrast, most data warehouses store data parameterized by time, so the user can query past data and detect temporal trends. Most operational systems store only recent data, such as data for the last 60 or 90 days. A data warehouse has a much longer time horizon and may store data for as long as 5 or 10 years. A data warehouse acquires its information from periodic feeds from operational applications. Thus there is some time lag between the data in the operational systems and the data in the data warehouse.

The enterprise model provides the foundation for the data warehouse by integrating the logical content of the various applications. A data warehouse is not helpful unless the data is integrated, so the user can query across application boundaries. The data warehouse is not normally used to mediate the integration of operational applications. The purpose of the data warehouse is to facilitate decision support.

Because of the broad scope (many applications) and long time (possibly 5 to 10 years), a data warehouse must handle much data. By necessity, the data warehouse developer must aggregate data to reduce storage and speed computation. Typically, as data ages, less detail is kept and the data is rolled up into more summarized forms. For example, recent data can be organized by day, older data by week, and the oldest data by month. Eventually, after 5 or 10 years, the data is migrated off-line into a data archive. You can also aggregate data by other criteria such as the business structure—for a department, division, or entire corporation.

19.6 CHAPTER SUMMARY

Integration is a critical capability for large, complex information systems. It is no longer sufficient to meet only the needs of individual applications, but applications must be coordinated to serve the broader goals of an organization. With the current state of the art, the focus is on integrating data—and this is, by itself, difficult to achieve. In the future, as technology progresses we can look forward to integration of objects with increased reuse of code and greater consistency in behavior.

Object-oriented modeling is a critical technology for achieving integration. With object-oriented models you can not only develop superior applications in less time, but you also produce the substrate for integration. Models allow you to compare applications and deeply understand their similarities and differences. An enterprise model is the net result of the reconciliation of the application models and essentially provides the blueprint for causing the applications to exchange data.

We recommend that you construct an enterprise model mostly bottom-up. You should successively rationalize applications with an enterprise model and add their information. We do not favor top-down construction, because there is too much risk that the modeling effort will become excessive and disconnected from reality.

As the enterprise model provides the blueprint for integration, the integration techniques provide the structure. There are three techniques for connecting applications: the master database approach (rewrite all applications so they access the same database), point-to-point integration (directly connect each pair of applications that must exchange data), and indirect integration (connect applications via an intermediary database). In practice, all three integration approaches are helpful, but the most significant technique is indirect integration. You can augment your integration architecture by implementing a small number of identity applications for important abstractions. An identity application simplifies integration by providing an anchor point that cuts across applications.

The integrated applications and enterprise model provide the foundation for constructing a data warehouse. A data warehouse is a database that is dedicated to the needs of decision-support applications. Essentially, a business can mine the data accumulated from the operational transaction-based systems to uncover opportunities for improvements.

Figure 19.13 lists the key concepts for this chapter.

data warehouse	integration architecture
decision-support application	integration technique
enterprise model	legacy data conversion
identity application	mobile client-server application
integration	operational application

Figure 19.13 Key concepts for Chapter 19

BIBLIOGRAPHIC NOTES

[Chu-83] describes an early effort to integrate VLSI circuit design applications. The database technology that they cite is crude, which is not surprising given the age of the paper. Nevertheless, their approach to integration is well considered and they summarize their enterprise model in the paper.

The PDXI project [Book-94] is an example of a large integration project. The purpose of the PDXI project has been to model chemical process engineering data and cause the models to be accepted as standards. Then chemical engineering vendors can write their software against a common target and exchange data. The PDXI project has been funded by a consortium of about 30 petrochemical companies under the auspices of the AIChE (American Institute of Chemical Engineers). The initial technical deliverable was 125 dense pages of OMT object models that serve as an enterprise model for integrating the individual applications. The current status is that the models are slowly moving toward standardization.

Our treatment in Section 19.2 was heavily influenced by [Batini-86]. This paper has many insights for dealing with integration that we have found helpful in our industrial work.

The integration of database applications is also addressed by the heterogeneous database literature [Elmagarmid-90]. A ***federated database*** is a collection of cooperating but autonomous component databases that are integrated. A ***multidatabase*** is a collection of loosely coupled databases that are not integrated. Sometimes full integration of application models is too difficult and it is better to reconcile just the portions of the application models needed for a particular query.

The federated and multidatabase literature emphasizes variations in database paradigms (relational, hierarchical, etc.), query languages, and transaction processing. We are not concerned with these physical mechanisms in this book, but instead focus on the conceptual aspect of integration—integration of models. Federated databases and multidatabases use a query language to define views that map between application databases. These views are functionally equivalent to the transformations that we use.

[Bischoff-97] provides helpful business-oriented information for deploying a data warehouse. [Inmon-93] also has much information on data warehouse technology.

REFERENCES

[Batini-86] C Batini, M Lenzirini and S Navathe. A comparative analysis of methodologies for database schema integration. *ACM Computing Surveys 18*, 4 (December 1986), 323–364.

[Bischoff-97] Joyce Bischoff and Ted Alexander. *Data Warehouse: Practical Advice from the Experts*. Englewood Cliffs, New Jersey: Prentice Hall, 1997.

[Book-94] N Book, O Sitton, R Motard, M Blaha, B Maia-Goldstein, J Hedrick, and J Fielding. The road to a common byte. *Chemical Engineering* (September 1994), 98–110.

[Chu-83] K Chu, JP Fishburn, P Honeyman, and YE Lien. Vdd—A VLSI design database system. *1983 Database Week: Engineering Design Applications, Proceedings of the Annual Meeting*, May 23–26, 1983, San Jose, California.

[Elmagarmid-90] Ahmed K Elmagarmid and Calton Pu. Guest editors' introduction to the special issue on heterogeneous databases. *ACM Computing Surveys 22*, 3 (September 1990), 175–178.

[Inmon-93] WH Inmon. *Building the Data Warehouse*. New York: Wiley-QED, 1993.

[Sheth-90] Amit Sheth and James Larson. Federated database systems for managing distributed, heterogeneous, and autonomous databases. *ACM Computing Surveys 22*, 3 (September 1990), 183–236.

[Tsichritzis-78] D Tsichritzis and A Klug, editors. The ANSI/X3/SPARC DBMS framework. *Information Systems 3*, 4 (1978).

20

Reverse Engineering

Most methodologies emphasize forward engineering—that is, building a new application from an analysis model. A robust, industrial-strength process for software development must broaden its focus to deal with existing software. You may replace a past system to simplify maintenance, permit porting to a new platform, or allow further enhancement. Often a new system lacks explicit requirements and is specified relative to an existing system. It is desirable to extract implicit requirements from the existing legacy system to seed development of the new system.

During reverse engineering, the developer takes a past design or an implementation that embodies a design and extracts the essential application problem content. The reverse engineer discards design optimizations and implementation decisions.

This chapter extends the OMT methodology to include processes for reverse engineering of hierarchical, network, and relational databases. We do not address the full reverse engineering problem, which also includes reverse engineering of programming code. Consequently, we only consider the object model and disregard the other two OMT models.

20.1 OVERVIEW

20.1.1 Re-engineering

Re-engineering is the process of redeveloping existing software systems. Re-engineering consists of a sequence of three steps: reverse engineering, consideration of new requirements, and then forward engineering. ***Reverse engineering*** is the process of taking an existing design and extracting the underlying logical intent.

The rationale for re-engineering is straightforward: New software is expensive to develop but old software can be costly to maintain and adapt to new uses. The goal of re-engineering is mechanically to reuse past development efforts in order to reduce maintenance expense and improve software flexibility. Re-engineering is applicable to diverse software such as programming code, databases, and inference logic, though we just discuss database reverse engineering in this chapter. Object-oriented models facilitate the re-engineering process because the same modeling paradigm is adept at representing abstract conceptual models and models with implementation decisions.

Reverse engineering is not a panacea for evolving from old software to new software. Reverse engineered models require much interpretation, and you should not use them to perpetuate past flaws. Instead, you should regard reverse-engineered models as merely one source of requirements for a new software system. Reverse engineering ensures that the new system addresses earlier concerns, facilitates migration of data from the old database to the new database, and speeds development.

20.1.2 Motivation for Database Reverse Engineering

As we mentioned earlier, a major motivation for reverse engineering is to elicit requirements for re-engineering. Reverse engineering of existing software can yield tentative requirements for the new replacement system. Reverse engineering ensures that the functionality of the existing system is not overlooked or forgotten. There are additional motives for database reverse engineering:

- **Software assessment**. We have derived substantial benefit from reverse engineering of databases from vendor software. Reverse engineering provides an unusual source of insight. The quality of the database design is an indicator of the quality of the software as a whole. An understanding of the concepts supported by the underlying database schema lets you judge functionality claims better.
- **Integration**. Reverse engineering facilitates integration of applications. A logical model of encompassed software is a prerequisite for integration.
- **Conversion of legacy data**. You must understand the logical correspondence between the old database and the new database before attempting to convert data.
- **Assessment of state of the art**. From our perspective as methodologists, reverse engineering provides candid insight about the state of the database design art—as practiced.

20.1.3 Reverse Engineering Strategies

There are multiple strategic issues for reverse engineering:

- **Sources of input**. A complete reverse engineering process must exploit information from several sources: database schema, examples of data, programming code, suggestive names, user interface screens and reports, test cases, manuals and documentation, and application understanding. The available information varies widely across problems, and you should exploit any information you can find. Data analysis often cannot prove a proposition, but the more data that is encountered, the more likely will be the conclusion.

  We believe reverse engineering of code and reverse engineering of databases will converge in the future. Code slicing [Weiser-84] appears to be a promising technique for restructuring procedural code into the object-oriented paradigm for combination with tentative models from database reverse engineering.

- **Database reverse engineering versus program reverse engineering**. Our experience has been that a database can be reverse engineered independently of programming code. Databases are explicitly declared and simpler and more amenable to study. Furthermore, the data are generally the most stable parts of applications. Usually it is easier to reverse engineer the database and then address programming code, rather than deal with both simultaneously or programming code first. Reverse engineering of a database alone often provides adequate input for modeling. This chapter only addresses reverse engineering of databases.
- **Alternative models from reverse engineering**. In general, the mapping between object models and database schemas is many-to-many. Various optimizations and design decisions can be used to forward engineer an object model into a database schema. Similarly, when reverse engineering a database, alternative interpretations of the structure and data can yield different object models. Usually, there is no obvious, single correct answer for reverse engineering. Multiple interpretations can all yield plausible results. You must thoroughly understand the application to reach firm conclusions.
- **Modeling tools**. A good way to begin reverse engineering is by entering the existing schema into a software modeling tool. You will often find associations in a degraded form, such as pointers or relational database foreign keys. Inheritance will also be degraded for non-object-oriented databases. You can use the flow of identity notation (Section 13.3.2) to capture identity dependencies in the design model. You can then scrutinize the dependencies to determine the underlying relationships.
- **Transformations**. In principle, you can use transformations (Chapter 10) to perform the reverse engineering steps. In practice, this is difficult because of lack of tool support. You can also use transformations to refine the model that results from reverse engineering.
- **Flawed designs**. A practical approach to reverse engineering must tolerate a wide range of problem styles; some database designers are more skilled than others. In one sense, the flawed database designs are the most relevant problems for reverse engineering because remedy of the flaws provides additional benefits. We have encountered some surprising design techniques that violate the norms of good design practice [Blaha-95]. Although the techniques have varied widely across problems, within a given problem we have often found a uniform strategy, especially for large schemas.
- **Reverse engineering skills**. Reverse engineering requires a different mix of skills than forward engineering. Basically, you must be proficient at both modeling and database design. Often you will have a large schema, and you must be able to recognize design idioms quickly and translate them into modeling constructs. This interpretation of database designs is complicated by the many optimizations and flaws that are found in practice.

  The reverse engineer must also have an ability to focus and cannot become mired in detail. You must be satisfied with quickly obtaining 80–90% of the available information. The remaining 10–20% of information is often not worth the effort that would be needed for recovery. Reverse engineering is incidental to the primary task of developing an application and you must not digress.

The next sections present our reverse engineering processes for hierarchical, network, and relational databases. As would be expected with any textbook, these processes are somewhat idealistic and simplified. Our steps are weakly ordered; in practice there is much iteration, backtracking, and reordering of steps. Nevertheless, these processes contain the key steps that we perform and form a useful starting point for more complex problems.

20.2 HIERARCHICAL DATABASES

20.2.1 Concepts

A ***hierarchical database*** is a database that is organized as a collection of inverted trees of records. The inverted trees may be of arbitrary depth. The record at the root of the tree has zero or more child records; the child records, in turn, serve as parent records for their immediate descendants. The parent-child relationship recursively continues down a tree. IBM's IMS product is the most prominent hierarchical DBMS.

The forced asymmetry of many-to-many associations is a major drawback of hierarchical databases. Figure 20.1 shows an object model and some sample data. The left side of Figure 20.2 shows a hierarchical database organized about company records and the right side shows an alternative hierarchical database organized about person records.

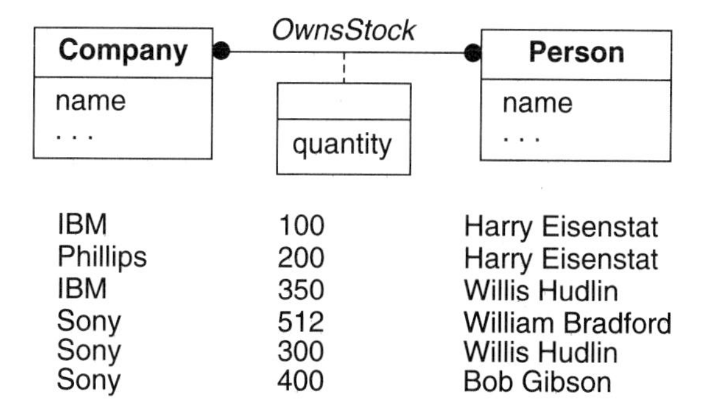

Figure 20.1 Sample object model and data

A data definition language (DDL) defines hierarchical database structure—record types, field types, pointers, and parent-child relationships. Programs must use another language, the data manipulation language (DML), to access the actual data. Hierarchical databases have two ways of accessing records: via the parent and via pointers. Each parent record usually has its child records, grandchild records, and other descendants stored in contiguous disc space.

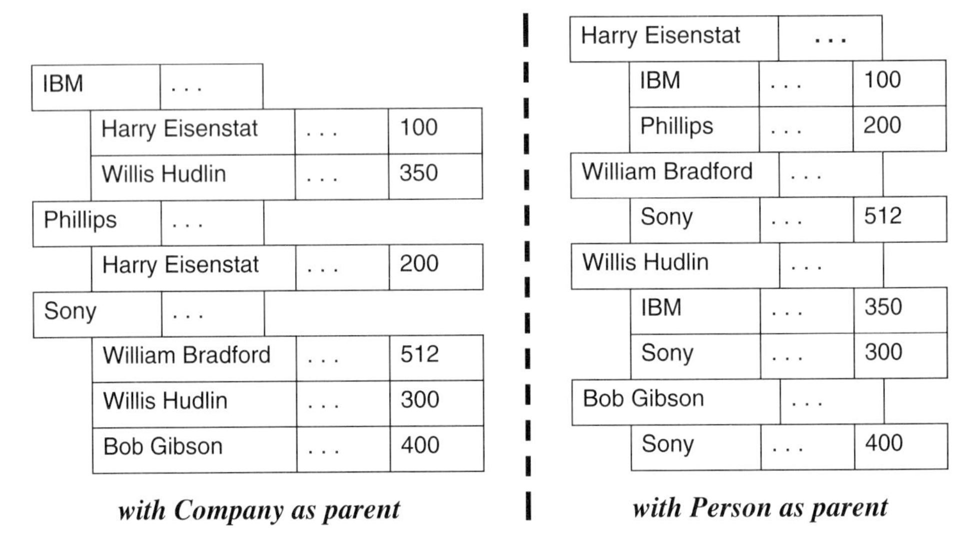

Figure 20.2 Sample hierarchical databases

Figure 20.3 shows a simplified object metamodel for hierarchical databases. A hierarchical database contains various record types, each of which provides a format for record instances. The instances of some record types are roots of trees. Applications can navigate a hierarchical database by starting at a root and successively navigating downward from parent to child until the desired record is found. Applications can interleave parent-child navigation with traversal of pointers that also may be embedded in a record. Searching down a hierarchical tree is very fast due to contiguous storage; traversal of pointers is a little slower but still fast; other types of queries require sequential search.

A record is a list of fields; each field contains a simple value or a pointer to a record. A field type defines the structure for values. Some combination of fields may form the key for a record relative to its parent. Few hierarchical DBMSs support null values or variable-length fields.

The hierarchical paradigm is impoverished for expressing complex information models. Often a natural hierarchy does not exist and it is awkward to impose a parent and child. Pointers partially compensate for this forced dominance, but it is still difficult to devise a suitable hierarchical schema for large models. Programs must directly navigate the physical data structures, so there is no data independence as with relational DBMSs. Consequently, hierarchical databases and their applications are difficult to maintain and extend to new unforeseen applications. Hierarchical DBMSs are obsolete and you should not use them for new applications, but you may encounter them with legacy applications.

20.2.2 Reverse Engineering Process

We recommend the following sequence of steps for reverse engineering a hierarchical database schema to an object model. The rigidity of hierarchical databases makes classes and as-

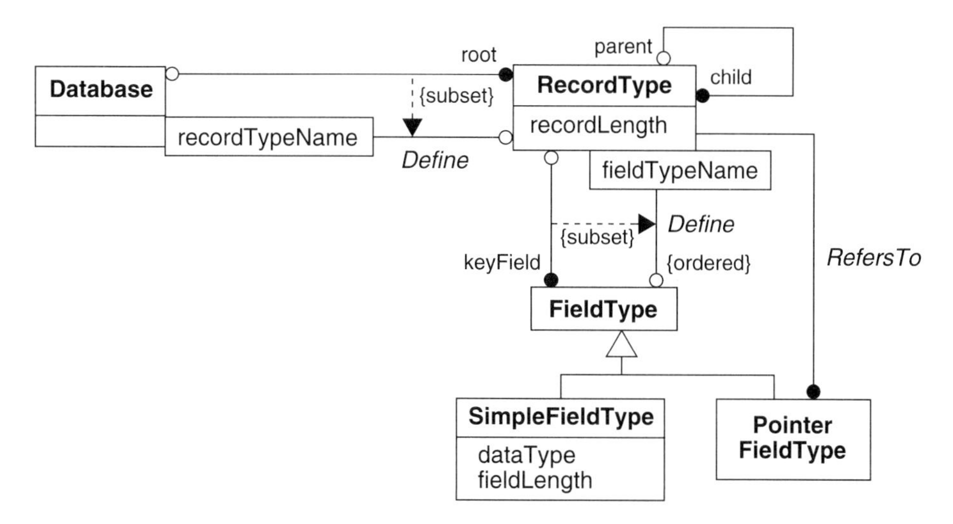

Figure 20.3 Simplified object metamodel for hierarchical databases

sociations straightforward to find. Generalization requires more effort. You can find generalizations by analyzing data and learning about the application.

Step 1. Prepare an initial object model.

- **Tentative classes**. Represent each record type as a tentative class. All fields of record types become attributes of classes.
- **Anonymous fields**. Beware of ***anonymous fields***, fields for which the data structure is not known to the database. A DBMS treats an anonymous field as just a sequence of bytes, and an application program must deduce the meaning. An anonymous field may correspond to several logical attributes. You will need to inspect programming code or consult with an application expert to determine the meaning of anonymous fields.
- **Blank fields**. Beware of blank fields that are not currently used but are merely added for future growth.

Step 2. Determine candidate keys and flow of identity (Section 13.3.2). Add corresponding tentative associations.

- **Candidate keys**. Note candidate keys. Each record type in a hierarchical database is either a root or a child of some parent. Each root record type has at least one explicit candidate key. A child record type may have candidate keys specified relative to a parent record type.
- **Candidate keys consisting entirely of one foreign key**. Remove the candidate key attribute(s) and add a one-to-one association.
- **Candidate keys combining one foreign key with nonforeign key attributes**. Remove the candidate key attributes and introduce a qualified association. The nonforeign key attributes are qualifiers.

- **Candidate keys consisting of two foreign keys**. Convert a tentative class to a many-to-many association. Any attributes of the class become link attributes of the association.
- **Candidate keys consisting of three foreign keys**. Convert a tentative class to a ternary association. We seldom encounter ternary associations in our analysis models. A ternary association in a reverse engineered model may indicate an optimized or inadequately conceived design. You should question ternary associations and try to decompose them into binary associations.
- **Minimum multiplicity**. Most hierarchical databases do not support nulls, so you cannot determine minimum multiplicity from inspection of the schema or data. You will have to rely on application understanding or inspection of programming code.
- **Maximum multiplicity**. You can determine maximum multiplicity from the schema.
- **Role names**. Note role names that arise through foreign key references.

Step 3. Add tentative associations for foreign keys that are not part of a candidate key. Note multiplicity and role names.

- **Foreign keys**. Add a one-to-many association for each foreign key that is not part of a candidate key. The class with the buried pointer is the "many" role; the referenced class is the "one" role. There is little need to search for foreign keys, because foreign keys are normally explicitly marked.
- **Disconnected classes**. Do not be surprised to find some classes that are free standing and unrelated.

Step 4. As appropriate, restate associations as aggregations.

- **Aggregations**. Apply your understanding of the application and restate associations as aggregations where appropriate. Aggregation is the *a-part-of* relationship.

Step 5. Discover generalizations.

- **One-to-one associations**. Look for clusters of one-to-one associations. Generalization is likely when several classes are interconnected with one-to-one associations.
- **Discriminators**. Look for a field of enumeration domain that indicates applicability of attributes and associations. Only some enumeration attributes will be discriminators; do not introduce a generalization just because you have found an enumeration domain. You must apportion attributes and associations among the superclass and subclasses.
- **Overloaded fields**. A field may be overloaded with multiple meanings, one for each generalization subclass.
- **Similarity**. Note patterns of similar attributes, similar names, and similar associations. You can use generalization to recognize and abstract the commonality.
- **Inheritance forests**. In the most general case there may be a forest of generalizations with multiple superclasses and intermediate levels.
- **Hierarchies**. Do not misinterpret hierarchies. A hierarchy need not be an aggregation or generalization. In fact, most hierarchies correspond to ordinary associations. In a hi-

erarchical database, physical hierarchies are the dominant paradigm for organizing records, and various relationships can be cast into this form.

Step 6. Discard design notation.

- **Flow of identity**. Drop flow of identity notation; the choice of path for deriving identity is a design decision.

20.2.3 Example

Figure 20.4 shows a reverse engineering example for a hierarchical database. We have chosen the same problem (a simple flight information database) to facilitate contrast between the three database paradigms (hierarchical, network, and relational) covered in this chapter.

We start with a hierarchical schema. For step 1 we prepare an initial object model restating record types as classes and fields as attributes. During step 2 we determine sources of identity and add the corresponding associations. The *key* clauses in the hierarchical schema explicitly state primary keys. We infer that *Airline.symbol* is a candidate key based on our understanding of the problem.

For step 3 we add tentative associations for foreign keys that are not part of candidate keys. The flight information database has no aggregations to note in step 4. During step 5 we introduce a generalization; the comment about employee type is highly suggestive. We discard flow of identity notation, resulting in the final object model.

20.3 NETWORK DATABASES

20.3.1 Concepts

A ***network database*** is a database that is organized as a collection of graphs of records that are related with pointers. Network databases represent data in a symmetric manner, unlike the hierarchical distinction of a parent and a child. Network database technology arose in response to the limitations of hierarchical DBMSs. A network is a more flexible data structure than a hierarchy and still permits efficient navigation. Most network DBMSs adhere to the CODASYL (COmmittee for DAta SYstem Languages) standard. Do not confuse *network* databases with communication *networks*. These are different uses of the term "network."

Figure 20.5 shows the network database that corresponds to the simple object model and data of Figure 20.1. The boxes denote records and the lines denote pointers. Note that networks directly represent many-to-many relationships.

A data definition language (DDL) defines network database structure—record types, field types, pointers, and indexes. Programs must use another language, the data manipulation language (DML), to access the actual data. Network databases support two ways of accessing records—via pointers and via indexes.

Figure 20.6 shows a simplified object metamodel for network databases. A database consists of one or more areas that physically partition the database for reasons of performance and security. Names for record types, set types, and indexes are unique within the

Initial hierarchical DBMS schema

```
Record = Airline, Parent = Root, Key = name
   field = name
   field = symbol
   Record = Flight, Parent = Airline, Key = flight_num + date
      field = flight_num
      field = date
      field = copilot   /* key to Employee */
      field = pilot   /* key to Employee */
Record = Employee, Parent = Root, Key = taxpayer_number
   field = name
   field = taxpayer_number
   field = airline   /* key to airline */
   field = employee_type   /* pilot or flight attendant */
   field = flight_rating
Record = Attendant_assignment, Parent = Root,
   Key = flight + flight_attendant
   field = flight   /* key to flight */
   field = flight_attendant   /* key to employee */
```

Step 1. Prepare an initial object model.

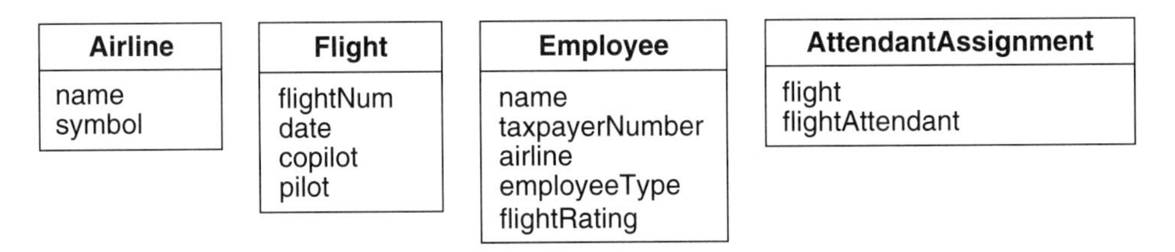

Step 2. Determine candidate keys. Add corresponding associations.

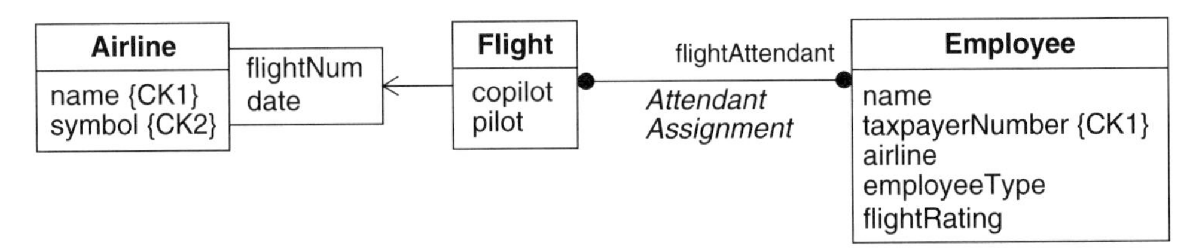

Figure 20.4 Example of reverse engineering for a hierarchical database

Step 3. Add associations for foreign keys that are not part of a candidate key.

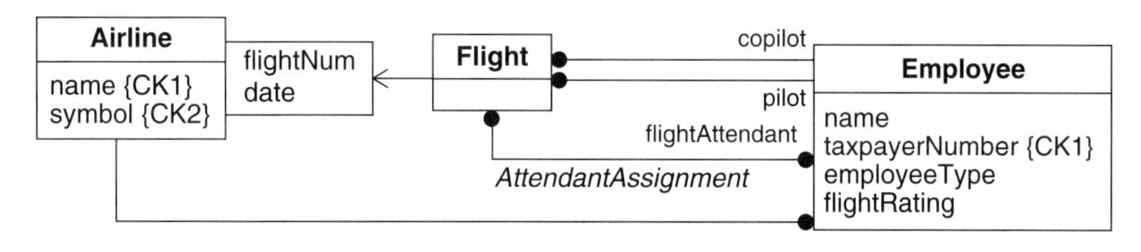

Step 4. As appropriate, restate associations as aggregations.

..... Not needed for this example .....

Step 5. Discover generalizations.

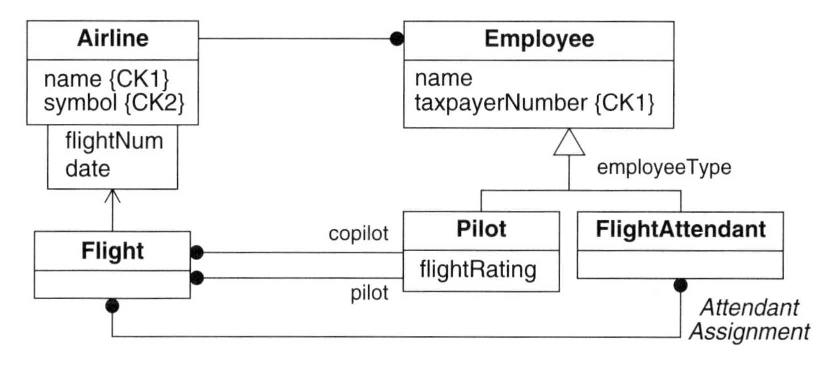

Step 6. Discard design notation.

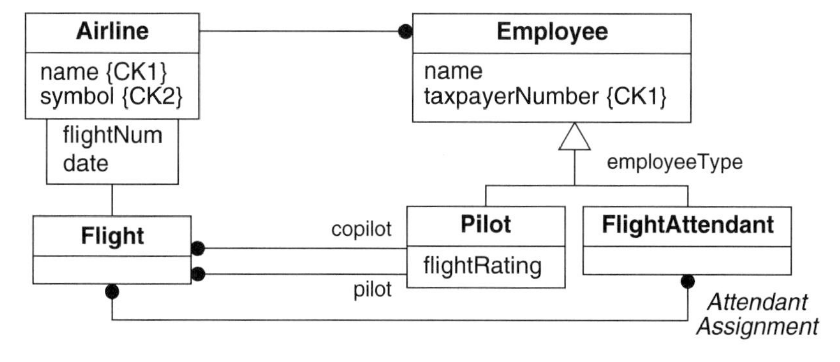

Figure 20.4 (continued) Example of reverse engineering for a hierarchical database

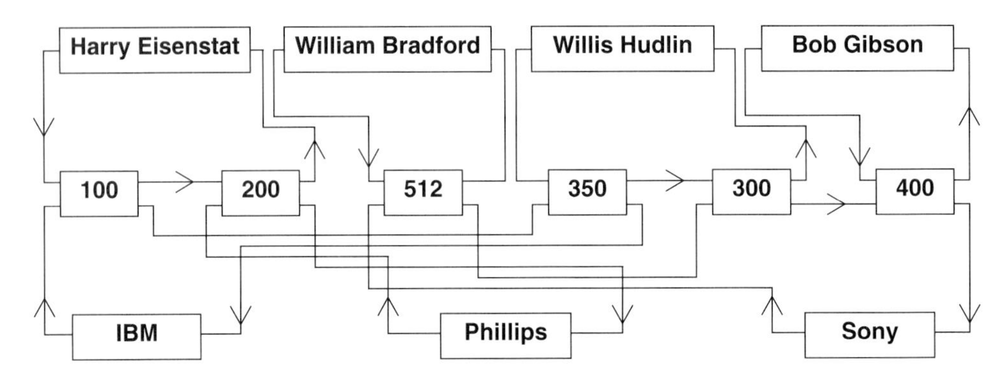

Figure 20.5 Sample network database

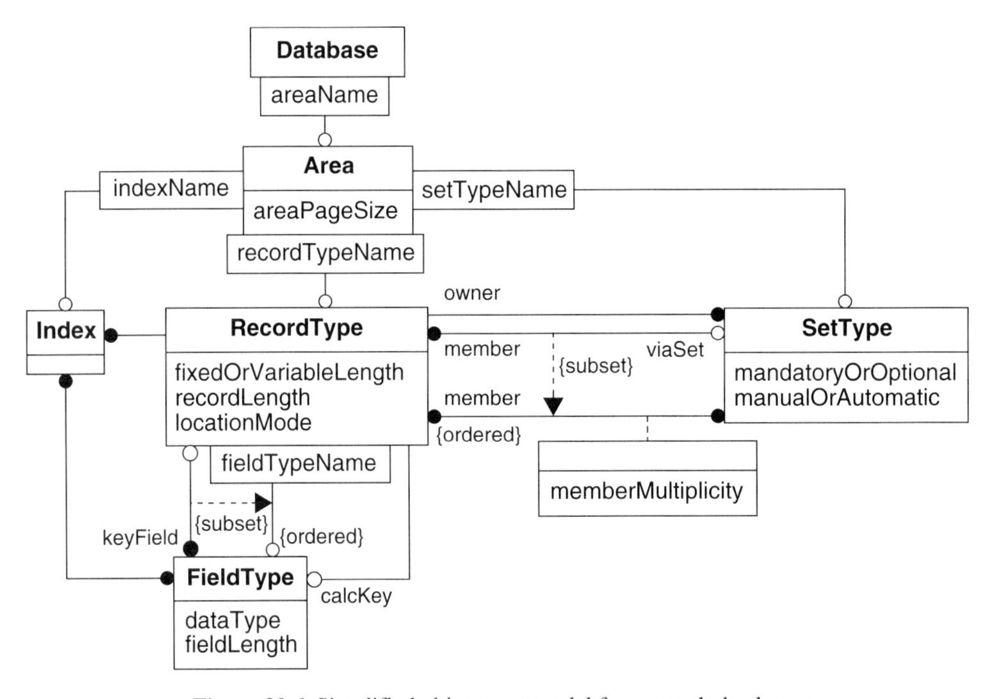

Figure 20.6 Simplified object metamodel for network databases

scope of an area. (The term "set type" is CODASYL jargon that refers to pointers.) Each record type, set type, and index must be assigned to a single area; their instances also cannot cross areas. All fields of an index must belong to the same record type.

Each record type provides a format for one or more record instances. A record is a list of values; most network DBMSs require that the values not be null. A record may be fixed length or variable with a maximum length. Each record has a location mode for physical placement. With a location mode of *calc*, the DBMS hashes a field to yield a physical page. When *via* is specified, the DBMS tries to place each member record (CODASYL jargon) on the same page as an owner record (CODASYL jargon). When *direct* is specified, the user must supply the actual database page number on which to place records. Records may be indexed to facilitate certain searches.

Sets (CODASYL jargon for chains of pointers) provide the mechanism for relating records. Each set has one record that serves as an *owner.* A set may involve multiple *member* records of the same or different types. Each record has a pointer chain for each set type in which it participates. By necessity, a record may participate at most once in a *via* set because of the collocation. In Figure 20.5 the boxes denote records and the lines denote sets.

There are several set membership options. A record may have a mandatory or optional owner. When a member record for an automatic set type is stored, it is automatically entered in all sets to which it belongs. In contrast, with a manual set type the DBMS programmer must write the code to connect a record to its sets.

Network databases are similar to hierarchical databases in that they also lack data independence. Application programs must directly navigate pointers. Consequently, network databases are difficult to maintain and extend. Network DBMSs are obsolete and you should not use them for new applications, but you may encounter them with legacy applications.

20.3.2 Reverse Engineering Process

We recommend the following sequence of steps for reverse engineering a network database schema to an object model. The rigidity of network databases makes classes and associations straightforward to find. Generalization requires more effort, because network databases provide no semantic support. You can find generalizations by analyzing data and learning about the application.

Step 1. Prepare an initial object model.

- **Tentative classes**. Represent each record type as a tentative class. All fields of record types become attributes of classes. Note candidate keys that are defined for record types.
- **Anonymous and blank fields**. Beware of anonymous and blank fields that are used for restructuring and future growth.

Step 2. Add tentative associations.

- **Tentative associations**. Represent each set type as one or more associations. A set type with one member type corresponds to one association; a set type with multiple member types corresponds to multiple associations.
- **Minimum multiplicity**. The minimum multiplicity for member roles is zero; some owners of a set type may not have any members. The minimum multiplicity for an owner role is zero if the set type is optional and one if the set type is mandatory.
- **Maximum multiplicity**. The maximum multiplicity for owner roles is “one”; the maximum multiplicity for members is “many.” You can be more precise if you have additional semantic information.

- **Role names**. Note role names that arise through owner and member references to a record type.
- **Disconnected classes**. Do not be surprised to find some classes that are free standing and unrelated.

Step 3. Refine associations.

- **Many-to-many and ternary associations**. Network databases cannot directly express many-to-many and ternary associations; they require an intermediate intersection record.
- **Qualified associations**. Change an association to a qualified association when a candidate key combines a foreign key with nonforeign key attributes.
- **Repeating fields**. An association is sometimes implemented as a repeating field in a record, subject to a maximum number of repetitions. When such a repeating field is encountered, replace it by the true underlying association.

Step 4. As appropriate, restate associations as aggregations.

- **Aggregations**. Apply your understanding of the application and restate associations as aggregations where appropriate. Aggregation is the *a-part-of* relationship.

Step 5. Discover generalizations.

- **One-to-one associations**. Look for clusters of one-to-one associations. Generalization is likely when several classes are interconnected with one-to-one associations.
- **Discriminators**. Look for a field of enumeration domain that indicates applicability of attributes and associations. Only some enumeration attributes will be discriminators; do not introduce a generalization just because you have found an enumeration domain. You must apportion attributes and associations among the superclass and subclasses.
- **Overloaded fields**. A field may be overloaded with multiple meanings, one for each generalization subclass.
- **Similarity**. Note patterns of similar attributes, similar names, and similar associations. You can use generalization to recognize and abstract the commonality.
- **Inheritance forests**. In the most general case there may be a forest of generalizations with multiple superclasses and intermediate levels.

Step 6. Discard design notation. Improve organization of the model.

- **Flow of identity**. Drop flow of identity notation; the choice of path for deriving identity is a design decision.
- **Tentative packages**. Areas may indicate closely coupled classes and associations that you can organize into packages. However, be aware that some areas may be performance tuning or database administration artifacts that are irrelevant to an analysis model.

20.3.3 Example

Figure 20.7 shows a reverse engineering example for a network database. This example also concerns the simple flight information database.

Initial network DBMS schema

```
Record = Airline, Area = system, Key = name
   field = name
   field = symbol
Record = Flight, Area = system,
   Key = airline_name + flight_num + date
   field = airline_name
   field = flight_num
   field = date
Record = Employee, Area = system, Key = taxpayer_number
   field = name
   field = taxpayer_number
   field = employee_type   /* pilot or flight attendant */
   field = flight_rating
Record = Attendant_assignment, Area = system,
   Key = airline_name + flight_num + date + taxpayer_number
   field = airline_name
   field = flight_num
   field = date
   field = taxpayer_number
Set = Airline__Employee, Area = system
   owner is Airline, member is Employee
Set = Pilot__Flight, Area = system
   owner = Employee, member = Flight
Set = Copilot__Flight, Area = system
   owner = Employee, member = Flight
Set = Flight__Attendant_assignment, Area = system
   owner = Flight, member = Attendant_assignment
Set = Flight_attendant__Attendant_assignment, Area = system
   owner = Employee, member = Attendant_assignment
Set = Airline__Flight, Area = system
   owner = Airline, member = Flight
```

Figure 20.7 Example of reverse engineering for a network database

We start with a network schema. For step 1 we prepare an initial object model restating record types as classes and fields as attributes. During step 2 we add tentative associations corresponding to set types. Then we refine the associations by recognizing many-to-many and qualified associations. The flight information database has no aggregations to note in step 4. During step 5 we use semantic judgment and introduce a generalization; the comment about employee type is highly suggestive. We discard flow of identity notation, resulting in the final object model.

Step 1. Prepare an initial object model.

Airline
name {CK1}
symbol {CK2}

Flight
airlineName {CK1}
flightNum {CK1}
date {CK1}

Employee
name
taxpayerNumber {CK1}
employeeType
flightRating

AttendantAssignment
airlineName {CK1}
flightNum {CK1}
date {CK1}
taxpayerNumber {CK1}

Step 2. Add tentative associations.

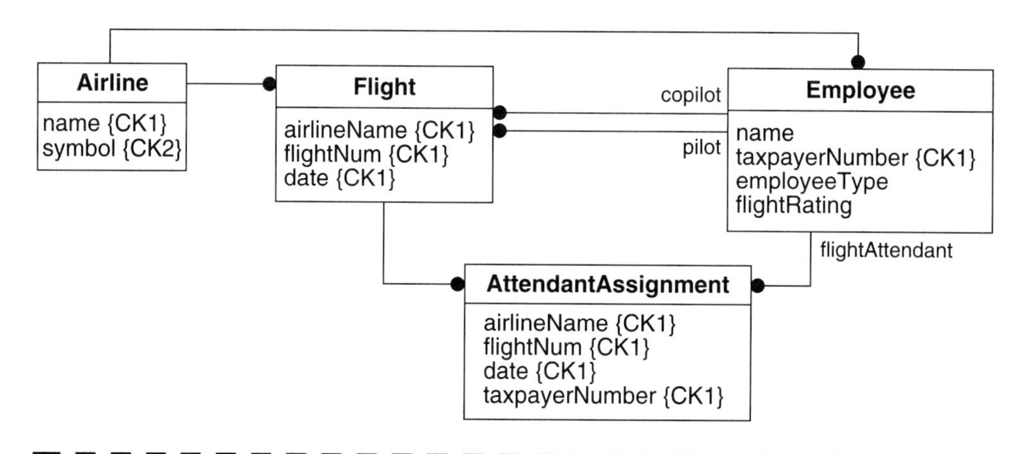

Step 3. Refine associations.

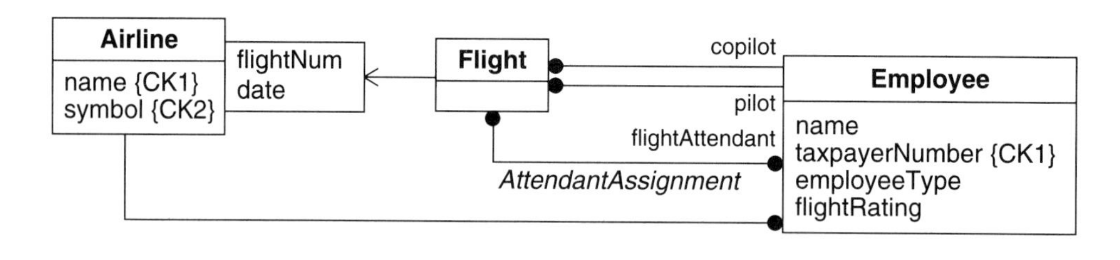

Step 4. As appropriate, restate associations as aggregations.

..... Not needed for this example .....

Figure 20.7 (continued) Example of reverse engineering for a network database

Step 5. Discover generalizations.

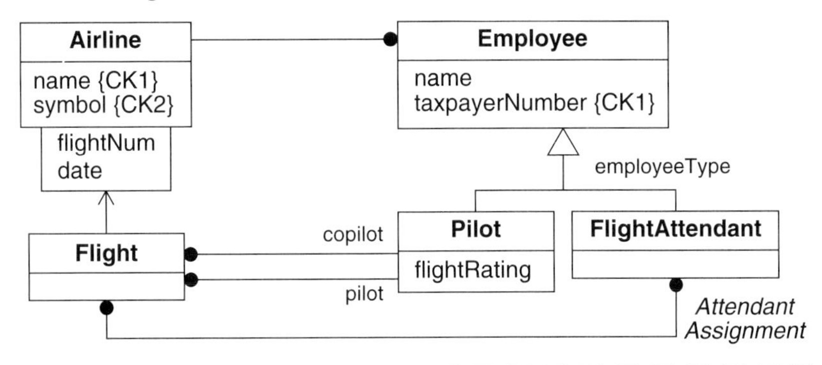

Step 6. Discard design notation. Improve organization of the model.

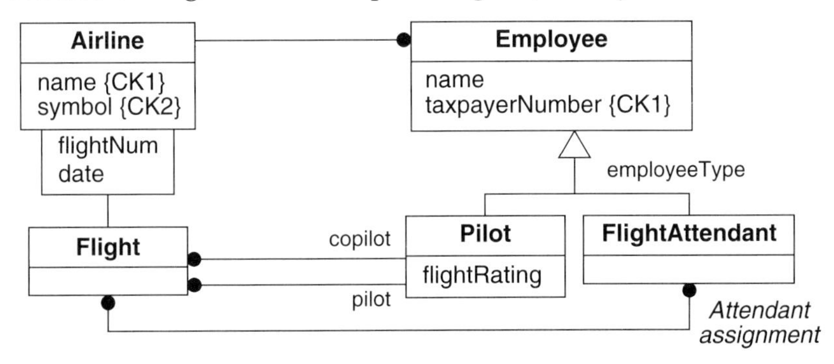

Figure 20.7 (continued) Example of reverse engineering for a network database

20.4 RELATIONAL DATABASES

20.4.1 Concepts

Section 13.1 explains basic relational database concepts.

20.4.2 Reverse Engineering Process

The effort required to reverse engineer a relational database depends on the difficulty of finding foreign keys and the depth of your application understanding. During reverse engineering you may have difficulty with representing constraints, such as mutually exclusive associations or complex derivations of identity. If a constraint is difficult to represent with a model, you should note it textually. Some relational database schema are difficult to represent with an object model. Our experience is that such schema are difficult to understand and likely to be flawed. We have identified the following sequence of steps for reverse engineering a relational database to an object model.

Step 1. Prepare an initial object model

- **Tentative classes**. Represent each table as a tentative class. All columns of tables become attributes of classes.
- **Anonymous and blank fields**. Beware of anonymous and blank fields. Such fields are sometimes used to facilitate later customization. For example, some vendors include user-defined fields.

Step 2. Determine candidate keys.

- **Candidate keys**. Look for unique indexes but keep in mind that some candidate keys may not be enforced by unique indexes. If you have a populated database, you can search for unique combinations of data. However, inferences from data are only suggestive and you must have application knowledge to assert a candidate key.
- **Primary keys**. Even if primary keys are specified, just record candidate keys. Many relational database designs reference candidate keys with foreign keys, even though this is a bad practice. (As explained in Chapter 13, foreign keys should only reference primary keys.)
- **Flaws in identity**. Beware of flaws in identity. Some applications are sloppy and do not define primary keys. We have also seen applications with primary key attributes that are null and extraneous attributes that are not required for uniqueness.

Step 3. Determine foreign key groups. If you are lucky, you will have a relational database with foreign key clauses specified as part of the schema and can skip this step.

- **Homonyms and synonyms**. Try to resolve homonyms, attributes with the same name that refer to different things, and synonyms, attributes with different names that refer to the same thing.
- **Finding foreign keys**. Matching attribute names, data types, and domains may suggest foreign keys. Foreign key attributes are often involved in join clauses in view definitions and may have a secondary index defined to facilitate traversal. You can refute some foreign key hypotheses by analyzing data. (See *inclusion dependencies* bullet.)
- **Foreign key groups**. Aside from homonyms and synonyms, the target of a foreign key reference may be ambiguous because of generalization. This is why this step refers to *foreign key groups*. During this step we do not attempt to determine specific reference-referent attribute pairs but merely groups of attributes within which foreign keys may be found.
- **Inclusion dependencies**. Use inclusion dependencies to detect foreign keys. For a correct schema, the foreign keys in one table must be a subset of the primary keys for the referent table. (All foreign keys must refer to some primary key.) You can often discover inclusion dependencies by a query of the form: *Select ... from A except B*. (This query may not work for databases with flaws in identity.)

Step 4. Refine tentative classes.

- **Partitioned classes**. Combine partitioned classes into a single class. Horizontally partitioned classes have the same schema. (Such classes must also have the same semantic intent. We presume that identical schema is a good indicator of semantic intent.) Vertical partitioning replicates the primary key and apportions other attributes. (In Chapter 18 we recommend against the use of vertical partitioning.) Distributed databases may use partitioning to disperse records.
- **Generic classes**. Do not be surprised if you encounter generic classes (Section 4.1.1) in your reverse-engineered model. Generic classes can sometimes help you simplify a model and capture concepts more succinctly.
- **Tables that are not classes**. Detect functions, constraints, and enumerations that are represented as tables. You should suspect classes that do not participate in foreign keys. Some tables are just implementation artifacts.

Step 5. Discover generalizations.

- **Foreign key groups**. Analyze large foreign key groups, particularly those with five, ten, or more cross-related attributes. Generalization is likely when several classes are interconnected with one-to-one associations. Common primary keys for the classes further increase the odds of generalization. Data analysis can increase confidence in the discovery of a generalization by revealing a subset of records.
- **Inclusion dependencies**. Inclusion dependencies can also suggest generalization: The candidate keys for one table are a subset of the candidate keys for another table.
- **Replicated attributes**. Look for many replicated attributes. A generalization may have been implemented by pushing superclass attributes down to each subclass.
- **Mutually exclusive attributes**. Similarly, look for data where a class has mutually exclusive groups of attributes. This may indicate an implementation of generalization where subclass attributes were pushed up to the superclass.
- **Inheritance forests**. Do not forget that there may be a forest of generalizations with multiple superclasses and intermediate levels. Data analysis can help distinguish multiple inheritance. As is true throughout reverse engineering, data analysis only yields hypotheses, and application understanding is required to reach firm conclusions.

Step 6. Discover associations.

- **Candidate keys consisting entirely of one foreign key**. Consider converting a tentative class to an association. This may indicate a one-to-one or a one-to-many association being implemented as a distinct table. You must apply application judgment to distinguish these associations from generalizations.
- **Candidate keys combining one foreign key with nonforeign key attributes**. Introduce a qualified association. This will find some, but not all, qualifiers.
- **Candidate keys consisting of two foreign keys**. Convert a tentative class to a many-to-many association.

- **Candidate keys consisting of three foreign keys**. Convert a tentative class to a ternary association. A ternary association in a reverse engineered model may indicate an optimized or poorly conceived design. You should question ternary associations and try to decompose them into binary associations.
- **Foreign keys**. The remaining associations are buried and manifest as foreign keys.
- **Minimum multiplicity**. Note minimum multiplicity. The most general case is a minimum multiplicity of zero. Often, you must have application knowledge to infer a more restrictive minimum multiplicity of one (or another number).
- **Maximum multiplicity**. Note maximum multiplicity. The most general case is a maximum multiplicity of "many" (unbounded). Often you must have application knowledge to infer a more restrictive upper limit of "one" (or another number).
- **Disguised nulls**. Beware of encodings of null values and default values.
- **Parallel attributes**. Even though it is not a good practice, many designers implement the "many" role of an association with parallel attributes (Section 13.7). Often you can replace parallel attributes with a qualified association.
- **Double-buried associations**. We have occasionally encountered schema with a one-to-one association between *A* and *B* buried in both classes. You must understand the application to be able to distinguish this situation from two distinct associations. You should represent double-buried associations with a single association.
- **Intermediate tables**. You may need to introduce some intermediate tables to model qualifiers and convey flow of identity properly.

Step 7. As appropriate, restate associations as aggregations.

- **Aggregations**. Apply your understanding of the application and restate associations as aggregations where appropriate. Aggregation is the *a-part-of* relationship.

Step 8. Discard design notation.

- **Flow of identity**. Drop flow of identity notation; the choice of path for deriving identity is a design decision.
- **Identifiers**. Remove any object identifiers that are an artifact of implementation and are not meaningful in the real world.

20.4.3 Example

Figure 20.8 shows a reverse engineering example for a relational database. This example also concerns the simple flight information database.

Figure 20.8 shows the initial relational schema. Steps 1 through 4 elicit the classes, foreign keys, and candidate keys; steps 2 and 3 are degenerate because the initial schema specifies candidate keys and foreign keys. For foreign keys we list the foreign key attribute, a colon, and the class that is the referent. In step 5 we introduce a generalization guided by our understanding of the problem and the suggestive comment about employee type. In step 6 we replace the remaining foreign keys with associations. Aggregation does not apply for this problem. We discard design notation to reach the final model.

Initial relational DBMS schema

```
Table = Airline
   column = airline_ID
   column = name
   column = symbol
   primary key (airline_ID)
   candidate key (name)
   candidate key (symbol)
Table = Flight
   column = flight_ID
   column = airline_ID
   column = flight_num
   column = date
   column = pilot
   column = copilot
   primary key (flight_ID)
   candidate key (airline_ID, flight_number, date)
   foreign key (airline_ID) references Airline
   foreign key (pilot) references Employee
   foreign key (copilot) references Employee
Table = Employee
   column = employee_ID
   column = name
   column = taxpayer_number
   column = employee_type   /* pilot or flight attendant */
   column = flight_rating
   column = airline_ID
   primary key (employee_ID)
   candidate key (taxpayer_number)
   foreign key (airline_ID) references Airline
Table = Attendant_assignment
   column = employee_ID
   column = flight_ID
   primary key (employee_ID, flight_ID)
   foreign key (employee_ID) references Employee
   foreign key (flight_ID) references Flight
```

Figure 20.8 Example of reverse engineering for a relational database

Step 1. Prepare an initial object model.

Step 2. Determine candidate keys.

Step 3. Determine foreign key groups.

Step 4. Refine tentative classes.

Flight
flightID {CK1}
airlineID {CK2}
flightNum {CK2}
date {CK2}
pilot:Employee
copilot:Employee

Employee
employeeID {CK1}
name
taxpayerNumber {CK2}
employeeType
flightRating
airlineID

AttendantAssignment
employeeID {CK1}
flightID {CK1}

Airline
airlineID {CK1}
name {CK2}
symbol {CK3}

Step 5. Discover generalizations.

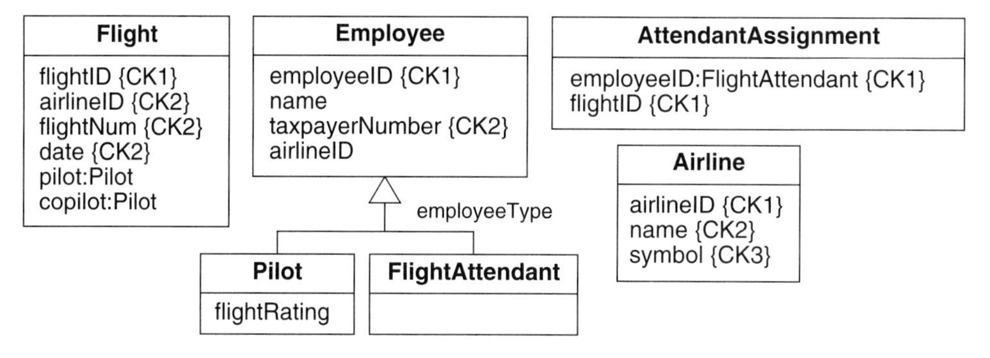

Step 6. Discover associations.

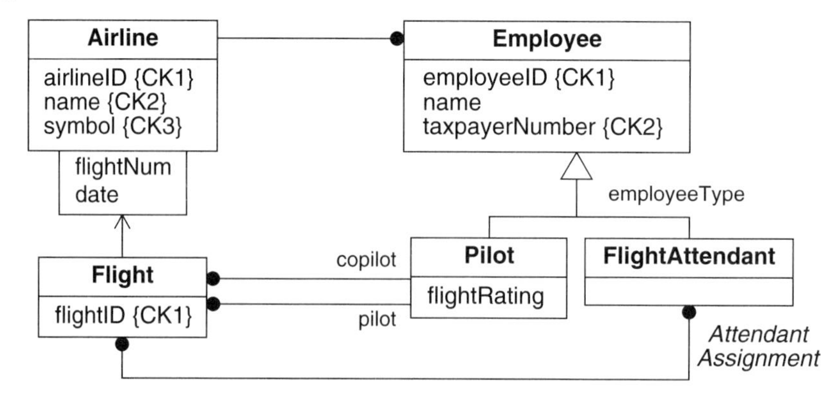

Figure 20.8 (continued) Example of reverse engineering for a relational database

Step 7. As appropriate, restate associations as aggregations.

..... Not needed for this example .....

Step 8. Discard design notation.

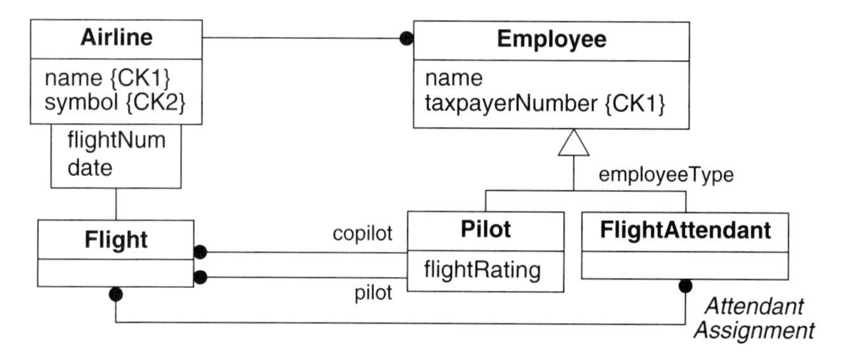

Figure 20.8 (continued) Example of reverse engineering for a relational database

20.5 CHAPTER SUMMARY

Reverse engineering is the process of taking an existing design and extracting the underlying logical intent. There are many motivations for performing reverse engineering. You can use the models that result from reverse engineering to seed the development of new systems. Reverse engineering is also helpful for assessing vendor software and integrating stand-alone applications. We have received ready industrial acceptance of database reverse engineering and frequently apply the technology in our consulting work.

During reverse engineering you should use any and all information sources you can find. Having said this, we must emphasize that reverse engineering is normally incidental to some other task, such as developing an application. Thus you must focus your efforts and only expend the resources justified by your motivation for performing reverse engineering. Often it is sufficient to acquire just 80–90% of the available information. This chapter has presented reverse engineering processes for hierarchical, network, and relational databases.

Figure 20.9 lists the key concepts for this chapter.

hierarchical database	relational database	transformation
network database	reverse engineering process	
re-engineering	reverse engineering strategies	

Figure 20.9 Key concepts for Chapter 20

BIBLIOGRAPHIC NOTES

[Date-82] provides further information on hierarchical and network database technology.

We have taken some of the information in this chapter from [Premerlani-94] and [Blaha-95]. [Blaha-95] presents a listing of the various odd designs we have encountered when reverse engineering relational databases. We would appreciate receiving email (blaha@acm.org, premerlani@acm.org) about documented examples of reverse engineering idiosyncracies from readers of this book.

Hainaut has written many excellent papers concerning data reverse engineering [Hainaut-93] [Hainaut-95]. Davis is also an active researcher in the area [Davis-95]. The Chikofsky and Cross paper [Chikofsky-90] is a seminal work documenting basic reverse engineering terminology.

If you are especially interested in reverse engineering, we recommend that you browse the proceedings of the IEEE Working Conferences on Reverse Engineering.

REFERENCES

[Blaha-95] Michael Blaha and William Premerlani. Observed idiosyncracies of relational database designs. *Second Working Conference on Reverse Engineering*, July 1995, Toronto, Ontario, 116–125.

[Chikofsky-90] Eliot J Chikofsky and James H Cross II. Reverse engineering and design recovery: A taxonomy. *IEEE Software* (January 1990), 13–17.

[Date-82] CJ Date. *An Introduction to Database Systems*. Reading, Massachusetts: Addison-Wesley, 1982.

[Davis-95] Kathi Hogshead Davis. August-II: A tool for step-by-step data model reverse engineering. *Second Working Conference on Reverse Engineering*, July 1995, Toronto, Ontario, 146–154.

[Hainaut-93] JL Hainaut, M Chandelon, C Tonneau, and M Joris. Contribution to a theory of database reverse engineering. *International Conference on Software Engineering, Workshop on Reverse Engineering*, May 1993, Baltimore, Maryland, 161–170.

[Hainaut-95] JL Hainaut, V Englebert, J Henrard, JM Hick, and D Roland. Requirements for information system reverse engineering support. *Second Working Conference on Reverse Engineering*, July 1995, Toronto, Ontario, 136–145.

[Premerlani-94] William Premerlani and Michael Blaha. An approach for reverse engineering of relational databases, *Communications ACM 37*, 5 (May 1994), 42–49.

[Weiser-84] M Weiser. Program slicing. *IEEE Transactions on Software Engineering 10*, 4 (July 1984), 352–357.

Appendix A

Glossary

abstract class (for a generalization) a superclass that has no direct instances. The descendant classes can also be abstract but the generalization hierarchy must ultimately terminate in subclasses with direct instances.

abstract operation an operation that specifies the signature while deferring implementation to subclasses.

abstraction the ability to focus on essential aspects of an application while ignoring details.

aggregation a kind of association in which a whole, the assembly, is composed of parts, the components. Aggregation is often called the "a-part-of" or "parts-explosion" relationship and may be nested to an arbitrary number of levels. Aggregation bears the transitivity and antisymmetry properties.

analysis the step of OMT development in which requirements are scrutinized and rigorously restated by constructing models.

API an acronym for application programming interface. A collection of methods that provide the functionality of an application.

array an indexed, ordered collection of objects with duplicates allowed.

assembly (for an aggregation) a class of objects that combine component objects.

association a description of a group of links with common structure and common semantics.

association class an association whose links can participate in subsequent associations.

attribute a named property that describes values.

attribute multiplicity the possible number of values for an attribute. Attribute multiplicity specifies whether an attribute may be single or multivalued and whether an attribute is optional or mandatory.

bag an unordered collection of objects with duplicates allowed.

bill-of-material (pertains to aggregation) a report that lists each part on a separate line. The lines of the report are ordered by traversing the components in depth-first order starting from the root assembly.

BNF grammar Backus-Naur Form. One of the most popular notations for expressing a context-free grammar.

CAD an acronym for computer-aided design.

candidate key (for an association) a combination of roles and qualifiers that uniquely identifies links within an association. The collection of roles and qualifiers in a candidate key must be minimal; no role or qualifier can be discarded from the candidate key without destroying uniqueness.

candidate key (for a class) a combination of one or more attributes that uniquely identifies objects within a class. The collection of attributes in a candidate key must be minimal; no attribute can be discarded from the candidate key without destroying uniqueness. No attribute in a candidate key can be null.

cardinality the count of elements that are in a collection.

CASE tool a tool for computer-aided software engineering. Software for software development. Sophisticated CASE tools can generate code from models, maintain traceability from analysis through design and implementation, and manage the progress of a project.

catalog aggregation an aggregation for which components are reusable across multiple assemblies. (Contrast with *physical aggregation.*)

class a description of a group of objects with similar properties (object attributes), common behavior (operations and state diagrams), similar relationships to other objects, and common semantics.

class attribute an attribute whose value is common to a group of objects in a class rather than peculiar to each instance. (Contrast with *object attribute.*)

class diagram a graphical representation that describes objects and their relationships. (Contrast with *instance diagram.*)

class library a collection of methods that have been carefully polished and organized and are suitable for reuse.

class operation an operation on a class rather than on instances of the class.

client-server computing an architecture for which a resource, the server, provides computation for multiple components, the clients.

collection an object that aggregates other objects; the other objects can be lesser collections and primitive objects.

component (for an aggregation) objects that comprise the assembly object.

compound qualifier two or more attributes that combine to refine the multiplicity of an association. The attributes that compose the compound qualifier are “anded” together.

conceptualization the first step of OMT development in which business analysts or users conceive an application and formulate tentative requirements.

concrete class (for a generalization) a class that can have direct instances.

concurrency the ability of multiple users to read and write simultaneously to a database.

configuration a set of mutually consistent versions.

constraint a functional relationship between modeling constructs such as classes, attributes, and associations.

context-free grammar a grammar that can be defined in terms of production rules whose right side can be substituted for the left side in any context.

CORBA the Common Object Request Broker Architecture. A standard for object-oriented client-server computing. The CORBA object request broker separates clients and servers and mediates their communication.

crash recovery the protection of the database against accidental loss of persistent data.

data dictionary (of a database) the definition of the database schema that is also stored in the database.

data dictionary (of OMT methodology) the definition of all modeling entities (classes, associations, attributes, operations, domains, and enumeration values) and explanation of the rationale for key modeling decisions.

data warehouse a database dedicated to the needs of decision-support applications. A data warehouse is distinct from operational, application-oriented databases. The data is placed on a common basis—for the same period of time, for the same portion of a corporation, and for the same currency (such as dollars, yen, marks).

database a permanent, self-descriptive repository of data that is stored in one or more files.

database management system (DBMS) the software for managing access to a database.

decision table a table for which the rows are individual rules and the columns are the attributes (antecedents and consequences) that are the subjects of the rules.

decision-support application an application that tends to have few updates and can involve unpredictable and lengthy queries.

declarative model a model that is stated as structural properties and constraints.

degree (of an association) the number of distinct roles for each link. The vast majority of associations are binary or qualified binary. Ternary associations occasionally occur, but we have rarely encountered an association of higher degree.

delegation an approach to object orientation for which types are represented by objects themselves. New objects are defined with respect to an existing object.

derived data data that can be completely determined form other data. Classes, attributes, and associations can all be derived. Do not confuse our use of the term "derived" with the C++ derived class. A C++ derived class refers to the subclass of a generalization; it has nothing to do with OMT's meaning of derived data.

detailed design the second phase of design in which the analysis models are augmented and transformed into a form amenable to implementation.

dictionary an unordered collection of object pairs with duplicates allowed. Each pair binds a key to an element. You can then use the key to look up the element.

directed graph a set of nodes and edges, where an edge originates at a source node and terminates in a sink node. The nodes in a directed graph can have any number of edges.

discriminator an attribute that has one value for each subclass. The value indicates which subclass further describes an object. The discriminator is simply a name for the basis of generalization.

distributed database a database that is built on top of a computer network rather than on a single computer.

distributed DBMS the software for managing access to a distributed database. A distributed DBMS lets application software be written as if it were accessing a centralized DBMS, by isolating an application from the details of distribution.

domain the named set of possible values for an attribute.

dynamic model the OMT model that describes temporal interactions between objects. This model is seldom important for database applications.

dynamic SQL database code that is not known until run time. (Contrast with *static SQL.*)

encapsulation the separation of external specification from internal implementation.

enterprise model a model that describes an entire organization or some major aspect of an organization. An enterprise model abstracts multiple applications, combining and reconciling their logical content.

Entity-Relationship (ER) approach the classical approach to modeling originated by Peter Chen. You note entities from the real world, describe them, and observe relationships among them. The OMT methodology is an entity-based approach.

enumeration domain a domain that has a finite set of values.

exclusive-or association a member of a group of associations that emanate from a class, called the source class. For each object in the source class exactly one exclusive-or association applies. An exclusive-or association relates the source class to a target class.

existence-based identity an approach in which a system-generated object identifier (also called an OID, a surrogate, or a pointer) identifies each object. (Contrast with *value-based identity.*)

extent (of a class) the set of objects for a class.

external control how an application is controlled at its outermost level and how it interacts with users and other applications. External control contrasts with internal control, which is not visible outside the application.

file type (relevant for implementation with files) an organization of the classes and associations of an application. The distinction between file and file type is analogous to that between object and class—a file is an instance of a file type.

flow of identity notation a notation for expressing propagation of value-based identity. This notation is sometimes helpful for relational database applications.

foreign key a reference to a candidate key. We strongly advise that all foreign keys refer only to primary keys, not to other candidate keys.

fourth-generation language (4GL) a framework for straightforward database applications that provides screen layout, specification of simple computations, and report preparation.

framework a skeletal structure of a program that must be elaborated to build a complete application.

functional model the OMT model that defines the computations that objects perform.

GemStone a popular object-oriented DBMS product.

generalization the relationship between a class (called the superclass) and one or more variations of the class (the subclasses). Generalization organizes classes by their similarities and differences, structuring the description of objects.

generic class a class that combines data and metadata.

hierarchical database a database that is organized as a collection of inverted trees of records. The inverted trees may be of arbitrary depth. IBM's IMS product is the most prominent hierarchical DBMS.

identity the property of an object which distinguishes each object from all others.

imperative model a model that is stated as a series of programming statements.

implementation the step when the design is translated into the actual programming language and database code.

index a data structure that locates objects according to attribute values. An index is typically implemented as an inverted tree with a wide fan-out at each node (often a factor of 50 or more).

Informix a popular relational DBMS product.

inheritance the mechanism that implements the generalization relationship.

instance an occurrence of a description. For example, an object is an instance of a class. A link is an instance of an association. A value is an instance of an attribute.

instance diagram a diagram that involves only objects, links, and values. (Contrast with *class diagram.*)

instantiation the relationship between an object and its class.

integration the technology for facilitating information exchange between applications.

intent (of a class) the structure and behavior of objects of a particular type.

layer a subsystem that builds on subsystems at a lower level of abstraction.

link a physical or conceptual connection between objects. A link is an instance of an association.

link attribute a named property of an association that describes a value held by each link of the association.

list an ordered collection of objects with duplicates allowed.

logical horizon (of a class) the set of classes reachable by one or more paths terminating in a combined multiplicity of "one" or "zero or one."

logical data independence (for a DBMS) the quality by which an application is made aware only of relevant portions of the database. This allows you to add new applications to the database without disrupting existing applications.

long transaction a series of DBMS commands that extend over a long period—hours, days, weeks, or even months. A long transaction contrasts with the short transaction supported by conventional DBMSs, which ordinarily resolves within a few seconds.

maximum multiplicity the upper limit on the possible number of related objects. The most common values are one and infinite.

metadata information about a model.

metamodel an abstraction of metadata.

method the implementation of an operation for a class.

methodology a combination of concepts and process to guide a practitioner through system development.

minimum multiplicity the lower limit on the possible number of related objects. The most common values are zero and one.

model an abstraction of some aspect of a problem; we express models with various kinds of diagrams.

modularity the organization of a system into groups of closely related objects.

MS-Access a popular relational DBMS product.

multiple inheritance a generalization for which a class may inherit attributes, operations, and associations from multiple superclasses.

multiplicity the number of instances of one class that may relate to a single instance of an associated class.

***n*-ary association** an association with three or more roles that cannot be restated as binary associations.

nested transaction a transaction placed within another transaction. A nested transaction can succeed or fail or at local level, but when grouped with additional DBMS commands, it can still fail within a broader context. This gives you finer control and lets you rollback to intermediate states.

network database a database that is organized as a collection of graphs of records that are related with pointers. Most network DBMSs adhere to the CODASYL (COmmittee for DAta SYstem Languages) standard.

normal form a guideline for relational database design that increases the consistency of data. As tables satisfy higher normal forms, they are more likely to store correct data.

null a special value denoting that an attribute value is unknown or not applicable.

object a concept, abstraction, or thing that can be individually identified and has meaning for an application. An object is an instance of a class.

object attribute a named property of a class that describes a value held by each object of the class. (Contrast with *class attribute*.)

object model the OMT model that characterizes the static structure of things. This model is dominant for database applications.

Object Navigation Notation (ONN) the OMT notation for navigating object models.

object orientation (OO) a strategy for organizing systems as collections of interacting objects that combine data and behavior.

object-oriented database a persistent store of objects created by an object-oriented programming language.

object-oriented DBMS a DBMS that manages the data, programming code, and associated structures that constitute an object-oriented database.

Objectivity a popular object-oriented DBMS product.

ObjectStore a popular object-oriented DBMS product.

ODMG Object Data Management Group. A group that is standardizing features across OO-DBMSs.

OMG Object Management Group. A group that is developing standards for object-oriented computing.

OMT an acronym for the Object Modeling Technique. The methodology that is the subject of this book.

ONN an acronym for the Object Navigation Notation.

operation a function or procedure that may be applied to or by objects in a class.

operational application an application that updates a database by posting numerous transactions as part of the routine business of an organization.

optimistic locks database locks that are acquired at the time of writing to the database. (Contrast with *pessimistic locks*.)

Oracle a popular relational DBMS product.

ordered association an association with a sequence of objects for a many role.

O2 a popular object-oriented DBMS product.

package a group of elements (classes, associations, generalizations, and lesser packages) with a common theme.

path a sequence of consecutive associations and generalization levels.

partition a subsystem that is in parallel to other subsystems.

pattern an excerpt of an object model with one or more parameters as placeholders for classes and associations.

persistent object an object that is stored in the database and can span multiple application executions. (Contrast with *transient object*.)

pessimistic locks database locks that are acquired before accessing objects (records) and released after object (record) access is complete. (Contrast with *optimistic locks*.)

physical aggregation an aggregation for which each component is dedicated to at most one assembly. (Contrast with *catalog aggregation*.)

physical data independence (for a DBMS) the quality by which a database insulates applications from changes in tuning mechanisms. This allows you to restructure a database for faster performance without affecting application logic.

polymorphism the ability of an operation to be applied to many classes.

primary key an arbitrarily chosen candidate key that is used to reference instances preferentially. Each class should normally have exactly one primary key.

primitive transformation a transformation that cannot be decomposed into lesser transformations.

private (referring to an attribute or method of a class) accessible by methods of the containing class only. (Contrast with *protected* and *public*.)

production rule a definition of one language element in terms of other elements in the same language.

promotion (of an association) treating an association as if it were a class.

propagation the automatic application of some property to a network of objects when the property is applied to some starting object.

protected (referring to an attribute or method of a class) accessible by methods of the containing class or any of its subclasses. (Contrast with *private* and *public*.)

pseudocode an informal language that provides sequence, conditionality, and iteration.

public (referring to an attribute or method of a class) accessible by methods of any class. (Contrast with *private* and *protected*.)

qualification cascade a series of consecutive qualified associations.

qualified association an association in which the objects in a "many" role are partially or fully disambiguated by an attribute called the qualifier.

qualifier an attribute that selects among target objects in a qualified association.

query a database command that retrieves a collection of objects by specifying predicates on the attribute values.

re-engineering the process of redeveloping existing software systems. Re-engineering consists of a sequence of three steps: reverse engineering, consideration of new requirements, and then forward engineering.

referential integrity a constraint for a relational database ensuring that the values referenced in other tables really exist.

reflexive association an association between objects of the same class. (Contrast with *symmetric association.*)

reification the promotion of something that is not an object into an object.

relational database a database in which the data is logically perceived as tables.

relational DBMS a DBMS that manages tables of data and associated structures that increase the functionality and performance of tables.

reverse engineering the process of taking an existing design and extracting the underlying logical intent.

role one end of an association. Each role may have an explicit name and a multiplicity.

secondary data data about attributes and classes that the OMT notation does not explicitly address. Secondary data provides relevant information, but exists in a realm apart from the essence of an application.

sequential-access ASCII file a file for which you must access each data item in turn. You cannot read or update an item in the middle of the file without accessing all prior items. The file uses only printable characters and is human readable.

schema the structure of the data in a database.

schema evolution the migration of data across changes in data structure.

security the protection of data against unauthorized read and write access.

set an unordered collection of objects without duplicates.

signature the argument types, result type, exception conditions, and the semantics of an operation.

single inheritance a generalization for which each subclass has a single superclass.

software engineering a systematic approach to software development that emphasizes thorough conceptual understanding prior to design and coding.

specialization the splitting of variations for a class. Specialization has the same meaning as generalization but takes a top-down perspective. In contrast, generalization takes a bottom-up perspective.

SQL (language) the standard language for accessing the data of a relational database.

static SQL database code that is known at compile time. (Contrast with *dynamic SQL.*)

stored procedure a method that is stored in a relational database.

structured domain a domain with important internal detail.

subclass a class that adds specific attributes, operations, state diagrams, and associations for a generalization.

superclass the class that holds common attributes, operations, state diagrams, and associations for a generalization.

Sybase a popular relational DBMS product.

symmetric association an association between objects of the same class that have interchangeable roles. (Contrast with *reflexive association*.)

system architecture the high-level plan or strategy for solving an application.

system design the first phase of design in which the developer devises a system architecture and establishes general design policies.

ternary association an association with three roles that cannot be restated as binary associations.

three-tier architecture an approach to client-server computing that cleanly separates data management, application functionality, and the user interface. The data management layer holds the database schema and data. The application layer holds the methods that embody the application logic. The user-interface layer manages the forms and reports that are presented to the user.

transaction a group of commands that succeed or fail as a single unit of work.

transaction modeling forecasting the mix of queries that will execute against a database and estimating query frequency, the number of read and write I/Os, the required response time, the number of concurrent users, and other data.

transformation a mapping from the domain of object models to the range of object models. You can think of a transformation as accepting a source object model pattern and yielding a target object model pattern.

transient object an object that exists only in memory and disappears when an application terminates execution. Thus a transient object is an ordinary programming object. (Contrast with *persistent object*.)

transitive closure (from graph theory) the set of nodes that are reachable by some sequence of edges.

tree a set of nodes, such that each node has at most one parent node and zero or more child nodes. A tree may not have any cycles. Exactly one node is designated the root of the tree.

trigger a database command that performs an action upon the occurrence of the specified event.

UML the Unified Modeling Language developed by Booch, Rumbaugh, and Jacobson for modeling applications. This book uses their notation.

undirected graph a set of nodes and edges, where an edge connects two nodes. The nodes in an undirected graph can have any number of edges.

use case a theme for interacting with a system. You should think of the various external entities that interact with the system and the different ways they each use the system.

value a piece of data. A value is an instance of an attribute.

value-based identity an approach in which some combination of attributes identifies each object. (Contrast with *existence-based identity.*)

Versant a popular object-oriented DBMS product.

version an alternative object relative to some base object.

view a table that the RDBMS dynamically computes from a query stored in the database.

Appendix B

BNF Grammar for the ONN

This appendix presents a BNF (Backus-Naur Form) grammar for the Object Navigation Notation (ONN). We are placing this appendix in the public domain so that you may freely copy it and disseminate it. We assume that you are familiar with the notion of a grammar and BNF notation.

We are using the BNF dialect of *Lex*, *Yacc*, and *Yacc++*. A semicolon terminates each production rule. We list the token to the left of the colon and the definition to the right. The vertical bar denotes alternation. Quotes enclose literal characters. We list primitive elements of the grammar in uppercase and use white space and indentation to improve readability.

Ideally, we would have liked the ONN grammar to be context free, to simplify the implementation of a parser using tools like *Lex*, *Yacc*, and *Yacc++*. However, there is some context dependence in the grammar. For example, when a parser encounters an *EntryPoint*, it must decide whether the *EntryPoint* is for an *ObjectOrSet* or for a *LinkOrSet* by examining the object model. In principle, the ONN grammar could be modified to make it context free, but it would not be as clear, expressive, and easy to read as its present form.

Chapter 5 explains the semantics and presents examples of the grammar.

```
ONNexpression    : ObjectOrSet | LinkOrSet | ValueOrSet ;


ObjectOrSet      : ObjectOrSet '.' Role
                 | ObjectOrSet '.' Role '[' Qualif '=' Value ']'
                 | ObjectOrSet '[' Filter ']'
                 | ObjectOrSet ':' SuperClass
                 | ObjectOrSet ':' SubClass
                 | LinkOrSet '.' Role
                 | '(' ObjectOrSet ')'
                 | EntryPoint ;
Role             : TargetRole | SourceRole ;
```

```
TargetRole      : Name ;
SourceRole      : '~' Name ;
Qualif          : Name ;
SuperClass      : Name ;
SubClass        : Name ;
EntryPoint      : Name ;

LinkOrSet       : '{' LinkClauses '}'
                | '{' LinkClauses ',' AssocName '}'
                | ObjectOrSet '@' Role
                | LinkOrSet '[' Filter ']'
                | '(' LinkOrSet ')'
                | EntryPoint ;
LinkClauses     : LinkClause
                | LinkClauses ',' LinkClause ;
LinkClause      : Role '=' ObjectOrSet ;
AssocName       : Name ;

ValueOrSet      : ObjectOrSet '.' ObjectAttrib
                | LinkOrSet '.' LinkAttrib
                | '(' ValueOrSet ')' ;
ObjectAttrib    : Name ;
LinkAttrib      : Name ;

Name            : A NAME, NORMALLY CONSISTING OF CHARACTERS AND
                  DIGITS AND BEGINNING WITH A CHARACTER
Value           : ANY STRING
Filter          : AN EXPRESSION THAT EVALUATES TO TRUE OR FALSE
```

Index

A

abstract class **56**
  notation for 56
abstract operation 56
abstraction 1, 3
aggregation **51–55**
  catalog **55**
  implementing
    with files 255
    with object-oriented database 349
    with relational database 287
  notation for 52
  OMT metamodel 79
  physical **54**, 311
  physical vs. catalog **53–54**
  vs. association 52, 134
  vs. generalization 57
airline flight reservation example
  basic object model 27–29
  exercise 116–117, 304, 363, 392
  extended object model 64–67
  traversal for 98–110
analysis 6, 118, **125–168**, 236–237
  building the dynamic model 148
  building the functional model 148–156
    choosing a paradigm 150–151
    specifying operations 151–156
    specifying use cases 149–150
  building the object model 126–145
    adding attributes 134–137
    finding associations 130–134
    finding classes 127–130
    organizing a model 145
    refining a model 140–145
    testing access paths 139
    using generalization 137–139
  data dictionary 145–147
  exercise 160–167
  inputs to 125
  outputs from 125–126
a-part-of. *See* aggregation
API 172, 191, 229
application
  decision support 5, 171, 432–433
  operational 5, 171, 432–433
application programming interface. *See* API
architecture **170–175**
  canonical architectures 202
  design principles 171–172
  generating candidates 173
  proposing criteria 173–174
  quantifying compliance 174–175
assembly **51**
assertion 229
association **17–26**, **48–51**
  conventions for 17
  degree 49
  derived 67

directed name 21
exclusive-or 50
finding 17, 130–134
implementing 199
with files 250–255
with object-oriented database 344–349, 372–385
with relational database 282–287
importance of 18
intersecting lines 29
notation for 17
OMT metamodel 79
ordered 19
qualified **24–26**, 50–51
resolving ambiguity 17, 22
role **21–22**
ternary 49
traversal of 18, 21, **98–101**, **102–103**
for object-oriented database 351–353
for relational database 295–297, 299–301
vs. aggregation 52
vs. generalization 56
association class **23–24**
exercise 37, 91, 116, 267
implementing
with files 253
with object-oriented database 349, 376–385
with relational database 285
importance of 24
notation for 23
traversal of 102, 110
for object-oriented database 353
for relational database 299
attribute **15–16**
attribute multiplicity 43–44
notation for 43
vs. multiplicity 44
class attribute 42–43
derived 67
finding 134–137
implementing
with files 249–250
link attribute 23
notation for 15
object attribute 15–16

B

Backus-Naur Form. *See* BNF grammar
batch processing 190
benefits
of client-server architecture 402
of context-free grammar 240
of distributed database 400
of domain 46
of framework 81
of generic class 77
of integration 416
of metamodeling 77
of modeling 2–4
of Object Navigation Notation 98
of object-oriented DBMS 186
of pattern 81
of relational DBMS 184
of reverse engineering 3, 437
bill-of-material 52
exercise 333
BNF grammar 240
exercise 92, 265–268
BOM. *See* bill-of-material
business process reengineering 5

C

candidate key **44**, **49–50**, **271**, 297
creating index as side effect 309
enforcing for distributed database 410
notation for 44, 50
ordering data in forms 316
specifying 222
cardinality 19
CASE tool 119
class **14–15**
abstract 56
concrete 56
conventions for 14
derived 67
finding 14, 127–130
implementing
with files 248–249
with object-oriented database 342
with relational database 282
meaning of term 15
notation for 14
OMT metamodel 79
repeated class box 29

suppressing attributes and operations 14, 20
vs. domain 46
class attribute **42–43**
conventions for 43
notation for 43
practical tip 43
class diagram 15, **27**
class icon 62, 143, 145
class library 188–189
class operation **42–43**
conventions for 43
notation for 43
client-server architecture **401–406**
benefits of 402
CORBA 404–406
disadvantages of 402
SQL 405–406
three tier 172, 275, 314, 402–403
two tier 275
cluster 311, 328
CODASYL 443
collection class 336–337
component **51**
conceptualization 5, 118, **121–124**, 236
concrete class **56**
notation for 56
concurrency 178
optimistic 181
pessimistic 181
constraint 4, 19, 25, 31, 44, 50, 55, **68**
implementing
with files 263
with relational database 307–309
in dynamic model 316
in functional model 68
in object model 68
notation for 68
conventions for
association 17
class 14
class attribute 43
class operation 43
link 17
name 14
object 13
package 62, 145
CORBA **403–406**, 414
crash recovery 178, 263

D

data dictionary
for documenting models 145–147
for object-oriented DBMS 337
for relational DBMS 271
data flow diagram **110–111**, 115
unsuitability for databases 110, 150
data independence 273
location 401, 412
logical 178, 271
migration 401
physical 178, 271
replication 401
data warehouse 171, **432–433**
database 3, 177–187
See also DBMS
advice for using 181
suitable applications 181
database management system. *See* DBMS
DBMS 3, 177–187
See also database
active 187
deductive 187
distributed 399–414
hierarchical 439–443
network 443–450
object-oriented 185–187, 334–367, 368–393
real-time 187
relational 182–184, 269–306, 307–333
decision table **111–113**
advanced issues 112–113
advice for using 150
application example 115
declarative model 5, 119, 176
delegation 15
Demeter, law of 228, 355
derived data **67**
advice for using 130, 133
exercise 72
for portfolio manager 326
notation for 67
detailed design 6, 119, **210–235**, 238
elaborating the functional model 224–229
elaborating the object model 222–224
evaluating design quality 230
exercise 233–235

transformations 210–222
development
attribute-based 6
entity-based 6
diagram
class 15, **27**
instance 15, **29–30**
diagram layout 32
disadvantages
of client-server architecture 402
of distributed database 400
of framework 81
of generic class 78
of metamodeling 77
of modeling 3
of object-oriented DBMS 185–186
of relational DBMS 183–184
discriminator **26**
implementing
with relational database 320
importance of 26
distributed data **406–412**
allocating fragments to sites 410–411
considering during design 230
fragmenting a database 407–410
providing location transparency 412
replicating data 411–412
distributed database 178, **399–414**
benefits of 400
client-server computing 401–406
disadvantages of 400
distributing data 406–412
domain **45–46**, 311
assigning 222
benefits of 46
enumeration domain **45**
implementing
with files 246–247
with object-oriented database 340–342
with relational database 278–282
notation for 45
structured domain 45
notation for 45
use of during development 46
vs. class 46
dynamic model 8, 148
implementing
with object-oriented database 349–350
with relational database 314–318

E

email address 214, 269, 458
encapsulation 1, 57, 109
vs. query optimization 228–229
enhanced pseudocode **96–98**
advice for using 151
example 107–108, 151–156
enterprise model **417–425**, 433
choosing a development approach 418
choosing a merge sequence 419
comparing application models 420
example 421–422
resolving discrepancies 422–424
verifying the model 424–425
Entity-Relationship approach 10
enumeration domain **45**, 137
implementing
with files 247
with object-oriented database 341
with relational database 278–281, 318
vs. generalization 46
ER. *See* Entity Relationship approach
exclusive-or association **50**
exercise 91
notation for 50
existence dependency 48, 282
extensibility 46
extent 15, 369–372
external control 175–177

F

files 177–180, **239–268**
exercise 265–268
implementing associations 250–255
implementing attributes 249–250
implementing classes 248–249
implementing domains 246–247
implementing generalizations 255
implementing identity 246
implementing the dynamic model 256
organizing data into files 241–243
other issues 263–264
saving and loading 256–262
selecting a file approach 243–245
suitable applications 180
summary for mappings 256

vs. object-oriented database 394–397
vs. relational database 394–397
finding
association 17, 130–134
attribute 134–137
class 14, 127–130
generalization 137–139
link 17
link attribute 23
object 13
object attribute 15
role 21
value 15
five-schema architecture 429
flow of identity notation **277–278**
foreign key **272**
implementing association 282–287, 308–309
implementing generalization 288–291, 308–309
importance of indexing 310
should reference primary key 276
forward engineering 5
fourth-generation language (4GL) 176, 191–192
advice for using 314–318
framework **80–81**
benefits of 81
disadvantages of 81
example 81
usefulness of dynamic SQL 329
vs. pattern 82
functional model 8, 68, **96–117**
building a functional model 148–156
data flow diagram 110–111
decision table 111–113
enhanced pseudocode 96–98
equations 113
implementing
with files 256–262
with object-oriented database 350–360, 385–387
with relational database 295–301, 318–330
Object Navigation Notation 98–110
practical tips 113–114
pseudocode 96–97

G

generalization **26–27**, **56–60**
See also inheritance
analysis vs. design 57
exercise 37
finding
bottom-up 138–139
top-down 138
implementing
with files 255
with object-oriented database 344
with relational database 288–291
importance of 26
notation for 26
OMT metamodel 79
traversal of 101
for object-oriented database 352
for relational database 297–299
vs. aggregation 57
vs. association 56
vs. enumeration domain 46
vs. instantiation 57
generic class **77–78**
benefits of 77
disadvantages of 78
exercise 89–90, 92
implementing
with relational database 329
grammar
context-free **239–240**
graph
directed 30–31, 83–84, 89
exercise 40, 117, 233, 304
undirected 85, 89

H

hierarchical database
concepts 439–440
reverse engineering 440–443

I

IDEF1X 115, 303
identifier 54
implementing
with files 247
with object-oriented database 340–341
with relational database 278, 320

practical tip 16
identity **13**, 46, 54, 60
checking during design 230
existence-based 193–195
implementing
with files 246
with relational database 276–278
value-based 193–195
imperative model 5, 119
implementation 6, 119, 238
with files 239–268
with object-oriented database 334–367, 368–393
with relational database 269–306, 307–333
index **181**
for ObjectStore 338
for relational DBMS 272, 282, **309–311**
inheritance
See also generalization
multiple **27**, **58–60**, 139, 197
implementing 255, 291
workarounds 60
single **26–27**, 56–58
in-memory data 177–180
suitable applications 180
instance **14**
instance diagram 15, **29–30**
exercise 37, 40, 71, 89–90, 90
importance of 29
instantiation **42**
notation for 42
vs. generalization 57
integration 4, 172, 213, 291, **415–435**
architecture 429–432
for distributed application 431
identity application 429–430
isolating flawed application 430
legacy data conversion 432
benefits of 416
enterprise model 417–425
technique 425–429
indirect 428–429
master database 425–426
point-to-point 426–427
integrity of data in database 4
checking during design 230
enforcing
with relational database 307–309, 322–326
invariant 115
inverse member (ObjectStore feature) 345
is-a. *See* generalization

J

join (in SQL language) 295, 327–329

K

key. *See* candidate, foreign, and primary key

L

large object model. *See* package
layer 171
legacy data 4, 432
Lex 240, 260–262, 264–265
life cycle
analysis 6, 118, 125–167, 236–237
conceptualization 5, 118, 121–124, 236
detailed design 6, 119, 210–235, 238
implementation 6, 119, 238
with files 239–268
with object-oriented database 334–367, 368–393
with relational database 269–306, 307–333
maintenance 6
reverse engineering 436–458
system design 6, 118–119, 169–209, 237–238
testing 6
link **17–18**
conventions for 17
finding 17
multiple links between objects 20
notation for 17
traversal of 18
vs. object 18
vs. value 16, 18
link attribute **23**
exercise 267
finding 23, 136
implementing
with files 253
with object-oriented database 349, 372–385

with relational database 285, 287
importance of 23
notation for 23
locking
avoid insert lock contention 328
avoid locks across user input 226, 349
for files 263
logical horizon **62–64**, 115
exercise 72, 90
long transaction **183**

M

maintenance 6
mathematical equations **113**
advice for using 151
metadata **75–79**, 179
metamodel **75–79**
benefits of 77
disadvantages of 77
example 76
exercise 90–91, 117
for OMT object model 79
for relational DBMS 271–273
metamodel driven interaction 192–193
method **16**
algorithmic code vs. lookup table 226
designing algorithms 226–227
determining owner 227–228
implicit argument 97
signature 56
methodology 5
comparison of 10, 35
mobile computing 431
model
dynamic model 8
functional model 8, 96–117
object model 8, 12–41, 42–74, 75–92
relationship between models 7
modularity 1
MS-Access relational DBMS **274–275**, 278, 311–326, 405
multiple inheritance. *See* inheritance, multiple
multiplicity **19–21**, 32, 134
maximum multiplicity **48**
minimum multiplicity **48**
notation for 19
OMT vs. UML 21
vs. cardinality 19

N

name 32
conventions for 14
scope of 21
uniqueness of 25
navigation. *See* Object Navigation Notation
network database
concepts 443–447
reverse engineering 447–450
NIAM 232
normal form **3**, 7, **273–274**, 303
violated by buried link attributes 287
violated by combination mapping 284
violated by pushing subclass attributes up 290
notation for
aggregation 52
association 17
exclusive-or 50
qualified 25
ternary 49
association class 23
attribute 15
attribute multiplicity 43
candidate key 44, 50
class 14
abstract 56
concrete 56
class attribute 43
class operation 43
constraint 68
derived data 67
domain 45
structured domain 45
flow of identity 277–278
generalization 26
instantiation 42
link 17
link attribute 23
multiplicity 19
OMT vs. UML 21
navigation 98–105
object 13
operation 16
abstract 56
package 61
pattern 82

role 21
value 15
null 44, 109
default policy for implementation 199
specifying 222–223

O

object **13–14**
conventions for 13
finding 13
notation for 13
vs. link 18
vs. value 16
object attribute **15–16**
finding 15
notation for 15
Object Data Management Group 4, 18, 199
object design 6, 119
See detailed design
Object Management Group 404
object model 8
advanced concepts 42–74
basic concepts 12–41
building an object model 126–145
implementing
comparison 394–395
with files 240–256
with object-oriented database 339–349, 368–385
with relational database 276–292, 307–312
metamodeling concepts 75–92
notation 34, 70–71, 88
practical tips 31–33, 68–69
Object Modeling Technique. *See* OMT
Object Navigation Notation **98–110**
benefits of 98
BNF grammar 470–471
combining constructs 105–108, 355–356
examples 151–156
exercise 116–117
filtering 101, 103–104, 301, 354
implementing
comparison 397
with files 256
with object-oriented database 351–356
with relational database 295–301
implementing with MS-Access 321
mathematical properties 109
nulls 109
operator
at sign 103
braces 102
brackets 101, 103
colon 101
dot 98, 99, 100, 102, 104
tilde 99
summary 105
for object-oriented database 356
for relational database 301
traversal from link to object 102, 299, 353
traversal from link to value 104, 301, 354–355
traversal from object to link 102–103, 300–301, 353
traversal from object to value 104, 301, 354
traversal of generalization 101, 297–299, 352
traversal of qualified association 100–101, 297, 352
traversal of simple association 98–99, 295–297, 351–352
object-oriented, meaning of **1**
object-oriented database 4, 185–187, **334–367**, **368–393**
See also object-oriented DBMS 185
creation and deletion methods 356–360
exercise 363–367, 392–393
extents 369–372
folding attributes into a related class 372–376
implementing associations 344–349
implementing classes 342–344
implementing domains 340–342
implementing generalizations 344
implementing OMT models 185
implementing the dynamic model 349–350
implementing the functional model 350–360
keys 368–369
promoting associations 376–385
software engineering issues 387–390
summary for mappings 349
summary for ONN mappings 356
using ObjectStore queries 385–387

vs. files 394–397
vs. relational database 394–397
object-oriented DBMS 4, 185–187, **334–367**, **368–393**
See also object-oriented database 185
advantage 186
disadvantage 185–186
suitable applications 186–187
support for associations 199
vs. relational DBMS 185
object-oriented development 2
ObjectStore object-oriented DBMS 334–367, 368–393
ODMG. *See* Object Data Management Group
OLE 403, 414
OMG. *See* Object Management Group
OMT (Object Modeling Technique) 5–8
OMT methodology
abridged object metamodel **79**
comparison with others 35
relationship to prior book 10, 110, 115
ONN. *See* Object Navigation Notation
OO-DBMS. *See* object-oriented DBMS
operation **16**, 129
abstract 56
notation for 56
class operation 42–43
notation for 16
signature 56
Oracle relational DBMS 190, 278, 311, 328, 405
ordered association 19
implementing
with files 254
with object-oriented database 344
with relational database 286
ORM 232

P

package **61–67**, 145
basis for organizing into file types 242
example 64–67
notation for 61
practical tip 62
partition 171
part-whole relationship. *See* aggregation
path **62**
pattern **81–88**, 140
benefits of 81
directed graph 83–84
exercise 92
homomorphism 86–88
in transformation 211
item description 85–86
notation for 82
tree 82–83
undirected graph 85
vs. framework 82
performance tuning 4, 226, 228–229, 271
for object-oriented database **388–389**
for relational database **309–311**, **327–328**
persistent object 335–336
pharmacy exercises
detailed design 233–235
implementation
with files 268
with object-oriented database 363–367, 392–393
with relational database 306, 332
system design 203–209
polymorphism **16**
portability 229
portfolio manager case study
analysis 126–157
conceptualization 122–124
detailed design 218–229
final object model with major operations 318
implementation
lessons learned 330–331
with files 242–262
with object-oriented database 340–360, 373–387
with relational database 276–301, 308–331
system design 170–200
postcondition 115
practical tips
advanced object modeling 68–69
basic object modeling 31–33
evaluating the quality of a design 230
functional modeling 113–114
ObjectStore code 389
SQL code 329–330
precondition 115

primary key **246**, **271**
product data manager (PDM) 182
propagation 54
pseudocode **96–98**
  method invocation 97
  method signature 97

Q

qualified association **24–26**, 50–51
  cascade 51
  compound qualifier 51, 103
  exercise 37, 267
  implementing
    with files 253–254
    with object-oriented database 349, 372–385
    with relational database 285, 287
  importance of 25
  notation for 25
  traversal of 100–101, 104, 110
    for object-oriented database 352
    for relational database 297
qualifier **24–26**
query language 3, 178
  vs. programming 172, 295, 330
    exercise 333
query optimization
  vs. encapsulation 228–229

R

rapid prototype 6, 119
RDBMS. *See* relational DBMS
reengineering 436–437
referential integrity 270, 288, **307–309**, 400
reification **78**, 172
  exercise 91
relational database 3, 182–184, **269–306**, **307–333**
  See also relational DBMS 182
  allocating storage 311
  defining constraints 307–309
  defining indexes 309–311
  exercise 304–306, 332–333
  hidden beneath an OO layer 192
  implementing associations 282–287
  implementing classes 282
  implementing domains 278–282
  implementing generalizations 288–291
  implementing identity 276–278
  implementing OMT models 183
  implementing the dynamic model 314–318
  implementing the functional model 295–301, 318–330
  physical implementation issues 328, 330, 332
  reverse engineering 451–455
  summary for mappings 291–292
  summary for ONN mappings 301
  vs. files 394–397
  vs. object-oriented database 394–397
relational DBMS 3, 182–184, **269–306**, **307–333**
  See also relational database 182
  advantage 184
  disadvantage 183–184
  metamodel 271–273
  suitable applications 184
  support for associations 199
  vs. object-oriented DBMS 185
relationship
  aggregation 51–55
  association 17–26, 48–51
  generalization 26–27, 56–60
  instantiation 42
replication 411–412, 414
reuse 2, 3, **187–189**
reverse engineering 126, 213, **436–458**
  benefits of 437
  hierarchical database 440–443
  network database 447–450
  relational database 451–455
  strategies 437–439
role **21–22**, 98–99
  derived 67
  finding 21, 129, 133
  implementing
    with object-oriented database 346–347
    with relational database 282
  importance of 22
  notation for 21
  source **99**
  target **98**
  traversing association 21

S

schema 3, 180
schema evolution 183, 213
  implementing
    with files 263–264
    with ObjectStore 338–339
secondary data **46–48**, 196–198
security 200, 230, 330
signature **56**
software engineering 1–3
specialization **26**
SQL language 182, 270, **273**
  advanced queries 332
  dynamic SQL 191, 328–329
  limitation 273, 324, 330
  referential integrity options 308
  static SQL 191, 328–329
  style guidelines 329–330
storage
  allocating for relational database 311
  estimating 223
stored procedure 191
structured domain **45**
  implementing
    with files 247
    with object-oriented database 341
    with relational database 281
  notation for 45
subclass **26**
superclass **26**
Sybase relational DBMS 191, 200
symmetric association
  implementing
    with files 254
    with object-oriented database 344
    with relational database 286
synonym 330
system design 6, 118–119, **169–209**, 237
  choosing a DBMS 182–187
  choosing data interaction 190–193
  choosing data management 177–182
  choosing external control 175–177
  choosing object identity 193–195
  dealing with secondary data 196–198
  dealing with time 195–196
  determining reuse 187–189
  devising an architecture 170–175
  exercise 202–209
  interplay with analysis 169
  specifying design defaults 198–200
systems engineering 202

T

table (in relational DBMS) 3
table subtraction 322
temporal data 195–196
ternary association **49**, 132
  exercise 267
  implementing
    with files 254
    with object-oriented database 349, 376–385
    with relational database 285
  notation for 49
  traversal of 102
    for object-oriented database 353
    for relational database 300
testing 6
three-schema architecture 426
traceability 2
transaction 178, 321, 324
  long 183
transaction modeling 226, 311, 350
transformation 119, **210–222**, 438
  bibliographic notes 231–232
  concepts 211–212
  examples 213–218
  exercise 233
  for portfolio manager 218–219
  in OMT methodology 213
  mathematical properties 219–222
transient object 335–336
transitive closure 52, 57, 320, 323–326
  exercise 117
traversal. *See* Object Navigation Notation
trigger 314
tuning. *See* performance tuning
type 15

U

UML 10, 21, 34, 70
Unified Modeling Language. *See* UML
use case 125, 126, 149–150
user interface 226, 229

V

value **15–16**
finding 15
notation for 15
vs. link 16, 18
vs. object 16
version 54, 183, 410
view **272**, 274, 288, 314, 330

W

Web site 233, 269, 312

Y

Yacc 240, 260–262, 264–265
Yacc++ 240, 260–262, 264–265